INDUSTRIAL HEMP

INDUSTRIAL HEMP
Food and Nutraceutical Applications

Edited by

MILICA POJIĆ
Principal Research Fellow, Institute of Food Technology (FINS), University of Novi Sad, Novi Sad, Serbia

BRIJESH K. TIWARI
Principal Research Officer, TEAGASC – Agriculture and Food Development Authority;
Professor, University College Dublin, Ireland

ELSEVIER

ACADEMIC PRESS
An imprint of Elsevier

Academic Press is an Imprint of Elsevier
125 London Wall, London EC2Y 5AS, United Kingdom
525 B Street, Suite 1650, San Diego, CA 92101, United States
50 Hampshire Street, 5th Floor, Cambridge, MA 02139, United States
The Boulevard, Langford Lane, Kidlington, Oxford OX5 1GB, United Kingdom

Notices
Knowledge and best practice in this field are constantly changing. As new research and experience broaden our understanding, changes in research methods, professional practices, or medical treatment may become necessary.

Practitioners and researchers must always rely on their own experience and knowledge in evaluating and using any information, methods, compounds, or experiments described herein. In using such information or methods they should be mindful of their own safety and the safety of others, including parties for whom they have a professional responsibility.

To the fullest extent of the law, neither the Publisher nor the authors, contributors, or editors, assume any liability for any injury and/or damage to persons or property as a matter of products liability, negligence or otherwise, or from any use or operation of any methods, products, instructions, or ideas contained in the material herein.

ISBN: 978-0-323-90910-5

For information on all Academic Press publications
visit our website at https://www.elsevier.com/books-and-journals

Publisher: Nikki P. Levy
Acquisitions Editor: Megan R. Ball
Editorial Project Manager: Franchezca Cabural
Production Project Manager: Bharatwaj Varatharajan
Cover Designer: Miles Hitchen

Typeset by STRAIVE, India

Working together
to grow libraries in
developing countries

www.elsevier.com • www.bookaid.org

Contents

7. The significance of industrial hemp knowledge management

Anamarija Koren, Milica Pojić, and Vladimir Sikora

8. Nutraceutical potential of industrial hemp

Viviana di Giacomo, Claudio Ferrante, Luigi Menghini,
and Giustino Orlando

9. Industrial hemp nutraceutical processing and technology

Saša Đurović, Rubén Domínguez, Mirian Pateiro, Nemanja Teslić,
José M. Lorenzo, and Branimir Pavlić

10. Industrial hemp foods and beverages and product properties

Lorenzo Nissen, Flavia Casciano, Elena Babini, and Andrea Gianotti

11. Industrial hemp-based dietary supplements and cosmetic products

Anna Bakowska-Barczak, Yussef Esparza, Harmandeep Kaur,
and Tomasz Popek

12. Industrial hemp by-product valorization

Vita Maria Cristiana Moliterni, Milica Pojić, and Brijesh Tiwari

Contributors

Darja Kocjan Ačko Department for Agronomy, Biotechnical Faculty, University of Ljubljana, Ljubljana, Slovenia

Elena Babini DiSTAL-Department of Agricultural and Food Sciences, Alma Mater Studiorum-University of Bologna, Cesena, Italy

Anna Bakowska-Barczak Radient Technologies Inc, Edmonton, AB, Canada

Flavia Casciano DiSTAL-Department of Agricultural and Food Sciences, Alma Mater Studiorum-University of Bologna, Bologna, Italy

Marne Coit Coit Consulting LLC, Raleigh, NC, United States

Viviana di Giacomo Department of Pharmacy, Medicinal Plant Unit (MPU), Botanic Garden "Giardino dei Semplici", "G. d'Annunzio" University of Chieti-Pescara, Chieti, Italy

Rubén Domínguez Centro Tecnológico de la Carne de Galicia, Parque Tecnológico de Galicia, Ourense, Spain

Saša Đurović Institute of General and Physical Chemistry, Belgrade, Serbia

Yussef Esparza BioNeutra, Edmonton, AB, Canada

Claudio Ferrante Department of Pharmacy, Medicinal Plant Unit (MPU), Botanic Garden "Giardino dei Semplici", "G. d'Annunzio" University of Chieti-Pescara, Chieti, Italy

Marko Flajšman Department for Agronomy, Biotechnical Faculty, University of Ljubljana, Ljubljana, Slovenia

Andrea Gianotti DiSTAL-Department of Agricultural and Food Sciences, Alma Mater Studiorum-University of Bologna, Bologna, Italy

Harmandeep Kaur Advanced BioInnovations Inc, AB, Edmonton, Canada

Jane Kolodinsky Department of Community Development and Applied Economics, University of Vermont, Burlington, VT, United States

Anamarija Koren Institute of Field and Vegetable Crops, National Institute of the Republic of Serbia, Novi Sad, Serbia

Hannah Lacasse Department of Community Development and Applied Economics, University of Vermont, Burlington, VT, United States

Yonghui Li Department of Grain Science and Industry, Kansas State University, Manhattan, KS, United States

José M. Lorenzo Centro Tecnológico de la Carne de Galicia, Parque Tecnológico de Galicia; Área de Tecnología de los Alimentos, Facultad de Ciencias de Ourense, Universidad de Vigo, Ourense, Spain

Sari Mäkinen Natural Resources Institute Finland, Jokioinen, Finland

Luigi Menghini Department of Pharmacy, Medicinal Plant Unit (MPU), Botanic Garden "Giardino dei Semplici", "G. d'Annunzio" University of Chieti-Pescara, Chieti, Italy

Vita Maria Cristiana Moliterni Council for Agricultural Research and Economics (CREA), Research Centre for Genomics and Bioinformatics, Fiorenzuola d'Arda, Piacenza, Italy

Eva Mrkvicová Department of Animal Nutrition and Forage Production, Faculty of Agri-Sciences, Mendel University in Brno, Brno, Czech Republic

Lorenzo Nissen CIRI-Interdepartmental Centre of Agri-Food Industrial Research, Alma Mater Studiorum-University of Bologna, Cesena, Italy

Markus Nurmi Natural Resources Institute Finland, Jokioinen, Finland

Giustino Orlando Department of Pharmacy, Medicinal Plant Unit (MPU), Botanic Garden "Giardino dei Semplici", "G. d'Annunzio" University of Chieti-Pescara, Chieti, Italy

Mirian Pateiro Centro Tecnológico de la Carne de Galicia, Parque Tecnológico de Galicia, Ourense, Spain

Leoš Pavlata Department of Animal Nutrition and Forage Production, Faculty of Agri-Sciences, Mendel University in Brno, Brno, Czech Republic

Branimir Pavlić Faculty of Technology, University of Novi Sad, Novi Sad, Serbia

Brandy Phipps Agriculture Research Development Program; Department of Agricultural and Life Sciences, Central State University, Wilberforce, OH, United States

Anne Pihlanto Natural Resources Institute Finland, Jokioinen, Finland

Milica Pojić Institute of Food Technology (FINS), University of Novi Sad, Novi Sad, Serbia

Tomasz Popek Radient Technologies Inc, Edmonton, AB, Canada

Biljana B. Rabrenović Faculty of Agriculture, University of Belgrade, Belgrade-Zemun, Serbia

Craig Schluttenhofer Agriculture Research Development Program, Central State University, Wilberforce, OH, United States

Vladimir Sikora Institute of Field and Vegetable Crops, National Institute of the Republic of Serbia, Novi Sad, Serbia

Ondřej Šťastník Department of Animal Nutrition and Forage Production, Faculty of Agri-Sciences, Mendel University in Brno, Brno, Czech Republic

Xiuzhi Sun Department of Grain Science and Industry, Kansas State University, Manhattan, KS, United States

Nemanja Teslić Institute of Food Technology, University of Novi Sad, Novi Sad, Serbia

Brijesh Tiwari Food Chemistry & Technology Department, Teagasc Food Research Centre, Ashtown, Dublin, Ireland

Vesna B. Vujasinović Faculty of Sciences, University of Novi Sad, Novi Sad, Serbia

Donghai Wang Department of Biological and Agricultural Engineering, Kansas State University, Manhattan, KS, United States

Weiqun Wang Department of Food, Nutrition, Dietetics, and Health, Kansas State University, Manhattan, KS, United States

Jikai Zhao Department of Biological and Agricultural Engineering, Kansas State University, Manhattan, KS, United States

1

Perspectives of industrial hemp cultivation

Brandy Phipps[a,b] *and Craig Schluttenhofer*[a]

[a]Agriculture Research Development Program, Central State University, Wilberforce, OH, United States [b]Department of Agricultural and Life Sciences, Central State University, Wilberforce, OH, United States

1.1 Introduction

Throughout history, mankind has relied upon diverse plants for food, textiles, shelter, animal feed, medicines, and recreation. Of all plants, *Cannabis sativa* is conceivably one of the most intricately intertwined with humanity and society. The fiber of Cannabis has been used for over 10,000 years in the production of textiles and woven goods (Long, Wagner, Demske, Leipe, & Tarasov, 2017). Seeds are consumed as a human food and fed to animals. Female flowers produce an array of pharmacologically active constituents, including the plant's well-known intoxicant used by 2.5% of the world population (WHO, 2021). The diverse and contradictory applications of Cannabis, perhaps unsurprisingly, account for the dichotomous nature of its relationship with humans: green goddess and devil weed, licit and illicit, medicine and banned drug.

The duality of Cannabis was nominal for most of its history with mankind. In the 20th century, Cannabis was differentiated into two distinct categories—marijuana and hemp—based on legal definitions and crop usage. The primary intoxicant in Cannabis is Δ^9-tetrahydrocannabinol (Δ^9-THC). Depending on the country, hemp is defined as Cannabis containing less than a specified concentration of Δ^9-THC ($< 0.3\%$ by dry weight for the United States, Canada, and European Union). In contrast, by definition, marijuana contains greater than the prescribed amount of Δ^9-THC. As the United States and other international laws refer to marijuana as the intoxicating version of Cannabis, the term will be used in this context. Marijuana is used medically or recreationally (also called "adult use"), whereas hemp serves as a food, feed, fiber, and medicine. Current estimates suggest that hemp has over 25,000 distinct uses (Johnson, 2014; Schluttenhofer & Yuan, 2017).

The chemistry of Δ^9-THC production underlies the differentiation of hemp and marijuana and is necessary for understanding the diversity of metabolites marketed. Fig. 1.1 depicts the biosynthetic pathway of cannabinoids. Biosynthesis of phytocannabinoids begins with the coacylation of a short chain starter lipid by acyl-activating enzyme 1 (Stout, Boubakir, Ambrose, Purves, & Page, 2012). In Cannabis, the starter unit is typically a hexanoate moiety but less frequently may be butanoate. The lipid-CoA molecule then undergoes addition of three malonyl-CoA by tetraketide synthase to produce 3,5,7-trioxododecanoyl-CoA or 3,5,7-trioxodecanoyl-CoA depending on the starter unit as hexanoate or butanoate, respectively (Gagne et al., 2012). Olivetolic acid cyclase cyclizes the acetaldehyde groups into a resorcinol ring, producing olivetolic acid or divarinic acid (Gagne et al., 2012). Prenyltransferase 4 then attaches a geranyl moiety to the resorcinol ring to produce cannabigerolic acid (CBGA) or cannabigerovarinic acid (CBGVA) (Fellermeier & Zenk, 1998; Luo et al., 2019). CBGA and CBGVA are derived from the olivetolic acid and divarinic acid, respectively. CBGA serves as the parent molecule for the pentyl series of cannabinoids (named for the 5-carbon alkyl chain on the resorcinol ring), whereas CBGVA leads to the propyl

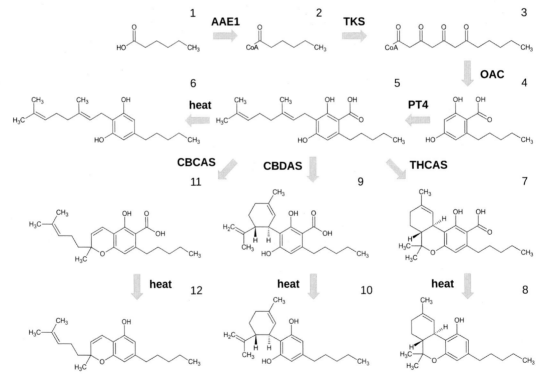

FIG. 1.1   *Cannabis sativa* biosynthetic pathway for the pentyl series of cannabinoids. Metabolites and enzymes are indicated by numbers and bold font, respectively. (1) Hexanoate; (2) hexanoyl-CoA; (3) 3,5,7-trioxododecanoyl-CoA; (4) olivetolic acid; (5) cannabigerolic acid; (6) cannabigerol; (7) Δ^9-tetrahydrocannabinolic acid; (8) Δ^9-tetrahydrocannabinol; (9) cannabidiolic acid; (10) cannabidiol; (11) cannabichromenic acid; and (12) cannabichromene. *AAE1*, acyl-activating enzyme; *TKS*, tetraketide synthase; *OAC*, olivetolic acid cyclase; *PT4*, prenyltransferase 4; *THCAS*, Δ^9-tetrahydrocannabinolic acid synthase; *CBDAS*, cannabidiolic acid synthase; and *CBCAS*, cannabichromenic acid synthase.

series (referencing the 3-carbon alkyl chain). The geranyl moiety CBGA and CBGVA can be cyclized by tetrahydrocannabinolic acid synthase (THCAS) into Δ^9-tetrahydrocannabinolic acid (Δ^9-THCA) or Δ^9-tetrahydrocannabivarinic acid (Δ^9-THCVA) (Merkus, 1971; Shoyama, Hirano, & Nishioka, 1984; Sirikantaramas et al., 2004; Taura, Morimoto, Shoyama, & Mechoulam, 1995). Alternatively, cannabidiolic acid synthase (CBDAS) can cyclize CBGA and CBGVA into cannabidiolic acid (CBDA) or cannabidivarinic acid (CBDVA), respectively (Merkus, 1971; Shoyama et al., 1984; Taura, Sirikantaramas, et al., 2007; Taura, Morimoto, & Shoyama, 1996). A third route is the cyclization of CBGA and CBGVA into cannabichromenic acid (CBCA) and cannabichromevarinic acid (CBCVA) by cannabichromenic acid synthase (Grassa et al., 2021; Morimoto, Komatsu, Taura, & Shoyama, 1997). Plants predominantly producing Δ^9-THCA or CBDA are referred to as type I and type III, respectively (de Meijer et al., 2003; Small & Beckstead, 1973). Type II plants have approximately a 1-to-1 mixture of Δ^9-THCA and CBDA. Type IV are predominantly CBGA, whereas type V lack cannabinoids (de Meijer & Hammond, 2005; de Meijer, Hammond, & Sutton, 2009; Fournier, Richez-Dumanois, Duvezin, Mathieu, & Paris, 1987; Mandolino & Carboni, 2004). Importantly, acidic cannabinoids decarboxylate into neutral cannabinoids upon heating. Δ^9-THCA, CBDA, and CBGA decarboxylate into Δ^9-THC, cannabidiol (CBD), and cannabigerol (CBG), respectively. Cannabinoids may undergo further degradation with storage, heating, and exposure to light (Doorenbos, Fetterman, Quimby, & Turner, 1971; Fairbairn, Liebmann, & Rowan, 1976; Lindholst, 2010; Zamengo et al., 2019). Δ^9-THC may be further oxidized into cannabinol (CBN), a compound that is less intoxicating (Hollister & Gillespie, 1975; Karniol, Shirakawa, Kasinski, Pfeferman, & Carlini, 1974).

As the name indicates, cannabinoids were originally thought to be solely produced by Cannabis. Later, additional plant species producing cannabinoids, or bioactive equivalents thereof, were identified. The endocannabinoid system consists of receptors and endogenous signal molecules found in vertebrates, with portions of the pathway in lower organisms (Elphick, 2012; Maccarrone, 2020; McPartland, Matias, Di Marzo, & Glass, 2006). The signaling molecules in the endocannabinoid system are long-chain fatty acids that target endogenous cannabinoid receptors. Moreover, further botanical studies identified additional plants that produce structural phytocannabinoids (i.e., those containing a resorcinol core with adjacent terpene moiety), including *Helichrysum umbraculigerum*, *Radula marginata*, *Radula perrottetii*, *Rhododendron collettianum*, and *Rhododendron dauricum* (Bohlmann & Hoffmann, 1979; Hakeem Said et al., 2017; Pollastro et al., 2017; Taura et al., 2014; Toyota et al., 2002). Additionally, select plant species produce cannabimimetic compounds (i.e., those which function on the endogenous human cannabinoid receptors but are not structural phytocannabinoids). Cannabimimetic compounds include falcarinol, *N*-benzyloleamide, and yangonin (Gertsch, Pertwee, & Di Marzo, 2010; Russo, 2016). Several phytocannabinoid compounds have been engineered into alternative biological platforms for production (Luo et al., 2019; Taura, Dono, et al., 2007), yet Cannabis remains the commercial source.

Traditionally, hemp primarily served as a fiber crop; however, increased applications in food and medicine spurred awareness and market diversification. With increased use of grain and metabolites, hemp is undergoing a renaissance, currently cultivated in over 47 countries with more approving production each year (Schluttenhofer & Yuan, 2017). This chapter will focus on the history of hemp grain use and medical applications, crop sustainability, and the current status of production.

1.2 History of hemp for food and medicine

Hemp has a long history of use as both food and medicine in communities across the globe. While the exact date that Cannabis began to be used by humans is uncertain, we know it was early in our history. Although evidence is lacking to support the assumption, some have speculated that early humans possessed Cannabis seeds over 70,000 years ago (Clarke & Merlin, 2013).

1.2.1 Cannabis in Asia

The *Li Qi* places hemp among the "five grains" of ancient China which included barley, rice, wheat, and soybeans. Hempseed remained a staple of the Chinese diet through the 10th century when other grains became more widespread (Li, 1974). Hempseeds have been found in relics from the Jin dynasty and the Han dynasty, stored with rice, millet, and wheat (Chang, 1977).

Legend says that in 2900 BC, Chinese emperor Fu Hsi referred to *ma* (the Chinese word for Cannabis) as a popular medicine with both *yin* and *yang*. Emperor Shen Nung also discovered the medicinal benefits of Cannabis in 2700 BC and taught his people how to cultivate the grain as food. He is thought to have authored *Shen-nung pen ts'ao ching* (*Divine Husbandman's Materia Medica*, also known as the *Materia Medica Sutra*), the earliest extant Chinese pharmacopeia. According to the text, while hemp contains both *yin* and *yang* properties, *ma-fen*—the flowers of the female plant—contain extensive *yin* energy and were prescribed for menstrual fatigue, malaria, and constipation, typically associated with *yin* deficiency (Li, 1974; Unschuld, 1986).

In the *Pen Ts'ao* and other ancient Chinese medical texts (i.e., *Ming'i Pieh'lu* by T'ao Hung in the 5th century), the character for hemp (*ma*) is subcategorized as either *ma-fen* or *ma-ze*, toxic and non-toxic, respectively. It was said that *ma-fen* taken in excess could cause hallucinations ("seeing devils") and taken long term could help one communicate with the spirits. The description of the taste and psychoactive properties of *ma-fen* suggest that the texts are referring to the properties of the resinous bract, rather than the seed itself (Mechoulam, 1986). Hua T'o, 2nd century Chinese surgeon, boiled Cannabis seeds in wine, which he used as an anesthetic during complicated abdominal surgeries (Julien, 1849; Li, 1974). The ethanol in the wine and the boiling likely served to extract and concentrate, respectively, the psychoactive compounds in the Cannabis.

Cannabis was prominently found in the Assyrian clay tablet pharmacopeia between 3000 and 2000 BC. Campbell-Thompson linguistically identified the Sumerian compound *ga'n-zi-gu`n-nu* as hashish, a mix of ground seeds, resinous extracts, or powdered leaves of Cannabis (Thompson, 1949).

The *Ebers Papyrus* (Bryan & Smith, 1930; Ebbell & Banov, 1937; Ghalioungui, 1987) is one of the oldest known complete medical texts (circa. 1500 BC), translated into over 800 sections and almost one thousand prescriptions with specific symptoms and diagnoses. Cannabis—*s˘ms˘mt*—is named in at least two remedies in the *Ebers Papyrus* [Formula No. 618 (plate # 78, lines10–11) and Formula No. 821 (plate # 96, lines 7–8)] (Manniche, 1989). Sicilian historian Diodorus Siculus (90–21 BC) reported the use of Cannabis by Egyptian women to relieve sorrow and bad humor (Siculus, 1933) and by the mid-13th century, the use of hashish in Egypt was widespread enough that it was taxed by the treasury (Marino, 2019).

In India, Cannabis has been considered one of the five sacred plants. Known as *bhang*, it was described in the *Atharva Veda* (written between 2000 and 1400 BC) as a sacred grass useful for disease prevention and extending life (Translation of Sacred Books of the East, 1897) (Müller, 1880). By the 10th century, *bhang* was considered a "food of the gods" with significant medicinal properties (Grierson, 1994). *Bhang* referred both to the plant itself and a drink made from the plant. One popular *bhang* recipe boils together Cannabis leaves, poppy seeds, pepper, ginger, caraway seed, cloves, cardamom, cinnamon, cucumber seed, almonds, nutmeg, rosebuds, sugar, and milk (Ball, 1910). Two other common Indian preparations of Cannabis include *ganja* and *charas*, with the latter being the most potent and made from the flowers at the height of resin production (Abel, 2013). While Cannabis was used both medicinally and recreationally, its primary function was religious, noted in Brahman, Tantric, Buddhist, and Hindu traditions (Aldrich, 1977; Avalon, 1972; Bharati, 1976; Bodhisattva, 1883; Campbell, 1893).

Persia had a collection of texts similar to the *Vedas*, known as the *Zend-Avesta*, thought to be authored by the prophet Zoroaster around 1200 BC. While much of the *Zend-Avesta* has been lost, the *Vendidad*—one of the few remaining books—refers to *bhanga* as a "good narcotic" (Darmesteter, 1880). Other Persian holy books, including the *Khorda Avesta* and the *Book of Arda* refer to *bhanga* in a similar way (Darmesteter, 1890; Horne, 1917).

1.2.2 Accounts of Cannabis in Europe

Herodotus of Halicarnassus first noted the use of Cannabis by societies north of the Mediterranean Sea. In his *The Histories* (written between 430 and 424 BC), he described the Scythian practice of inhaling hemp smoke as part of burial rituals (Rawlinson & Komroff, 2016). Scythians were nomads who lived in what is now southern Russia. An archaeological find in 1929 in the Pazyryk Valley of central Siberia confirmed the use of Cannabis smoke inhalation as a common ritual in Scythian culture (Artamonov, 1965; Rudenko, 1970; Wollner, 2001).

Pliny the Elder (79 AD)—the Roman nobleman, historian, scientist, and author of *Naturalis Historia*—referred to the use of *Cannabis* for pain and gout; and, referred to as *Kannabis* (Holland, 1601) various preparations (*emera* and *agria*) used to treat inflammation, edema, and gout were described by Greek botanist and physician Pedanius Dioscorides (90 AD) in *The Materia Medica* (Dioscorides, Gunther, & Goodyer, 1959; Dioscorides, Osbaldeston, & Wood, 2000).

Sara Benetowa (a.k.a. Sula Benet) suggested that language for Cannabis existed as far back as the Old Testament of *The Bible*, prior to the settling of the Scythians in what is modern-day Israel and was used in religious ceremonies and for intoxication (Benet, 1936, 1975; Benetowa, 1967). Recent work identified traces of cannabinoids on altars at a Judahite shrine (Arie, Rosen, & Namdar, 2020), confirming the presence of Cannabis within the region during the 8th century BC.

Early western European medical and botanical texts refer to Cannabis as hemp, prior to when it was thought to have been introduced as hashish by Napoleonic troops to France in the early 1800s (Crawford, 2002). In 5th–11th century England, two types of medicinal hemp were referred to—"manured" hemp, used for coughs and jaundice and "bastard" (wild-growing) hemp, used against nodes and tumors (Schultes, Hofmann, & Rätsch, 1992).

The *Herbarium* (11th century, rpt. 1984, CXVI, p.148) recommends hemp extracted in fat for breast health and *The Vertuous Boke of Distillacioun* suggested its use for headaches and "excess heat" (DeVriend, 1984). Other herbal texts from the 1500s to 1600s citing medical applications of Cannabis include William Turner's *The New Herball* (Turner et al., 1551); John Gerard's *The Herball or Generall Historie of Plantes* (Gerard, 1633); John Burton's *The Anatomy of Melancholy* (Burton, 1621); and Nicholas Culpeper's *The English Physitian*—later known as *Culpeper's Complete Herbal* (Culpeper, 1653). Hemp (primarily its "seed" and oil) was recommended for a variety of conditions, including worms, hot, dry coughs, jaundice, colic, gout, joint inflammation, depression, sleep disorders, and burns. The *Edinburgh New Dispensary*, originally published in 1791, also recommended hemp for skin inflammation, cough, and what was likely cystitis (Duncan, 1805). The *Doron Medicum* referred to the seed of hemp oil for joint pain and the French pox (Salmon, 1683). In France, Napoleonic troops returning from their invasion of Egypt in 1799 (Green, 2002) brought hashish home with them. Regular imports soon followed, and soon Cannabis as hashish could be purchased in most pharmacies in France. By the late 1800s, spurred in part by British colonization of India and the resin imported from this new colony, even Queen Victoria of England was known to use Cannabis for menstrual cramps. During this time, Victorian doctors throughout Europe and the United States were using *Cannabis indica* extracts to treat a variety of illnesses including epilepsy (Boire & Feeney, 2007; British Medical Association, 1997).

1.2.3 Cannabis makes its way to the Americas

In 1772, the Mexican—New Spain at the time—priest José Antonio Alzate y Ramírez wrote a newspaper article supporting the medical use of *pipiltzintzintlis* (Dierksmeier, 2020). By 1850, Cannabis was listed in the *United States Pharmacopeia* (USP, 1850). At that time, Cannabis was employed as a treatment for neuralgia, tetanus, typhus, cholera, rabies, dysentery, alcoholism, opiate addiction, anthrax, leprosy, incontinence, gout, convulsive disorders, tonsillitis, insanity, excessive menstrual bleeding, and uterine bleeding, among others. Cannabis tinctures and extracts were sold in pharmacies, and its attributes were published in medical journals such as *The Lancet* (Birch, 1889). Due to changing ideologies and policies, however, in 1942 Cannabis was officially removed from the *United States Pharmacopeia* (USP, 1942).

1.2.4 Cannabis in modern times

Hundreds of different compounds with potential biological activity, including more than 120 terpenoids; 100 cannabinoids; 50 hydrocarbons; 34 glycosidic compounds; 27 nitrogenous compounds; 25 non-cannabinoid phenolics; 22 fatty acids; 21 simple acids; 18 amino acids; 13 simple ketones; 13 simple esters and lactones; 12 simple aldehydes; 11 proteins, glycoproteins, and enzymes; 11 steroids; 9 trace elements; 7 simple alcohols; 2 pigments; and vitamin K have been identified in *Cannabis* (Hazekamp, Fischedick, Llano Díez, Lubbe, & Ruhaak, 2010).

The first attempt to successfully identify a cannabinoid was achieved by Wood, Spivey, and Easterfield (1899) who isolated cannabinol (CBN, $C_{21}H_{26}O_2$) from the exuded resin of

Indian *charas*. Its structure was elucidated first by Cahn (1932) and completed by Jacob and Todd (1940), Ghosh, Todd, and Wilkinson (1940), and Adams, Baker, and Wearn (1940). Subsequently, CBD was isolated (Adams, Pease, & Clark, 1940) and its structure reported in 1963 (Mechoulam & Shvo, 1963). A year later, Δ^9-THC was structurally characterized by the same group (Gaoni & Mechoulam, 1964).

A high affinity, stereoselective, and pharmacologically distinct cannabinoid receptor was identified in the rat brain in 1988 (Devane, Dysarz, Johnson, Melvin, & Howlett, 1988), with the human receptor encoded (Matsuda, Lolait, Brownstein, Young, & Bonner, 1990) and named CB1. CB2, a homolog that shares 48% amino acid identity, was cloned in 1993 (Munro, Thomas, & Abu-Shaar, 1993).

In 1961 the Single Convention on Narcotic Drugs, the United Nations drug control system, limited "the production, manufacture, export, import, distribution of, trade in, use and possession" of Cannabis "exclusively to medical and scientific purposes" (UNDOC, 1961). It was included under the strictest controls in the Convention—Schedule I and Schedule IV. Schedule I includes substances whose properties give rise to dependence and presents a serious risk of abuse. Schedule IV includes substances such as heroin, with extremely harmful properties and minimal evidence for medical use. For a full history of the criminalization and subsequent decriminalization of Cannabis, see Bewley-Taylor, Blickman, and Jelsma (2014).

The use of hemp as food and medicine has experienced a resurgence in recent years, as countries across the globe have legalized its use. In December 2018, the United States Food and Drug Administration completed its evaluation and approved the use of the following hempseed-derived food ingredients—hulled hempseed, hempseed protein powder, and hempseed oil—as generally recognized as safe (GRAS) for human consumption. In February 2021, an application for the first approval of hemp products as GRAS for animal feed (chickens) was submitted for approval (Hemp Feed Coalition, 2021).

There is no single European Union law on Cannabis use; each Member State determines its own response to drug use offenses. Member State laws vary, including full sanctions with penalties, decriminalization, depenalization, and full legalization. For a full review, see EMCDDA (2018).

The cultivation of *C. sativa* L. varieties is permitted in the European Union provided they are registered in the "Common Catalogue of Varieties of Agricultural Plant Species" and the Δ^9-THC content in the plant does not exceed a specific concentration. In 2019, the European Commission's Working Group of Novel Foods reclassified CBD as a Novel Food for the purposes of the European Union Novel Food Catalogue (Norwinski, Pippins, Willcocks, & Williams, 2019). As a result of this change, the European Commission now regards all extracts of hemp and derived products containing cannabinoids (including CBD) as novel. Hempseeds, flour, and seed oil without added cannabinoids remain excluded from the categorization of Novel Foods in the Catalogue.

In late 2020, the European Parliament voted in favor of the increase of Δ^9-THC level from 0.2% back to 0.3%, bringing the levels into agreement with current US and Canadian hemp laws (Fernández & Mirizzi, 2020). The regulations could also lead to marketing standards for hemp products, providing businesses new market opportunities (e.g., via trade), reduced costs, and clarity of product requirements.

1.3 The role of hemp for sustainable agriculture

1.3.1 The concept of sustainability

Global food and fiber security requires implementation of long-term sustainable agricultural practices. Sustainability encompasses the concepts of environment, society, and economics (Maynard et al., 2020; Yunlong & Smit, 1994). Environmental sustainability addresses biodiversity, efficient land use, water conservation, and air quality protection. The second pillar of sustainability is social impact, which focuses on the enhancement of human capital through education, beneficial governmental frameworks, and policy. Economic sustainability defines the long-term financial and market benefits of a crop and its production system without negatively impacting environmental and social facets. Collectively, these principles cooperate to form a sustainable agriculture system. Agriculture sustainability is best accomplished by developing a robust plan prior to implementation (Smith & McDonald, 1998). Hemp is oft touted and marketed as a sustainable crop. Sustainable utilization of hemp requires understanding how to maximize its environmental, social, and economic advantages.

Hemp is grown for fiber, grain, and metabolites (i.e., **fiber hemp**, **grain hemp**, and **metabolite hemp**, respectively) (Schluttenhofer & Yuan, 2017). Fiber, grain, and metabolite hemp crops vary in their production practices and possess distinct sustainable features. Fiber and grain hemp production is similar to large-scale forage and small grain crops, respectively. In contrast, metabolite hemp parallels production of tobacco and other horticulture crops. Each type of hemp crop has discrete and unique environmental, social, and economic sustainability outcomes.

1.3.2 Environmental sustainability

The environmental benefits of a crop depend on the species, its cultivation practices, tissues harvested, crop post-harvest processing method, and application in consumer products. Hemp crops have been proposed to have positive environmental impacts on crop production practices, soil health, water quality, and wildlife (Montford & Small, 1999). The different uses of hemp contribute to varying levels of sustainability. Table 1.1 evaluates the environmental sustainability of fiber, grain, and metabolite hemp production.

Understanding the cultivation of a crop provides insights into its sustainable and unsustainable attributes. Fiber hemp is directly seeded at a high population density (270 plants/m^2) to produce tall slender stemmed plants greater than 3–5 m in height. Plant height and density suppress weed growth. When male plants are flowering, the crop is mowed and allowed to field ret (alternatively water retted). Stalks are baled and shipped to a decortication facility for separation into fiber and hurd. Grain hemp is directly sown into the field at moderate population densities (160 plants/m^2) and produces plants 1–2 m in height (Fig. 1.2A). Plants are allowed to flower and set seed. Mature seeds are collected with a combine harvester and dried in a grain bin for storage. Grain may be used whole, dehulled, or pressed into oil that leaves a residual seed cake. In contrast, metabolite hemp is started indoors as clones or seedling transplants. Transplants are set into the field at a density of 2500–7500 plants/ha (Fig. 1.2B). Plastic mulch and drip tape may be used for weed control and irrigation, respectively. Any male plants are rogued, leaving females to produce a seedless flower crop. Plants typically

TABLE 1.1 The ability of different types of hemp crops to provide environmental services.

	Parameter	Fiber hemp	Grain hemp	Metabolites hemp
Crop production	Weed suppression	Canopy closure easily suppresses weeds	Good stand establishment prevents most weed problems	Plastics or manual/ mechanical removal used to control weeds in row
	Minimal pesticide use	Pesticides are generally not required	Pesticides are generally not required and residues are considered contaminants of grain	Limited pesticides occasionally used to control insects and diseases, but residues are considered contaminants of floral biomass
Soil	Low fertility requirement	Crop needs supplemental nitrogen to attain height necessary for profitable yield	Crop needs supplemental nitrogen for growth and grain fill	Plants are typically placed on continuous fertigation; crop needs supplemental nitrogen for growth prior to flowering
	Nutrient cycling	Retting stalks returns nutrients to field; carbonaceous material is removed	Crop residue returns nutrients to field; grain and nutrients are removed	Minimal nutrient returned to field; bulk removal of whole fresh plants removes nutrients
	Remediation of organic contaminants	Usable crop even with organic pollutant uptake	Organic pollutants contaminate crop	Organic pollutants contaminate crop
	Remediation of heavy metals	Usable crop even with uptake of heavy metals	Heavy metals contaminate crop	Heavy metals contaminate crop
Wildlife	Supports wildlife	Provides habitat for mammals, birds, and insects; pollen is food for insects and pollinators	Provides habitat for mammals, birds, and insects; pollen is food for insects and pollinators; seeds relished by birds and mammals	Provides some habitat for insects
Water	Water conservation	Crop is typically not irrigated	Crop is typically not irrigated	Crop is usually drip irrigated
	Protection of water quality	Crop and dew retting protects against water runoff; water retting can lead to microbial contaminated water	Crop and residues protect against water runoff	Raised beds may lead to surface water runoff between rows if a cover crop is not used

FIG. 1.2   Field production of grain (A) and metabolite hemp (B). *Photos courtesy of Craig Schluttenhofer.*

reach 1 m wide by 1–2 m in height at flowering. Whole plants or flowers are harvested and dried for extraction. Alternatively, metabolite hemp plants may be grown in greenhouses or indoor production systems. Such production systems require extensive external sources of soil, water, nutrients, and lighting.

Environmental sustainability incorporates crop production practices which support health of ecosystems (clean air, soil health, water quality, biodiversity, and energy conservation). Hemp crops vary in their planting density and height. The density and height of fiber plants, and to a lesser degree grain plants, suppress weed growth (Sandler & Gibson, 2019). Conversely, metabolite hemp necessitates extensive weed management in the form of manual weeding, mowing, cultivation, or plastic mulches. Herbicides may be used to eliminate weeds from fiber, grain, and metabolite hemp fields prior to planting but are generally not used while the crop is growing. Few other pesticides are typically used on the crop, despite susceptibility to insects and pathogens (McPartland, Clarke, & Watson, 2000). The limited use of pesticides is driven by (1) few registered pesticides for use on hemp crops, (2) poorly established economic thresholds of pest damage, and (3) consumer demand for contaminant-free products. The latter criterion is particularly important for grain and metabolite hemp where consumers are eating or otherwise taking the product into their bodies.

Soil health is measured by its physical, chemical, and biological properties (Cardoso et al., 2013). While information remains insufficient, different hemp crops probably have varying nutrient demands and impact on soil fertility (Wylie, Ristvey, & Fiorellino, 2021). In fiber crops, minerals are recycled back to the soil during the retting process, leaving primarily the carbonaceous stalks that are harvested. In contrast, harvesting grain and flower biomass removes mineral-rich tissues from the agroecosystem. The nutrient-dense grain is rich in minerals such as phosphorus, copper, iron, magnesium, and manganese (Callaway, 2004). Harvesting grain removes minerals from the agroecosystem. Grain production may partially return minerals to the soil through leaf and stalks left to decay in the field. Nutrient recycling in metabolite hemp production systems depends on-farm practices. Harvesting whole plants generates less in-field crop residues, rendering less nutrient-recycling into the system, versus only collecting flowers which returns minerals through left-behind leaves and stalks.

Anthropogenic practices have polluted the environment with heavy metal and organic toxins. Bioremediation rids pollutants from the environment via degradation or accumulation in biomass removed from the contaminated site followed by proper disposal. Hemp can bioremediate heavy metals and toxins from the soil (Angelova, Ivanova, Delibaltova, & Ivanov, 2004; Campbell, Paquin, Awaya, & Li, 2002; Citterio et al., 2003; Di Candito, Ranalli, & Dal Re, 2004). However, the bioremediation properties of hemp are problematic for producers growing for grain and metabolites, as accumulation and concentration of contaminants may transfer into consumer products. Concerns of consumer safety restrict production of grain and metabolite crops to non-polluted fields, limiting the beneficial environmental impact. In contrast, fiber production can take place on polluted sites as the terminal products (i.e., fiber and hurd) are utilized in non-consumed industrial products (e.g., absorbents, composites, cordage, fiberboard, insulation). Thus fiber plants may bioremediate the environment while providing a customer-safe product.

Environmentally sustainable management of water resources is essential when discussing agronomic practices. Fiber hemp is considered to require less water than cotton, producing a more sustainable natural fiber source (Schumacher, Pequito, & Pazour, 2020). Grain and fiber hemp crops typically rely on precipitation and are not irrigated, conserving water resources. Metabolite hemp crops, on the other hand, are oft irrigated. Drip irrigation systems, used for metabolite hemp, conserve more water than other watering methods (i.e., over-head irrigation, flooding furrows, etc.).

Contamination of water with soil sediment reduces water quality. When growing, the density of fiber and grain crops protects the soil surface, limiting erosion and contamination of water sources with sediment. Upon harvest, the stubble and residues continue to protect against water erosion. The raised beds and plastic used in metabolite hemp production can deflect rainfall into inter-row space, concentrating water and enhancing the potential for surface runoff. Use of cover crops between rows can reduce risks of water erosion and is practiced by some metabolite hemp growers (Fig. 1.2B). After harvesting metabolite hemp, crop or cover crop residue may be left in the field to protect against surface runoff. Plastic mulches, if used, must be managed for disposal. Complete removal of conventional plastic mulches from the field leaves rows of exposed soil prone to erosion. Biodegradable plastic mulches are incorporated into the soil via tillage, thereby exposing a bare soil to precipitation and increasing chances of water erosion.

There are two methods for preparing fiber hemp for processing (i.e., decortication)—dew retting and water retting. Dew retting involves cutting fiber hemp plants and leaving them to lay in the field a number of days until microbes partially decay the stalk, releasing the fibers. Water retting involves placing fresh-cut hemp plants in ponds to allow microbes to partially decay stalks and release fibers from the hurd. While water-retting hemp is considered to produce a higher quality fiber, it can lead to contamination of water from excessive microbial growth (Paridah, Basher, Saiful Azry, & Ahmed, 2011). Managing the microbe-laden wastewater, without damaging the environment, remains a challenge for utilizing the water-retting method.

Healthy ecosystems support wildlife survival and sustain biodiversity. Hemp crops provide ecological benefits to wildlife in the form of habitat (i.e., food, shelter, water, and space). The height and density of fiber and grain hemp stalks shelter insects, birds, reptiles, and small mammals. In contrast, metabolite hemp crops, with limited canopy cover and intensive

FIG. 1.3   Hemp grain serves as food for wildlife. (A) A mouse eating spilled grain. (B) A hemp research plot using netting to protect grain from consumption by birds. *Photos courtesy of Craig Schluttenhofer.*

management, primarily provide shelter to insects. Grain hemp, high in protein and fats, is particularly valuable to wildlife. Birds relish hempseeds and can be found energetically devouring grain from crops nearing maturity (Fig. 1.3A). Indeed, hemp grain has long been used as a bird feed. Frugivores and omnivores, such as mice and other small mammals, are also fond of hempseeds and can be found scavenging seeds dropped on the ground or in storage (Fig. 1.3B). The favorability of hemp grain to numerous species poses a possibility of utilizing this crop in conservation plantings.

The suitability of hemp to support the environment depends on the type of crop. Fiber hemp possesses multiple qualities promoting environmental sustainability. Grain hemp has more sustainable features than metabolite hemp, but less than fiber crops. Overall, metabolite hemp has a limited number of environmentally sustainable attributes. Further work is needed to fully understand the environmental impacts of the various hemp crops.

1.3.3 Societal and human sustainability

Sustainability includes the ability of a practice to better the human condition, improve well-being of citizens, build community, and enhance equitable governance. The long and complex sociopolitical history of Cannabis leads to hemp having multiple impacts, both positive and negative, on human and social sustainability.

The aging and declining number of farmers create a significant challenge to agriculture across the globe (Eistrup, Sanches, Muñoz-Rojas, & Pinto Correia, 2019; NASS, 2017; O'Meara, 2019; Zou, Mishra, & Luo, 2018). Interestingly, in the United States, legalization of hemp, along with a desire to produce the crop, has stimulated people to enter the agricultural workforce. Importantly, a portion of these new producers are younger and from historically underrepresented demographics in farming, helping revitalize and diversify the population of US farmers. The hemp industry strives to be diverse and inclusive. Hemp industry stakeholders have initiated programs to assist disadvantaged populations in succeeding within the space. Further work is needed to continue and expound upon this momentum.

Hempseed is nutrient dense, with roughly 25% protein, 35% oil, 25% carbohydrates, 10% moisture, and 5% minerals (Callaway, 2004; Galasso et al., 2016; Rodriguez-Leyva & Pierce, 2010), with variations between genotypes (Irakli et al., 2019). Seeds are consumed whole or dehulled (hearts). Hearts have a healthy balance of omega-6 to omega-3 polyunsaturated fatty acids (2.5:1) and easily digestible protein (Dubois, Breton, Linder, Fanni, & Parmentier, 2007) with sufficient essential amino acid levels (Callaway, 2004; Dubois et al., 2007; House, Neufeld, & Leson, 2010; Kriese et al., 2004; Tang, Ten, Wang, & Yang, 2006; Wang, Tang, Yang, & Gao, 2008). Pressing seeds produces edible oil and high protein seed cake used for flour. Small quantities (50 mg) of hempseeds provide at least half of the recommended daily intake of several minerals—including copper, magnesium, and zinc—and provide > 100% of the daily recommended intake of vitamins A, D, and E (Andrews et al., 2018). Hemp has strong prebiotic activity, the ability to support probiotic growth and increase the content of some bioactive compounds important for health (Nissen, di Carlo, & Gianotti, 2020). Fermented hemp flour—as a gluten-free replacement for bread—has been shown to possess increased concentration of antimicrobial compounds and a larger spectrum of bioactive volatile organic compounds compared to standard products (Nissen, Bordoni, & Gianotti, 2020). Pasta fortified with hempseed raw materials resulted in increased protein, total dietary fiber, and ash and fat content in the final pasta product (Teterycz, Sobota, Przygodzka, & Łysakowska, 2021). The nutrients available in hempseed may have protective effects against chronic diseases, including cardiovascular disease and metabolic syndrome (Rodriguez-Leyva & Pierce, 2010; Schwab et al., 2006) and there are ongoing clinical trials investigating the efficacy of hemp food products as nutritional therapy for hypertension (Samsamikor, Mackay, Mollard, & Aluko, 2020). These recent findings suggest that hemp grain products have the potential for use in fortification and formulation of novel functional foods. Increasing production of hemp grain could not only contribute to environmental sustainability in agriculture but also have significant social impact—especially in low-income, rural, and other socially and economically disadvantaged populations—by increasing production and access to nutrient-dense, lower cost sources of high-quality protein, fats, vitamins, minerals, and other important health-related bioactive compounds.

Chronic pain is a major cause of disability globally with a prevalence rate of 15%–30% in the general adult population. Unfortunately, reports indicate that less than 70% of patients are receiving adequate pain relief from current pharmacological options (Gatchel, McGeary, McGeary, & Lippe, 2014). Growing chronic pain management needs in patients along with the consequences of the opioid epidemic require effective, innovative, and safer alternatives to treat pain. Cannabis has been used as medicine for thousands of years, but its criminalization in the middle of the 20th century arrested much of the research into its compounds and their potential pain management benefits. The three major recognized medicinal components of the Cannabis plants are cannabinoids, terpenoids, and flavonoids. Evidence from animal studies suggests that CBD exerts its analgesic effects through its various interactions with and modulation of the endocannabinoid, inflammatory, and nociceptive systems (Friedman, French, & Maccarone, 2019; Kumar, Chambers, & Pertwee, 2001; Spanagel, 2020; Starowicz & Finn, 2017). Cannabinoids and other non-cannabinoid bioactive compounds in hemp have significant potential for use as primary or adjuvant treatment for pain management. For a full review of the pharmacodynamics, pharmacokinetics, animal models, and human clinical trials of Cannabis for pain management, see these references (Capano, Weaver, & Burkman,

2020; Urits et al., 2020; Vučković, Srebro, Vujović, Vučetić, & Prostran, 2018). Establishing the non-intoxicating bioactive compounds as effective potential treatments for chronic pain will have extensive social impact by increasing quality of life and productivity for a significant number of people worldwide.

Despite the positive sociopolitical impacts, hemp has human and social sustainability shortfalls. Individuals entering agriculture face challenges to access the financial and material resources necessary for success (Carlisle et al., 2019). Systemic challenges continue to hinder minority participation in agriculture (Horst & Marion, 2019). The law re-establishing the legal production of hemp in the United States contained language that many deem inequitable. Specifically, it prohibits individuals with a drug-related felony within the last 10 years from obtaining hemp licenses. This language, along with the disproportional impact of the US "War on Drugs" on minority populations (Vitiello, 2019), has led to fewer minority individuals able to obtain licenses and enter into the US hemp industry. Industry recognizes this challenge, and some companies have implemented initiatives to promote diversity within the hemp workforce.

Conflicts occur between metabolite and grain/fiber hemp growers. To maximize cannabinoid yield, growers seek to produce unpollinated (*sensimilla*, Spanish for "without seeds") flowers. Grain and fiber hemp production results in abundant pollen production. Fiber hemp is oft harvested when male plants are flowering. Grain hemp obligatorily requires pollination for seed set. Metabolite hemp producers take issue with pollen near their fields as it can pollinate their crop. Pollinated metabolite hemp flowers (i.e., containing seeds) are undesirable because of their perceived reduction in cannabinoid content. In addition, production of extracts from pollinated flowers requires extra purification steps, thereby increasing processing costs. The increased effort needed for processing may lead extractors to reject crops containing pollinated flowers.

Challenges of metabolite hemp producers with neighbors are not limited to other growers. Metabolite hemp crops have caused issues with residential neighbors. Metabolite hemp plants are strongly aromatic, especially when flowering, with a scent that may be distinguishable over considerable distances. This property has caused producers near residential housing to face complaints and lawsuits over the odor (Boteler, 2020).

Hemp legalization impacts services of local law enforcement. As plant material resembles marijuana, possession of legal hemp products, notably flowers, causes significant challenges to police. Only analytical tests suffice to confirm the identity of Cannabis material as either hemp or marijuana, with tests of suspect material backlogging existing facilities. If jurisdictions legalizing possession of hemp flower are inadequately prepared with proper analytical equipment, it may create *de facto* legalization (Borchardt, 2019).

Perhaps the greatest challenge facing hemp production is its definition. Hemp is Cannabis with less than a specified Δ^9-THC content. In Cannabis there is a continuum of Δ^9-THC levels from 0% to more than 35%. Despite intended use of the crop or best management practices, plants may exceed the Δ^9-THC limits imposed for hemp (colloquially, "hot hemp"). Producers in excess of permitted Δ^9-THC levels are subject to varying degrees of penalties. While this is less problematic for grain, fiber, and CBG hemp growers using well-bred genetics, CBD producers remain constantly at risk. Hemp producers growing for CBD face a challenging management dichotomy. The value of a metabolite crop depends upon the percent of cannabinoid per pound of dry weight. To maximize profits, growers tend to push the allowed Δ^9-THC limit to accumulate the greatest total CBD content [total $CBD = CBD + (0.8777 \times CBDA)$]. However, CBDAS generates Δ^9-THCA as a by-product (recall Δ^9-THCA decarboxylates into Δ^9-THC) (Zirpel, Kayser, & Stehle, 2018). With a 0.3% total Δ^9-THC content [total Δ^9-THC = Δ^9-THC + (0.877 × Δ^9-THCA)] tolerance level, this limits yield to 7% or 8% CBD (Toth et al., 2020). A grower's miscalculation during harvesting time could lead to excess Δ^9-THC accumulation and a non-marketable crop, plants that require destruction, and possible criminal charges.

1.3.4 Economic

Hemp is widely considered a higher-value crop, although the reality is less certain. Crop prices are subject to typical marketplace price fluctuations. Hemp, particularly that which is organically produced, may bring greater returns relative to nonorganic crops. Despite recent collapse in prices, metabolite hemp remains the most lucrative crop, followed by grain and fiber hemp. Nonetheless, the ability to find a buyer for the crops or products remains a limiting step for many hemp producers. Metabolite hemp competes with some medical and, to a lesser degree, recreational marijuana products. Fiber hemp competes with other natural fiber crops, many of which are cheaply produced worldwide. The combination of healthy oils and protein content of hemp hearts makes grain hemp a unique product that does not have many comparable crops with which to compete. Hempseed oil competes with soybean, canola, and flax seed for oil products high in polyunsaturated fats. Hempseed protein competes with other proteins, primarily soybean, pea, brown rice, and those derived from animals (whey, casein, and egg).

The environmental, social, and economic sustainability features of hemp, combined with the diversity of its crops, lead to a diverse array of sustainable industrial and consumer products. Hemp products comprised of fibers or hurd constitute reduced carbon, carbon-neutral and carbon-negative products. Observations suggest that including the term "hemp" or "made with hemp" adds value to a product. While the extent and rationale are unknown, consumers seem willing to pay more for products containing hemp. This may be beneficial for supporting economic sustainability of the hemp industry.

Most hemp grains end up dehulled into hemp hearts or expelled to produce hempseed oil, both food products. The seed cake is further processed to develop protein products—including protein powder/flour, concentrate, and isolate—also used in the food market. Industrial applications of hemp lag behind the food market. Hempseed oil produces a drying oil for protecting wooden floors and furniture. Hempseed oil can be mixed with paints and lubricants. While the European Union allows incorporation of hemp into livestock feeds (EFSA, 2011), groups in the United States are still working on approval for this application.

Fiber hemp produces sustainable products and may serve as a renewable biofuel crop. Fiber hemp yields range from 7 to 34 Mg/ha, comparable with biomass production of other bioenergy crops (Schluttenhofer & Yuan, 2017). Hemp hurds can make "hempcrete," a hurd and lime mixture used as insulation (Arrigoni et al., 2017; Elfordy, Lucas, Tancret, Scudeller, & Goudet, 2008). Fibers mixed with plastic resins produce an array of biocomposites (Chimeni, Dubois, & Rodrigue, 2018; Mohanty, Wibowo, Misra, & Drzal, 2004; Pil, Bensadoun, Pariset, & Verpoest, 2016; Yuanjian & Isaac, 2007). A product called HempWood®, comprised of compressed hemp stalks bound with natural adhesives, serves as a wood replacement for use in flooring, cabinetry, and furniture (Mirski, Boruszewski, Trociński, & Dziurka, 2017).

Hemp has a key economic disadvantage compared to other crops. Due to its relationship with marijuana, regulations require licenses for hemp production and crop testing. Licensing fees reduce farm profit margins through direct costs and lost time managing paperwork. Maintaining less than the legal limit of Δ^9-THC necessitates careful management by the producer. In addition to tests for compliance from the approving agency, growers may need to monitor their crops with further commercial testing. Analytical tests for compliance are pricey, negatively impacting grower profit margins. Canada has implemented a solution to expensive Δ^9-THC testing. When and where consistent compliance has been recorded over a number of years, a variety may be exempted from testing. While this method may be easily practiced for grain and fiber production, compliance for metabolite crops (i.e., CBD) possesses a greater challenge.

1.4 Current status of hemp cultivation and processing

After widespread prohibition of Cannabis in the 1960s decreased global production, a renewed interest in grain and fiber production occurred in the 1990s. During this era, Australia, Canada, and parts of the European Union restarted hemp production. Currently, the hemp industry is undergoing a second era of revitalization. In 2014, the United States initiated a pilot program to study the feasibility of hemp. Recently, Columbia and Ecuador adopted legislation allowing hemp production, with other countries are in the process of passing policies.

The current status of world hemp production remains in question. Statistics from the Food and Agriculture Organization (FAO) indicate North Korea and China are the largest producer of fiber (21,496 ha) and grain (12,603 ha), respectively (Table 1.2) (FAOSTAT, 2019). Cumulatively, the largest hemp producer is North Korea with 21,496 ha of hemp, followed by China (16,618 ha) and France (14,550 ha). However, FAO data does not include statistics for Canada or the United States, key hemp-producing countries. Additionally, the FAO does not include data for metabolite hemp crops. In 2019, estimates for Canada and US hemp production were 37,436 and 59,110 ha, respectively (Government of Canada, 2020; Olson & Thornsbury, 2020). In contrast to FAO estimates, a US Department of Agriculture Foreign Agriculture Service report on hemp in China placed production estimates at 66,700 ha in 2019, with half being for fiber (Mcgrath, 2020). These values place China, United States, and Canada as the top world hemp producers.

1.4.1 Grain

The grain hemp market largely remains restricted to Canada and China. Compared to other countries, Canada and China have large-scale dehulling and oil-expelling facilities that enable production of hemp grain products. The majority of Canadian hemp production occurs in the south central region (Manitoba, Saskatchewan) of the country (Canada, 2020). Seventy percent of Canadian hemp grain products are exported to the United States (Lupescu, 2019). The dependence on Canada for hemp grain products was a significant driving factor for US entry into the hemp industry in 2014. US growers were eager to establish a domestic hemp supply, capitalize on the high value grain, and strengthen the US agricultural economy. The nascent US grain hemp industry relies on foreign genetics. Northern US growers use Canadian grain

TABLE 1.2   Global hemp production by country from the FAOSTAT (2019).

Country	Product	Area (ha)
North Korea	Hemp fiber	21,496
France	Hemp fiber	14,550
Lithuania	Hemp fiber	6000
Chile	Hemp fiber	4381
China	Hemp fiber	4015
Germany	Hemp fiber	3600
Russia	Hemp fiber	3102
Austria	Hemp fiber	2010
Netherlands	Hemp fiber	1880
Poland	Hemp fiber	1830
Romania	Hemp fiber	1430
Ukraine	Hemp fiber	1333
Bulgaria	Hemp fiber	1010
Italy	Hemp fiber	910
Czech Republic	Hemp fiber	400
Greece	Hemp fiber	360
Ireland	Hemp fiber	310
Latvia	Hemp fiber	200
Spain	Hemp fiber	170
Hungary	Hemp fiber	110
Belgium	Hemp fiber	100
Slovakia	Hemp fiber	100
Turkey	Hemp fiber	28
South Korea	Hemp fiber	16
Japan	Hemp fiber	1
China	Hempseed	12,603
Russia	Hempseed	5992
Chile	Hempseed	3323
Ukraine	Hempseed	1176
Iran	Hempseed	197
Turkey	Hempseed	48

hemp varieties and some European genetics. Lower US latitudes have stronger success with southern European hemp genetics. The rapid shift to metabolite hemp production, however, has hindered the establishment of a US grain hemp industry. Europe has genetics suited for grain production but hosts limited processing facilities.

1.4.2 Cannabinoids

For decades, marijuana breeders focused on increasing Δ^9-THC (ElSohly et al., 2000; Mehmedic et al., 2010). In the mid-2000s, marijuana breeders developed cultivars with elevated levels of CBD. A cultivar high in CBD, "Cannatonic," won the Cannabis Cup in 2008, driving further interest in developing new CBD-rich cultivars. CBD-rich plants gained popularity in the medical marijuana space for treating diverse conditions, despite lack of scientific support. In particular, medical marijuana patients valued CBD-rich cultivars for the lack of or reduced intoxication, allowing for its use throughout the day. In 2013 the CNN documentary *Weed*, featuring neurosurgeon Dr. Sanjay Gupta, spurred a wave of interest in Cannabis and CBD (Gupta, 2013). The documentary covers the treatment of Charlotte Figi, a young girl with Dravet Syndrome, with a CBD-rich cultivar of Cannabis (later name "Charlotte's Web").

Passage of the *Agriculture Act of 2014* created a pilot program that opened the door for US states to explore the feasibility of domestic hemp production. The definition of hemp in this landmark legislation only specified hemp as having less than 0.3% Δ^9-THC—no usage requirement or specification to utilize only approved varieties. Prior to 2014, few hemp-derived cannabinoid products were on the market. US Cannabis companies quickly realized the opportunity for producing CBD in hemp instead of the more strictly regulated and limited access marijuana. The success of the pilot programs, specifically the cannabinoid market, led to language in the *Agriculture Improvement Act of 2018* that officially legalized hemp production in the United States (Schluttenhofer & Yuan, 2019).

An interest in cannabinoids has precipitated a dramatic change within the hemp industry. In 2011, Switzerland redefined its hemp laws to raise the THC limit to 1.0%. Reports indicate limited changes took place in the market over the next several years (EMCDDA, 2020). In 2015, CBD-rich genetics (>5% total CBD by dry weight) were being grown and extracts sold in the US market. Metabolite hemp genetics were derived from marijuana accessions instead of traditional grain and fiber varieties (Grassa et al., 2021). CBD-rich genetics were developed by selecting Cannabis plants that contain the CBDAS alleles instead of THCAS. In 2016, the metabolite hemp market began a rapid expansion in Europe, spreading to over 25 countries by 2020 (Zobel, Notari, Schneider, & Rudmann, 2019). By 2019, 94% of US growers were producing hemp for metabolites, primarily CBD (McVey et al., 2019).

Type III, CBD-rich progeny of marijuana crosses were rapidly deployed into the US market. Early CBD-rich cultivars included type III selections of "Cannatonic," "ACDC," "Cherry," "Boax," "Otto II," and "Cherry Wine." These parents have subsequently been crossed to marijuana plants to produce hundreds of CBD-rich hemp cultivars. As the CBD market expanded, companies sought to diversify their offerings of cannabinoid products. In 2019, the first type IV CBG-rich genetics (>5% total CBG by dry weight) were introduced to the market. New cultivars producing other cannabinoids (CBDV, CBC, and CBN) continue to be released.

With the development of the global hemp-derived cannabinoid market came a wide range of new products. Arrival of CBD-rich genetics led to the production of diverse hemp extracts.

Extracts include full spectrum, broad spectrum, and isolates. Full-spectrum products contain up to the legal limit of Δ^9-THC. Broad-spectrum products have undergone a separation process, typically distillation or chromatography, to remove the Δ^9-THC. Broad-spectrum products, particularly those which have not been diluted, are sometimes called distillates. Isolates are refined and purified to contain only a single compound, usually in the form of a crystalline powder.

Production of extracts requires a number of steps, depending upon the final product. Dried floral material is extracted with supercritical carbon dioxide (scCO$_2$) or ethanol to produce a raw extract. In addition to cannabinoids, this extract contains terpenes, phenolics, lipids, and chlorophylls. Raw extracts may be refined by a process called "winterization." Winterization involves dissolving the extract in ethanol and chilling the solution to precipitate out lipids. Winterized solutions are then filtered to remove precipitates. Chlorophylls may also be removed by winterization or through photodegradation methods to generate a transparent/translucent product. Distillation or chromatography may then be used to remove Δ^9-THC or other cannabinoid fractions. The same steps may also be used to purify a desired cannabinoid to produce the isolate. During extraction, heat generated intentionally or as a result of processing leads to decarboxylation of cannabinoids into the neutral form (e.g., CBD). Upon completion of processing, full-spectrum extracts must be diluted to reduce the Δ^9-THC below legal limits. While either method may be used, ethanol extraction generally creates full- or broad-spectrum products. Conversely, scCO$_2$ is used for production of broad-spectrum products and isolates.

Other less common hemp products include formulations more frequently found within the marijuana market—rosins, live resins, hashes, and concentrates. Rosins are produced by using a heated hydraulic press to compress the flower which forces out the resinous extracts. Extraction of fresh material and maintenance of low temperatures during process inhibits cannabinoid decarboxylation producing a "live resin" (i.e., retains cannabinoids in the acidic form, e.g., CBDA). Shatters, crumbles, and budders are produced via hydrocarbon (e.g., butane) solvent extraction and then vacuum purged to remove any residual solvent. Hash products are created by physical removal of the trichomes followed by applied pressure, with or without gentle heating, to form a solid mass. Hemp-derived versions of these products are available to consumers in the United States and Europe.

Extracts and isolates are used in wellness, cosmetics, and food products. Full- and broad-spectrum extracts are sold as "oils" (an extract), tinctures, or capsules. Isolates have more diverse applications. In addition to crystalline powders, extracts, tinctures, and capsules, isolates are incorporated into a wide variety of foods (e.g., chocolate, honey, gummies, protein bars), beverages (e.g., coffee, energy drinks, teas), and personal care products (e.g., bath salts, salves, soaps, transdermal patches). Extracts and isolates are used to make e-liquids for vape devices. Vaping extracts or isolates, dissolved in a carrier solvent (typically, a combination of glycerin and propylene glycol), serve as a common method for consuming cannabinoid products. Versions of extracts and isolates for pets are also available. Cannabinoid-infused clothing has been developed for managing joint and body pain. Many products lack scientific evidence supporting their efficacy.

Expansion of market options oft requires novel formulation of products. Corporate desire for CBD-infused beverages required identifying a method for dissolving a hydrophobic molecule into a polar solvent (i.e., water). Despite cannabinoids' hydrophobicity, water-soluble products have been developed. Creating water-soluble cannabinoid formulations involves either using additives to enhance solubility or suspending them in the solution as micelles. Water-soluble products are now widely used in beverages, foods, and personal care products.

Traditionally, smoked Cannabis products were the domain of the marijuana industry. Over the last decade, the United States and Europe have developed markets for smoked hemp flower. In Europe and the United States, referred to as "Cannabis light" and "smokable hemp," respectively, cured hemp flowers for smoking command a premium price. High-quality smokable hemp production requires proper plant production, drying, curing, and trimming of flowers, akin to processing for marijuana.

The US policy created a unique opportunity for the metabolite hemp sector to flourish. The United States continues to dominate the hemp-derived cannabinoid market. European countries and Canada have had challenges with accessing CBD-rich germplasm due to regulations requiring growers to use cultivars on the European Commission or Health Canada approved lists, respectively. Switzerland and Italy, with their higher tolerances for Δ^9-THC levels, are better positioned to capitalize on the metabolite hemp market. Legalization of marijuana in Canada allowed processors to produce and sell CBD (Schluttenhofer, 2018). CBD in Canada may be produced (1) from any Cannabis plant growing under license of the *Cannabis Regulations* or (2) in hemp, if plants are licensed per *Industrial Hemp Regulations* and come from the list of approved varieties. A 2020 report indicates that China, at least regionally, has begun producing CBD products (Mcgrath, 2020). The extent of Chinese CBD production remains uncertain.

1.4.3 Improving safety of cannabinoid products

As hemp metabolites usually serve as a medicinal product, there are rising safety concerns about contaminants. Over the past half decade, the US market has progressed from only testing for Δ^9-THC and CBD levels in products to routinely testing for additional cannabinoids, terpenes, pesticide residues, and heavy metals. Hemp producers and processors now have options of testing a given crop or product for cannabinoids, terpenes, moisture content, pesticides, heavy metals, microbes, mycotoxins, residual solvents, and foreign objects. Aside from tests for levels of Δ^9-THC for regulatory compliance, no tests are currently required in the United States, and few companies conduct all tests for a given product. Continuing to sell hemp as health products requires consistent safety testing using industry standard methods. Although various organizations are developing standardized methods for Cannabis analyses, currently each lab uses their own protocols. Companies selling cannabinoid products may provide "certificates of analysis" indicating test results for a given lot of goods to consumers. However, fraudulent, stolen, and old certificates of analysis have been observed. Further work is needed to advance the safety of hemp products.

In 2019, some e-cigarette users developed breathing complications, with some cases resulting in death (Jonas & Raj, 2020; Knopf, 2019). The condition, termed e-cigarette or vaping product use-associated lung injury (EVALI), was associated with vaping (Lewis et al., 2019). Later studies linked vitamin E acetate in vape products with EVALI (Wu & O'Shea, 2020). The EVALI event precipitated three key shifts within the hemp space. The first change involved companies quickly removing vitamin E acetate from products. Companies began advertising products as vitamin E acetate-free in attempts to reassure customers of product safety. The second change was a shift of consumers away from vape products and toward "smoked" hemp. After the EVALI epidemic, consumer demand for cured hemp flower products increased. Finally, US states have started implementing policies controlling contaminants in Cannabis e-cigarette products (Schaneman, 2020).

1.4.4 The cannabinoid market and oversupply

Hemp is a prolific producer of specialized metabolites, particularly cannabinoids. As with tobacco, limited acreage is sufficient to supply market demand. Until 2019, the high value to metabolite hemp provided growers with potentially large profit margins on small acreage production. Recent price crashes resulted from an oversupply of metabolite hemp. Table 1.3 indicates the kilogram of cannabinoid needed to fill demand for a given percentage of the world population. The cannabinoid yield (Table 1.4) and percent recovery are used to calculate the number of hectares needed to supply the market, according to the following equation:

$$\text{Number of ha needed} = (\text{cannabinoids need}) / (\text{cannabinoid yield} \times \%\text{recovery})$$

where

$$\%\text{recovery} = 100 - \%\text{losses}$$

and

$$\%\text{losses} = (\%\text{farm losses} + \%\text{extraction losses})$$

In the United States, approximately 14%–28% of people utilize CBD products on a regular basis, with a majority of survey respondents using them for analgesic, anxiolytic, or soporific properties (Acosta, 2019; Berger, 2021; Brightfield Group, 2017; Gill, 2019). Research studies have found that doses of CBD ranging from 300 to 1000 mg/day significantly reduce mental health disorders (Bonaccorso, Ricciardi, Zangani, Chiappini, & Schifano, 2019; Crippa et al., 2009; Khan et al., 2020; Larsen & Shahinas, 2020; Skelley, Deas, Curren, & Ennis, 2020). In contrast, online guides suggest doses of CBD in the range of 5–100 mg/day (DailyCBD, 2021; Fiorenzi, 2021; Moltke, 2019; Simone, 2020).

Assuming 1% of the world population consumed 30 mg CBD/d, the estimate for the world need would be 854,100 kg of CBD per year. US' growers can typically reach around 7% CBD while maintaining legal compliance for THC (Toth et al., 2020), which gives 78.5 kg CBD/ha. Typical extraction processes lose 20%–30% of the material. The on-farm losses include loss of cannabinoids due to handling, drying, stripping, storage, and degradation. Assuming farm and extraction losses of 20% and 30%, respectively, an estimated 21,760 ha are needed to supply this market. For comparison, the United States alone produced 36,421 ha of hemp for CBD in 2019, an estimated 1.67 times the market need for CBD.

1.4.5 Controversy of new cannabinoid products

In the United States, there is disparity in states obeying and rejecting federal prohibition of marijuana. The ample supply of CBD, ease of derivation into Δ^8-THC or Δ^{10}-THC, and illicit nature of marijuana in states created a market of hemp-derived intoxicants. Similarly, Δ^9-THC

TABLE 1.3 The amount of cannabinoid needed (kg) gives a percent of the world population taking a specified dose for a year.

CBD (mg/day)	% World population							
	0.01%	0.10%	1.00%	5.00%	10.00%	25.00%	50.00%	100.00%
1	285	2847	28,470	142,350	284,700	711,750	1,423,500	2,847,000
10	2847	28,470	284,700	1,423,500	2,847,000	7,117,500	14,235,000	28,470,000
20	5694	56,940	569,400	2,847,000	5,694,000	14,235,000	28,470,000	56,940,000
30	8541	85,410	854,100	4,270,500	8,541,000	21,352,500	42,705,000	85,410,000
40	11,388	113,880	1,138,800	5,694,000	11,388,000	28,470,000	56,940,000	113,880,000
50	14,235	142,350	1,423,500	7,117,500	14,235,000	35,587,500	71,175,000	142,350,000
60	17,082	170,820	1,708,200	8,541,000	17,082,000	42,705,000	85,410,000	170,820,000
70	19,929	199,290	1,992,900	9,964,500	19,929,000	49,822,500	99,645,000	199,290,000
80	22,776	227,760	2,277,600	11,388,000	22,776,000	56,940,000	113,880,000	227,760,000
90	25,623	256,230	2,562,300	12,811,500	25,623,000	64,057,500	128,115,000	256,230,000
100	28,470	284,700	2,847,000	14,235,000	28,470,000	71,175,000	142,350,000	284,700,000
150	42,705	427,050	4,270,500	21,352,500	42,705,000	106,762,500	213,525,000	427,050,000
200	56,940	569,400	5,694,000	28,470,000	56,940,000	142,350,000	284,700,000	569,400,000
250	71,175	711,750	7,117,500	35,587,500	71,175,000	177,937,500	355,875,000	711,750,000
300	85,410	854,100	8,541,000	42,705,000	85,410,000	213,525,000	427,050,000	854,100,000
350	99,645	996,450	9,964,500	49,822,500	99,645,000	249,112,500	498,225,000	996,450,000
400	113,880	1,138,800	11,388,000	56,940,000	113,880,000	284,700,000	569,400,000	1,138,800,000
450	128,115	1,281,150	12,811,500	64,057,500	128,115,000	320,287,500	640,575,000	1,281,150,000
500	142,350	1,423,500	14,235,000	71,175,000	142,350,000	355,875,000	711,750,000	1,423,500,000
600	170,820	1,708,200	17,082,000	85,410,000	170,820,000	427,050,000	854,100,000	1,708,200,000
700	199,290	1,992,900	19,929,000	99,645,000	199,290,000	498,225,000	996,450,000	1,992,900,000
800	227,760	2,277,600	22,776,000	113,880,000	227,760,000	569,400,000	1,138,800,000	2,277,600,000

900	256,230	2,562,300	25,623,000	128,115,000	256,230,000	640,575,000	1,281,150,000	2,562,300,000
1000	284,700	2,847,000	28,470,000	142,350,000	284,700,000	711,750,000	1,423,500,000	2,847,000,000
1100	313,170	3,131,700	31,317,000	156,585,000	313,170,000	782,925,000	1,565,850,000	3,131,700,000
1200	341,640	3,416,400	34,164,000	170,820,000	341,640,000	854,100,000	1,708,200,000	3,416,400,000
1300	370,110	3,701,100	37,011,000	185,055,000	370,110,000	925,275,000	1,850,550,000	3,701,100,000
1400	398,580	3,985,800	39,858,000	199,290,000	398,580,000	996,450,000	1,992,900,000	3,985,800,000
1500	427,050	4,270,500	42,705,000	213,525,000	427,050,000	1,067,625,000	2,135,250,000	4,270,500,000

TABLE 1.4 The cannabinoid yield (kg/ha) produced at a dried flower yield with a given cannabinoid percentage.

Yield (kg flower/ha)	Percent cannabinoid													
	1%	2%	3%	4%	5%	6%	7%	8%	9%	10%	11%	12%	13%	14%
280	2.8	5.6	8.4	11.2	14	16.8	19.6	22.4	25.2	28	30.8	33.6	36.4	39.2
560	5.6	11.2	16.8	22.4	28	33.6	39.2	44.8	50.4	56	61.6	67.3	72.9	78.5
841	8.4	16.8	25.2	33.6	42	50.4	58.8	67.3	75.7	84.1	92.5	100.9	109.3	117.7
1121	11.2	22.4	33.6	44.8	56	67.3	78.5	89.7	100.9	112.1	123.3	134.5	145.7	156.9
1401	14	28	42	56	70.1	84.1	98.1	112.1	126.1	140.1	154.1	168.1	182.1	196.1
1681	16.8	33.6	50.4	67.3	84.1	100.9	117.7	134.5	151.3	168.1	184.9	201.8	218.6	235.4
1961	19.6	39.2	58.8	78.5	98.1	117.7	137.3	156.9	176.5	196.1	215.8	235.4	255	274.6
2242	22.4	44.8	67.3	89.7	112.1	134.5	156.9	179.3	201.8	224.2	246.6	269	291.4	313.8
2522	25.2	50.4	75.7	100.9	126.1	151.3	176.5	201.8	227	252.2	277.4	302.6	327.8	353.1
2802	28	56	84.1	112.1	140.1	168.1	196.1	224.2	252.2	280.2	308.2	336.3	364.3	392.3

can be derivatized into CBN, a mild intoxicant. The intoxicant Δ^9-tetrahydrocannabidivarin (THCV) can also be extracted from hemp selected for production of propyl-type cannabinoids. The legality of Δ^8-THC, Δ^{10}-THC, CBN, and THCV remains uncertain. Traditionally, advocates have promoted hemp as non-intoxicating to foster the crop's legalization and adoption. Uses of intoxicating metabolites from hemp undermine advocacy and have created a division within the hemp industry—proponents for both permitting and prohibiting hemp-derived intoxicants. Supporters are capitalizing on economic benefits and expansion of the saturated metabolite hemp market. Opponents take ethical issues with using hemp to provide products akin to marijuana, suggesting these compounds should be restricted to that sector. Uncertainty is further driven by the lack of health studies on non-Δ^9-THC intoxicants. Without clear laws regarding intoxicating compounds, other countries allowing hemp-derived metabolite products may come to face a similar scenario.

1.5 Challenges and opportunities

1.5.1 Challenges

Despite worldwide progress in legalization and market expansion, significant challenges still exist for the hemp industry. Many of the issues countries continue to struggle with are in the emerging metabolite hemp sector. Problem areas include legality of compounds derived from hemp, stability of genetics, overlap with pharmaceutical applications, rapid diversification of products, and product safety.

The primary challenge facing the metabolite hemp sector is overproduction. Currently, supply far exceeds demand, driving down crop prices. Expansion of the market is needed to allow growth of the sector. While widely marketed in the United States and allowed by law, the FDA continues to evaluate data and debate on allowing hemp-derived metabolites as herbal supplements. One of the challenges faced by the FDA is an approved pharmaceutical version of CBD (Jamie & Rod, 2018; Mead, 2019; Rubin, 2018). Per regulations, the presence of an existing pharmaceutical drug precludes marketing the compound as an herbal supplement. Other countries face similar struggles with the legality of CBD and hemp-derived metabolites. Clarification of local, federal, and international regulations can provide certainty to the metabolite hemp market, allowing possible expansion. How regulations develop for hemp-derived intoxicating compounds may also have considerable impact on the metabolite hemp market.

Another major issue with metabolite hemp production is the limited availability of stable genetics. European and Canadian growers have few approved CBD-rich varieties. US' growers do not have to comply with a standard list of varieties and therefore have access to CBD-rich genetics. This makes new genetics quickly available to growers compared to traditional variety development practices. While improving, many of the seeds sold in the United States lack sufficient breeding to stabilize important production traits (e.g., chemistry, flowering time, plant architecture). "Feminized" seeds to produce uniform female populations come with additional challenges in the quality of crosses used to produce the seed. Some growers have lost thousands of dollars buying feminized seeds which were not fully feminized (i.e., only female plants). Growers should seek seed that has been evaluated for its quality of feminization. Clones from selected female plants have been used to overcome issues with feminized seeds; however, they are more expensive. With reducing returns, clones further narrow a grower's profit margin. In addition to conventionally bred seeds, new regulations from the Association of Seed Certifying Agencies (AOSCA) allow variety certification for clones, feminized seed, and transplants. AOSCA certification confirms a variety meets specified purity and identity standards, which provides growers' confidence in the approved material.

The primary challenge of the grain hemp sector is lack of processing infrastructure. Currently, Canada is the primary grain hemp producing country. Canadian companies equipped with dehulling and oil expelling facilities provide growers market outlets for the grain. Outside of Canada, such facilities are limited. Even in the United States, which is the largest hemp grain consumer, few grain hemp processing companies exist, hindering development of this market sector. Expansion of infrastructure, globally and domestically, is needed to expand the world hemp grain market.

The grain market has the opportunity to expand if seeds are allowed in more products. Food and feed regulators are concerned that trace levels of cannabinoids in products will transfer to the consumer. While hemp hearts, hempseed oil, and hemp protein powder are GRAS in the United States, there remain concerns about their use in animal feed. Animals may bioaccumulate cannabinoids to levels exceeding daily dose requirements. The lack of data supporting appropriate levels for the daily dose requirements further hinder approval of these products for feed applications in the United States. European countries allow hemp grain products in animal feed, provided the content does not exceed specified thresholds so as to meet daily dose (EFSA, 2011).

1.5.2 Opportunities

Hemp can be, and is, a vector for significant global changes. For as many challenges as the global hemp industry faces, there is a corresponding number of opportunities. Opportunities can be consolidated into two broad categories: improving sustainability and enhancing human health.

With a growing world population and increased demand for food, fiber, and medicine—along with declining water, soil, and space resources—methods for increasing sustainability are essential for the future. Hemp has a strong potential to positively impact environmental, human, societal, and economic sustainability. The beneficial environmental impacts to soil health, water quality, conservation of resources, and wildlife can provide a diverse array of more sustainable consumer products while protecting the planet. Recent changes in perception of hemp around the globe have led to more sustainable human and social policies, although further work remains. The diversity of crops and products produced by hemp enhances market opportunities for farmers and companies. Large-scale production of hemp may diversify crop markets and positively impact prices for other crops by reducing tendency for domestic overproduction. Further exploration of hemp crops as they relate to sustainability is highly warranted.

Plants have long served as a source of medicines. Hemp provides new market options for development of pharmaceutical drugs and herbal supplements, both directly and indirectly. Cannabis-derived CBD, CBG, CBDV, and other cannabinoids are actively being explored as pharmaceuticals. As some regulatory antagonism exists between availability of pharmaceuticals and herbal supplements, a careful path must be established to retain both market opportunities. Metabolites from non-Cannabis species may have valuable agricultural, industrial, or medicinal applications. Historically, metabolites have been expensive to extract and isolate, making them cost prohibitive for commercial use. The hemp-derived cannabinoid market has brought rapid developments in the scale and speed of scCO$_2$, ethanol, solvent (propane and butane), and solvent-less (rosin presses, etc.) extraction processes. To minimize costs, the hemp industry has focused on using large-scale scCO$_2$ and ethanol extraction methods. CO$_2$ and ethanol are sustainable solvents as they can be captured from industrial process. Ethanol, in particular, has been named an environmentally friendly solvent (Capello, Fischer, & Hungerbühler, 2007). Additionally, there have been key advancements in large-scale purification of extracts into single compounds. Such technologies and infrastructure now make it feasible to extract and isolate metabolites from other species in a potentially cost-effective manner. Overall, there are opportunities for the hemp industry to diversify into new crops and increase the availability of inexpensive and environmentally sustainable metabolite products for use as agriculture, industrial, and medicinal chemicals.

The foods people eat impact their overall health and well-being. Hemp grain is a nutrient-dense food rich in easily digestible protein and healthy polyunsaturated fats. The high nutrient density of hemp grain can potentially help mitigate global epidemics of chronic diseases (i.e., cardiovascular disease, diabetes, and some cancers) (Albracht-Schulte et al., 2018; Fabian, Kimler, & Hursting, 2015; Jang & Park, 2020; Simopoulos, 2016). Regions and countries suffering inadequate nutrition could benefit from consumption of hemp grain. The GRAS status of hemp hearts, hempseed oil, and hemp protein provides opportunities for increased use of ingredients within commercial food products. Increasing grain hemp production can produce nutrient-dense foods while protecting the environment. Further work is needed to better understand the health implications of hemp grain.

1.6 Conclusion

Hemp has a long history of use as a source of food, fiber, and medicine. In the 20th century, hemp production declined due to its association with marijuana. Within the last few decades, hemp has reemerged as an agricultural crop in countries across the globe. Recent changes in perception have led to a renewed interest in this ancient crop. The resurgence of interest in hemp is timely as the crop can fulfill important niches in environmental, human, societal, and economic sustainability. While the crop poses tremendous potential for improving agriculture, the hemp industry faces many challenges. Limitations span the crop life-cycle, from production to consumer products. The rapidly expanding and diversifying metabolite hemp sectors possess unique difficulties in simultaneously managing access to new products while complying with local, national, and international regulations. Expansion of the hemp grain market requires greater regional access to processing equipment and demand for products. Despite the challenges, the hemp industry faces inordinate opportunities to improve sustainability and human health. The rapid changes in the industry are anticipated to result in advances of hemp research, production, processing, consumer products, sustainability, and health within the coming years.

Acknowledgments

This work was supported by the United States Department of Agriculture fund number NI201445XXXXG018-0001.

References

Abel, E. L. (2013). *Marihuana: The first twelve thousand years*. New York: McGraw Hill.

Acosta. (2019). *New Acosta report finds 28 percent of consumers use CBD products daily or as-needed*. Retrieved from https://www.prnewswire.com/news-releases/new-acosta-report-finds-28-percent-of-consumers-use-cbd-prod-ucts-daily-or-as-needed-300915291.html.

Adams, R., Baker, B., & Wearn, R. (1940). Structure of cannabinol. III. Synthesis of cannabinol, 1-hydroxy-3-n-amyl-6, 6, 9-trimethyl-6-dibenzopyran1. *Journal of the American Chemical Society, 62*(8), 2204–2207.

Adams, R., Pease, D. C., & Clark, J. H. (1940). Isolation of cannabinol, cannabidiol and quebrachitol from red oil of Minnesota wild hemp. *Journal of the American Chemical Society, 62*(8), 2194–2196. https://doi.org/10.1021/ja01865a080.

Albracht-Schulte, K., Kalupahana, N. S., Ramalingam, L., Wang, S., Rahman, S. M., Robert-McComb, J., & Moustaid-Moussa, N. (2018). Omega-3 fatty acids in obesity and metabolic syndrome: A mechanistic update. *Journal of Nutritional Biochemistry, 58*, 1–16. https://doi.org/10.1016/j.jnutbio.2018.02.012.

Aldrich, M. R. (1977). Tantric Cannabis use in India. *Journal of Psychedelic Drugs, 9*(3), 227–233. https://doi.org/10.1080/02791072.1977.10472053.

Andrews, K. W., Gusev, P. A., McNeal, M., Savarala, S., Dang, P. T. V., Oh, L., … Douglass, L. W. (2018). Dietary supplement ingredient database (DSID) and the application of analytically based estimates of ingredient amount to intake calculations. *Journal of Nutrition, 148*(Suppl. 2), 1413S–1421S. https://doi.org/10.1093/jn/nxy092.

Angelova, V., Ivanova, R., Delibaltova, V., & Ivanov, K. (2004). Bio-accumulation and distribution of heavy metals in fibre crops (flax, cotton and hemp). *Industrial Crops and Products, 19*(3), 197–205. https://doi.org/10.1016/j.indcrop.2003.10.001.

Arie, E., Rosen, B., & Namdar, D. (2020). Cannabis and frankincense at the Judahite shrine of Arad. *Tel Aviv, 47*(1), 5–28. https://doi.org/10.1080/03344355.2020.1732046.

Arrigoni, A., Pelosato, R., Melià, P., Ruggieri, G., Sabbadini, S., & Dotelli, G. (2017). Life cycle assessment of natural building materials: The role of carbonation, mixture components and transport in the environmental impacts of hempcrete blocks. *Journal of Cleaner Production, 149*, 1051–1061. https://doi.org/10.1016/j.jclepro.2017.02.161.

Artamonov, M. (1965). Frozen tombs of the Scythians. *Scientific American, 212*(5), 100–109.

Avalon, A. (1972). *Tantra of the great liberation (Mahanirvana tantra)*. New York: Dover.

Ball, M. (1910). The effects of haschisch not due to cannabis indica. *Therapeutic Gazette, 34*, 777–780.

Benet, S. (1936). *Konopie w wierzeniach i zwyczajach ludowych*. Warszawa: Nakl. Towarzystwa Naukowego Warszawskiego.

Benet, S. (1975). Early diffusion and folk uses of hemp. In *Cannabis and culture* (pp. 39–50). De Gruyter Mouton.

Benetowa, S. (1967). Tracing one word through different languages. In G. Andrews, & S. Vinkenoog (Eds.), *The book of grass: An anthology of Indian hemp* (pp. 15–18). New York: Grove Press.

Berger, K. (2021). CBD statistics 2021. *The Checkup*. Retrieved from https://www.singlecare.com/blog/news/cbd-statistics/.

Bewley-Taylor, D., Blickman, T., & Jelsma, M. (2014). *The rise and decline of cannabis prohibition. The history of cannabis in the UN drug control system and opions for reform*. Amsterdam: Jubels. Retrieved from https://www.tni.org/files/download/rise_and_decline_web.pdf.

Bharati, A. (1976). *The tantric tradition*. London: Rider and Co.

Birch, E. A. (1889). The use of Indian hemp in the treatment of chronic chloral and chronic opium poisoning. *Lancet, 1*(625), 25.

Bodhisattva, A. (1883). In S. Beal (Ed.), *Vol. 19. The Fo-sho-hing-tsan-king: A life of Buddha*. Oxford: Clarendon Press.

Bohlmann, F., & Hoffmann, E. (1979). Cannabigerol-ähnliche verbindungen aus Helichrysum umbraculigerum. *Phytochemistry, 18*(8), 1371–1374. https://doi.org/10.1016/0031-9422(79)83025-3.

Boire, R., & Feeney, K. (2007). *Medical Marijuana Law*. Ronin Publishing.

Bonaccorso, S., Ricciardi, A., Zangani, C., Chiappini, S., & Schifano, F. (2019). Cannabidiol (CBD) use in psychiatric disorders: A systematic review. *Neurotoxicology, 74*, 282–298. https://doi.org/10.1016/j.neuro.2019.08.002.

Borchardt, J. (2019). No, Ohio did not accidentally legalize marijuana. But a new hemp legalization law is causing problems for prosecutors. *Cincinnati Enquirer*. Retrieved from https://www.cincinnati.com/story/news/2019/08/09/ohios-hemp-law-causing-problems-prosecuting-marijuana-crimes/1954669001/.

Boteler, C. (2020). Odor of industrial hemp farm has Baltimore County residents fuming. *The Baltimore Sun*. Retrieved from https://www.baltimoresun.com/maryland/baltimore-county/cng-co-to-industrial-hemp-20200217-offz-6gi6rrhmfjcqmzqhcdnl5a-story.html.

Brightfield Group. (2017). *Understanding Cannabidiol*. Retrieved from https://daks2k3a4ib2z.cloudfront.net/595e80a3d32ef41bfa200178/59946dd86c6b200001c5b9cb_CBD_-_HelloMD_Brightfield_Study_-_Expert_Report_-_FINAL.pdf.

British Medical Association. (1997). *Therapeutic uses of cannabis*. CRC Press.

Bryan, C. P., & Smith, G. E. (1930). *The papyrus Ebers: Translated from the German version*. London: G. Bles.

Burton, R. (1621). *The anatomy of melancholy*. Project Gutenburg.

Cahn, R. S. (1932). Cannabis indica resin. Part III. The constitution of cannabinol. *Journal of the Chemical Society*, 1342–1353.

Callaway, J. C. (2004). Hempseed as a nutritional resource: An overview. *Euphytica, 140*(1), 65–72. https://doi.org/10.1007/s10681-004-4811-6.

Campbell, J. M. (1893). On the religion of hemp. In *Vol. 2. Indian Hemp Drugs Commission Report* (pp. 250–252).

Campbell, S., Paquin, D., Awaya, J. D., & Li, Q. X. (2002). Remediation of benzo[a]pyrene and chrysene-contaminated soil with industrial hemp (*Cannabis sativa*). *International Journal of Phytoremediation, 4*(2), 157–168. https://doi.org/10.1080/15226510208500080.

Capano, A., Weaver, R., & Burkman, E. (2020). Evaluation of the effects of CBD hemp extract on opioid use and quality of life indicators in chronic pain patients: A prospective cohort study. *Postgraduate Medicine, 132*(1), 56–61. https://doi.org/10.1080/00325481.2019.1685298.

Capello, C., Fischer, U., & Hungerbühler, K. (2007). What is a green solvent? A comprehensive framework for the environmental assessment of solvents. *Green Chemistry, 9*(9), 927–934. https://doi.org/10.1039/B617536H.

Cardoso, E. J. B. N., Vasconcellos, R. L. F., Bini, D., Miyauchi, M. Y. H., Santos, C. A. D., Alves, P. R. L., … Nogueira, M. A. (2013). Soil health: Looking for suitable indicators. What should be considered to assess the effects of use and management on soil health? *Scientia Agricola, 70*, 274–289.

Carlisle, L., de Wit, M. M., DeLonge, M. S., Calo, A., Getz, C., Ory, J., … Press, D. (2019). Securing the future of US agriculture: The case for investing in new entry sustainable farmers. *Elementa: Science of the Anthropocene, 7*. https://doi.org/10.1525/elementa.356.

Chang, K.-C. (1977). *Food in Chinese culture: Anthropological and historical perspectives*. New Haven: Yale University Press.

Chimeni, D. Y., Dubois, C., & Rodrigue, D. (2018). Polymerization compounding of hemp fibers to improve the mechanical properties of linear medium density polyethylene composites. *Polymer Composites, 39*(8), 2860–2870. https://doi.org/10.1002/pc.24279.

Citterio, S., Santagostino, A., Fumagalli, P., Prato, N., Ranalli, P., & Sgorbati, S. (2003). Heavy metal tolerance and accumulation of Cd, Cr and Ni by *Cannabis sativa* L. *Plant and Soil, 256*(2), 243–252. https://doi.org/10.1023/a:1026113905129.

Clarke, R. C., & Merlin, M. D. (2013). *Cannabis evolution and ethnobotany* (1st ed.). University of California Press.

Crawford, V. (2002). A homelie herbe. *Journal of Cannabis Therapeutics, 2*(2), 71–79. https://doi.org/10.1300/J175v02n02_05.

Crippa, J. A., Zuardi, A. W., Martín-Santos, R., Bhattacharyya, S., Atakan, Z., McGuire, P., & Fusar-Poli, P. (2009). Cannabis and anxiety: A critical review of the evidence. *Human Psychopharmacology: Clinical and Experimental, 24*(7), 515–523. https://doi.org/10.1002/hup.1048.

Culpeper, N. (1653). *The complete herbal*. Project Gutenberg.

DailyCBD. (2021). *CBD dosage calculator: How much CBD should I take?*. Retrieved from https://dailycbd.com/en/cbd-dosage/.

Darmesteter, J. (1880). In F. M. Müller (Ed.), *Vol. 4. The Zend-Avesta: The Vendîdâd*. Oxford: Clarendon Press. Translated by James Darmesteter.

Darmesteter, J. (1890). *Avestâ Khorda Avestâ: Book of common prayer*. Oxford: Oxford University Press.

de Meijer, E. P. M., Bagatta, M., Carboni, A., Crucitti, P., Moliterni, V. M. C., Ranalli, P., & Mandolino, G. (2003). The inheritance of chemical phenotype in *Cannabis sativa* L. *Genetics, 163*(1), 335–346.

de Meijer, E. P. M., & Hammond, K. M. (2005). The inheritance of chemical phenotype in Cannabis sativa L. (II): Cannabigerol predominant plants. *Euphytica, 145*(1), 189–198. https://doi.org/10.1007/s10681-005-1164-8.

de Meijer, E. P. M., Hammond, K. M., & Sutton, A. (2009). The inheritance of chemical phenotype in Cannabis sativa L. (IV): Cannabinoid-free plants. *Euphytica, 168*(1), 95–112. https://doi.org/10.1007/s10681-009-9894-7.

Devane, W. A., Dysarz, F. A., Johnson, M. R., Melvin, L. S., & Howlett, A. C. (1988). Determination and characterization of a cannabinoid receptor in rat brain. *Molecular Pharmacology, 34*(5), 605–613.

DeVriend, H. J. (1984). *The Old English Herbarium and Medicina de Quadrupedibus*. Oxford University Press.

Di Candito, M., Ranalli, P., & Dal Re, L. (2004). Heavy metal tolerance and uptake of Cd, Pb and Tl by hemp. *Advances in Horticultural Science, 18*(3), 138–144.

Dierksmeier, L. (2020). Forbidden herbs: Alzate's defense of pipiltzintzintlis. *Colonial Latin American Review, 29*(2), 292–315. https://doi.org/10.1080/10609164.2020.1755941.

Dioscorides, Gunther, R. T., & Goodyer, J. (1959). *The Greek herbal of Dioscorides*. New York: Hafner Publishing.

Dioscorides, P., Osbaldeston, T. A., & Wood, R. P. A. (2000). *De materia medica: Being an herbal with many other medicinal materials written in Greek in the first century of the common era*. Johannesburg: Hafner Publishing.

Doorenbos, N. J., Fetterman, P. S., Quimby, M. W., & Turner, C. E. (1971). Cultivation, extraction, and analysis of Cannabis sativa L*. *Annals of the New York Academy of Sciences, 191*(1), 3–14. https://doi.org/10.1111/j.1749-6632.1971.tb13982.x.

Dubois, V., Breton, S., Linder, M., Fanni, J., & Parmentier, M. (2007). Fatty acid profiles of 80 vegetable oils with regard to their nutritional potential. *European Journal of Lipid Science and Technology, 109*(7), 710–732. https://doi.org/10.1002/ejlt.200700040.

Duncan, A. (1805). *The Edinburgh new dispensatory: Containing, I. The elements of pharmaceutical chemistry II. The materia medica; or, the natural, pharceutical and medicinal history of different substancs employed in medicine III. The pharmaceutical preparations and compositions; including complete and accurate translations of the octavo editions of the London pharmacopoeia, published in 1791; Dublin pharmacopoeia, published in 1794; and the new edition of the Edinburgh pharmacopoeia, published in 1803* (1st ed.). Worcester: Isaiah Thomas, Jun.

Ebbell, B., & Banov, L., Jr. (1937). *The papyrus Ebers: The greatest Egyptian medical document*. Copenhagen: Levin & Munksgaard.

EFSA. (2011). Scientific opinion on the safety of hemp (Cannabis genus) for use as animal feed. *EFSA Journal, 9*(3), 2011.

Eistrup, M., Sanches, A. R., Muñoz-Rojas, J., & Pinto Correia, T. (2019). A "Young farmer problem"? Opportunities and constraints for generational renewal in farm management: An example from Southern Europe. *Land, 8*(4), 70.

Elfordy, S., Lucas, F., Tancret, F., Scudeller, Y., & Goudet, L. (2008). Mechanical and thermal properties of lime and hemp concrete ("hempcrete") manufactured by a projection process. *Construction and Building Materials, 22*(10), 2116–2123. https://doi.org/10.1016/j.conbuildmat.2007.07.016.

Elphick, M. R. (2012). The evolution and comparative neurobiology of endocannabinoid signalling. *Philosophical Transactions of the Royal Society, B: Biological Sciences, 367*(1607), 3201–3215. https://doi.org/10.1098/rstb.2011.0394.

ElSohly, M. A., Ross, S. A., Mehmedic, Z., Arafat, R., Yi, B., & Banahan, B. F. (2000). Potency trends of delta9-THC and other cannabinoids in confiscated marijuana from 1980–1997. *Journal of Forensic Sciences, 45*(1), 24–30.

EMCDDA. (2018). *Cannabis legislation in Europe: An overview*. Luxembourg: Publications Office of the European Union. Retrieved from https://www.emcdda.europa.eu/system/files/publications/4135/TD0217210ENN.pdf.

EMCDDA. (2020). *Low-THC cannabis products in Europe*. Retrieved from Luxembourg.

Fabian, C. J., Kimler, B. F., & Hursting, S. D. (2015). Omega-3 fatty acids for breast cancer prevention and survivorship. *Breast Cancer Research, 17*(1), 62. https://doi.org/10.1186/s13058-015-0571-6.

Fairbairn, J. W., Liebmann, J. A., & Rowan, M. G. (1976). The stability of cannabis and its preparations on storage. *Journal of Pharmacy and Pharmacology, 28*(1), 1–7. https://doi.org/10.1111/j.2042-7158.1976.tb04014.x.

FAOSTAT. (2019). *Crops and livestock products*. Retrieved from http://www.fao.org/faostat/en/#data/TP.

Fellermeier, M., & Zenk, M. H. (1998). Prenylation of olivetolate by a hemp transferase yields cannabigerolic acid, the precursor of tetrahydrocannabinol. *FEBS Letters, 427*(2), 283–285. https://doi.org/10.1016/S0014-5793(98)00450-5.

Fernández, V. T., & Mirizzi, F. (2020). *The EU Parliament endorses the increase of THC level for industrial hemp on the field in a key vote on the Common Agricultural Policy [Press release]*.

Fiorenzi, R. (2021). *CBD dosage guide: How much should you take?*. Retrieved from https://startsleeping.org/cbd-dosage/#dosage.

Fournier, G., Richez-Dumanois, C., Duvezin, J., Mathieu, J.-P., & Paris, M. (1987). Identification of a new chemotype in Cannabis sativa: Cannabigerol-dominant plants, biogenetic and agronomic prospects. *Planta Medica, 53*(03), 277–280.

Friedman, D., French, J. A., & Maccarrone, M. (2019). Safety, efficacy, and mechanisms of action of cannabinoids in neurological disorders. *Lancet Neurology*, *18*(5), 504–512. https://doi.org/10.1016/S1474-4422(19)30032-8.

Gagne, S. J., Stout, J. M., Liu, E., Boubakir, Z., Clark, S. M., & Page, J. E. (2012). Identification of olivetolic acid cyclase from *Cannabis sativa* reveals a unique catalytic route to plant polyketides. *Proceedings of the National Academy of Sciences*, *109*(31), 12811–12816. https://doi.org/10.1073/pnas.1200330109.

Galasso, I., Russo, R., Mapelli, S., Ponzoni, E., Brambilla, I. M., Battelli, G., & Reggiani, R. (2016). Variability in seed traits in a collection of *Cannabis sativa* L. genotypes. *Frontiers in Plant Science*, *7*. https://doi.org/10.3389/fpls.2016.00688.

Gaoni, Y., & Mechoulam, R. (1964). Isolation, structure, and partial synthesis of an active constituent of hashish. *Journal of the American Chemical Society*, *86*(8), 1646–1647. https://doi.org/10.1021/ja01062a046.

Gatchel, R. J., McGeary, D. D., McGeary, C. A., & Lippe, B. (2014). Interdisciplinary chronic pain management: Past, present, and future. *American Psychologist*, *69*(2), 119–130. https://doi.org/10.1037/a0035514.

Gerard, J. (1633). *The herball or Generall historie of plantes. Gathered by Iohn Gerarde of London Master in Chirurgerie very much enlarged and amended by Thomas Iohnson citizen and apothecarye of London.* London: Adam Islip Ioice Norton and Richard Whitakers.

Gertsch, J., Pertwee, R. G., & Di Marzo, V. (2010). Phytocannabinoids beyond the Cannabis plant – Do they exist? *British Journal of Pharmacology*, *160*(3), 523–529. https://doi.org/10.1111/j.1476-5381.2010.00745.x.

Ghalioungui, P. (1987). *The Ebers papyrus: A new English translation, commentaries and glossaries.* Cairo: Academy of Scientific Research and Technology.

Ghosh, R., Todd, A., & Wilkinson, S. (1940). Cannabis indica. Part V. The synthesis of cannabinol. *Journal of the Chemical Society*, 1393–1396.

Gill, L. (2019). *CBD goes mainstream.* Retrieved from https://www.consumerreports.org/cbd/cbd-goes-mainstream/.

Government of Canada. (2020). *Industrial hemp licensing statistics.* Retrieved from https://www.canada.ca/en/health-canada/services/drugs-medication/cannabis/producing-selling-hemp/about-hemp-canada-hemp-industry/statistics-reports-fact-sheets-hemp.html.

Grassa, C. J., Weiblen, G. D., Wenger, J. P., Dabney, C., Poplawski, S. G., Timothy Motley, S., … Schwartz, C. J. (2021). A new Cannabis genome assembly associates elevated cannabidiol (CBD) with hemp introgressed into marijuana. *New Phytologist*, 1–15. https://doi.org/10.1111/nph.17243.

Green, J. (2002). *Cannabis.* Running Press Book Publishers.

Grierson, G. (1994). On references to the hemp plant occurring in Sanskrit and Hindi literature. In *Vol. 7. Indian Hemp Drugs Commission Report* (pp. 1893–1894).

Gupta, S. (Writer). (2013). *Weed: Special report by S. Gupta.* CNN.

Hakeem Said, I., Rezk, A., Hussain, I., Grimbs, A., Shrestha, A., Schepker, H., … Kuhnert, N. (2017). Metabolome comparison of bioactive and inactive Rhododendron extracts and identification of an antibacterial cannabinoid(s) from Rhododendron collettianum. *Phytochemical Analysis*, *28*(5), 454–464. https://doi.org/10.1002/pca.2694.

Hazekamp, A., Fischedick, J. T., Llano Díez, M., Lubbe, A., & Ruhaak, R. L. (2010). Chemistry of Cannabis. In H.-W. Liu, & L. Mander (Eds.), *Vol. 3. Comprehensive natural products II: Chemistry and biology* (pp. 1033–1084). Elsevier.

Hemp Feed Coalition. (2021). Hemp Feed Coalition submits the first ingredient application in the United States for hemp to be federally approved as a feed ingredient for poultry. *The State of Hemp as Animal Feed.* Retrieved from https://hempfeedcoalition.org/2021/02/06/hemp-feed-coalition-submits-the-first-ingredient-application-in-the-united-states-for-hemp-to-be-federally-approved-as-a-feed-ingredient-for-poultry/.

Holland, P. (1601). *Pliny's natural history in thirty-seven books.* Leicester Square: George Barclay.

Hollister, L. E., & Gillespie, H. (1975). Interactions in man of delta-9-tetrahydrocannabinol; II. Cannabinol and cannabidiol. *Clinical Pharmacology & Therapeutics*, *18*(1), 80–83. https://doi.org/10.1002/cpt197518180.

Horne, C. F. (1917). *The sacred books and early literature of the east.* New York: Parks, Austin, and Lipscomb, Inc.

Horst, M., & Marion, A. (2019). Racial, ethnic and gender inequities in farmland ownership and farming in the U.S. *Agriculture and Human Values*, *36*(1), 1–16. https://doi.org/10.1007/s10460-018-9883-3.

House, J. D., Neufeld, J., & Leson, G. (2010). Evaluating the quality of protein from hemp seed (*Cannabis sativa* L.) products through the use of the protein digestibility-corrected amino acid score method. *Journal of Agricultural and Food Chemistry*, *58*(22), 11801–11807. https://doi.org/10.1021/jf102636b.

Irakli, M., Tsaliki, E., Kalivas, A., Kleisiaris, F., Sarrou, E., & Cook, C. M. (2019). Effect of genotype and growing year on the nutritional, phytochemical, and antioxidant properties of industrial hemp (*Cannabis sativa* L.) seeds. *Antioxidants*, *8*(10), 491.

Jacob, A., & Todd, A. (1940). Cannabidiol and cannabol, constituents of Cannabis indica resin. *Nature*, *145*, 350.

Jamie, C., & Rod, K. (2018). Regulatory status of cannabidiol in the United States: A perspective. *Cannabis and Cannabinoid Research*, 3(1), 190–194. https://doi.org/10.1089/can.2018.0030.

Jang, H., & Park, K. (2020). Omega-3 and omega-6 polyunsaturated fatty acids and metabolic syndrome: A systematic review and meta-analysis. *Clinical Nutrition*, 39(3), 765–773. https://doi.org/10.1016/j.clnu.2019.03.032.

Johnson, R. (2014). *Hemp as an agricultural commodity (RL32725)*. Washington, DC.

Jonas, A. M., & Raj, R. (2020). Vaping-related acute parenchymal lung injury: A systematic review. *Chest*, 158(4), 1555–1565. https://doi.org/10.1016/j.chest.2020.03.085.

Julien, M. S. (1849). *Chirurgie Chinoise – Substance anesthetique employee en Chine, dans le commencement du III siecle de notre ere, pour paralyser momentanement la sensibilite*. Retrieved from https://char-fr.net/Compte-rendu-des-seances-de-l,223.html.

Karniol, I. G., Shirakawa, I., Kasinski, N., Pfeferman, A., & Carlini, E. A. (1974). Cannabidiol interferes with the effects of Δ9-tetrahydrocannabinol in man. *European Journal of Pharmacology*, 28(1), 172–177. https://doi.org/10.1016/0014-2999(74)90129-0.

Khan, R., Naveed, S., Mian, N., Fida, A., Raafey, M. A., & Aedma, K. K. (2020). The therapeutic role of Cannabidiol in mental health: A systematic review. *Journal of Cannabis Research*, 2(1), 2. https://doi.org/10.1186/s42238-019-0012-y.

Knopf, A. (2019). FDA: Avoid THC-containing vaping products. *Alcoholism & Drug Abuse Weekly*, 31(35), 5–6. https://doi.org/10.1002/adaw.32481.

Kriese, U., Schumann, E., Weber, W. E., Beyer, M., Brühl, L., & Matthäus. (2004). Oil content, tocopherol composition and fatty acid patterns of the seeds of 51 *Cannabis sativa* L. genotypes. *Euphytica*, 137(3), 339–351. https://doi.org/10.1023/b:euph.0000040473.23941.76.

Kumar, R. N., Chambers, W. A., & Pertwee, R. G. (2001). Pharmacological actions and therapeutic uses of cannabis and cannabinoids. *Anaesthesia*, 56(11), 1059–1068. https://doi.org/10.1111/j.1365-2044.2001.02269.x.

Larsen, C., & Shahinas, J. (2020). Dosage, efficacy and safety of cannabidiol administration in adults: A systematic review of human trials. *Journal of Clinical Medicine Research*, 12(3), 129–141. https://doi.org/10.14740/jocmr4090.

Lewis, N., McCaffrey, K., Sage, K., Cheng, C.-J., Green, J., Goldstein, L., … Dunn, A. (2019). E-cigarette use, or vaping, practices and characteristics among persons with associated lung injury – Utah, April–October 2019. *MMWR. Morbidity and Mortality Weekly Report*, 68(42), 953–956. https://doi.org/10.15585/mmwr.mm6842e1.

Li, H.-L. (1974). An archaeological and historical account of Cannabis in China. *Economic Botany*, 28(4), 437–448. https://doi.org/10.1007/bf02862859.

Lindholst, C. (2010). Long term stability of cannabis resin and cannabis extracts. *Australian Journal of Forensic Sciences*, 42(3), 181–190. https://doi.org/10.1080/00450610903258144.

Long, T., Wagner, M., Demske, D., Leipe, C., & Tarasov, P. E. (2017). Cannabis in Eurasia: Origin of human use and Bronze Age trans-continental connections. *Vegetation History and Archaeobotany*, 26(2), 245–258. https://doi.org/10.1007/s00334-016-0579-6.

Luo, X., Reiter, M. A., d'Espaux, L., Wong, J., Denby, C. M., Lechner, A., … Keasling, J. D. (2019). Complete biosynthesis of cannabinoids and their unnatural analogues in yeast. *Nature*, 567(7746), 123–126. https://doi.org/10.1038/s41586-019-0978-9.

Lupescu, M. (2019). *Canada: Industrial hemp production trade and regulation*. Retrieved from https://apps.fas.usda.gov/newgainapi/api/report/downloadreportbyfilename?filename=Industrial%20Hemp%20Production%20Trade%20and%20Regulation_Ottawa_Canada_8-26-2019.pdf.

Maccarone, M. (2020). Missing pieces to the endocannabinoid puzzle. *Trends in Molecular Medicine*, 26(3), 263–272. https://doi.org/10.1016/j.molmed.2019.11.002.

Mandolino, G., & Carboni, A. (2004). Potential of marker-assisted selection in hemp genetic improvement. *Euphytica*, 140(1), 107–120. https://doi.org/10.1007/s10681-004-4759-6.

Manniche, L. (1989). *An ancient Egyptian herbal*. Austin: University of Texas Press.

Marino, D. (2019). *Hashish and food: Arabic and European medieval dreams of edible paradises* (pp. 190–213). Leiden, The Netherlands: Brill.

Matsuda, L. A., Lolait, S. J., Brownstein, M. J., Young, A. C., & Bonner, T. I. (1990). Structure of a cannabinoid receptor and functional expression of the cloned cDNA. *Nature*, 346(6284), 561–564.

Maynard, D. D. C., Vidigal, M. D., Farage, P., Zandonadi, R. P., Nakano, E. Y., & Botelho, R. B. A. (2020). Environmental, social and economic sustainability indicators applied to food services: A systematic review. *Sustainability*, 12(5), 1804.

Mcgrath, C. (2020). *China: 2019 hemp annual report*. Retrieved from https://apps.fas.usda.gov/newgainapi/api/Report/DownloadReportByFileName?fileName=2019%20Hemp%20Annual%20Report_Beijing_China%20-%20Peoples%20Republic%20of_02-21-2020.

McPartland, J. M., Clarke, R. C., & Watson, D. P. (2000). *Hemp diseases and pests: Management and biological control: An advanced treatise*. CABI.

McPartland, J. M., Matias, I., Di Marzo, V., & Glass, M. (2006). Evolutionary origins of the endocannabinoid system. *Gene, 370*, 64–74. https://doi.org/10.1016/j.gene.2005.11.004.

McVey, E., Cowee, M., Nichols, K., Drotleff, L., Huhn, K., & Madrid, C. (2019). In J. Stelton-Holtmeier (Ed.), *Annual hemp & CBD industry factbook*. Retrieved from https://mjbizdaily.com/bizbooks/2019-hemp-factbook/3/#zoom=z.

Mead, A. (2019). Legal and regulatory issues governing Cannabis and Cannabis-derived products in the United States. *Frontiers in Plant Science, 10*(697). https://doi.org/10.3389/fpls.2019.00697.

Mechoulam, R. (1986). The pharmacohistory of Cannabis sativa. In R. Mechoulam (Ed.), *Cannabinoids as therapeutic agents* (pp. 1–19). Taylor&Francis.

Mechoulam, R., & Shvo, Y. (1963). Hashish-I The structure of cannabidiol. *Tetrahedron, 19*(12), 2073–2078.

Mehmedic, Z., Chandra, S., Slade, D., Denham, H., Foster, S., Patel, A. S., … ElSohly, M. A. (2010). Potency trends of Δ9-THC and other cannabinoids in confiscated cannabis preparations from 1993 to 2008. *Journal of Forensic Sciences, 55*(5), 1209–1217. https://doi.org/10.1111/j.1556-4029.2010.01441.x.

Merkus, F. W. H. M. (1971). Cannabivarin and tetrahydrocannabivarin, two new constituents of Hashish. *Nature, 232*(5312), 579–580.

Mirski, R., Boruszewski, P., Trociński, A., & Dziurka, D. (2017). The possibility to use long fibres from fast growing hemp (*Cannabis sativa* L.) for the production of boards for the building and furniture industry. *BioResources, 12*(2), 3521–3529.

Mohanty, A. K., Wibowo, A., Misra, M., & Drzal, L. T. (2004). Effect of process engineering on the performance of natural fiber reinforced cellulose acetate biocomposites. *Composites Part A: Applied Science and Manufacturing, 35*(3), 363–370. https://doi.org/10.1016/j.compositesa.2003.09.015.

Moltke, J. (2019). *CBD oil for anxiety: Can cannabidiol treat anxiety?*. Retrieved from https://www.netdoctor.co.uk/healthy-living/mental-health/a25601198/cbd-oil-anxiety/.

Montford, S., & Small, E. (1999). A comparison of the biodiversity friendliness of crops with special reference to hemp (*Cannabis sativa* L.). *Journal of the International Hemp Association, 6*, 53–63.

Morimoto, S., Komatsu, K., Taura, F., & Shoyama, Y. (1997). Enzymological evidence for cannabichromenic acid biosynthesis. *Journal of Natural Products, 60*(8), 854–857. https://doi.org/10.1021/np970210y.

Müller, F. M. (1880). *The sacred books of the East*. Oxford: Clarendon Press.

Munro, S., Thomas, K. L., & Abu-Shaar, M. (1993). Molecular characterization of a peripheral receptor for cannabinoids. *Nature, 365*(6441), 61–65. https://doi.org/10.1038/365061a0.

NASS. (2017). *Farm producers*. Retrieved from https://www.nass.usda.gov/Publications/Highlights/2019/2017Census_Farm_Producers.pdf.

Nissen, L., Bordoni, A., & Gianotti, A. (2020). Shift of volatile organic compounds (VOCs) in gluten-free hemp-enriched sourdough bread: A metabolomic approach. *Nutrients, 12*(4), 1050.

Nissen, L., di Carlo, E., & Gianotti, A. (2020). Prebiotic potential of hemp blended drinks fermented by probiotics. *Food Research International, 131*. https://doi.org/10.1016/j.foodres.2020.109029, 109029.

Norwinski, E. J., Pippins, R., Willcocks, J., & Williams, A. (2019). *EU regulation of CBD in foods and cosmetics*. Retrieved from https://www.arnoldporter.com/en/perspectives/publications/2019/04/eu-regulation-of-cannabidiol-in-foods.

Olson, D. W., & Thornsbury, S. (2020). *U.S. acres planted with industrial hemp expanded rapidly in 2019*. Retrieved from https://www.ers.usda.gov/data-products/chart-gallery/gallery/chart-detail/?chartId=96988.

O'Meara, P. (2019). The ageing farming workforce and the health and sustainability of agricultural communities: A narrative review. *Australian Journal of Rural Health, 27*(4), 281–289. https://doi.org/10.1111/ajr.12543.

Paridah, M. T., Basher, A. B., Saiful Azry, S., & Ahmed, Z. (2011). Retting process of some bast plant fibres and its effect on fibre quality: A review. *BioResources, 6*(4), 5260–5281.

Pil, L., Bensadoun, F., Pariset, J., & Verpoest, I. (2016). Why are designers fascinated by flax and hemp fibre composites? *Composites Part A: Applied Science and Manufacturing, 83*, 193–205. https://doi.org/10.1016/j.compositesa.2015.11.004.

Pollastro, F., De Petrocellis, L., Schiano-Moriello, A., Chianese, G., Heyman, H., Appendino, G., & Taglialatela-Scafati, O. (2017). Amorfrutin-type phytocannabinoids from Helichrysum umbraculigerum. *Fitoterapia, 123*, 13–17. https://doi.org/10.1016/j.fitote.2017.09.010.

Rawlinson, G., & Komroff, M. (2016). *History of Herodotus*. Hansebooks GmbH.

Rodriguez-Leyva, D., & Pierce, G. N. (2010). The cardiac and haemostatic effects of dietary hempseed. *Nutrition and Metabolism, 7*(1), 32. https://doi.org/10.1186/1743-7075-7-32.

Rubin, R. (2018). The path to the first FDA-approved cannabis-derived treatment and what comes next. *Journal of the American Medical Association*, *320*(12), 1227–1229. https://doi.org/10.1001/jama.2018.11914.

Rudenko, S. I. (1970). *Frozen tombs of Siberia*. University of California Press.

Russo, E. B. (2016). Beyond cannabis: Plants and the endocannabinoid system. *Trends in Pharmacological Sciences*, *37*(7), 594–605. https://doi.org/10.1016/j.tips.2016.04.005.

Salmon, W. (1683). *Doron medicum, or, A supplement to the new London dispensatory: In III books: Containing a supplement I. To the materia medica, II. To the internal compound medicaments, III. To the external compound medicaments: Compleated with the art of compounding medicines*. London: T. Dawks, T. Bassett, J. Wright and R. Chiswell.

Samsamikor, M., Mackay, D., Mollard, R. C., & Aluko, R. E. (2020). A double-blind, randomized, crossover trial protocol of whole hemp seed protein and hemp seed protein hydrolysate consumption for hypertension. *Trials*, *21*, 354.

Sandler, L. N., & Gibson, K. A. (2019). A call for weed research in industrial hemp (*Cannabis sativa* L). *Weed Research*. https://doi.org/10.1111/wre.12368.

Schaneman, B. (2020). *Oregon cannabis regulators ban additives first seen in vaping crisis, require more disclosure*. Retrieved from https://mjbizdaily.com/oregon-cannabis-regulators-ban-additives-first-seen-in-vaping-crisis/#:~:text=Squalene%20and%20squalane,lipoid%20pneumonia%2C%20among%20other%20ailments.

Schluttenhofer, C. (2018). Canada begins a great ganja experiment. *Science*, *361*(6401), 460. https://doi.org/10.1126/science.aau5323.

Schluttenhofer, C., & Yuan, L. (2017). Challenges towards revitalizing hemp: A multifaceted crop. *Trends in Plant Science*, *22*(11), 917–929. https://doi.org/10.1016/j.tplants.2017.08.004.

Schluttenhofer, C., & Yuan, L. (2019). Hemp hemp hooray for cannabis research. *Science*, *363*(6428), 701–702. https://doi.org/10.1126/science.aaw3537.

Schultes, R., Hofmann, A., & Rätsch, C. (1992). *Plants of the gods: Their sacred, healing, and hallucinogenic powers*. Rochester, Vermont: Inner Traditions – Bear & Company.

Schumacher, A. G. D., Pequito, S., & Pazour, J. (2020). Industrial hemp fiber: A sustainable and economical alternative to cotton. *Journal of Cleaner Production*, *268*. https://doi.org/10.1016/j.jclepro.2020.122180, 122180.

Schwab, U. S., Callaway, J. C., Erkkilä, A. T., Gynther, J., Uusitupa, M. I., & Järvinen, T. (2006). Effects of hempseed and flaxseed oils on the profile of serum lipids, serum total and lipoprotein lipid concentrations and haemostatic factors. *European Journal of Nutrition*, *45*(8), 470–477.

Shoyama, Y., Hirano, H., & Nishioka, I. (1984). Biosynthesis of propyl cannabinoid acid and its biosynthetic relationship with pentyl and methyl cannabinoid acids. *Phytochemistry*, *23*(9), 1909–1912. https://doi.org/10.1016/S0031-9422(00)84939-0.

Siculus, D. (1933). *The library of history*. *Vol. 12*. Harvard University Press.

Simone, D. (2020). *New to CBD? This is how much to take the first time*. Retrieved from https://greatist.com/health/how-much-cbd-should-i-take-the-first-time.

Simopoulos, A. P. (2016). An increase in the omega-6/omega-3 fatty acid ratio increases the risk for obesity. *Nutrients*, *8*(3), 128. https://doi.org/10.3390/nu8030128.

Sirikantaramas, S., Morimoto, S., Shoyama, Y., Ishikawa, Y., Wada, Y., Shoyama, Y., & Taura, F. (2004). The gene controlling marijuana psychoactivity: Molecular cloning and heterologous expression of Δ1-tetrahydrocannabinolic acid synthase from *Cannabis sativa* L. *Journal of Biological Chemistry*, *279*(38), 39767–39774. https://doi.org/10.1074/jbc.M403693200.

Skelley, J. W., Deas, C. M., Curren, Z., & Ennis, J. (2020). Use of cannabidiol in anxiety and anxiety-related disorders. *Journal of the American Pharmacists Association*, *60*(1), 253–261. https://doi.org/10.1016/j.japh.2019.11.008.

Small, E., & Beckstead, H. D. (1973). Cannabinoid phenotypes in Cannabis sativa. *Nature*, *245*(5421), 147–148. https://doi.org/10.1038/245147a0.

Smith, C. S., & McDonald, G. T. (1998). Assessing the sustainability of agriculture at the planning stage. *Journal of Environmental Management*, *52*(1), 15–37. https://doi.org/10.1006/jema.1997.0162.

Spanagel, R. (2020). Cannabinoids and the endocannabinoid system in reward processing and addiction: From mechanisms to interventions. *Dialogues in Clinical Neuroscience*, *22*(3), 241–250. https://doi.org/10.31887/DCNS.2020.22.3/rspanagel.

Starowicz, K., & Finn, D. P. (2017). Cannabinoids and pain: Sites and mechanisms of action. In D. Kendall, & S. P. H. Alexander (Eds.), *Vol. 80. Advances in pharmacology* (pp. 437–475). Academic Press.

Stout, J. M., Boubakir, Z., Ambrose, S. J., Purves, R. W., & Page, J. E. (2012). The hexanoyl-CoA precursor for cannabinoid biosynthesis is formed by an acyl-activating enzyme in *Cannabis sativa* Trichomes. *Plant Journal*, *71*(3), 353–365. https://doi.org/10.1111/j.1365-313X.2012.04949.x.

Tang, C.-H., Ten, Z., Wang, X.-S., & Yang, X.-Q. (2006). Physicochemical and functional properties of hemp (*Cannabis sativa* L.) protein isolate. *Journal of Agricultural and Food Chemistry*, 54(23), 8945–8950. https://doi.org/10.1021/jf0619176.

Taura, F., Dono, E., Sirikantaramas, S., Yoshimura, K., Shoyama, Y., & Morimoto, S. (2007). Production of Δ1-tetrahydrocannabinolic acid by the biosynthetic enzyme secreted from transgenic Pichia pastoris. *Biochemical and Biophysical Research Communications*, 361(3), 675–680. https://doi.org/10.1016/j.bbrc.2007.07.079.

Taura, F., Iijima, M., Lee, J.-B., Hashimoto, T., Asakawa, Y., & Kurosaki, F. (2014). Daurichromenic acid-producing oxidocyclase in the young leaves of *Rhododendron dauricum*. *Natural Product Communications*, 9(9). https://doi.org/10.1177/1934578x1400900928, 1934578X1400900928.

Taura, F., Morimoto, S., & Shoyama, Y. (1996). Purification and characterization of cannabidiolic-acid synthase from Cannabis sativa L: Biochemical analysis of a novel enzyme that catalyzes the oxidocyclization of cannabigerolic acid to cannabidiolic acid. *Journal of Biological Chemistry*, 271(29), 17411–17416. https://doi.org/10.1074/jbc.271.29.17411.

Taura, F., Morimoto, S., Shoyama, Y., & Mechoulam, R. (1995). First direct evidence for the mechanism of DELTA 1-tetrahydrocannabinolic acid biosynthesis. *Journal of the American Chemical Society*, 117(38), 9766–9767.

Taura, F., Sirikantaramas, S., Shoyama, Y., Yoshikai, K., Shoyama, Y., & Morimoto, S. (2007). Cannabidiolic-acid synthase, the chemotype-determining enzyme in the fiber-type *Cannabis sativa*. *FEBS Letters*, 581(16), 2929–2934. https://doi.org/10.1016/j.febslet.2007.05.043.

Teterycz, D., Sobota, A., Przygodzka, D., & Łysakowska, P. (2021). Hemp seed (*Cannabis sativa* L.) enriched pasta: Physicochemical properties and quality evaluation. *PLoS One*, 16(3). https://doi.org/10.1371/journal.pone.0248790, e0248790.

Thompson, R. C. (1949). *A dictionary of Assyrian botany*. London: British Academy.

Toth, J. A., Stack, G. M., Cala, A. R., Carlson, C. H., Wilk, R. L., Crawford, J. L., … Smart, L. B. (2020). Development and validation of genetic markers for sex and cannabinoid chemotype in *Cannabis sativa* L. *GCB Bioenergy*, 12(3), 213–222. https://doi.org/10.1111/gcbb.12667.

Toyota, M., Shimamura, T., Ishii, H., Renner, M., Braggins, J., & Asakawa, Y. (2002). New bibenzyl cannabinoid from the New Zealand liverwort *Radula marginata*. *Chemical and Pharmaceutical Bulletin*, 50(10), 1390–1392. https://doi.org/10.1248/cpb.50.1390.

Turner, W., Coombe, J., Coombe, R., Coombe, E., Gyvken, J., Mierdman, S., … Morwood, R. D. (1551). *A new herball, wherin are conteyned the names of herbes in Greke, Latin, Englysh, Duch, Frenche, and in the potecaries and herbaries Latin: With the properties degrees, and natural places of the same*. London: Steven Myerdman.

UNDOC. (1961). Narcotic drugs and psychotropic substances. In *Single convention on narcotic drugs of 1961 as amended by the 1972 protocol* (p. 70). Geneva, Switzerland: United Nations (Chapter VI).

Unschuld, P. U. (1986). *Medicine in China: A history of pharmaceutics*. Berkerley, United States: University of California Press.

Urits, I., Gress, K., Charipova, K., Habib, K., Lee, D., Lee, C., … Viswanath, O. (2020). Use of cannabidiol (CBD) for the treatment of chronic pain. *Best Practice & Research. Clinical Anaesthesiology*, 34(3), 463–477. https://doi.org/10.1016/j.bpa.2020.06.004.

USP. (1850). *Pharmacopoeia of the United States of America* (3rd ed.). Philadelphia: Lippincott, Grambo, & Co.

USP. (1942). *Pharmacopoeia of the United States of America* (12th ed.). Easton, Pennsylvania: Mack Printing Company.

Vitiello, M. (2019). Marijuana legalization, racial disparity, and the hope for reform. *Lewis & Clark Law Review*, 23, 789.

Vučković, S., Srebro, D., Vujović, K. S., Vučetić, Č., & Prostran, M. (2018). Cannabinoids and pain: New insights from old molecules. *Frontiers in Pharmacology*, 9(1259). https://doi.org/10.3389/fphar.2018.01259.

Wang, X.-S., Tang, C.-H., Yang, X.-Q., & Gao, W.-R. (2008). Characterization, amino acid composition and in vitro digestibility of hemp (*Cannabis sativa* L.) proteins. *Food Chemistry*, 107(1), 11–18. https://doi.org/10.1016/j.foodchem.2007.06.064.

WHO. (2021). *Cannabis*. Retrieved from https://www.who.int/teams/mental-health-and-substance-use/alcohol-drugs-and-addictive-behaviours/drugs-psychoactive/cannabis.

Wollner, F. (2001). The hemp ritual of the Scythians. In *Marijuana medicine: A world tour of the healing and visionary powers of cannabis* (p. 56). Healing Arts Press.

Wood, T. B., Spivey, W. N., & Easterfield, T. H. (1899). III.—Cannabinol. Part I. *Journal of the Chemical Society, Transactions*, 75, 20–36.

Wu, D., & O'Shea, D. F. (2020). Potential for release of pulmonary toxic ketene from vaping pyrolysis of vitamin E acetate. *Proceedings of the National Academy of Sciences*, 117(12), 6349–6355. https://doi.org/10.1073/pnas.1920925117.

Wylie, S. E., Ristvey, A. G., & Fiorellino, N. M. (2021). Fertility management for industrial hemp production: Current knowledge and future research needs. *GCB Bioenergy, 13*(4), 517–524. https://doi.org/10.1111/gcbb.12779.

Yuanjian, T., & Isaac, D. H. (2007). Impact and fatigue behaviour of hemp fibre composites. *Composites Science and Technology, 67*(15), 3300–3307. https://doi.org/10.1016/j.compscitech.2007.03.039.

Yunlong, C., & Smit, B. (1994). Sustainability in agriculture: A general review. *Agriculture, Ecosystems & Environment, 49*(3), 299–307. https://doi.org/10.1016/0167-8809(94)90059-0.

Zamengo, L., Bettin, C., Badocco, D., Di Marco, V., Miolo, G., & Frison, G. (2019). The role of time and storage conditions on the composition of hashish and marijuana samples: A four-year study. *Forensic Science International, 298,* 131–137. https://doi.org/10.1016/j.forsciint.2019.02.058.

Zirpel, B., Kayser, O., & Stehle, F. (2018). Elucidation of structure-function relationship of THCA and CBDA synthase from *Cannabis sativa* L. *Journal of Biotechnology, 284,* 17–26. https://doi.org/10.1016/j.jbiotec.2018.07.031.

Zobel, F., Notari, L., Schneider, E., & Rudmann, O. (2019). *Cannabidiol (CBD): Situation analysis.* Retrieved from https://idpc.net/fr/publications/2019/02/cannabidiol-cbd-analyse-de-situation.

Zou, B., Mishra, A. K., & Luo, B. (2018). Aging population, farm succession, and farmland usage: Evidence from rural China. *Land Use Policy, 77,* 437–445. https://doi.org/10.1016/j.landusepol.2018.06.001.

2

Industrial hemp breeding and genetics

Marko Flajšman and Darja Kocjan Ačko

Department for Agronomy, Biotechnical Faculty, University of Ljubljana, Ljubljana, Slovenia

2.1 Introduction

Cannabis sativa L. is a crop with a history of cultivation of several millennia (Fike, 2016). Despite the fact that genetic and morphological differences between many groups of cannabis make the taxonomic classification difficult and that the classification of the species is not uniform, a lay classification into two groups of cannabis has become established in legislation and in everyday use, specifically according to the content of the psychoactive cannabinoid Δ^9-THC (delta-9-tetrahydrocannabinol)—plants (varieties, populations, lines, genotypes, ecotypes) containing over 0.2% Δ^9-THC by dry weight (in Canada and some EU countries this limit is 0.3% or higher) are classified as so-called drug-type or medical cannabis. All other cannabis groups with the Δ^9-THC content below 0.2% are called fiber-type cannabis or industrial hemp or just hemp and can be grown outdoors as a crop (Fike, 2016; Small & Cronquist, 1976). This is mainly a legal and not a strict taxonomic classification. Besides, this is also not a classification by use as the names "medical" or "industrial" suggests, as recently industrial hemp (containing less than 0.2% Δ^9-THC and more CBD (cannabidiol) or any other cannabinoid) has also been used for medical purposes and phytocannabinoids are produced from dry inflorescences using an extraction process for use in the treatment of various diseases and conditions. Medical cannabis is used exclusively for medical purposes (Janatova et al., 2018). There are old varieties and new lines of hemp obtained in breeding processes that have a high potential for successful production of seeds and fibers; however, because they contain Δ^9-THC above the permitted limit—generally not more than 1%—their production outdoors for industrial purposes is not allowed, and the varieties and lines have not been preserved (Berenji, Sikora, Fournier, & Beherec, 2013). 1% Δ^9-THC by dry weight is the limit that causes psychotropic effects upon ingestion (Chait et al., 1988; Grotenhermen & Karus, 1998), so the current content limit of 0.2% Δ^9-THC by dry weight, which is the dividing line between permitted and prohibited varieties, appears unreasonably set to the detriment of agricultural production and exploitation of hemp.

In medical cannabis, there has been systematic cross-breeding between many different populations and varieties in recent history, so genetic diversity has greatly increased (Onofri & Mandolino, 2017). However, new trends in medicinal cannabis cultivation from cuttings (clones) of parent plants reduce the significance of traditional cross-breeding and cultivation from seeds, limiting the formation of new allelic combinations, thus greatly reducing the genetic variability of medical cannabis varieties (Clarke & Merlin, 2017). On the other hand, breeding of hemp is done exclusively using seeds, i.e., by cross-breeding different varieties and groups of cannabis (Clarke & Merlin, 2017). The primary objective of breeders is to fix the desired properties by increasing homozygosity at the locus or loci for a particular trait, but at the same time maintaining heterozygosity at other loci in the genome, and thus the vigor of such a variety (Clarke & Merlin, 2017). Breeding of hemp has been intensive in Europe in recent years, as evidenced by the growing number of varieties registered on the EU variety list; 12 varieties were registered in 1995, 45 varieties in 2004, and 46 varieties in 2008; 51 varieties were in the EU Plant variety database in 2013, and the database currently includes 75 varieties of C. sativa L. (Plant variety database—European Commission, 2021).

The chapter reviews the breeding of exclusively industrial hemp. First, some general features of plant morphology, species taxonomy, and genetics are presented. Second, the contents of hemp breeding are initiated by an overview of some milestones of hemp breeding in Europe and continued with a description of breeding objectives, methods used, and obstacles to hemp improvement. Finally, the breeding paths of some well-known European varieties are presented.

2.2 Some features of C. *sativa* L. as a plant

2.2.1 Plant description

C. *sativa* L. is an annual herbaceous plant that occurs in nature in a dioecious sexual form (female and male flowers appear on separate plants, Fig. 2.1A and B), with a small proportion (one per thousand to one per million dioecious plants) of monoecious plants (Fig. 2.1C) in the population (male and female flowers are on the same plant) (Berenji et al., 2013). Only approximately 15% of flowering plant species have unisexual flowers that develop only carpels or stamens and are thus called monoecious or dioecious (Renner, 2014).

It is a short-day plant that is highly dependent on photoperiod; flowering initiation is also highly dependent on temperature (Salentijn, Petit, & Trindade, 2019). Cannabis forms a taproot with many fine roots. The stems of the cannabis plant are herbaceous, erect, 0.5–5 m tall, often hollow at the bottom, and composed of 5–20 nodes. The leaves of the plant are opposite or alternate on the stem, composed in palmate form of 3–13 leaflets (basally have more leaflets and apically less). Male plants are more slender and taller compared to female plants and they also flower earlier. On the other hand, female plants are much more robust, shorter, and flower later than male plants (Kocjan Ačko, 1999). The flowers of the male plants have a perianth consisting of five sepals (green appearance) and five free stamens (yellow). Female flowers consist of two carpels surrounded by a perigonal bract. The carpel is consisting of filamentous style and a brush-like stigma (Leme, Schönenberger, Staedler, & Teixeira, 2020; Spitzer-Rimon, Duchin, Bernstein, & Kamenetsky, 2019). The female flowers are of great importance because phytocannabinoids and terpenes are synthesized there, which are used as pharmacologically active compounds (Spitzer-Rimon et al., 2019). These secondary metabolites are predominantly produced in glandular trichomes (Fig. 2.2) covering perigonal bracts and subordinate leaves within an inflorescence (Andre, Hausman, & Guerriero, 2016).

FIG. 2.1   (A) Male plant, (B) female plant, (C) monoecious plant.

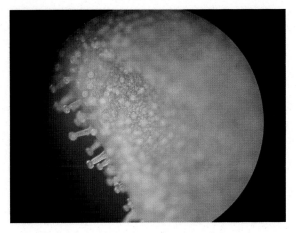

FIG. 2.2   Glandular trichomes on perigonal bract of a female "Finola" flower.

2.2.2 Taxonomy

C. *sativa* L. belongs to the genus *Cannabis* and the family Cannabaceae, which includes 10 genera and about 120 species (Jin et al., 2020; Yang et al., 2013). Genus *Humulus* is the sister genus of Cannabis and its closest relative, which includes three dioecious species: *H. yunnanensis* Hu, *H. japonicus* Merr., and *H. lupulus* L. (Kovalchuk et al., 2020).

The origin of C. *sativa* L. is in Central Asia, with Hindustani and European–Siberian as two centers of origin (Zeven & Zhukovsky, 1975), from where it spread across the globe within the last 6000 years (Small, 2015). It then evolved separately and in isolation in some regions for centuries or even millennia, either as a cultivated plant or in wild populations. This is probably the reason for the formation of populations that differ significantly both morphologically and genetically (Salentijn, Zhang, Amaducci, Yang, & Trindade, 2015). Therefore the taxonomy of *C. sativa* L. is still not universally accepted. Some authors advocate the classification of cannabis genotypes into several species, e.g., *C. sativa*, *C. indica*, and *C. ruderalis* (e.g., Hillig, 2005; McPartland & Guy, 2004). Population-based studies by Lynch et al. (2016), Sawler et al. (2015), and Soorni, Fatahi, Haak, Salami, and Bombarely (2017) generated GBS (genotyping by sequencing) data for ~ 400 samples of hemp and marijuana lines. Their results showed that the lines segregated at the genome-wide level into drug- and fiber-type populations. Probably the other most widely accepted alternative is the categorization of cannabis as a monotypic genus consisting of a single species (*C. sativa* L.) with several subspecies and varieties (e.g., *C. sativa* subsp. *sativa* var. *sativa*, *C. sativa* subsp. *indica* var. *indica*, *C. sativa* subsp. *sativa* var. *spontanea*) (e.g., de Meijer, 2014; Small, 2015; Small & Cronquist, 1976). This theory was also recently supported by genetic data from Zhang et al. (2018), in which five chloroplast DNA (cpDNA) markers were applied to 645 individuals from multiple wild populations, landraces, and official cultivars from five different world continents. Regardless of classification, the fact remains that plants from all species/subspecies and groups can freely pollinate with each other (Small, 1972).

Another classification, based on phytocannabinoid profiles rather than evolutionary relationships that is useful for breeding programs, is chemical classifications. Based on the content of the three major phytocannabinoids (cannabigerol—CBG, cannabidiolic acid—CBDA, and tetrahydrocannabinolic acid—THCA), which are the most perspective cannabinoids medicinally and commercially, cannabis plants are classified into five chemotypes. Chemotype I plants secrete high concentrations of THCA and only low amounts of CBDA (and CBGA), meaning the ratio between THCA and CBDA is well above 1 ($\gg$ 1). Chemotype II plants have an intermediate THCA/CBDA ratio (usually 0.5–2.0). Chemotype I and II are both drug-type cannabis. Chemotype III plants have high CBDA content and low to very low amounts of THCA, implying a low THCA/CBDA ratio ($\ll$ 1.0). Plants of chemotype IV contain CBGA as the main cannabinoid and there are almost no cannabinoids in chemotype V plants. The latter three chemotypes are classified as fiber-type cannabis (Aizpurua-Olaizola et al., 2016; De Meijer & Hammond, 2005; Small & Beckstead, 1973).

2.2.3 Genetics

The cannabis genome is diploid ($2n = 20$) and contains nine autosomes and a pair of sex chromosomes that are heteromorphic (female XX and male XY). The haploid genome size is estimated to be 818 Mb for the male and 843 Mb for the female. The male genome is larger due to the larger Y chromosome, which explains the sex difference in genome size (Braich, Baillie, Jewell, Spangenberg, & Cogan, 2019; Sakamoto, Akiyama, Fukui, Kamada, & Satoh, 1998; Van Bakel et al., 2011). The plastid genome is estimated to be 153,871 bp (Vergara, White, Keepers, & Kane, 2016) and the mitochondrial genome size is 415,545 bp (White, Vergara, Keepers, & Kane, 2016).

The cannabis genome has been difficult to resolve due to its high heterozygosity (estimated at 12.5%–40.5%) and abundance of repetitive DNA sequences (Lynch et al., 2016). We now know that approximately 70% of the cannabis genome consists of repetitive sequences (Gao et al., 2020; Kovalchuk et al., 2020; Laverty et al., 2019; Pisupati, Vergara, & Kane, 2018), which has further complicated genome assembly in the past. The first draft genome of cannabis was sequenced in 2011 from the cultivar marijuana "Purple Kush" using short-read sequencing and was unable to resolve low-complexity repeat-rich regions (Van Bakel et al., 2011), resulting in an incomplete genome assembly (size of 534 Mb), which was, however, very valuable due to the first insight into cannabis DNA sequences.

Recently, third-generation single-molecule sequencing (long-read sequencing) led to the assembly of some cannabis genomes at the chromosome level. The genome of cultivar "CBDRx" (a female individual) was sequenced using Oxford Nanopore technology and has an assembly size of 876.148 Mb (Grassa et al., 2018). Laverty et al. (2019) extended the research of Van Bakel et al. (2011) by sequencing the genomes of "Purple Kush" (female) and "Finola" (male) using PacBio single-molecule sequencing and determined an assembly size of 891.965 Mb and 1009.67 Mb for female and male, respectively. Recently, Gao et al. (2020) also used PacBio single-molecule sequencing to assemble the 812.525 Mb genome of a wild cannabis plant and McKernan et al. (2020) used the same sequencing platform to determine the genome sequences of a father (1009.156 Mb)-mother (876.736 Mb)-daughter (999.122 Mb) trio from the cultivar "Jamaican Lion." To date, 12 different cannabis cultivars have been used for genome and transcriptome assembly, as indicated by publicly available data. The version of the "CBDRx" assembly (v.2.0; GenBank acc. no. GCA_900626175.2) is currently the most complete and contiguous chromosome-level assembly with updated chromosome renumbering and has therefore been proposed as the reference cannabis genome (Hurgobin et al., 2021).

2.3 Breeding objectives

The breeding objective is the most important starting point in the breeding of cultivated plants. In the case of hemp, the breeding objectives can vary greatly due to the specific use of various plant parts. As cannabis use has changed throughout history, breeding objectives have changed as needed. The common objectives in breeding of all varieties of hemp, regardless of the purpose of use, are a low level of Δ^9-THC, which must be below the statutory limit of 0.3% or 0.2% (depends on the legislation), maximum resistance to diseases and pests, and maximum crop stability in different growing conditions. Other morphological and physiological traits—e.g., length of growing season, length of flowering, degree of monoecism or dioecism, plant height, content and composition of other (than THC) cannabinoids, seed and fiber yield, etc.—are inherent in varieties that are selectively bred according to the purpose of use (Salentijn et al., 2015).

2.3.1 Fibers

In the cultivation of hemp for fiber, the objective is to have the highest possible fiber content. The genetic potential of dioecious varieties in terms of fiber content is estimated at 38%–40% of fiber in the stem, but in practice it is possible to realistically achieve 28%–30% of fibers from the stem of dioecious hemp plants (Bócsa & Karus, 1998).

Adequate fiber quality is also important, with the objective to achieve the best possible ratio between primary and secondary fibers in favor of the former. For paper and textile industries, fibers with a high cellulose content, low rate of lignification, and few cross-links of pectins and other structural elements in the cell wall of fiber-forming cells are desirable

(Mandolino & Carboni, 2004). In the preparation of biocomposites, where fibers are mixed into a synthetic carrier, the surface properties of fibers, their fineness, and tensile strength are very important (Gamelas, 2013; Placet, 2009).

Weather and growing conditions, as well as stem processing and fiber extraction, have been shown to have a greater impact on fiber quality than the genotype itself (Müssig, 2003). Therefore hemp production method and fiber extraction have a greater potential for improving the quality of obtained hemp fibers than the breeding of new varieties (Finta-Korpelova & Berenji, 2007). Nevertheless, several varieties have been developed in the Netherlands whose fibers are more resistant to mechanical processing of stems before and during extraction—i.e., in retting and friction processing—which can lead to improved fiber quality and to easier and cheaper extraction (Salentijn et al., 2015).

2.3.2 Seed

The seed yield of hemp is low, from a few 100 kg/ha to approximately of 1.5 t/ha (Flajšman & Kocjan Ačko, 2019; Flajšman, Kocjan Ačko, & Čeh, 2018), so one of the more important goals in hemp breeding is to increase the seed yield. This objective was partially achieved by development of monoecious varieties, which usually have a higher seed yield than dioecious varieties (Berenji et al., 2013). Because seed yield is a quantitative trait inherited by many genes at multiple loci, progress in selective breeding using traditional breeding methods is slow (Clarke & Merlin, 2017). Therefore, in terms of seed yield, hemp remains uncompetitive in comparison to other crops, e.g., cereals.

Another aspect of improving seed content is its nutritional composition. Hempseeds have a beneficial nutritional composition, as they are a rich source of protein, unsaturated fatty acids, fiber, and some vitamins and minerals (Kušar, Flajšman, Kocjan Ačko, Pravst, & Čeh, 2018). In breeding of hemp for use in consumption, the main goal is to increase the protein and oil content in seeds, and to improve the fatty acid composition of oil. Hemp oil contains about 80% of unsaturated fatty acid, which gives it a low oxidative stability and therefore makes it perishable at room temperature—which breeding aims to improve (Bielecka et al., 2014). Another objective of breeding is to improve the already favorable ratio between linoleic and α-linolenic acid (Ranalli, 2004) and to increase its gamma-linolenic acid content (Berenji et al., 2013). The synthesis pathway of proteins and fatty acids is encoded by fewer genes, so the selection of suitable lines and breeding of varieties is slightly less demanding than in the case of yield increase (Clarke & Merlin, 2017).

2.3.3 Inflorescences for cannabinoid extraction

In 2001 legislation was introduced in the EU that defined the upper permitted limit of Δ^9-THC in industrial hemp to 0.2%. Since then, hemp breeding has focused more intensively on lines with very little or no Δ^9-THC (Grassi & McPartland, 2017).

Recently, varieties of hemp with high cannabinoids content (except Δ^9-THC, which must remain below 0.2% by dry weight) are becoming more interesting, primarily CBD. The goal is also to develop varieties that would have a high content of only a certain cannabinoid, e.g., CBG (cannabigerol), THCV (tetrahydrocannabivarin), and CBC (cannabichromene), for use in pharmaceutical industry for the preparation of active substances

(Mandolino & Carboni, 2004). Hemp is becoming attractive for cultivation for the purpose of obtaining cannabinoids because it is permitted to be grown outdoors, where it can reach its full potential in terms of expression (quantitatively and qualitatively) of secondary metabolites under favorable weather and growing conditions. The genetic background of chemotype inheritance (= the cannabinoid profile of the plant) is well explained (de Meijer, 2014; de Meijer et al., 2003; Weiblen et al., 2015). However, the content of cannabinoids produced is influenced by several different genes, environmental factors, and plant sex (Grassi & McPartland, 2017), making breeding and precise selection of plants with desired traits, which result solely from genotype, difficult.

2.4 Hemp breeding methods

C. sativa L. is an allogamous plant (anemophilous), so breeding methods for cross-pollinating plants are most commonly used in breeding processes.

2.4.1 Mass selection

Mass selection is carried out by preserving the seeds from the best plants, which are then sown again the following year; the procedure is repeated year after year (Brown, Caligari, & Campos, 2014). In the past, farmers have (unknowingly) created landraces from wild populations using repetitive mass selection. Farmers increased population homozygosity by sowing seeds from plants with agronomically desirable traits, while heterozygosity on loci for resistance and vigor of plants was enhanced through natural selection (Clarke & Merlin, 2017). Mass selection is an effective method for improving simple qualitative traits (e.g., plant height, seed size, cannabinoid content) that are usually encoded by a small number of genes and therefore have high heredity. However, if precise mass selection is carried out over a long period of time, it may also result in progress in more complex quantitative traits that have a low degree of heredity. For hemp, mass selection is successfully used to improve highly hereditary traits such as, for example, fiber content of stems and cannabinoid content of trichome glands (Hennink, 1994).

Mass selection can be undertaken using plants grown from seeds of several parent plants, as well as using plants that are offspring of a single parent plant, which ultimately results in a genetically more homogeneous breeding line. Mass selection in hemp is more effective if we allow only selected male plants to flower and pollinate female plants, and then to store the seeds of only selected female plants. This method was used for selected breeding and is still used for maintaining several European dioecious varieties (e.g., "Carmagnola," "Kompolti," "Lovrin," and "Novosadska konoplja") (Clarke & Merlin, 2017).

2.4.2 Cross-breeding

Using cross-breeding method, breeders select as parents two lines that are as uniform as possible (all plants within the line must be as morphologically identical as possible) but genetically different. Recently, varieties have also been used as parent lines. After crossing, the progeny is selected according to fiber content, stem and seed yield, sex, phenological development, stem quality, low Δ9-THC content, resistance to diseases, pests, and lodging. Improved offspring lines were developed using a selection of offspring from a single maternal plant that was pollinated by several different paternal plants (half-sib family). The selection was based on the general combining ability of different combinations of crosses (Salentijn et al., 2015).

2.4.3 Hybrid breeding by crossing varieties or synthetic lines

Cannabis breeders noticed that planned crosses of genotypes that are significantly different contribute to a greater vigor of offspring, which they attributed to the effect of heterosis or hybrid vigor. *C. sativa* L. is a species that has no pure lines (i.e., homozygous genotypes), so true hybrids produced by crossing pure lines have not yet been developed (Berenji et al., 2013; Sharon Reikhav, personal communication). Therefore, in the case of hemp, hybrid breeding is called cross-breeding: (a) between genetically different varieties, which are as homogeneous as possible; several varieties were selectively bred, the first developed in this way are "Uniko-B," "Kompolti Hybrid TC," and "YunMa 3" (Salentijn et al., 2015); (b) between different homogeneous lines which are repeatedly self-pollinated. A synthetic line is an artificial population of plants that has been selected using mass selection and then maintained in isolation for a long time (monoecious plants have been self-pollinated over several generations) in order to maximize genetic uniformity. The best lines (usually between 4 and 10) are then crossed with each other to develop a so-called hybrid seed. This approach was developed in France and is used to improve stem pith and seed yield (Berenji et al., 2013).

Hybrid hempseed production is similar as with true hybrids, where a maternal component, which is male-sterile and does not produce pollen, and a paternal component that serves as a pollinator are needed. With dioecious varieties, we have female plants that can serve as a maternal component, but the problem is that the sown seeds of the dioecious variety produce 50% male plant that must be removed before flowering, which is practically impossible to do on a large scale, i.e., for purposes of obtaining large quantities of seeds, where plants are sown over a large area. The described problem was solved by Bócsa (1967), when he discovered that F1 offspring of a crossing between dioecious female plants and monoecious plants, which act as the pollinator line, are 70%–85% female, 10%–15% monoecious plants, and only 1%–2% male plants. The F1 generation was named unisexual plants. In the process of producing the hybrid seed, these F1 offspring served as the maternal line, from which male plants still needed to be removed, but to a much lesser extent. Finta-Korpelova and Berenji (2007) described in more detail the production of three-line hybrid seed in Hungary; the crossing of dioecious female plants and monoecious plants takes place on an area of 1 ha, where the ratio between maternal and paternal component must be 3:1 or lower. In this way, 400–500 kg of F1 seeds (50–100 g per female plant) can be obtained in the first year. In the second year, the F1 seed serves as the maternal component for sowing in rows on 200–250 ha and by sowing the paternal component next to it, the seed of the three-line hybrid is produced. This is a dioecious line, where male plants serve as pollenizers. In terms of sex, such a three-line hybrid consists of 50% female and 50% male plants. In the second year, 0.5–1 tonnes of three-line hybrid seed per hectare can be obtained from female plants of the maternal component, which is intended for commercial production and is sufficient for sowing 3000 ha, if the sowing quantity is 80 kg/ha (the purpose of production of such hemp hybrids was for fibers). By selecting the genotype of the paternal component (monoecious line) in the first year of breeding and by selecting the paternal component in the last year of crossing (dioecious line), breeders affected the traits of the three-line hybrid seed. The pioneer of this type of hybrid seed production was Dr. Ivan Bócsa from the Fleischmann Rudolf Agricultural Research Institute in Kompolt, Hungary (Finta-Korpelova & Berenji, 2007).

Modern French hemp varieties are monoecious varieties consisting of 50% female and 50% monoecious plants (Berenji et al., 2013). The production process is slightly different from the Hungarian one, as the French varieties are two-line open-pollinated hybrids (Fig. 2.3).

Production of F1 unisex generation in the first year is done in the same way as in Hungary, i.e., by crossing dioecious female plants (male plants must be removed before flowering) and monoecious plants, which are the pollinating line. In the second year, F1 plants are backcrossed with a monoecious paternal component of the same line, leading to the production of F2 seeds. In both years, remaining male plants and unwanted phenotypes, as well as atypical plants (e.g., atypical stem morphology), need to be removed from the plant population of monoecious paternal components and F1 offspring (Fig. 2.4). In the third year, F2 plants pollinate freely with each other, and their yield is a two-line hybrid seed (50% female and 50% monoecious plants) intended for commercial production (Berenji et al., 2013; Finta-Korpelova & Berenji, 2007). This is the method of simple production of large quantities of monoecious variety seeds, with less effort needed than in the case of traditional reproduction and preservation of monoecious varieties (Berenji et al., 2013). This method has also been used to obtain several modern Polish monoecious cultivars (Poniatowska, Wielgus, Szalata, Ozarowski, & Panasiewicz, 2019) and very recently (in March 2021) the first Slovenian two-line open-pollinated monoecious hybrid named "Fiona" was also registered.

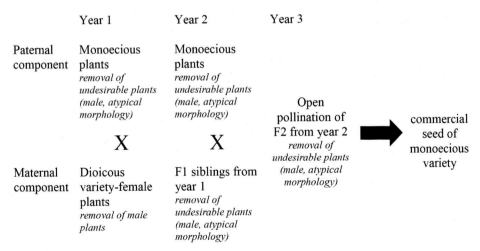

FIG. 2.3 Scheme for producing seed of two-line open-pollinated hybrids and breeding of new monoecious varieties. *Adapted from Berenji, J., Sikora, V., Fournier, G., & Beherec, O. (2013). Genetics and selection of hemp. In P. Bouloc (Ed.), Hemp: Industrial production and uses (pp. 48–72). London: CABI.*

FIG. 2.4   An example of two plants with atypical stem morphology; left—a plant stem with too much robust lateral shoots; right—a plant with atypically curved stem.

2.4.4 Mutation breeding

Mutation breeding mimics the natural process of spontaneous mutations in the plant's genetic information using various ways (physical and chemical methods) (Liang, 2012). Monoecious cannabis plants are thought to result from spontaneous mutation in dioecious plants. The yellow cannabis stems, which were used in breeding in the past and which were clearly distinguished from other varieties by the color of the stem, were also the result of a spontaneous natural mutation (Berenji et al., 2013). Hemp varieties produced in this way are not officially registered. Grassi and McPartland (2017) mentioned that di Candilo et al. (2000) used ^{60}Co gamma-ray irradiated pollen of Carmagnola and Fibranova varieties to produce "Red Petiole" (plants had red-colored petioles—leaf stalks) and "Yellow Apex" (plants had yellow-colored leaflets on top of the plant) varieties. Because the traits described were incompletely dominant, they did not survive after several generations of reproduction and the green color returned. The "Ermo" variety was also formed by pollinating plants of the "Ermes" variety with pollen from a low-cannabinoid male plant irradiated with ^{60}Co gamma rays (Grassi & McPartland, 2017). None of the varieties described is on the EU Plant variety database today. Flajšman, Kocjan Ačko, and Bohanec (2017) irradiated seeds of three varieties of industrial hemp and determined the radiation dose that would be suitable for use in hemp breeding purposes. The work in hemp mutation breeding was not continued.

2.4.5 C. *sativa* L. genome research and breeding using molecular markers

Molecular markers for genes involved in the biosynthetic pathways of specific metabolites or those responsible for expressing important agronomic traits would allow early plant selection in their juvenile stage, well before flowering, and would be of great help to breeders (de Meijer, 2004). In order to use molecular markers in breeding of plants, the genome of the plant under study must be at least partially known.

The publication of draft genomes and transcriptome in 2011 represented significant progress in the field of genome research on *C. sativa* L. Van Bakel et al. (2011) used Illumina sequencing to determine the genome sequence of female plants of medical cannabis (variety "Purple Kush") and industrial hemp (variety "Finola"). It was discovered that there are many differences between varieties; e.g., medical cannabis had the *THCAS* gene (Δ^9-tetrahydrocannabinolic acid synthase), but not the *CBDAS* gene (cannabidiolic acid synthase), which are responsible for the synthesis of THCA and CBDA. In industrial hemp, the situation is the opposite. Thus the theory that the cannabis chemotype is encoded by two closely related loci was partially confirmed (several pseudogenes for both enzymes were also discovered). Medical cannabis was found to have many copies of the *AAE3* gene, which could be involved in cannabinoid biosynthesis. Further genome analysis, which looked at additional differences between medical cannabis and industrial hemp, included "USO-31" (industrial hemp variety) and "Chemdawg" (medical cannabis variety). It was discovered that medical cannabis and industrial hemp from the study had significant genetic differences. The cannabis transcriptome (determined from roots, stems, shoots, and flowers of the "Purple Kush" variety) contains more than 30,000 genes. When comparing the expression of genes of the biosynthetic enzymatic pathway of cannabinoids between varieties "Purple Kush" and "Finola," it was discovered that the genes of this pathway are 15 times higher expressed in medical cannabis. The more expressed genes in variety "Purple Kush" also include a number of transcription factors that influence many developmental processes in trichome glands (Van Bakel et al., 2011). The published genome of *C. sativa* L. was the foundation for identification of 55 genes for synthase enzymes in the terpenes biosynthesis pathway, which were also tissue specifically expressed (Allen et al., 2019).

Weiblen et al. (2015) discovered a quantitative trait locus (QTL) that is associated with the total amount of cannabinoids and their ratio (especially the ratio of THCA to CBDA) in a plant.

Sawler et al. (2015) used 14,031 markers for single nucleotide polymorphism (SNP) to successfully distinguish industrial hemp (43 samples) from medical cannabis (81 samples), with only two samples according to origin and name did not correspond to the established genetic profile.

Based on a cross between medical cannabis ("Purple Kush" variety) and industrial hemp ("Finola" variety), Laverty et al. (2019) produced a physical and genetic map. They assembled 10 chromosomes from the sequence data. They found that the genes for the synthesis of the two main cannabinoids (*THCAS* and *CBDAS*) are unrelated and located at two different loci, but very close to chromosome 6. Chromosomes 5, 9, and 10 are telocentric and one of them could be the sex chromosome. The map is not yet complete, as not all known cannabis transcriptomes can be mapped to it, including markers for determining the male sex.

Grassa et al. (2018) also compiled 10 chromosomes from the obtained data on the nucleotide sequence of cannabis. They published a precise genetic map obtained by sequencing 100 different cannabis genomes of parents and offspring of one family of cross-breeding, which helped them reveal the genetic mechanism responsible for the relationship between CBDA and THCA in cannabis plants. They found that the cannabis genome contained a large number of copies of genes for synthase enzymes that synthesize cannabinoids. They hypothesized that the ratio of cannabinoids ultimately depended on the number of copies of genes of a particular synthase. Furthermore, synthase genes are controlled by regulatory elements from other regions of the genome and have a greater impact on the amount of cannabinoids produced than synthases themselves. Loci with *CBDAS* genes (not loci with *THCAS* genes) are likely to have a major impact on the CBDA/THCA ratio.

Salentijn et al. (2019) reviewed the potential genes that control flowering time in cannabis, determine sex and fiber quality. By knowing the genetic mechanisms that control the described traits, it is possible to improve breeding methods (use of molecular markers) and shorten the time to develop new improved varieties.

Despite extensive research efforts in discovering the genetic background of *C. sativa* L., few useful research results have crossed the line of basic research and established themselves as an aid in real breeding purposes. The utility of molecular markers for breeding has so far been demonstrated in sex determination (Kolenc & Čerenak, 2017; Mandolino, Carboni, Forapani, Faeti, & Ranalli, 1999; Törjék et al., 2002) and in chemotype determination. Pacifico et al. (2006) developed the PCR marker B1080/B1192, which was able to reliably determine the chemotype of the plant. However, it turned out that genetic chemotype determination can also lead to deviations (Weiblen et al., 2015), suggesting the complexity of the system of gene inheritance and expression for the synthetic pathway of cannabinoids. Welling, Liu, Shapter, Raymond, and King (2016) used two SCAR markers to predict cannabinoid profiles of 22 different cannabis accessions with 98% rate of success.

There is still no marker system associated with monoecious plants (monoecious plants have the same karyotype as dioecious female plants), despite the fact that monoecious varieties are gaining in importance in terms of extensive breeding and widespread production. There have also been no developed markers associated with, for example, growth period and response to the photoperiod, to the presence of trichomes on flowers, and to fiber quality (Onofri & Mandolino, 2017). Such markers would be particularly useful in dioecious varieties for the selection of male plants that do not carry important economic traits (seed yield, cannabinoids), and their contribution is seen only with expressed traits of next-generation female plants (Clarke & Merlin, 2017). The reason for slow progress in the development of marker systems for use in breeding lies mainly in the fact that in *C. sativa* L. the genetic variability is very high due to the allogamy and cross-pollination of *C. sativa* L. plants in the past (Salentijn et al., 2015).

2.4.6 Difficulties in breeding of C. *sativa* L.

Problems in breeding of *C. sativa* L. stem from its characteristics, such as dioecism, allogamy, and cross-pollination. *C. sativa* L. is not a plant that is easy to breed, although it produces a lot of both pollen and seeds. Successful breeding of wind-pollinated allogamous plants requires first selecting plants with the desired traits, obtaining breeding lines by repeated selection, and then crossing different breeding lines and testing their offspring in the field (Posselt, 2010). In dioecious hemp, female plants are the ones that are most useful (fibers, seeds, inflorescences), while male plants, which are essential for crossing because of their pollen, are used only for fibers. Therefore the selection of male cannabis plants is difficult, as they cannot be selected according to the phenotype and useful value of male plants (the exception is selection based on fiber content, known as the Bredemann method), but only the female population of offspring shows the good or bad traits of the paternal parental line. Dioecious plant also cannot be self-pollinated naturally (Clarke & Merlin, 2017). On the other hand, monoecious varieties are problematic primarily for maintenance and seed multiplication, as they have a natural tendency to return to the dioecious state, requiring great effort in removing unwanted male plants in propagating seed material (Berenji et al., 2013). Cannabis pollen can be carried over long distances by wind (Small & Antle, 2003), so individual lines need to be cultivated in isolation to avoid unwanted pollination, as hemp plants, regardless of origin, are freely pollinated with each other with fully fertile offspring (Small, 1972).

The development of a new variety is a lengthy process and can take more than 20 years (Poniatowska et al., 2019). Quantitative traits (yield, content of cannabinoids, etc.) are usually inherited with many alleles at different loci in the genome, with a single gene having little effect. Therefore the heredity of such traits is low and progress in breeding is slow, which makes any progress in cannabis breeding more difficult (Clarke & Merlin, 2017). To start breeding a new variety, where we use target-oriented cross-breeding, it is necessary to have a large population of plants from which we can select plants with the desired traits and then use them for further cross-breeding. A large population of plants is needed so that most important allelic forms can be passed on to offspring; for monoecious lines, it is necessary to use at least 1000 selected plants for one parental component, and for dioecious lines, at least 2000 selected plants are required (Crossa, Hernandez, Bretting, Eberhart, & Taba, 1993). This means that the initial population must be much larger.

2.5 Hemp breeding in Europe

2.5.1 Milestones of hemp breeding in Europe

In the first three decades of the 20th century, varieties were bred in Europe using selections from old landrace populations according to the length of the growing season, height, diameter, and weight of stems and, in some cases, the seed yield. L. H. Dewey is considered one of the first breeders of industrial hemp, having published a work in 1927 in which he described production of late varieties from Chinese landraces (Dewey, 1927). One of the parental components was also the Ferrara landrace (Ranalli, 2004). The selection according to the morphological traits described before is simple and relatively quickly results in population uniformity. However, using these methods, qualitative traits such as the fiber content of the stem did not improve. The first turning point in hemp breeding was the discovery that the fiber content is inherited from both the maternal and paternal side in dioecious hemp. Bredemann (1924) was the first to use the male plant selection method to increase the fiber content in hemp stems, and increased fiber content over the next 30 years by a factor of three, from 12%–15% to 25%–35% (Berenji et al., 2013). The famous breeders from this early period are Rudolf Fleischmann from Hungary; Otto Heuser from Germany; and Vedenski, Grishko, and Malusha from Russia (Ranalli, 2004). Another major turning point in the selective breeding of industrial hemp was the discovery of hermaphroditic monoecious plants (male and female flowers are found on the same plant), which could also be self-pollinated and thus used to develop more uniform new varieties (Grishko, Levchenko, & Seletski, 1937; Grishko &

Malusha, 1935). Since then, the methods of self-pollination and cross-breeding of monoecious and dioecious plants have become more widely used and have led to the development of modern varieties. Neuer and Sengbusch (1943) were the first to fix the trait for monoecism. Their work led to the development of variety "Fibrimon," which is probably the first officially registered monoecious variety (Bredemann, Garber, Huhnke, & von Sengbusch, 1961) and represents the basis for many modern varieties. The development of monoecious lines led to the discovery that the F1 offspring of cross-breeding of dioecious female and monoecious plants could be used as a maternal component in the production of hemp hybrids (Bócsa, 1967), leading to the registration of varieties "Uniko-B" in 1969 and "Kompolti Hybrid TC" in 1983. It should be noted that these are not true hybrids that would result from cross-breeding pure (genetically uniform, homogeneous) lines, but cross-breeds of plants that may otherwise originate from an officially registered variety, but the plants are genetically heterogeneous. An important milestone in the hemp breeding is also the development of variety "Finola" (registered in the EU in 1999; Clarke & Merlin, 2017), which flowers regardless of the length of the day (autoflowering). Additionally, the variety is short and therefore more suitable for machine seed harvesting, which is a rarity among hemp varieties (Callaway & Laakkonen, 1996).

2.5.2 Breeding of some European varieties of hemp

Cultivation of hemp in Europe before the 20th century was based on the use of local landraces, which were slowly improved by growers themselves through mass selection. However, in the 20th century, breeders began to use foreign genetic material for cross-breeding to increase fiber and seed yields. Crossings between Mediterranean and Central Russian ecotypes represent the main selective breeding lines from which most European modern varieties emerged (Clarke & Merlin, 2017).

2.5.2.1 Dioecious varieties

Many of today's dioecious varieties of hemp, which are used to produce fibers, have a common ancestor, namely the "Carmagnola" variety (Clarke & Merlin, 2017).

"Carmagnola" is a cannabis landrace from the northern part of Italy and is the oldest population in Italy. It was selectively bred into a variety using mass selection. It originates in a Chinese broad-leaflet hemp landrace (Clarke & Merlin, 2017). In the past, it was spread throughout Italy, where ecotypes and then varieties of "Bolognese," "Tuscany," and "Ferrarese" were obtained using mass selection (Ranalli, 2004). The "CS" variety ("Carmagnola Selezionata") was obtained from "Carmagnola" in the 1970s. "Carmagnola" was also one of the parents in the development of varieties "Fibranova" (the other parent was the Bredemann "Eletta" variety) and "Eletta Campana" (the other parent was probably the German variety "Fibridia") (de Meijer, 1995). Varieties "CS," "Fibranova," and "Eletta Campana" are still listed on the EU Plant variety database today.

Most Hungarian varieties are dioecious. An old Hungarian variety was "Fleischmann hemp" or "F-hemp," developed in 1931 by Rudolf Fleischmann by crossing inbred landraces of "Bologna" and "Ferrara" which are of Italian origin (Grassi & McPartland, 2017). The "Kompolti" variety was probably developed from the Italian variety "Carmagnola" by selection of plants with a higher fiber content in stems (Clarke & Merlin, 2017). "Uniko-B" is a one-line hybrid (registered in 1969), where the maternal components were dioecious plants of the "Kompolti" variety and the paternal component was the monoecious variety "Fibrimon 21." The F1 offspring were almost entirely female plants, some monoecious plants, and very few male plants. F1 offspring were then pollinated with each other and the F2 generation (commercial seed) contained about 30% of male plants, and the variety was suitable for fiber production (Bócsa & Karus, 1998). "Kompolti Hybrid TC" (Fig. 2.5) is a three-line hybrid developed according to the scheme described in Section 2.4.3. The maternal component was female dioecious plants of the Chinese variety "Kinai," which were crossed with monoecious plants of variety "Kinai." The result was called F1 "Kinai Uniszex." In the second year, F1 "Kinai Uniszex" was used as the parental component and crossed with male plants of the "Kompolti" variety. The F2 offspring were a new variety of "Kompolti Hybrid TC," consisting of 50% female and 50% male plants (Ranalli, 2004).

In the territories of former Yugoslavia (especially Serbia and Croatia), primarily foreign varieties were used in the past for cultivation, and breeding started only after 1950 (Bouloc & Berenji, 2013). Two domestic varieties were used, namely "Flajsmanova" (these are adapted plants of "Fleischmann hemp" from Hungary) and "Novosadska konoplja," which was bred using mass selection from "Fleischmann hemp" in the 1960s (de Meijer, 1995). Since 2002, the Serbian variety list includes three varieties: "Marina" (dioecious), "Helena" (monoecious), and "Diana" (F1 hybrid) (Bouloc & Berenji, 2013). Varieties "Marina" and "Helena" were recently registered with the EU Plant variety database as well.

"Finola" is a special variety among all industrial hemp varieties because of its characteristic ability to start flowering independently of the photoperiod (Callaway & Laakkonen, 1996). It is an early variety, bred in Finland, and is intended for seed production in more northern latitudes. It is widely grown in Canada for seed and oil production (Small & Marcus, 2002).

FIG. 2.5   Plants of variety "Kompolti Hybrid TC" at the field.

"Finola" is a cross between two very similar, early northern Russian narrow-leaflet hemp land-races, which were kept at the Vavilov Research Institute in St. Petersburg, Russia (Clarke & Merlin, 2017). It is a low and early variety intended for mechanical production. It was registered in the EU in 1999 (Clarke & Merlin, 2017). It can achieve high seed yields, even up to 1.7 t/ha (Callaway & Laakkonen, 1996). In Italy, it has been used in recent years as one of the parental components in crossing with the "Carmagnola" variety and other varieties derived from "Carmagnola" in order to introduce the monoecious traits into the offspring and to increase seed yield (Grassi & McPartland, 2017).

2.5.2.2 Monoecious varieties

Almost all modern European monoecious varieties are derived from the "Fibrimon" variety, which was bred in Germany between 1951 and 1955. "Fibrimon" was developed from a single monoecious plant that was accidentally found in the Central Russian landrace. The plant was self-pollinated and a line was developed which was then crossed with female plants from a high-fiber line from Germany and with late-flowering cultivated populations from Italy and Turkey (Clarke & Merlin, 2017; de Meijer, 1995). Cannabis loses vigor through self-pollination (Clarke & Merlin, 2017). This could be the reason that monoecious varieties are usually less lush than dioecious varieties, e.g., they achieve a lower final height, a lower stem diameter, and a lower biomass yield (Berenji et al., 2013).

Poland developed its monoecious varieties by cross-breeding and selection from other European varieties. The oldest varieties are "Bialobrzeskie" and "Beniko." "Bialobrzeskie" was developed by multiple crossing of dioecious and monoecious varieties and lines ("LKCSD" x "Kompolti"), with the offspring then crossed with the monoecious "Fibrimon" variety. This was followed by several years of selection—after 12 years of selection, the variety was registered in 1967. The monoecious variety "Beniko" (registered in 1985) had been selectively bred for 21 years. The development of this variety was based on the crossing of two monoecious varieties ("Fibrimon 21" x "Fibrimon 24") and then on individual selection. Both described varieties further served as one of the parents in the development of varieties "Tygra," "Wojko," and "Rajan." The last registered variety is "Henola" from 2017, which is a cross between varieties "Zołotonowska 13" (monoecious) and "Zenica" (dioecious) (Poniatowska et al., 2019).

All French varieties are descendants of the monoecious "Fibrimon" variety, which was brought to France in the late 1960s from Germany, where its selective breeding had started. Varieties "Fibrimon 21," "Fibrimon 24," and "Fibrimon 56" (they are no longer on the variety list today) were selected directly from "Fibrimon" and differed from each other during flowering. "Ferimon" is also a variety that was selected directly from "Fibrimon" and is still on the variety list. "Fedora 19" was developed as a two-line hybrid by crossing female dioecious plants of the Russian variety "JUS-9" and monoecious plants of the "Fibrimon 21" variety, followed by back-crossing of the F1 generation with monoecious plants of the "Fibrimon 21" variety. Today, the "Fedora 17" variety (selected from the "Fedora 19" variety and containing less Δ^9-THC) is on the variety list. The same scheme was used to develop varieties "Félina 34" (the maternal component was female dioecious plants of the Hungarian variety "Kompolti"), "Fédrina 74" and "Futura 77" (the maternal component of both varieties was female dioecious plants of the German variety "Fibridia"), with variety "Fibrimon 24" used twice as the pollinating line (de Meijer, 1995). The "Fédrina 74" variety is no longer on the variety list, and the "Félina 34" and "Futura 77" varieties have been replaced by the newer "Félina 32" and "Futura 75" varieties, which contain less Δ^9-THC. In the early 1980s in France, lines began to be intensively selected for their Δ^9-THC content. After 2001, they managed to select the lines that contained only 0.05% Δ^9-THC by dry weight. This is the origin of modern varieties with a very low content of this cannabinoid, e.g., "Fedora 17," "Félina 32," and "Futura 75." Lines with Δ^9-THC levels below the detection limit due to natural mutation were also selected from individual plants. Varieties "Santhica 23," "Santhica 27," and "Santhica 70" were developed from such selections (Berenji et al., 2013).

2.6 Conclusions and future perspectives

Industrial hemp is a plant with many uses and has been selectively bred for such purposes. Many varieties from the past are no longer used and have been replaced by newer varieties, which have primarily low Δ^9-THC content, but are distinguished by other useful traits intended for each type of production according to their ultimate purpose of use. The growing number of varieties on the EU Plant variety database clearly shows that there is intensive breeding of hemp in Europe.

Hemp has been systematically bred for the past 100 years. Despite many recent studies in the field of genetics of *C. sativa* L., e.g., publication of genome and transcriptome (Gao et al., 2020; Grassa et al., 2018; Laverty et al., 2019; Van Bakel et al., 2011) and the use of many of markers to study the genetic variability of different groups of cannabis plants and to determine sex and chemotype (Onofri & Mandolino, 2017), there are no available effective molecular methods for marker-assisted selection, with the exception of sex and chemotype markers. Traditional approaches in the breeding of allogamous plants—such as mass selection, cross-pollination, self-pollination, and development of hybrids—are still the most widely used, but they do not result in "true" hybrids. Therefore progress on improving important economic properties of hemp, such as fiber, seed, and cannabinoid yields, is slow.

Hemp is already being grown for medical purposes due to the importance of cannabinoids in medicine (Booth & Bohlmann, 2019; Deiana, 2017; García-Tejero et al., 2019; Janatova et al., 2018) and is likely to be grown on an even larger scale in the future. Therefore breeders face an important challenge: how to develop varieties that contain as much target cannabinoid as possible (e.g., CBD, CBG, THCV, CBC, etc.), but keeping Δ^9-THC below the permitted limit. Another aspect of development of varieties for medical use is to obtain plants with a suitable qualitative and quantitative ratio between certain cannabinoids.

The use of industrial hemp for medical purposes will also have to be followed by its cultivation. Pollinated female plants synthesize lower quantities of cannabinoids because they divert the primary metabolites to seed maturation after flowering (Chandra, Lata, Khan, & ElSohly, 2017). Therefore, in cultivation for the purpose of cannabinoid production, male and monoecious plants are not desirable, as they pollinate female plants with their pollen and thus reduce the yield of cannabinoids. To this end, it would be necessary to obtain feminized industrial hempseeds, which would allow cultivation of only female plants without the lengthy and laborious removal of male or monoecious plants.

One of the bigger challenges is certainly also the development of a pure cannabis line and derived true hybrids.

The protein and oil of hempseeds are known for their remarkable nutritional value and health benefits. Therefore increasing the protein and oil content in the seeds remains one of the main goals of industrial hemp cultivation for food use. In particular, improving the amino acid and fatty acid composition of hempseed oil and protein will have to be closely monitored in selection programs to achieve this important breeding objective.

The milestones in the breeding of industrial hemp in the 20th century (selection of male plants before flowering according to fiber content, development of monoecious varieties, schemes of developing "hybrid" hemp varieties, discovery and use of photoperiod independence in cannabis genetic material) led to the development of many new varieties, selectively bred in accordance with the intended use. Landraces that have historically been shaped by human selection and environmental selection pressures have been the foundation for new crossings and the development of modern dioecious varieties, such as "Kompolti Hybrid TC" and "Finola," and monoecious varieties, such as "Futura 75," "USO-31," and "Henola." However, there are still many new varieties yet to come in order to meet specific production purposes. Due to the high diversity of the cannabis genome, new methods of breeding will likely be prominent, including genetic engineering (Russo, 2019).

References

Aizpurua-Olaizola, O., Soydaner, U., Öztürk, E., Schibano, D., Simsir, Y., Navarro, P., et al. (2016). Evolution of the cannabinoid and terpene content during the growth of Cannabis sativa plants from different chemotypes. *Journal of Natural Products, 79*(2), 324–331.

Allen, K. D., McKernan, K., Pauli, C., Roe, J., Torres, A., & Gaudino, R. (2019). Genomic characterization of the complete terpene synthase gene family from *Cannabis sativa*. *PLoS One, 14*(9), e0222363.

Andre, C. M., Hausman, J. F., & Guerriero, G. (2016). Cannabis sativa: The plant of the thousand and one molecules. *Frontiers in Plant Science, 7*, 19.

Berenji, J., Sikora, V., Fournier, G., & Beherec, O. (2013). Genetics and selection of hemp. In P. Bouloc (Ed.), *Hemp: Industrial production and uses* (pp. 48–72). London: CABI.

Bielecka, M., Kaminski, F., Adams, I., Poulson, H., Sloan, R., Li, Y., et al. (2014). Targeted mutation of $\Delta 12$ and $\Delta 15$ desaturase genes in hemp produce major alterations in seed fatty acid composition including a high oleic hemp oil. *Plant Biotechnology Journal, 12*(5), 613–623.

Bócsa, I. (1967). Kender fajtahibrid előállításához szükséges unisexuális (hímmentes) anyafajta nemesítése. *Rostnövények,* 3–7 (in Hungarian).

Bócsa, I., & Karus, M. (1998). *The cultivation of hemp. Botany, varieties, cultivation and harvesting.* Sebastopol, USA: Hemptech.

Booth, J. K., & Bohlmann, J. (2019). Terpenes in *Cannabis sativa*—From plant genome to humans. *Plant Science, 284*, 67–72.

Bouloc, P., & Berenji, J. (2013). Hemp production outside the EU – North America and Eastern Europe. In P. Bouloc (Ed.), *Hemp: Industrial production and uses* (pp. 268–277). London: CABI.

Braich, S., Baillie, R. C., Jewell, L. S., Spangenberg, G. C., & Cogan, N. O. (2019). Generation of a comprehensive transcriptome atlas and transcriptome dynamics in medicinal cannabis. *Scientific Reports, 9*(1), 1–12.

Bredemann, G. (1924). Beitrage zur Hanfzuchtung II. Anslese faserreicher. Mannchen zur Befruchtung durch Faserbestimmung an der lebeden Pflanze vor deu Blute. *Angewandte Botanik, 6*, 348–360 (in German).

Bredemann, G., Garber, K., Huhnke, W., & von Sengbusch, R. (1961). Die Züchtung von monözischen und diözischen, faserertragreichen Hanfsorten Fibrimon und Fibridia. *Zeitschrift für Pflanzenzüchtung, 46*, 235–245 (in German).

Brown, J., Caligari, P., & Campos, H. (2014). *Plant breeding* (2nd ed.). West Sussex: John Wiley and Sons Ltd.

Callaway, J. C., & Laakkonen, T. T. (1996). Cultivation of *Cannabis* oil seed varieties in Finland. *Journal of the International Hemp Association, 3*, 32–34.

Chait, L. D., Evans, S. M., Grant, K. A., Kamien, J. B., Johanson, C. E., & Schuster, C. R. (1988). Discriminative stimulus and subjective effects of smoked marijuana in humans. *Psychopharmacology, 94*, 206–212.

Chandra, S., Lata, H., Khan, I. A., & ElSohly, M. A. (2017). *Cannabis sativa* L.: Botany and horticulture. In S. Chandra, H. Lata, & M. A. ElSohly (Eds.), *Cannabis Sativa L. – Botany and biotechnology* (pp. 76–100). Cham, Switzerland: Springer International Publishing.

Clarke, R. C., & Merlin, M. D. (2017). *Cannabis* domestication, breeding history, present-day genetic diversity, and future prospects. *Critical Reviews in Plant Sciences, 35*(5–6), 293–327.

Crossa, J., Hernandez, C. M., Bretting, P., Eberhart, S. A., & Taba, S. (1993). Statistical genetic considerations for maintaining germplasm collections. *Theoretical and Applied Genetics, 86*, 673–678.

de Meijer, E. P. M. (1995). Fibre hemp cultivars: A survey of origin, ancestry, availability and brief agronomic characteristics. *Journal of the International Hemp Association, 2*(2), 66–73.

de Meijer, E. P. M. (2004). The breeding of *Cannabis* cultivars for pharmaceutical end uses. In G. W. Guy, B. A. Whittle, & P. Robson (Eds.), *The medicinal uses of cannabis and cannabinoids* (pp. 55–69). London: Pharmaceutical Press.

de Meijer, E. P. M. (2014). The chemical phenotypes (chemotypes) of *Cannabis*. In R. G. Pertwee (Ed.), *Handbook of cannabis* (pp. 89–110). London: Oxford University Press.

de Meijer, E. P. M., Bagatta, M., Carboni, A., Crucitti, P., Moliterni, V. C., Ranalli, P., et al. (2003). The inheritance of chemical phenotype in *Cannabis sativa* L. *Genetics, 163*(1), 335–346.

De Meijer, E. P. M., & Hammond, K. M. (2005). The inheritance of chemical phenotype in *Cannabis sativa* L.(II): Cannabigerol predominant plants. *Euphytica, 145*(1), 189–198.

Deiana, S. (2017). Potential medical uses of cannabigerol: A brief overview. In V. R. Preedy (Ed.), *Handbook of cannabis and related pathologies* (pp. 958–967). Academic Press.

Dewey, L. H. (1927). Hemp varieties of improved type are result of selection. In *Yearbook of the Department of Agriculture* (pp. 358–361). U.S. Dept. of Agriculture, Extension Service.

di Candilo, M., di Bari, V., Giordano, I., Grassi, G., Pentagelo, A., & Ranalli, P. (2000). Due nuovi genotipi di canapa da fibra: Descrizione morfo-produttiva. *Sementi Elette, 46*(1), 25–31 (in Italian).

Fike, J. (2016). Industrial hemp: Renewed opportunities for an ancient crop. *Critical Reviews in Plant Sciences, 35*(5–6), 406–424.

Finta-Korpelova, Z., & Berenji, J. (2007). Trends and achievements in industrial hemp (*Cannabis sativa L.*) breeding. *Bulletin for Hops, Sorghum & Medicinal Plants, 39*(80), 63–75.

Flajšman, M., & Kocjan Ačko, D. (2019). Influence of growing conditions on stem yield and morphological characteristics of 12 hemp varieties (*Cannabis sativa* L.) in years 2018 and 2019. *Hop Bulletin, 26*, 119–128 (in Slovenian).

Flajšman, M., Kocjan Ačko, D., & Bohanec, B. (2017). Determination of dosage effect of X-ray RD30 from leaf area on three varieties of hemp. In B. Čeh (Ed.), *New challenges in agronomy 2017: Proceedings of symposium, Laško, 2017* (pp. 140–146). Ljubljana: Slovenian Society of Agronomy (in Slovenian).

Flajšman, M., Kocjan Ačko, D., & Čeh, B. (2018). Characteristics of common hemp varieties that are grown in Slovenia. *Hop Bulletin, 25*, 44–58 (in Slovenian).

Gamelas, J. A. F. (2013). The surface properties of cellulose and lignocellulosic materials assessed by inverse gas chromatography: A review. *Cellulose, 20*, 2675–2693.

Gao, S., Wang, B., Xie, S., Xu, X., Zhang, J., Pei, L., et al. (2020). A high-quality reference genome of wild *Cannabis sativa*. *Horticulture Research, 7*(1), 1–11.

García-Tejero, I. F., Zuazo, V. D., Sánchez-Carnenero, C., Hernández, A., Ferreiro-Vera, C., & Casano, S. (2019). Seeking suitable agronomical practices for industrial hemp (*Cannabis sativa* L.) cultivation for biomedical applications. *Industrial Crops and Products, 139*, 111524.

Grassa, C. J., Wenger, J. P., Dabney, C., Poplawski, S. G., Motley, S. T., Michael, T. P., et al. (2018). A complete *Cannabis* chromosome assembly and adaptive admixture for elevated cannabidiol (CBD) content. *BioRxiv*, 458083.

Grassi, G., & McPartland, J. M. (2017). Chemical and morphological phenotypes in breeding of *Cannabis sativa* L. In S. Chandra, H. Lata, & M. A. ElSohly (Eds.), *Cannabis sativa L. – Botany and biotechnology* (pp. 137–160). Cham, Switzerland: Springer International Publishing.

Grishko, N. N., Levchenko, V. I., & Seletski, V. I. (1937). Question of sex in hemp, the production of monoecious forms and of varieties with simoultaneous ripening of both sexes. *Vszesoy. Nauchno-Issled Inst Konopli, 5*, 73–108 (in Russian).

Grishko, N. N., & Malusha, K. V. (1935). Probleme und Richtlinien in Hanfzuchtung. *Trudy po Prikladnoi Botanike, 4*, 61–67 (in Russian).

Grotenhermen, F., & Karus, M. (1998). Industrial hemp is not marijuana: Comments on the drug potential of fiber *Cannabis*. *Journal of the International Hemp Association, 5*, 96–101.

Hennink, S. (1994). Optimization of breeding for agronomic traits in fiber hemp (*Cannabis sativa* L.) by study of parent-offspring relationships. *Euphytica, 78*, 69–76.

Hillig, K. W. (2005). Genetic evidence for speciation in *Cannabis* (*Cannabaceae*). *Genetic Research and Crop Evolution, 52*(2), 161–180.

Hurgobin, B., Tamiru-Oli, M., Welling, M. T., Doblin, M. S., Bacic, A., Whelan, J., et al. (2021). Recent advances in *Cannabis sativa* genomics research. *New Phytologist, 230*(1), 73–89.

Janatova, A., Frankova, A., Tlustoš, P., Hamouz, K., Božik, M., & Klouček, P. (2018). Yield and cannabinoids contents in different cannabis (*Cannabis sativa* L.) genotypes for medical use. *Industrial Crops and Products, 112*, 363–367.

Jin, J. J., Yang, M. Q., Fritsch, P. W., van Velzen, R., Li, D. Z., & Yi, T. S. (2020). Born migrators: Historical biogeography of the cosmopolitan family *Cannabaceae*. *Journal of Systematics and Evolution, 58*(4), 461–473.

Kocjan Ačko, D. (1999). *Pozabljene poljščine*. Ljubljana: Kmečki glas (in Slovenian).

Kolenc, Z., & Čerenak, A. (2017). Aplication of sex molecular markers in hemp plant (*Cannabis sativa* L.). *Hop Bulletin, 24*, 121–128 (in Slovenian).

Kovalchuk, I., Pellino, M., Rigault, P., van Velzen, R., Ebersbach, J., Ashnest, J. R., et al. (2020). The genomics of *Cannabis* and its close relatives. *Annual Review of Plant Biology, 71*, 713–739.

Kušar, A., Flajšman, M., Kocjan Ačko, D., Pravst, I., & Čeh, B. (2018). Nutritional composition of hemp in relation to the variety. *Hop Bulletin, 25*, 76–84 (in Slovenian).

Laverty, K. U., Stout, J. M., Sullivan, M. J., Shah, H., Gill, N., Holbrook, L., et al. (2019). A physical and genetic map of Cannabis sativa identifies extensive rearrangements at the THC/CBD acid synthase loci. *Genome Research, 29*(1), 146–156.

Leme, F. M., Schönenberger, J., Staedler, Y. M., & Teixeira, S. P. (2020). Comparative floral development reveals novel aspects of structure and diversity of flowers in Cannabaceae. *Botanical Journal of the Linnean Society, 193*(1), 64–83.

Liang, Q. (2012). Foreword. In Q. Y. Shu, B. P. Forster, & H. Nakagawa (Eds.), *Plant mutation breeding and biotechnology*. Oxfordshire: CABI. 608 pp.

Lynch, R. C., Vergara, D., Tittes, S., White, K., Schwartz, C. J., Gibbs, M. J., et al. (2016). Genomic and chemical diversity in *Cannabis*. *Critical Reviews in Plant Sciences, 35*(5–6), 349–363.

Mandolino, G., & Carboni, A. (2004). Potential of marker-assisted selection in hemp genetic improvement. *Euphytica, 140*, 107–120.

Mandolino, G., Carboni, A., Forapani, S., Faeti, V., & Ranalli, P. (1999). Identification of DNA markers linked to the male sex in dioecious hemp (*Cannabis sativa* L.). *Theoretical and Applied Genetics, 98*, 86–92.

McKernan, K. J., Helbert, Y., Kane, L. T., Ebling, H., Zhang, L., Liu, B., et al. (2020). Sequence and annotation of 42 cannabis genomes reveals extensive copy number variation in cannabinoid synthesis and pathogen resistance genes. *BioRxiv*. https://doi.org/10.1101/2020.01.03.894428.

McPartland, J. M., & Guy, G. W. (2004). The evolution of *Cannabis* and coevolution with the cannabinoid receptor – A hypothesis. In G. W. Guy, B. A. Whittle, & P. Robson (Eds.), *The medicinal uses of Cannabis and cannabinoids* (pp. 71–101). London: Pharmaceutical Press.

Müssig, J. (2003). Quality aspects in hemp fiber production influence of cultivation, harvesting and retting. *Journal of Industrial Hemp, 8*(1), 11–32.

Neuer, H. V., & Sengbusch, R. V. (1943). Die Geschlechtsvererbung bei Hanf und die Züchtung eines monöcischen Hanfes. *Der Züchter, 15*(3), 49–62 (in German).

Onofri, C., & Mandolino, F. (2017). Genomics and molecular markers in *Cannabis sativa* L. In S. Chandra, H. Lata, & M. A. ElSohly (Eds.), *Cannabis sativa L. – Botany and biotechnology* (pp. 319–342). Cham, Switzerland: Springer International Publishing.

Pacifico, D., Miselli, F., Micheler, M., Carboni, A., Ranalli, P., & Mandolino, G. (2006). Genetics and marker-assisted selection of the chemotype in *Cannabis sativa* L. *Molecular Breeding, 17*, 257–268.

Pisupati, R., Vergara, D., & Kane, N. C. (2018). Diversity and evolution of the repetitive genomic content in *Cannabis sativa*. *BMC Genomics, 19*(1), 1–9.

Placet, V. (2009). Characterization of the thermo-mechanical behaviour of Hemp fibres intended for the manufacturing of high performance composites. *Composites Part A: Applied Science and Manufacturing, 40*(8), 1111–1118.

(2021). *Plant variety database—European Commission*. http://ec.europa.eu/food/plant/plant_propagation_material/plant_variety_catalogues_databases/search/public/index.cfm?event=SearchVariety&ctl_type=A&species_id=240&variety_name=&listed_in=0&show_current=on&show_deleted. (Accessed 10 March 2021).

Poniatowska, J., Wielgus, K., Szalata, M., Ozarowski, M., & Panasiewicz, K. (2019). Contribution of polish agrotechnical studies on *Cannabis sativa* L. to the global industrial hemp cultivation and processing economy. *Herba Polonica, 65*(2), 37–50.

Posselt, U. K. (2010). Breeding methods in cross-pollinated species. In B. Boller (Ed.), *Vol. 5. Fodder crops and amenity grasses. Handbook of plant breeding* (pp. 39–87). New York: Springer Science and Business Media.

Ranalli, P. (2004). Current status and future scenarios of hemp breeding. *Euphytica, 140*(1–2), 121–131.

Renner, S. S. (2014). The relative and absolute frequencies of angiosperm sexual systems: Dioecy, monoecy, gynodioecy, and an updated online database. *American Journal of Botany, 101*(10), 1588–1596.

Russo, E. B. (2019). The case for the entourage effect and conventional breeding of clinical cannabis: No "strain," no gain. *Frontiers in Plant Science, 9*, 1969.

Sakamoto, K., Akiyama, Y., Fukui, K., Kamada, H., & Satoh, S. (1998). Characterization; genome sizes and morphology of sex chromosomes in hemp (*Cannabis sativa* L.). *Cytologia, 63*(4), 459–464.

Salentijn, E. M., Petit, J., & Trindade, L. M. (2019). The complex interactions between flowering behavior and fiber quality in hemp. *Frontiers in Plant Science, 10*, 614.

Salentijn, E. M., Zhang, Q., Amaducci, S., Yang, M., & Trindade, L. M. (2015). New developments in fiber hemp (*Cannabis sativa* L.) breeding. *Industrial Crops and Products, 68*, 32–41.

Sawler, J., Stout, J. M., Gardner, K. M., Hudson, D., Vidmar, J., Butler, L., et al. (2015). The genetic structure of marijuana and hemp. *PLoS One, 10*(8), e0133292.

Small, E. (1972). Interfertility and chromosomal uniformity in *Cannabis. Canadian Journal of Botany, 50*(9), 1947–1949.

Small, E. (2015). Evolution and classification of *Cannabis sativa* (marijuana, hemp) in relation to human utilization. *Botanical Review, 81*(3), 189–294.

Small, E., & Antle, T. (2003). A preliminary study of pollen dispersal in *Cannabis sativa* in relation to wind direction. *Journal of Industrial Hemp, 8*(2), 37–50.

Small, E., & Beckstead, H. D. (1973). Cannabinoid phenotypes in *Cannabis sativa. Nature, 245*(5421), 147–148.

Small, E., & Cronquist, A. (1976). A practical and natural taxonomy for *Cannabis. Taxon, 25*, 405–435.

Small, E., & Marcus, D. (2002). Hemp: A new crop with new uses for North America. In J. Janick, & A. Whipkey (Eds.), *Trends in new crops and new uses* (pp. 284–326). Alexandria: ASHS Press.

Soorni, A., Fatahi, R., Haak, D. C., Salami, S. A., & Bombarely, A. (2017). Assessment of genetic diversity and population structure in Iranian cannabis germplasm. *Scientific Reports, 7*(1), 1–10.

Spitzer-Rimon, B., Duchin, S., Bernstein, N., & Kamenetsky, R. (2019). Architecture and florogenesis in female *Cannabis sativa* plants. *Frontiers in Plant Science, 10*, 350.

Törjék, O., Bucherna, N., Kiss, E., Homoki, H., Finta-Korpelová, Z., Bócsa, I., et al. (2002). Novel male-specific molecular markers (MADC5, MADC6) in hemp. *Euphytica, 127*(2), 209–218.

Van Bakel, H., Stout, J. M., Cote, A. G., Tallon, C. M., Sharpe, A. G., Hughes, T. R., et al. (2011). The draft genome and transcriptome of *Cannabis sativa. Genome Biology, 12*(10), 1–18.

Vergara, D., White, K. H., Keepers, K. G., & Kane, N. C. (2016). The complete chloroplast genomes of *Cannabis sativa* and *Humulus lupulus. Mitochondrial DNA, 27*(5), 3793–3794.

Weiblen, G. D., Wenger, J. P., Craft, K. J., ElSohly, M. A., Mehmedic, Z., Treiber, E. L., et al. (2015). Gene duplication and divergence affecting drug content in *Cannabis sativa. New Phytologist, 208*(4), 1241–1250.

Welling, M. T., Liu, L., Shapter, T., Raymond, C. A., & King, G. J. (2016). Characterisation of cannabinoid composition in a diverse *Cannabis sativa* L. germplasm collection. *Euphytica, 208*(3), 463–475.

White, K. H., Vergara, D., Keepers, K. G., & Kane, N. C. (2016). The complete mitochondrial genome for *Cannabis sativa. Mitochondrial DNA, 1*(1), 715–716.

Yang, M. Q., van Velzen, R., Bakker, F. T., Sattarian, A., Li, D. Z., & Yi, T. S. (2013). Molecular phylogenetics and character evolution of *Cannabaceae. Taxon, 62*(3), 473–485.

Zeven, A. C., & Zhukovsky, P. M. (1975). *Dictionary of cultivated plants and their centres of diversity.* Wageningen: Pudoc.

Zhang, Q., Chen, X., Guo, H., Trindade, L. M., Salentijn, E. M., Guo, R., et al. (2018). Latitudinal adaptation and genetic insights into the origins of *Cannabis sativa* L. *Frontiers in Plant Science, 9*, 1876.

3

Legal and regulatory oversight of hemp cultivation and hemp foods

Marne Coit

Coit Consulting LLC, Raleigh, NC, United States

3.1 Introduction

Foods made from hemp include items such as hulled hempseed (sometimes referred to as "hemp hearts"), flour made out of ground hempseed, and hempseed oil. Hempseed discussed here refers to sterile hempseed that cannot be planted but rather is used as a source of food (Fig. 3.1).

As hemp and hemp-derived products have gained popularity in recent years, the study of hemp law has emerged as a specialized area of law in its own right. Hemp law is a vast body of law that encompasses everything that impacts hemp as it moves along the supply chain, from seed sales and sales contracts at the farm level to laws regarding banking and transportation to laws that regulate processed consumer products. The legal issues related to hemp are extremely broad, and to cover all of them would be beyond the scope of this chapter. Since the focus of this book is on hemp for food, the scope of this chapter will be limited to the laws related to hemp cultivation and how hemp is regulated as a food product; however, it is outside the scope of this chapter to cover cannabidiol (CBD) that is added to foods.

It is also important to note that the legal and regulatory framework for hemp is extremely complex. In part, this is due to the fact that many countries outlawed the production of this crop for years, in some cases for more than a half of a century. After being prohibited for so long, the laws in many countries are now changing to allow hemp to be grown again. However, there are still growing pains in reestablishing this crop, and as is often the case, the law—although evolving rapidly—still sometimes lags behind where society wants to be with hemp.

It is not feasible to write a chapter on the "global law" as it relates to this topic, as every country regulates hemp very differently. Even within a single country, regulations can vary between localities. Instead, three geographic areas have been chosen as a representative sampling to illustrate what hemp laws look like in different parts of the world. In particular, this chapter will look in depth at the laws in the United States, Canada, and the European Union (EU).

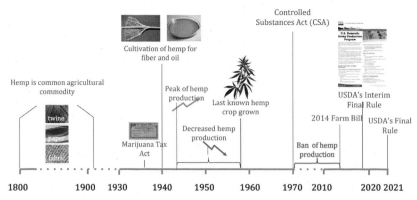

FIG. 3.1   Legal history of industrial hemp in the United States.

3.2 Hemp regulation in the United States

3.2.1 History of the law

The United States, like many other countries, has a complex history with hemp. Hemp had commonly been grown and utilized in the United States in the 1800s and into the mid-1900s. It was grown for a variety of purposes such as fabrics, twine, and paper (Coit, 2018). During this time the United States Department of Agriculture (USDA), which is the federal administrative agency that is responsible for regulating agricultural crops, treated hemp in the same way that it treated other agricultural commodities. That is, it provided assistance to hemp growers with production, compiled statistical analysis of the crop, and published crop reports (Coit, 2018). In 1943 an estimated 150 million pounds of hemp was produced in the United States, which is thought to be the peak of production. However, after this time production dwindled quickly, until 1958 when the last known hemp crop was grown in the United States (Coit, 2018).

There are a variety of reasons given as to why hemp cultivation was prohibited. In part, it was due to agricultural innovations. For example, harvesting equipment for cotton advanced during this time that mechanized harvesting, transitioning away from manual labor. As a result, it became easier to harvest cotton, which made it more of a competitor against hemp to be used as fiber for items such as clothing.

Societal attitudes toward drugs and drug use were also shifting during this time. This shift was reflected in the laws that were passed. In 1937 the US Congress passed the Marijuana Tax Act, which was the first time legislation had been passed that regulated marijuana (Coit, 2018). The purpose of this law—as the name implies—was to impose a tax on the sale of marijuana. While it did not ban the cultivation or sale of marijuana outright, it did make it much more difficult to grow and more burdensome to purchase or sell. It prohibited the possession and sale of marijuana for individuals, and limited it to medicinal use, which was highly regulated.

One important note is that this statute made a clear legal distinction between marijuana and hemp. This was the first time that this distinction was made in US law, and it has carried through to the present day in terms of how hemp and marijuana are categorized and regulated by the federal government, so it has had long-term implications. The Marijuana Tax Act specifically applied to marijuana. It did not regulate hemp production or the sale of hemp products. Actually, in the 1940s the US government encouraged the cultivation of hemp to be used for both fiber and oil during World War II (Coit, 2018).

Another major shift in the law came in 1970, when the US Congress passed the Controlled Substances Act (CSA). Under this law, the US Drug Enforcement Agency (DEA) was given jurisdiction to regulate *Cannabis sativa*. Significantly, unlike the Marijuana Tax Act, the CSA did not make a distinction between hemp and marijuana. This lack of distinction between the two has created confusion and complexities that carry through to those working in hemp in the present day.

Under the CSA, drugs are categorized according to both their potential for abuse and whether there is an acceptable medical use for that drug. There are five categories of drugs, known as "schedules." Marijuana was placed in Schedule I, which is the most restrictive of the five, meaning that it has been deemed to have a high potential for abuse and that there are no currently accepted medical uses (Schedules of Controlled Substances, n.d.-a, n.d.-b).

Similar to the Marijuana Tax Act, the CSA did not explicitly ban the cultivation of hemp. However, it implemented such rigorous controls that it would have been very cumbersome for farmers to comply with the regulations. For example, under the CSA, if one wanted to import cannabis seed, one would have needed to register with the DEA to do so. In addition, a permit from the DEA was required in order to obtain that seed (Coit, 2018).

With the legal restraints discussed previously in place again, it was very difficult to obtain seed and grow hemp. To further complicate the matter, there was a debate as to whether hemp fell within the scope of DEA's authority. The CSA defined the term "marihuana" as "all parts of the plant *Cannabis sativa* L., whether growing or not; the seeds thereof; … and every compound, manufacture, salt, derivative, mixture, or preparation of such plant, its seeds or resin. **Such term does not include** the mature stalks of such plant, fiber produced from such stalks, oil or cake made from the seeds of such plant … **or the sterilized seed of such plant which is incapable of germination**" (emphasis added) (Schedules of Controlled Substances, n.d.-a, n.d.-b).

The DEA contended that this definition included all categories of *C. sativa*, which included hemp, and therefore the authority to regulate hemp fell within their jurisdiction. However, others argued that the portion of the definition in bold, before, was to be read such that the definition of marijuana excluded hemp, and therefore it remained outside of the scope of DEA's authority. For example, hulled hempseeds that are sold as a food product are sterile and not capable of germination. The argument was that they should, according to the bolded part of the definition before, be excluded from the CSA's definition of marijuana and therefore outside of the DEA's jurisdiction.

At best, the laws discussed before made it cumbersome for farmers in the United States to cultivate hemp. At worst, they made it not only illegal but a criminal act to cultivate hemp. As a result, as mentioned before, the last known field of hemp was planted in the United States in 1958. It was not until 2014—more than a half of a century later—that this started to change. It is also worth noting that these laws treated hemp as one category and did not make a distinction between hemp grown for various end uses such as grains, fiber, or CBD. This is a paradigm that has carried on to how hemp is regulated currently.

3.2.2 Current law

3.2.2.1 *Cultivation*

The first major shift in hemp regulation was the passage of the 2014 Farm Bill by the US Congress. This piece of legislation was the first move toward the end of the period that has become known as prohibition in hemp cultivation.

The federal farm bill is an omnibus piece of legislation that is typically renewed every 3–5 years. The first farm bill was passed in 1933 with the purpose of addressing issues that arose during the Great Depression, such as providing subsidies to farmers, as well as provisions related to conservation of farmland. It has been expanded over time and now represents the main legislation that regulates farming and agriculture in the United States.

In the present day, there are typically 12 "titles" in the farm bill, with each title representing a different subject area. In 2014 there was language that was put into Title 12, which is the title for "miscellaneous" subjects; language was inserted that permitted hemp to be grown on a limited basis. Specifically, it allowed hemp to be grown by state departments of agriculture or academic institutions, as long as this was permitted in the state they were located in. In other words, individual states first had to pass legislation allowing hemp to be grown within their borders, in what were known as "pilot programs." This meant that, although hemp was now legal federally, individual states could decide to not allow it, meaning that farmers in those states would not be able to grow hemp.

Another restriction was that states and state departments of agriculture were limited to growing hemp specifically for the purpose of researching cultivation and/or marketing of hemp. By their nature, pilot programs are meant to be utilized on a temporary basis, as a way to test something. In this case, it was a means to reintroduce hemp production in the United States, but on a somewhat limited basis.

Unlike other agricultural commodities, when farmers grew hemp under a state pilot program, they were required to apply for a cultivation license. The specific rules in each state differ, but common information needed to get a license included the location of where the hemp would be grown, the varieties of hemp to be grown, whether it would be grown indoors or outdoors, etc. In addition, some states included a requirement that growers report back some of the data regarding their hemp grown at the end of the year, as part of the research purpose of the pilot program. There is also a fee that is typically associated with receiving a hemp license, which is often based on how much is grown and whether it is grown indoors or outdoors.

One of the most significant features of the language in the 2014 Farm Bill was the definition of hemp. It states that *[t]he term "industrial hemp" means the plant Cannabis sativa L. and any part of such plant, whether growing or not, with a delta-9 tetrahydrocannabinol concentration of not more than 0.3 percent on a dry weight basis* (Agricultural Act, 2014). Notably, the 2014 Farm Bill did not clarify that industrial hemp was exempt from the definition of marijuana under the CSA. Therefore, even after the passage of the 2014 Farm Bill, although hemp was legal at a federal level, it was still considered to be a Schedule I drug and under DEA's authority. As a result, hemp farmers needed to register with the DEA if they wanted to grow hemp (or a state pilot program could do so on their behalf) which complicated the process of purchasing seed, because Schedule I drugs are not permitted to be transported across state lines. This restriction applied whether the seed was being transported in a vehicle, mailed via the United States Postal Service, or via another carrier. The punishment for violating this law was a fine and/or prison sentence. The practical result was a limitation on the sources and ways through which farmers in states with pilot programs could obtain seed.

It took another 4 years until a legislative fix was implemented to address this issue. In the 2018 Farm Bill, the US Congress removed hemp from the definition of marijuana in the CSA (Agricultural Improvement Act, 2018a, 2018b).

The result was that hemp was no longer considered to be a Schedule I drug, which significantly lowered the burden on growers to obtain seed. The farm bill also clarified that it was legal for hemp to travel in interstate commerce, meaning that it could be transported more easily across state lines (Agricultural Improvement Act, 2018a, 2018b). This clarification was important for growers as well as the rest of the supply chain, as it made it clear that moving hemp between states was permitted. This point had been open to different interpretations previously.

In addition, the 2018 Farm Bill also established shared authority between the United States Department of Agriculture (USDA) and individual states to regulate hemp. The USDA is the federal administrative agency charged with regulating agriculture, food, natural resources, and rural development. The 2018 Farm Bill directed the USDA to promulgate regulations related to hemp production. Furthermore, the farm bill said that states and tribal nations could create their own plans for regulating hemp, which would have to be approved by the USDA. If states and tribal nations chose not to create their own plans, then farmers who grew hemp within that state or tribal nation would have to comply with the USDA's regulations. This provision provided a path for farmers in every state and tribal nation to grow hemp. Overall, the changes made in the 2018 Farm Bill solidified hemp as a legal commodity and established some clear legal guidance in areas that had been left murky under the 2014 Farm Bill.

In terms of the legal process, once the US Congress creates a statute, such as the farm bill, then the next step is for the appropriate federal agency to develop regulations, which provide more detailed guidance on the subject matter. In this case, the next step in the process was when USDA published the Interim Final Rule (IFR) to establish the US Domestic Hemp Production Program (DHPP) in October 2019. As stated in the IFR, the purpose was to "help expand production and sales of domestic hemp, benefiting both U.S. producers and consumers" (USDA, 2019).

The regulations provided procedures that states and tribal governments needed to follow if they decided to submit a plan to USDA for approval. For example, state or tribal plans were required to contain the following information: licensing requirements for growers, a way to track land that was used for the production of hemp, how the governing agency would conduct compliance testing for THC, a plan for disposal of crops that tested above the allowable THC level. In addition, the IFR required that any lab that was conducting compliance testing for THC was required to register with the DEA.

Other key provisions included acceptable ways to dispose of hemp that tested over the legally allowable limit of 0.3% THC (as was set in the farm bill), as well as required sampling procedures that needed to be conducted on hemp crops to ensure that they were in compliance with the allowable THC limit. When a crop did test above 0.5% THC on a dry weight basis, then IFR stated that the hemp licensee would be considered to have a negligent violation. If a licensee acquired three negligent violations within a 5-year period, then that grower would not be allowed to grow hemp again for 5 years (USDA, 2019).

As far as the compliance test for THC, the IFR stated in relevant part:

> State and Tribal plans must incorporate procedures for sampling and testing hemp to ensure the cannabis grown and harvested does not exceed the acceptable hemp THC level. Sampling procedures, among other requirements, must ensure that a representative sample of the hemp production is physically collected and delivered to a DEA-registered laboratory for testing. Within 15 days prior to the anticipated harvest of cannabis plants, a Federal, State, local, or Tribal law enforcement agency or other Federal, State or Tribal designated person shall collect samples from the flower material from such cannabis plants for delta-9 tetrahydrocannabinol concentration level testing. If producers delay harvest beyond 15 days, the plant will likely have a higher THC level at harvest than the sample that is being tested. This requirement will yield the truest measurement of the THC level at the point of harvest. Accepting that a pre-harvest inspection is best to identify suspicious plants and activities, and that the sample should be taken as close to harvest as possible, the time was selected based on what would be a reasonable time for a farmer to harvest an entire field. This 15-day post-sample harvest window was also designed to allow for variables such as rain and equipment delays. We are requesting comments and information regarding the 15-day sampling and harvest timeline.
>
> USDA (2019)

In the legal process, once an interim rule is published, then it is opened for public comment and feedback. The administrative agency then takes the comments, reviews them, and makes adjustments to the rule accordingly. There was considerable feedback on a number of requirements laid out in the interim final rule. In particular, the DEA registration requirement for labs, the rules around negligent violations, and the harvest window period. The USDA made adjustments based on this feedback, which was evidenced by the changes made in the final rule (FR), which was published in January 2021 and became effective on March 22, 2021.

In particular, per the USDA, the FR made the following changes:

- **"Negligent violation** – producers must dispose of plants that exceed the acceptable hemp THC level. However, if the plant tests at or below the negligent threshold stated in the rule, producer will not have committed a negligent violation. The final rule raises the negligence threshold from .5 percent to 1 percent and limits the maximum number of negligent violations that a producer can receive in a growing season (calendar year) to one.
- **Disposal and remediation of non-compliant plants** – the final rule allows for alternative disposal methods for non-compliant plants that do not require using a DEA reverse distributor or law enforcement and expands the disposal and remediation measures available to producers. AMS will provide acceptable remediation techniques in a separate guidance document.
- **Testing using DEA-registered laboratories** – there are an insufficient number of DEA-registered laboratories to test all the anticipated hemp that will be produced in 2020 and possibly 2021. The DEA has agreed to extend the enforcement flexibility allowing non-DEA registered labs to test hemp until January 1, 2023 and is processing lab registration applications quickly to get more labs testing hemp DEA-registered.
- **Timing of sample collection** – the IFR stated a 15-day window to collect samples before harvest. The FR extends this requirement to 30 days before harvest.

- **Sampling method** – stakeholders requested that samples may be taken from a greater part of the plant or the entire plant. They also requested sampling from a smaller number of plants. The FR allow states and tribes to adopt a performance-based approach to sampling in their plans. The plan must be submitted to USDA for approval. It may take into consideration state seed certification programs, history of producer compliance and other factors determined by the State or Tribe.
- **Extent of Tribal Regulatory Authority over the Territory of the Indian Tribe** – the IFR did not specifically address whether a tribe with an approved USDA plan could exercise primary regulatory authority over the production of hemp across all its territory or only lands over which it has inherent jurisdiction. The final rule provides that a tribe may exercise jurisdiction and therefore regulatory authority over the production of hemp throughout its territory regardless of the extent of its inherent regulatory authority" (USDA, 2021).

With the final rule, the USDA made a number of significant changes and addressed some of the concerns that had been raised during the public comment period.

While the provisions of the final rule went into effect in March 2021, states and tribal governments were permitted to continue to operate under their respective pilot programs until the end of 2021 per federal law. At that time, all pilot programs that were started under the 2014 Farm Bill were set to expire, which also means that all licenses issued under those pilot programs would expire. In their place, states and tribal governments were allowed to choose one of two options: (a) submit a state or tribal plan to the USDA to be approved and continue to regulate their own programs or (b) opt to have the USDA regulate the program in lieu of that particular state or tribal government.

As of the time of this writing, it is not clear yet which option all of the states and tribal governments will choose. But the beginning of 2022 will mark a big shift in how hemp cultivation is regulated in the United States. While it was first allowed to be cultivated in 2014 for the first time in decades, in the years since there have been a lot of changes and uncertainty. Now that the USDA's final rule is in place, hopefully 2022 will also mark a period of increased certainty for the hemp industry as the law in this area becomes more stable.

3.2.2.2 *Regulation of hemp as a food*

While the USDA was given authority under the 2018 Farm Bill to regulate the cultivation of hemp, the Food & Drug Administration (FDA) is the federal administrative agency that was given authority to regulate hemp as a food. When hemp was legalized federally, the US Congress did not establish a new framework specifically to regulate hemp foods; rather, authority was given to the FDA to regulate hemp within the existing regulatory framework for foods.

Much of the FDA's authority to regulate food and food products generally is derived from the Federal Food, Drug and Cosmetic Act (FD&C Act). This statute allows the FDA to regulate foods that travel in interstate commerce. In relevant part, the FDA has authority to regulate food and food products, as well as the packaging, labels, and labeling of food. The FD&C Act defines the term "food" as "articles used for food or drink for man … and articles used for components of any such article" FDA, (1938a–f). It also provides authority for the FDA to regulate the labeling of food products, which includes "written, printed, or graphic matter" that is on the container or wrapper of the food product (FDA, (1938a–f)). Furthermore, it prohibits food from being adulterated or misbranded (FDA, (1938a–f)).

Under the FD&C Act, food is considered to be adulterated "[i]f it bears or contains any poisonous or deleterious substance which may render it injurious to health" (FDA, (1938a–f)). It is considered to be misbranded under the statute if the labeling of the food product is false or misleading (FDA, (1938a–f)). Again, hemp falls under these same provisions, the same as other foods and food products.

Under the FD&C Act, food additives generally need to go through a premarket review in order to be approved by the FDA prior to being sold in the marketplace. Food additives are substances that have intentionally been added to food (Federal Food, Drug, and Cosmetic Act, 1938). The exception to the premarket review requirement is for food additives that are considered to be Generally Recognized As Safe (GRAS). The FDA has found that foods derived from hempseeds are considered to be GRAS (FDA, (1938a–f)).

The FDA has clarified its position on hemp food on its website as follows, in response to a question about whether hulled hempseed, protein powder made from hemp, and hempseed oil could be used as food:

> In December 2018, FDA completed its evaluation of three generally recognized as safe (GRAS) notices for the following hemp seed-derived food ingredients: hulled hemp seed, hemp seed protein powder, and hemp seed oil. FDA had no questions regarding the company's conclusion that the use of such products as described in the notices is safe. Therefore, these products can be legally marketed in human foods for the uses described in the notices, provided they comply with all other requirements. These GRAS notices related only to the use of these ingredients in human food …
>
> Hemp seeds are the seeds of the *Cannabis sativa* plant. The seeds of the plant do not naturally contain THC or CBD. The hemp seed-derived ingredients that are the subject of these GRAS notices contain only trace amounts of THC and CBD, which the seeds may pick up during harvesting and processing when they are in contact with other parts of the plant. Consumption of these hemp seed-derived ingredients is not capable of making consumers "high."
>
> The GRAS conclusions can apply to ingredients for human food marketed by other companies, if they are manufactured in a way that is consistent with the notices and they meet the listed specifications. Some of the intended uses for these ingredients include adding them as source of protein, carbohydrates, oil, and other nutrients to beverages (juices, smoothies, protein drinks, plant-based alternatives to dairy products), soups, dips, spreads, sauces, dressings, plant-based alternatives to meat products, desserts, baked goods, cereals, snacks and nutrition bars. Products that contain any of these hemp seed-derived ingredients must declare them by name on the ingredient list.
>
> FDA (1938)

While it is legal to sell hempseed-derived foods for human consumption, food makes up approximately 17% of the overall US hemp sales, which were valued at $818,000,000 as of 2017. The three largest categories of hemp product sales were hemp-derived CBD, personal care, and industrial applications, with the category of food coming in fourth place in terms of sales. This indicates that while it is legal, there may be more room for growth for hemp food sales in the United States (Johnson, 2021).

3.3 Hemp regulation in the European Union

3.3.1 History of hemp in the European Union

The European Union (EU) is currently composed of 27 member countries. These are countries that have decided to join together for economic and political purposes. These partnerships first started after World War II. Countries may join the EU based on acceptance of a set of conditions for membership which include the rule of law, freedom of expression, and media and regional cooperation, among other criteria.

In terms of how the legal system is set up within the EU, there are a number of different types of legal authority. Depending on the classification of the type of law, it may automatically apply to and be binding upon all member countries, or it may pass the EU legislative body but then need to be adopted by individual member countries if they so choose (European Commission, 2021).

Because of how the EU is structured, it means that the history of hemp and hemp cultivation differs between member states. Similar to the United States, some countries in the EU have a history of hemp cultivation which was then banned for a number of years, and the specifics vary from country to country.

Germany, for example, prohibited the cultivation of cannabis in 1982 but reversed this ban specifically for hemp in 1996 and has permitted hemp to be grown since that time (Raymunt, 2020). France started to permit the cultivation of hemp for uses such as food in 1990. There are additional restrictions, such as a list of approved varieties that may be grown. In addition, hemp that is cultivated must meet the standards laid out for compliance testing and must not contain more than 0.2% THC.

3.3.2 Cultivation

Hemp that is cultivated in the EU represents a significant portion of the world's overall supply of hemp. Up to 25% of hemp that is cultivated worldwide is grown in the EU (New Frontier Data Europe, 2019). Furthermore, it is estimated that as of 2019, 60% of the hemp crops grown in the EU were used for food (New Frontier Data Europe, 2019).

In general, hemp is permitted to be grown and utilized as a food in the European Union (EU). However, individual Member States are permitted to prohibit it or regulate it within their own jurisdictions, so there is some variation in different parts of the EU.

There are general requirements that apply to Member States if they do permit cultivation. For example, any varieties that are cultivated must be registered with the EU's "Common Catalogue of Varieties of Agricultural Plant Species." In addition, since 1999 the allowable legal limit for THC had been set at 0.2% (EU Novel food catalogue, 2018). There was a movement to increase the THC limit from 0.2% to 0.3% so that it would be more in line with the limits in other countries around the world. One rationale to do this was to expand the range of varieties that could be bred and grown. The European Parliament voted in favor of this increase in October 2020, and it was approved by the European Council in December of 2021. It is set to go into effect starting in January 2023.

Agriculture in the EU is supported by the Common Agricultural Policy (CAP). If a particular crop is considered to be an agricultural crop under the CAP, then growers may be eligible to receive income support through direct payments. Hemp is classified as an agricultural crop under CAP, so hemp growers are eligible for direct payments (Raymunt, 2020).

3.3.3 Food

Hemp is permitted to be sold as a food in the EU. Under the legal structure of the EU, if a food or food product does not have a history of being marketed and consumed prior to May

15, 1997, then it is automatically considered to be a "novel food" (European Commission, 2015). Foods that are considered to be novel must meet additional regulatory burdens under the Novel Food Regulation. However, if a food was on the market and consumed widely prior to that date, then it is not subject to the Novel Food Regulations (European Commission, 2015). Hemp products utilized for food such as seeds, oil made from hempseeds, and hempseed flour are not considered to be novel and therefore are permitted to be sold in the market and are not subject to the Novel Food Regulations (EU Novel food catalogue, 2018).

The European Industrial Hemp Association (EIHA) is a pan-European member-based organization that represents farmers, processors, and manufacturers in the hemp industry. The mission of EIHA is "to steer and promote hemp farming, processing and trading in the European Union. We aim at developing a single and safe common market of high-quality hemp products, inspired by the principle of social, environmental and economic sustainability" (EIHA, (2021a-c)). According to the EIHA, the hemp industry is growing quickly in the EU. It is estimated that as of 2018, hemp was grown on 50.081 ha, which is a 3.3% increase from 2017, and a 70% increase compared to the 5-year average (EIHA, 2018). As of 2018, France was the largest hemp producer within the EU, with Italy following as the second largest producer and the Netherlands the third largest (EIHA, 2018).

The EIHA also serves as an advocacy group that addresses policy issues relevant to its membership. For example, one issue that the organization has worked on that relates to hemp-derived food products is harmonized standards for THC levels that are allowed in foods. According to the EIHA, "[h]armonized legislation should be applied in all member states. This will guarantee consumer safety and the further expansion of the hemp food industry, attract direct and indirect investments and create new jobs" (EIHA, 2018).

The call for harmonized standards stems from the fact that different member nations have their own set of standards—or no standards. Germany was the first country in the EU to set standards for the amount of THC that is allowable in foods, with some other EU members adopting those standards (Banas et al., 2017).

The EIHA argues that the standard for the EU should be similar to the one used in Canada, where there is a list of approved hemp cultivars that can be grown. The approved cultivars are known to produce 0.3% or less of THC, and therefore do not require compliance testing at the time the crop is harvested. The result, the EIHA argues, is a reduction in the amount of time, testing, money, and other barriers for hemp crops, which in turn is beneficial for the hemp industry and allows it to grow more quickly (Banas et al., 2017).

Hemp and hemp-derived foods are an increasing share of the food market in the EU. As of September 2017, this market segment was valued at 40 million euros within the EU. Worldwide, the hemp food market accounted for 200 million euros (Banas et al., 2017).

3.4 Hemp regulation in Canada

3.4.1 Background

Similar to other countries discussed, hemp cultivation was also prohibited in Canada for about 60 years (Johnson, 2018). It was banned in 1938 under the Opium and Narcotic Drug Act (Government of Canada, 2021a). Hemp cultivation was permitted again during WWII, mainly for the purposes of producing fiber at a time when there was a limited supply of fiber from other sources. However, the prohibition was implemented again after the war (Government of Canada, 2021a).

Legalization of hemp in Canada started first with licenses issued for research purposes. Research licenses were first issued in 1994, followed a few years later by licenses for commercial cultivation in 1998 (Johnson, 2018).

Canada now has a strong hemp export market. For example, approximately 90% of the hemp products (this includes all hemp products, not just hemp ingredients/hemp foods) imported into the United States originate from Canada. Canada also supplies the United States with the majority of the dehulled seeds that are used for food in the United States (Johnson, 2018).

3.4.2 Cultivation

Similar to other countries, Canada has a law on the books that prohibits cannabis. Cannabis is considered to be a controlled substance under Canada's Controlled Drugs and Substances Act (CDSA). However, there is an exemption under this statute that allows for industrial hemp to be grown for commercial purposes under the Industrial Hemp Regulations (IHR). The IHR went into effect in March 1998 and lays out the licensing and compliance testing requirements for the cultivation of hemp. In addition, the IHR defines industrial hemp as Cannabis plants and plant parts, of any variety, that contains 0.3% tetrahydrocannabinol (THC) or less in the leaves and flowering heads (Government of Canada, 2021a).

In order to grow hemp in Canada, one must apply for and receive a license from the Office of Controlled Substances. Under the IRH, a license holder is entitled to cultivate and sell industrial hemp, as well as to possess seed or grain in order to clean or process it. The ability to transport hemp is also included in the list of activities that licensees may engage in (IHR, 2019) (Government of Canada, 2021a). License applicants must provide information regarding their name, address, the location of where the hemp will be grown, the variety to be grown, the number of hectares cultivated, and the location of each site where hemp will be grown. In addition, growers applying for a cultivation license must either declare that they own the land that the hemp will be grown on or else provide written documentation of consent from the land owner if the farmland is being rented (IHR, 2019).

In terms of the required compliance testing that must be completed, a representative sample of dried flower and leaves must be collected and tested at a laboratory that uses validated testing methods (IHR, 2019) (Government of Canada, 2021c). License holders are required to maintain a portion of the sampled material for up to 1 year (IHR, 2019) (Government of Canada, 2021c).

There is also a requirement regarding the maintenance of equipment that is used for hemp cultivation. The IRH provides that "[a] holder of a license that authorizes cultivation must maintain all equipment that is used to sow or harvest industrial hemp in a manner that prevents its contamination and avoids the inadvertent dissemination of the seed" (IHR, 2019) (Government of Canada, 2021c).

3.4.3 Food

When it comes to food and food products made from hemp, there is a distinction made in Canadian law between how non-viable seeds are regulated versus how derivatives from industrial hemp are regulated. Hemp foods and food products in Canada are regulated and must follow applicable statutes and regulations, the same as other food products. However, hemp foods that are considered to be derivatives from industrial hemp are subject to additional regulatory requirements. For example, hulled hempseed, which is considered to be non-viable seed, is excluded from the definition of industrial hemp. Therefore hulled hempseeds do not fall under the authority of the CDSA and do not require a license to sell or transport. They are not regulated any differently than other foods in Canada.

However, derivatives of hempseed are considered to be within the definition of industrial hemp and therefore require authorization under the CDSA, as well as needing to comply with other food-related laws. For example, hempseed oil and hemp flour would both be considered derivatives of industrial hemp and therefore fall under the CDSA (Government of Canada, 2018).

The main statutory authority for regulating food in Canada is the Food and Drugs Act (FDA). It is similar to the FD&C Act in the United States in that it serves to ensure the safety of food that is sold and allows for the regulation of food labels in order to provide accurate information to consumers and to prevent fraud. Misleading and false food labeling and advertising is not permitted. The statute provides that "[n]o person shall label, package, treat, process, sell or advertise any food in a manner that is false, misleading or deceptive or is likely to create an erroneous impression regarding its character, value, quantity, composition, merit or safety" (Food and Drugs Act, 1985). The definition of "foods" that are covered by the FDA includes "items manufactured, sold or represented for use as a food or drink." It also includes "ingredients" that are mixed with foods (Food and Drugs Act, 1985) (Government of Canada, 2021b).

A significant portion of Canada's hemp products is exported. As of 2018, Canada exported approximately 5400 metric tons of hempseed. This was valued at about $50 million USD, and about 70% of these products were exported to the United States (USDA Foreign Agricultural Service, 2019).

3.5 Conclusion and future perspectives

As can be seen from the examples discussed before, hemp and hemp-based foods are allowed in places that permit hemp to be grown. However, a parallel that runs through each is the fact that it has been prohibited from being cultivated for years, oftentimes decades. Different countries and geographic regions have started to allow it to be grown again, with additional countries coming online in the future. As hemp becomes a crop that is grown more commonly, it is likely that the markets for hemp foods will also continue to grow and increase.

Although this chapter focuses on three geographic areas, more countries across the globe are also lifting restrictions on hemp cultivation. As this happens, the supply of hemp available for food will increase, with a corresponding interest by consumers likely as more learn about hemp, including its use as a food source. In order to sustain continued growth, harmonization of standards between countries would ease the burden of trade between countries and increase the availability of hemp foods in order to meet increasing demand.

Also of note is a pattern that has emerged in that once hemp is legalized, many farmers will grow hemp varieties that are suited for CBD content. Initially it seems to be more profitable than hemp grown for grain or even for fiber. For example, this occurred in the United States in

2019, when there was an oversupply and then corresponding decrease in prices for farmers. Now that prices for CBD have fallen and are moving closer to the prices for food, there may be more farmers that grow hemp for food, particularly as consumer demand increases.

References

Agricultural Act. (2014). *7 USC §5940(a)(2)*. https://www.govinfo.gov/content/pkg/USCODE-2015-title7/pdf/USCODE-2015-title7-chap88-subchapVII-sec5940.pdf.

Agricultural Improvement Act. (2018a). *7 USC §10114*. https://www.congress.gov/bill/115th-congress/house-bill/2/text.

Agricultural Improvement Act. (2018b). *7 USC §12619(a)*. https://www.congress.gov/bill/115th-congress/house-bill/2/text.

Banas, B., Beitzke, B., Carus, M., Iffland, K., Kruse, D., Sarmento, L., et al. (2017). *Position paper of the European Industrial Hemp Association (EIHA) on: Reasonable guidance values for THC (tetrahydrocannabinol) in food products*. https://eiha.org/wp-content/uploads/2021/01/17-09-18-THC-Position-paper_EIHA.pdf.

Coit, M. (2018). The fate of industrial hemp in the 2018 Farm Bill: Will our collective ambivalence finally be resolved? *Journal of Food Law and Policy, 14*(1), 12–23.

(2018). *EU Novel food catalogue*. https://ec.europa.eu/food/safety/novel_food/catalogue/search/public/index.cfm.

European Commission. (2015). *Regulation (EC) No 258/97 of the European Parliament and of the Council of 27 January 1997 concerning novel foods and novel food ingredients. Regulation (EU) 2015/2282*. https://eur-lex.europa.eu/legal-content/EN/TXT/PDF/?uri=CELEX:01997R0258-20090807&from=EN.

European Commission. (2021). *Types of EU law*. https://ec.europa.eu/info/law/law-making-process/types-eu-law_en.

European Industrial Hemp Association (EIHA). (2018). *Hemp cultivation & production in Europe in 20184*. https://eiha.org/wp-content/uploads/2020/10/2018-Hemp-agri-report.pdf.

European Industrial Hemp Association (EIHA). (2021a). *History, mission & vision*. https://eiha.org/history-mission-vision/.

European Industrial Hemp Association (EIHA). (2021b). *About hemp: Hemp in Europe*. https://eiha.org/about-hemp-hemp-in-europe/.

European Industrial Hemp Association (EIHA). (2021c). *Tetrahydrocannabinol (THC) residues in food*. https://eiha.org/thc-in-food/.

Government of Canada. (2021a). *Hemp and the hemp industry: Frequently asked questions*. https://www.canada.ca/en/health-canada/services/drugs-medication/cannabis/producing-selling-hemp/about-hemp-canada-hemp-industry/frequently-asked-questions.html#a23.

Government of Canada. (2021b). Food and Drugs Act, R.S.C., 1985, c. F-27 (1985). https://laws-lois.justice.gc.ca/eng/acts/f-27/FullText.html.

Government of Canada. (2021c). Industrial hemp regulations, SOR/2018-145(2018). https://laws-lois.justice.gc.ca/eng/regulations/SOR-2018-145/page-1.html#h-851200.

Johnson, R. (2018). *Hemp as an agricultural commodity*. Congressional Research Service. https://sgp.fas.org/crs/misc/RL32725.pdf.

Johnson, R. (2021). *Hemp-derived cannabidiol (CBD) and related hemp extracts*. U.S. Congressional Research Service. https://sgp.fas.org/crs/misc/IF10391.pdf.

New Frontier Data Europe. (2019). *Hemp cultivation in Europe*. New Frontier Data. https://newfrontierdata.com/cannabis-insights/developed-global-markets-hemp-acreage-comparison/.

Raymunt, M. (2020). *Hemp cultivation in Europe: Key market details and opportunities*. Hemp Industry Daily. https://hempindustrydaily.com/wp-content/uploads/2020/07/hemp-in-europe-2020-FINAL.pdf.

Schedules of Controlled Substances. (n.d.-a). 21 U.S.C. §802(16) https://www.deadiversion.usdoj.gov/21cfr/21usc/802.htm.

Schedules of Controlled Substances. (n.d.-b). 21 U.S.C. §812(b). https://www.deadiversion.usdoj.gov/21cfr/21usc/812.htm.

United States Department of Agriculture (USDA). (2019). Establishment of a domestic hemp production program. *Federal Register, 84*(211), 58522–58564.

United States Department of Agriculture (USDA) (2021). Establishment of a domestic hemp production program. Federal Register, 5596-5691.

U.S. Food and Drug Administration (FDA). (1938a). *Federal Food, Drug, and Cosmetic Act, 21 USC § 342(a)(1)*. http:// uscode.house.gov/view.xhtml?req=Federal+Food%2C+Drug%2C+and+Cosmetic+Act+21+USC&f=tree-sort&fq=true&num=82&hl=true&edition=prelim&granuleId=USC-prelim-title21-section342.

U.S. Food and Drug Administration (FDA). (1938b). *Federal Food, Drug, and Cosmetic Act, 21 USC § 343(a)*. http:// uscode.house.gov/view.xhtml?req=Federal+Food%2C+Drug%2C+and+Cosmetic+Act+21+USC&f=tree-sort&fq=true&num=84&hl=true&edition=prelim&granuleId=USC-prelim-title21-section343-1.

U.S. Food and Drug Administration (FDA). (1938c). *Federal Food, Drug, and Cosmetic Act, 21 USC §201(s) and 21 USC §409*. https://www.accessdata.fda.gov/scripts/cdrh/cfdocs/cfcfr/CFRSearch.cfm?CFRPart=1&showFR=1.

U.S. Food and Drug Administration (FDA). (1938d). *Federal Food, Drug, and Cosmetic Act, 21 USC §321(f)*. http:// uscode.house.gov/view.xhtml?req=Federal+Food%2C+Drug%2C+and+Cosmetic+Act+21+USC&f=tree-sort&fq=true&num=62&hl=true&edition=prelim&granuleId=USC-prelim-title21-section321.

U.S. Food and Drug Administration (FDA). (1938e). *Federal Food, Drug, and Cosmetic Act, 21 USC §321(m)*. http:// uscode.house.gov/view.xhtml?req=Federal+Food%2C+Drug%2C+and+Cosmetic+Act+21+USC&f=tree-sort&fq=true&num=62&hl=true&edition=prelim&granuleId=USC-prelim-title21-section321.

U.S. Food and Drug Administration (FDA). (1938f). *Federal Food, Drug, and Cosmetic Act, 21 USC §331(b)*. http:// uscode.house.gov/view.xhtml?req=Federal+Food%2C+Drug%2C+and+Cosmetic+Act+21+USC&f=tree-sort&fq=true&num=68&hl=true&edition=prelim&granuleId=USC-prelim-title21-section331.

USDA Foreign Agricultural Service. (2019). *Industrial hemp production trade and regulation*. Retrieved 8 August 2021 https://apps.fas.usda.gov/newgainapi/api/report/downloadreportbyfilename?filename=Industrial%20 Hemp%20Production%20Trade%20and%20Regulation_Ottawa_Canada_8-26-2019.pdf.

Nutritional and chemical composition of industrial hemp seeds

Jikai Zhao[a], Weiqun Wang[b], Yonghui Li[c], Xiuzhi Sun[c], and Donghai Wang[a]

[a]Department of Biological and Agricultural Engineering, Kansas State University, Manhattan, KS, United States [b]Department of Food, Nutrition, Dietetics, and Health, Kansas State University, Manhattan, KS, United States [c]Department of Grain Science and Industry, Kansas State University, Manhattan, KS, United States

4.1 Introduction

Industrial hemp (*Cannabis sativa* L.) is a versatile, sustainable, and drought-resistant crop (Adesina, Bhowmik, Sharma, & Shahbazi, 2020; Jami, Karade, & Singh, 2019). Because hempseed contains the intoxicating component delta-9-tetrahydrocannabinol (THC), legislative restraints in the past decades have markedly enfeebled the economic importance of industrial hemp cultivation and processing in Western Europe, the United States, and Canada, to inhibit its abuse and illegal utilization for drug production (Cherney & Small, 2016). Therefore industrial hemp is often recognized as an underdeveloped and underutilized crop. Recently, the loosening of legislation has reintroduced the commercial interests of industrial hemp, allowing legal cultivation and research of various industrial hemp varieties with a THC content below 0.3% (Zhao et al., 2020). Regulatory changes are leading to a resurgent commercial exploration of this crop for food, feed, pharmaceutical, and industrial applications (Crini, Lichtfouse, Chanet, & Morin-Crini, 2020; Farinon, Molinari, Costantini, & Merendino, 2020; Xu, Li, et al., 2021; Xu, Zhao, et al., 2021). Indeed, hempseed is of great interest with important nutritional and functional features of its bioactive compounds (Farinon et al., 2020; Leonard, Zhang, Ying, & Fang, 2020; Wang & Xiong, 2019).

The emerging interest in the use of hempseed in the food and animal industries is triggered by its high nutritional value of components (Crescente et al., 2018; Zhou, Wang, Lou, & Fan, 2018). Among them, hempseed protein contains a large number of essential amino acids with excellent digestibility and functionality. In particular, the hempseed edestin is of high biological value due to its structural similarity to serum globulins and its amino acid composition on behalf of all essential amino acids (Wang & Xiong, 2019). The beneficial effects of hempseed are also attributed to the high content of lipids with a unique and perfectly balanced ratio between polyunsaturated and saturated fatty acids for human nutrition (Farinon et al., 2020; Zhou et al., 2018). The unsaponifiable portion is a crucial source of interesting compounds, including β-sitosterol, campesterol, phytol, cycloartenol, and γ-tocopherol (Farinon et al., 2020). Hempseed also contains carbohydrates, dietary fibers, and minerals. These specific constituents previously mentioned have been reported to play an antihypertensive and hemostatic role in human health (Crescente et al., 2018; Farinon et al., 2020; Leonard et al., 2020; Xu, Li, et al., 2021; Xu, Zhao, et al., 2021). It also should be mentioned that anti-nutritional compounds such as phytic acid, trypsin inhibitors, condensed tannins, cyanogenic glycosides, and saponins in hempseed products (Galasso et al., 2016; Mattila, Pihlava, et al., 2018; Pojić et al., 2014; Russo, 2013; Russo & Reggiani, 2015) have been reported to reduce the bioavailability of nutrients and interfere with metabolism, leading to impaired gastrointestinal and metabolic functions (Farinon et al., 2020).

Hempseed contains a spectrum of complex organic macromolecules, including proteins, oil, carbohydrates, and polyphenols. Its incorporation as a raw material could therefore lead to the development of multi-stream processes for producing food ingredients, platform nutrients, or pharmaceutical chemicals. This book chapter provides an overview of the recent advances in the nutritional and chemical composition and functional properties of hempseed. Additionally, secondary metabolites and anti-nutritive compounds derived from hempseed products are elucidated.

4.2 Major and minor hemp nutrients

A summary of representative studies on the chemical composition of hempseed from various cultivars is presented in Table 4.1. The composition of hempseed is of great interest, particularly concerning nutritional characteristics. Hempseed contains approximately 21.3%–32.0% protein, 25.4%–35.9% oil, 27.8%–38.8% fiber, and 3.7%–6.3% ash (Table 4.1). In addition, there are few studies of other components, including carbohydrates (32.5%–38.1%) and dietary fibers (27.6%–33.8%) (Lan et al., 2019; Mattila, Pihlava, et al., 2018; Xu, Li, et al., 2021; Xu, Zhao, et al., 2021). The chemical composition and nutrient contents of hempseed showed significant variation among hemp cultivars, which can be directly correlated with the genotypes and environmental (temperature, rainfall, soil, etc.) conditions (Lan et al., 2019; Xu, Li, et al., 2021; Xu, Zhao, et al., 2021). On the other hand, hempseed can be differentiated in terms of its natural antioxidants and bioactive components such as peptides, phenolic compounds, tocopherols, carotenoids, and phytosterols, mainly due to the environmental and agronomic factors and, to a lesser extent, by genetic variability (Irakli et al., 2019). In the following subsections, these nutritional components are discussed in detail.

4.2.1 Hempseed proteins

4.2.1.1 Crude protein

It has been demonstrated that the hempseed proteins are typically positioned in the inner layer of the seed and only a low amount of total proteins are located in the hull

TABLE 4.1 Main chemical composition of hempseed (%, dry basis) from different sources.

Cultivar	Dry matter	Crude protein	Crude fat	Crude fiber	Ash	References
Alyssa	98.87	24.1	30	37.3	5.6	Vonapartis, Aubin, Seguin, Mustafa, and Charron (2015)
Anka	93.97	23.8	28.8	38.8	5.7	Vonapartis et al. (2015)
CanMa	93.93	26.4	30.4	35	5.3	Vonapartis et al. (2015)
CFX1	92.82–95.80	26.0–27.1	29.8–35.9	33.2–33.93	5.1–6.1	Lan et al. (2019), Vonapartis et al. (2015), Xu, Li, et al. (2021), Xu, Zhao, et al. (2021)
CFX2	93.85–95.90	27.4–28.1	29.5–35.5	32.47–32.7	5.3–5.5	Lan et al. (2019), Vonapartis et al. (2015)
CRS1	94.02–96.80	25.7–32.0	29.5–35.2	28.78–36.2	4.9–6.0	Lan et al. (2019), Vonapartis et al. (2015), Xu, Li, et al. (2021), Xu, Zhao, et al. (2021)
Delores	94.01–95.80	24.3–24.5	26.9–34.0	37.5–37.53	5.6	Lan et al. (2019), Vonapartis et al. (2015)
Finola	91.80–95.60	23.3–28.0	30.6–33.0	29.8–34.5	4.7–5.8	House, Neufeld, and Leson (2010), Vonapartis et al. (2015)
Jutta	93.81	24.6	27.6	38.1	5.5	Vonapartis et al. (2015)
Yvonne	94.01	24.8	28.6	35.2	5.4	Vonapartis et al. (2015)
X59	95.8	28.1	32.8	34.91	5.1	Lan et al. (2019)
Grandi	95.8	28	34.9	33.42	5.3	Lan et al. (2019)
Picolo	95.8	28	34.9	33.05	5.2	Lan et al. (2019), Xu, Li, et al. (2021), Xu, Zhao, et al. (2021)
Katani	91.80–95.70	25.9–30.1	33.2–35.4	28.85–34.89	5.3–6.0	Lan et al. (2019), Xu, Li, et al. (2021), Xu, Zhao, et al. (2021)
Canda	91.30–95.70	26.3–26.8	28.9–35.4	36.04–36.55	5.7–6.0	Lan et al. (2019), Xu, Li, et al. (2021), Xu, Zhao, et al. (2021)
Joey	92.90–95.80	25.2–30.2	28.0–34.1	34.54–36.7	5.4	Lan et al. (2019), Xu, Li, et al. (2021), Xu, Zhao, et al. (2021)
USO31	91.80–95.30	21.9–28.1	25.6–30.0	33.0–34.79	3.7–5.8	House et al. (2010), Xu, Li, et al. (2021), Xu, Zhao, et al. (2021)
USO14	91.20–95.60	21.3–24.0	25.4–31.4	31.9–36.2	2.4–5.2	House et al. (2010)
Crag	95.10–96.0	23.3–27.5	31.7–33.0	27.8–32.3	4.7–5.1	House et al. (2010)
Fedora17	91.4	26.5	30.2	35.02	6.3	Xu, Li, et al. (2021), Xu, Zhao, et al. (2021)
Helena	91.8	27.5	29.2	35.64	5.7	Xu, Li, et al. (2021), Xu, Zhao, et al. (2021)
Hlukouskii51	91.2	29.3	29.9	32.84	6.1	Xu, Li, et al. (2021), Xu, Zhao, et al. (2021)
Felina32	91.9	27	32.9	32.36	5.8	Xu, Li, et al. (2021), Xu, Zhao, et al. (2021)
Futura75	90.08	27.3	21	34.25	5.7	Xu, Li, et al. (2021), Xu, Zhao, et al. (2021)
Tygra	91.9	29	29.8	33.19	6.2	Xu, Li, et al. (2021), Xu, Zhao, et al. (2021)
Hlesia	91.7	27.1	30.7	34.55	5.8	Xu, Li, et al. (2021), Xu, Zhao, et al. (2021)
Range	90.60–98.87	21.3–32.0	25.4–35.9	27.8–38.8	3.7–6.3	

(Mattila, Mäkinen, et al., 2018). Therefore the removal of hempseed hull can lead to a relative increase in the protein content of resultant products, which might be explained as protein concentration via removing some section of the whole seed that totally or almost lacks in protein (Farinon et al., 2020; Pojić et al., 2014). House et al. (2010) compared four sources of hempseed products to evaluate their nutritional value and found that de-hulled hempseed contained approximately 1.5-fold higher protein (30.3%–38.7%) than the common hempseed (21.3%–27.2%).

It should be highlighted that more than 180 proteins have been identified in hempseed with the two major proteins of legumin-type globulin edestin and globular-type albumin (Aiello et al., 2016). Edestin protein in hempseed characterized by crystallographic techniques showed comparable structure as soy glycinin (Patel, Cudney, & McPherson, 1994). The edestin protein contains six identical subunits, and each is composed of an acidic and a basic subunit allied by one disulfide bond (Tang, Ten, Wang, & Yang, 2006; Yin et al., 2008). The molecular weights of edestin, acidic subunit, and basic subunit were reported to be around 300, 33, and 20 kDa, respectively (Tang et al., 2006; Wang, Tang, Yang, & Gao, 2008). Differences in edestin content among hemp cultivars are directly related to the variation in genotypes and sources of raw material. For example, Mamone, Picariello, Ramondo, Nicolai, and Ferranti (2019) reported that edestin accounted for about 70% of total hemp protein isolate, whereas Tang et al. (2006) observed that edestin consisted of about 82% of total protein isolate. Albumin fraction accounts for about 25% of hempseed storage protein and contains fewer disulfide bonds among its proteins with a flexible structure (Wang & Xiong, 2019). The functional traits of these two major proteins have been reported to differ. In this regard, Malomo and Aluko (2015a, 2015b) compared the structural and functional characteristics between water-soluble albumin and salt-soluble globulin of hemp protein and demonstrated that albumin exhibited notably higher solubility and foaming capacity than globulin at a range of pH from 3 to 9, but emulsion-forming ability was comparable for both protein fractions.

4.2.1.2 Amino acid profile

In terms of amino acids of hempseed edestin and albumin proteins, Kim and Lee (2011) isolated edestin protein via acid precipitation and gel filtration chromatography and analyzed the N-terminal amino acid sequence of the first seven and six amino acid residues of an acidic and a basic subunit, respectively. They reported that the seven amino acid residues in the acidic subunit showed a sequence of Ile-Ser-Arg-Ser-Ala-Val-Tyr in the N-terminus, whereas two constituents of the basic subunit had the same N-terminus of Gly-Leu-Glu-Glu-Thr-Phe. In another study, Malomo and Aluko (2015a, 2015b) quantitatively analyzed the amino acid composition of hempseed albumin and globulin fractions and found that globulin contained higher hydrophobic (Ala, Cys, Val, Met, Ile, and Leu) and aromatic (Tyr, Phe, and Trp) amino acids (Table 4.2). The differences in amino acids between the isolated albumin and globulin can be attributed to the genomic forms, isolation technologies, and post-treatment conditions (Dapčević-Hadnađev et al., 2019; Galves et al., 2019; Potin, Lubbers, Husson, & Saurel, 2019; Russo & Reggiani, 2015; Shen, Gao, Xu, Rao, & Chen, 2020; Teh, Bekhit, Carne, & Birch, 2014; Wang, Jin, & Xiong, 2018), therefore, resulting in various functional properties (Dapčević-Hadnađev, Hadnađev, Lazaridou, Moschakis, & Biliaderis, 2018; Malomo, He, & Aluko, 2014; Raikos, Duthie, & Ranawana, 2015).

TABLE 4.2 Amino acid composition (%) of hempseed albumin and globulin fractions.[a]

Amino acids	Protein fractions	
	Albumin	Globulin
Histidine	3.7	3.9
Isoleucine	2.0	2.9
Leucine	4.1	5.6
Lysine	7.4	3.7
Methionine	1.7	4.1
Phenylalanine	1.3	3.3
Proline	3.8	3.9
Serine	5.1	5.7
Threonine	4.6	2.6
Tryptophan	0.2	0.3
Tyrosine	2.0	3.4
Valine	2.9	3.4
Alanine	3.9	2.8
Arginine	12.8	16.1
Asparagine	7.9	9.5
Cysteine	3.2	3.3
Glutamine	20.4	21.5
Glycine	8.3	4.1

[a] The data were obtained from Malomo and Aluko (2015a).

The amino acid profiles of hempseed from different sources are summarized in Table 4.3. Regardless of which source, hempseed protein is composed of all the essential amino acids needed by humans. The non-essential amino acids of whole hemp seed are characterized by higher levels of arginine (mean=2.28%) and glutamic acid (mean=3.74%). The former has been investigated by several clinical trials for its significant role in ammonia detoxification, fetal growth, and reducing insulin resistance (Wu et al., 2009), and the latter is recognized to be used as an immune-stimulating agent (Poo et al., 2010). In the case of essential amino acids, whole hempseed contains higher leucine (mean=1.49%) and lower tryptophan (mean=0.23%). The specific ratio of around 3.0 between arginine/lysine is significantly higher than those of casein (0.5) and soy protein isolate (1.4) (Tang et al., 2006; Wang et al., 2008), indicating hempseed protein is particularly competitive as a nutritional and bioactive ingredient since this ratio is a determinant factor of the cholesterolemia and atherogenesis. Whole hempseed has excellent amounts of sulfur-containing cysteine (mean=0.41%) and methionine (mean=0.56%). Moreover, the amino acid score for hempseed proteins has been calculated

TABLE 4.3 Amino acid content (%, dry basis) of hempseed from different sources.[a]

Amino acids	Whole hempseed (n=11)		De-hulled hempseed (n=6)		Hempseed meal (n=9)		Hempseed hull (n=3)	
	Range	Mean	Range	Mean	Range	Mean	Range	Mean
Essential								
His	0.48–0.67	0.55±0.06	0.83–1.14	0.97±0.11	0.70–1.14	0.93±0.19	0.11–0.40	0.25±0.15
Iso	0.54–0.93	0.80±0.11	0.83–1.56	1.29±0.35	1.05–1.77	1.45±0.23	0.24–0.49	0.39±0.14
Leu	1.19–1.73	1.49±0.16	1.74–2.39	2.14±0.28	1.82–3.16	2.35±0.45	0.43–0.96	0.71±0.27
Lys	0.76–1.02	0.86±0.09	1.21–1.31	1.26±0.05	1.03–1.76	1.32±0.27	0.16–0.47	0.33±0.16
Met	0.45–0.71	0.56±0.08	0.75–1.10	0.94±0.12	0.53–1.32	0.88±0.25	0.05–0.30	0.18±0.12
Phe	0.68–1.34	1.03±0.16	0.94–1.64	1.43±0.30	1.37–2.09	1.62±0.30	0.44–0.58	0.53±0.09
Thr	0.77–1.34	1.01±0.22	1.13–1.41	1.27±0.11	1.08–1.74	1.35±0.23	0.22–0.47	0.36±0.13
Trp	0.15–0.37	0.23±0.06	0.27–0.45	0.38±0.07	0.26–0.55	0.39±0.10	0.02–0.09	0.06±0.04
Val	0.99–1.49	1.14±0.14	1.52–1.95	1.78±0.19	1.52–2.38	1.91±0.30	0.30–0.91	0.60±0.31
Nonessential								
Tyr	0.47–0.81	0.68±0.11	1.03–1.64	1.28±0.22	0.66–1.63	1.15±0.28	0.33–0.46	0.40±0.07
Pro	0.79–1.12	0.90±0.10	1.08–2.10	1.62±0.41	1.06–1.98	1.59±0.32	0.31–1.23	0.69±0.48
Ser	0.95–1.44	1.19±0.17	1.49–1.90	1.70±0.17	1.34–2.25	1.73±0.32	0.24–0.56	0.42±0.16
Ala	0.81–1.12	0.96±0.09	1.32–1.71	1.52±0.14	1.05–2.05	1.61±0.32	0.21–0.51	0.40±0.17
Arg	1.96–2.76	2.28±0.26	4.04–5.31	4.55±0.45	2.93–5.37	3.91±0.89	0.28–1.82	0.94±0.80
Asp	2.14–2.72	2.39±0.18	3.10–4.06	3.66±0.37	3.04–4.80	3.66±0.67	0.54–1.23	0.90±0.35
Cys	0.35–0.57	0.41±0.06	0.57–0.73	0.65±0.07	0.53–0.93	0.70±0.15	0.11–0.23	0.18±0.06
Glu	3.28–4.21	3.74±0.30	3.28–4.21	3.74±0.30	4.69–7.98	6.03±1.24	0.53–1.76	1.19±0.62
Gly	0.95–1.23	1.06±0.10	1.36–1.78	1.61±0.15	1.17–2.14	1.66±0.35	0.22–0.52	0.41±0.16

[a] The data were obtained from House et al. (2010); n is the sample number; the maximum and minimum values of the obtained data range are exhibited; the mean±standard deviation is reported; Ala, alanine; Arg, arginine; Asp, asparagine; Cys, cysteine; Glu, glutamine; Gly, glycine; His, histidine; Iso, isoleucine; Leu, leucine; Lys, lysine; Met, methionine; Phe, phenylalanine; Pro, proline; Ser, serine; Trp, tryptophan; Thr, threonine; Tyr, tyrosine; Val, valine.

in several studies (House et al., 2010; Tang et al., 2006; Wang et al., 2008), where it reached an agreement that lysine was one of the limiting amino acids. However, due to the variation in raw sources of hempseed proteins, the ranking of limiting amino acids was reported to differ: sulfur-containing amino acid (methionine and cysteine) with an amino acid score of 0.65 (Tang et al., 2006) and 0.62 (Tang et al., 2006) was the first limiting amino acid followed by lysine; House et al. (2010) observed that lysine (amino acid score, 0.62) was the first limiting amino acid followed by tryptophan (amino acid score, 0.87) and leucine (amino acid score, 0.94). It has been recognized that the essential amino acids of hempseed are somewhat comparable to other high-quality proteins such as casein and soy protein (Callaway, 2004; Tang et al., 2006).

Here, it was observed that after de-hulling and oil extraction the amino acid contents in the resultant products can be significantly enhanced (House et al., 2010), because of the removal of the components that totally or almost lacks in protein, which is evidenced by that de-hulled hempseed and hempseed meal had higher amino acids than the whole hempseed, and hempseed hull showed lower amino acids than the whole hempseed (Table 4.3).

4.2.2 Hempseed fat

4.2.2.1 Fatty acid

The hempseed fat, namely oil, has been reported to account for 25.4%–35.9% of whole hempseed (Table 4.1). Based on the literature, the fatty acid composition was only described in the hempseed oil instead of the whole hempseed. Therefore it is necessary to consider the impact of specific methods used for oil extraction on its profile and properties (Aladić et al., 2015; Da Porto, Decorti, & Natolino, 2015; Da Porto, Decorti, & Tubaro, 2012; Devi & Khanam, 2019; Grijó, Piva, Osorio, & Cardozo-Filho, 2019; Rezvankhah, Emam-Djomeh, Safari, Askari, & Salami, 2018; Subratti, Lalgee, & Jalsa, 2019). In this section, we focused mainly on the fatty

TABLE 4.4   Fatty acid profile (%, dry basis) of hempseed oil.

Fatty acid	Source 1 ($n=13$)	Source 2 ($n=10$)	Source 3 ($n=20$)	Source 4 ($n=19$)	Source 5 ($n=10$)
Palmitic acid	8.11–8.45	6.66–6.98	5.98–8.60	7.10–9.10	6.09–7.81
Palmitoleic acid	0.12–0.15	n.a.	0.10–0.24	n.a.	n.a.
Heptadecanoic acid	0.06–0.07	n.a.	n.a.	n.a.	n.a.
Stearic acid	2.12–2.49	2.08–2.82	2.57–4.61	2.10–2.80	2.31–3.96
Oleic acid	10.77–14.00	9.38–13.00	9.20–16.80	10.60–17.90	12.21–18.78
Linoleic acid	55.97–57.34	55.56–56.58	46.10–58.2	51.60–54.20	53.86–58.99
α-Linolenic acid	13.28–15.47	14.69–17.27	12.60–28.4	10.50–15.10	12.28–18.88
γ-Linolenic acid	2.43–4.50	2.56–4.49	0.49–4.54	1.90–4.50	3.51–6.22
Arachidic acid	0.64–0.77	0.64–0.78	0.52–1.02	n.a.	n.a.
Eicosenoic acid	0.33–0.38	n.a.	n.a.	n.a.	n.a.
Eicosadienoic acid	0.07–0.11	0.75–1.29	n.a.	n.a.	n.a.
Erucic acid	n.a.	1.31–2.55	n.a.	n.a.	n.a.
Stearidonic acid	0.63–0.96	n.a.	n.a.	n.a.	n.a.
Behenic acid	0.24–0.33	n.a.	0.15–0.42	n.a.	0.42–0.66
Lignoceric acid	0.10–0.15	n.a.	n.a.	n.a.	n.a.
ω-6/ω-3	3.6–4.2	n.a.	2.8–4.5	3.9–5.5	3.2–5.0
References	Xu, Li, et al. (2021), Xu, Zhao, et al. (2021)	Vonapartis et al. (2015)	Galasso et al. (2016)	Irakli et al. (2019)	Lan et al. (2019)

n is the sample number; the maximum and minimum values of the obtained data range are shown; n.a., not available.

acid composition obtained from hempseed oil rather than the whole hempseed because of a lack of literature data, as presented in Table 4.4. Significant variation in the fatty acid profile of hempseed oil was observed within the same study, which can primarily be associated with the intrinsic genotype of hempseed and external environmental conditions such as geography, climatic conditions, and local agronomic factors (Anwar, Latif, & Ashraf, 2006; Chen et al., 2010; Faugno et al., 2019). Overall, Table 4.4 shows that hempseed oil is categorized by higher percentages of polyunsaturated fatty acids, which was in accordance with the previous study (Vonapartis et al., 2015), in which hempseed oil was found to contain up to 90% unsaturated fatty acids. As presented in Table 4.4, linoleic acid (46.10%–58.99%) is the dominant fatty acid in hempseed oil of all analyzed genotypes. The second is α-linolenic acid, accounting for 10.50%–18.88% of hempseed oil. Both of them are recognized as the essential fatty acids since humans are incapable of synthesizing them directly (Farinon et al., 2020). Moreover, linoleic acid is the precursor to synthesize dihomo-γ-linolenic acid and arachidonic acid, whereas α-linolenic acid is the platform chemical for the conversion of eicosapentaenoic acid (Leonard et al., 2020). These two bioactive essential fatty acids play an important role in regulating physiological, metabolic, and inflammatory processes to prevent cardiovascular diseases, obesity, and diabetes mellitus through the synthesis of prostaglandins, anti-inflammatory eicosanoids, leukotrienes, and skin integrity (Kapoor & Huang, 2006; Sokoła-Wysoczańska et al., 2018).

Hempseed oil is rich in both ω-6 and ω-3 essential fatty acids, with the optimal ratio between 2:1 and 3:1 good for human metabolism. The ω-6/ω-3 fatty acid ratio has been highlighted as a crucial index to evaluate an optimal state of health by Simopoulos (2008), where the high ω-6/ω-3 ratio in the diet would induce cardiovascular disease, cancer, obesity, and inflammatory and autoimmune diseases. It was also reported that the optimal ratio of ω-6/ω-3 may differ for specific diseases: a ratio of 2.5/1 decreased rectal cell proliferation in patients with colorectal cancer, a ratio of 2–3/1 suppressed inflammation in patients with rheumatoid arthritis, and a ratio of 5/1 showed a beneficial effect on patients with asthma. Strikingly, the ω-6/ω-3 ratio in hempseed was reported to fall in the range of 3.2–5.0 (Lan et al., 2019), 3.6–4.2 (Xu, Li, et al., 2021; Xu, Zhao, et al., 2021), 3.9–5.5 (Vonapartis et al., 2015), and 2.8–4.5 (Galasso et al., 2016) (Table 4.4), highly depending on the cultivar of hempseed and environmental condition. In addition to linoleic acid and α-linolenic acid, among ω-6 and ω-3 fatty acids, hempseed oil is also composed of their corresponding biologic metabolites γ-linolenic acid (0.49%–6.22%) and stearidonic acid (0.63%–0.96%). The former can function as an anti-inflammatory molecule since it can be rapidly converted into dihomo-γ-linolenic acid in the human body (Farinon et al., 2020). In terms of saturated fatty acids, palmitic acid ranges from 5.98% to 9.10%, followed by stearic acid (2.10%–4.61%), behenic acid (0.15%–0.66%), lignoceric acid (0.10%–0.15%), and heptadecanoic acid (0.06%–0.07%).

4.2.2.2 The unsaponifiable matter

Based on the International Organization for Standardization (ISO) 18609:2000 definition, the unsaponifiable matter consists of substances dissolved in the (hempseed) oil which are unable to be saponified by alkali hydroxides, but are soluble in the organic solvents. It includes sterols, aliphatic and terpenic alcohols, and tocopherols (Farinon et al., 2020; Leonard et al., 2020). Their profiles in hempseed oil have been reported in detail by Montserrat-De La Paz, Marín-Aguilar, García-Giménez, and Fernández-Arche (2014), where the unsaponifiable

TABLE 4.5 The tocopherol and phytosterol profiles of hempseed unsaponifiable matter.

Unsaponifiable matter	Whole seed (mg/100 g)					Oil (mg/100 g)		
Total tocopherol	n.a.	n.a.	n.a.	97.13	80.28	n.a.	22.14–38.52	41.3–102.0
α-Tocopherol	0.78–3.05	0.35	n.a.	0.43	3.22	2.78	1.83–3.53	2.2–9.9
β-Tocopherol	0.11–0.25	n.a.	n.a.	n.a.	0.81	n.a.	n.a.	0.4–0.8
γ-Tocopherol	15.97–28.20	0.55	n.a.	89.26	73.38	56.41	17.16–33.26	36.2–89.8
δ-Tocopherol	0.51–2.49	n.a.	n.a.	3.55	2.87	n.a.	1.89–3.49	1.5–7.6
Total phytosterol	n.a.	n.a.	124	n.a.	279.37	n.a.	n.a.	n.a.
β-Sitosterol	n.a.	53.61	79.7	n.a.	190.51	n.a.	n.a.	n.a.
Stigmasterol	n.a.	2.47	3.4	n.a.	10.02	n.a.	n.a.	n.a.
Campesterol	n.a.	11.54	7.3	n.a.	50.57	n.a.	n.a.	n.a.
References	Kriese et al. (2004)	Siano et al. (2019)	Vecka et al. (2019)	Rezvankhah et al. (2018)	Montserrat-De La Paz et al. (2014)	Teh and Birch (2013)	Chen et al. (2010)	Matthäus, Schumann, Brühl, and Kriese (2005)

Where more than one cultivar has been analyzed, the maximum and minimum values of the obtained data range are listed; *n.a.*, not available.

matter accounted for about 2.0% of hempseed oil and the most relevant compounds are tocopherols and phytosterols. Table 4.5 presents the tocopherol and phytosterol profiles of the unsaponifiable matter based on the whole seed and its oil. In terms of tocopherol, γ-tocopherol, ranging from 0.6 to 28.2 mg/100 g of whole seed and from 17.2 to 89.8 mg/100 g of oil, is the most abundant isomer, followed by α-tocopherol (0.4–3.1 mg/100 g seed; 0.4–10.0 mg/100 g oil), δ-tocopherol (0.5–2.5 mg/100 g seed; 1.5–7.6 mg/100 g oil), and β-tocopherol (0.1–0.3 mg/100 g seed; 0.4–0.8 mg/100 g oil). Among them, γ-tocopherol is the most bioactive antioxidant in lipids (Da Porto, Natolino, & Decorti, 2015), and it can cooperate with other antioxidant compounds such as polyphenols to protect the hempseed oil from oxidation ascribed to their capacity to scavenge free radicals (Farinon et al., 2020). The total tocopherol amount in the hempseed oil ranges from 22.14 to 102.0 mg/100 g of oil (Table 4.5). The extraction technique was found to have an essential influence on the tocopherol amount of the hempseed oil (Rezvankhah et al., 2018). However, under the same oil extraction condition, genotype played a dominant role in the tocopherol profiles of the oil, which was demonstrated by Matthäus et al. (2005) and Kriese et al. (2004), where the hempseed oil composition from more than 50 hemp genotypes was investigated.

Phytosterols, referred to as plant sterol and stanol esters, are fat-soluble organic molecules found in plant cell membranes, and they are identified to have a comparable structure to cholesterol (Farinon et al., 2020). It has been reported that phytosterols are capable of diminishing cholesterol solubility and excluding it from lipid micelles, thereby lowering the cholesterol intestinal absorption and final cholesterol content in the blood (Kritchevsky & Chen, 2005). Based on the limited literature data (Table 4.5), it still can be found that β-sitosterol (53.6–79.7 mg/100 g seed; 190.5 mg/100 g oil) is the dominant phytosterol as compared to stigmasterol (2.5–3.5 mg/100 g seed; 10.0 mg/100 g oil) and campesterol (7.3–11.5 mg/100 g seed; 50.6 mg/100 g oil). In a previous study, Vecka et al. (2019) compared 11 nuts (Brazil nut, cashew, chestnut, coconut, hazelnut, macadamia, peanut, pecan, pine, pistachio, and walnut), 7 seeds (almond, hemp, linseed, musk melon, pumpkin, sesame, and sunflower), and apricot kernel to evaluate their nutritional value in terms of sterols and fatty acids, and concluded that β-sitosterol (79.7 mg/100 g seed) in whole hempseed is higher than that in linseed (55.4 mg/100 g seed). In addition to tocopherol and phytosterol, the unsaponifiable matter of hempseed oil may also contain a limited amount of other compounds such as carotenoid, chlorophyll, phytol, etc., while the relevant reports were unavailable.

4.2.3 Hempseed carbohydrate and dietary fiber

4.2.3.1 Carbohydrate

Carbohydrate is defined as a large group of organic compounds or biomolecules consisting of carbon, hydrogen, and oxygen atoms, usually with a hydrogen-oxygen atom ratio of 2:1. It includes starch, soluble sugars, cellulose, and hemicellulose in the hempseed (Schultz et al., 2020). Only a few studies have analyzed the total carbohydrate content of the whole hempseed. Mattila, Mäkinen, et al. (2018) compared the proximate composition of some commercial protein-rich seeds and reported that among them hempseed contained a relatively lower amount of total carbohydrates—about 34.4% for whole seed. Callaway (2004) observed that the hemp cultivar Finola had a total carbohydrate content of 27.6% in the seed. The carbohydrate content in the hempseed meal is relatively higher than in the whole seed due to oil removal. Lan et al. (2019) investigated the carbohydrate content of 10 varieties of hemp flours and found that the carbohydrate content ranged from 32.47% to 37.53%. Siano et al. (2019) characterized the flour and seeds of Fedora cultivar hemp and the results showed that the carbohydrate content in flour (41.6%) is significantly higher than in the whole seed (38.1%). Xu, Li, et al. (2021) and Xu, Zhao, et al. (2021) studied the physicochemical, nutritional, and antioxidant properties of 13 hempseed varieties and reported that starch content ranged from 1.69% to 1.97%. In another study, Vonapartis et al. (2015) characterized the composition of 10 hemp cultivars and found that these seeds contained 16.5%–18.1% cellulose and 8.0%–9.4% hemicellulose.

4.2.3.2 Dietary fiber

Dietary fiber is a well-known term that has been defined by the Codex Alimentarius Commission as the carbohydrate polymers which cannot be hydrolyzed by endogenous enzymes in the small intestine of humans (Saura-Calixto, 1998). It is the edible portion of plant material that includes polysaccharides, oligosaccharides, lignin, and associated plant substances. Therefore there will be a significant overlap between dietary fiber and carbohydrate in the hempseed. Indeed, Callaway (2004) found a total dietary fiber content of 27.6% (5.4% digestible fiber; 22.2% non-digestible fiber) in hempseed, indicating that the entire carbohydrate fraction is entailed in dietary fiber as previously mentioned. From the research by Mattila, Mäkinen, et al. (2018), the total dietary fiber of hempseed was 33.8 g/100 g of seed, amounting up to 98% of the total carbohydrate. In addition, Xu, Li, et al. (2021) and Xu, Zhao, et al. (2021) reported 28.78%–36.55% of crude fiber among 13 hempseed varieties. Of further importance, hempseed flour was also reported to have 0.16% of soluble fiber and 25.49% of insoluble fiber (Multari et al., 2016). These results suggest that hempseed is a low-starch and high-fiber food matrix (Farinon et al., 2020).

4.2.4 Hempseed minerals

The total mineral content of hempseed is often reflected by the ash content (3.7%–6.3%), which is composed of naturally inorganic compounds. Minerals recognized as dietary micronutrients act physiological and structural indispensable roles to maintain optimal human health (Farinon et al., 2020). The main reported macro-elements in hempseed or hemp flour are phosphorous (P, 890–1170 mg/100 g), potassium (K, 252–2821 mg/100 g), magnesium (Mg, 237–694 mg/100 g), calcium (Ca, 90–955 mg/100 g), and sodium (Na, 6.8–27 mg/100 g), while micro-elements are iron (Fe, 4–240 mg/100 g), manganese (Mn, 4–15 mg/100 g), zinc (Zn, 4–11 mg/100 g), copper (Cu, 0.5–2 mg/100 g), and cadmium (Cd, 0.0015–0.4 mg/100 g), as presented in Table 4.6. Herein, the mineral profiles of hempseed were reported to differ because of the environmental and agronomic conditions (such as climate, soil, and fertilizer composition, etc.) as well as the hemp cultivars (Lan et al., 2019; Mattila, Mäkinen, et al., 2018; Siano et al., 2019). The comprehensive

TABLE 4.6   Mineral composition (mg/100 g) of different hempseed cultivars.

P	K	Mg	Ca	Na	Fe	Mn	Zn	Cu	Cd	References
n.a.	252	268	94	6.8	10	4	5	0.5	n.a.	Siano et al. (2019)
1160	859	483	145	12	14	7	7	2	n.a.	Callaway (2004)
1170	921	496	127	n.a.	4	11	7	1.9	0.0015	Mattila, Mäkinen, et al. (2018)
n.a.	463–2821	237–694	144–955	n.a.	113–240	6–11	4–9	n.a.	0.1–0.4	Mihoc, Pop, Alexa, and Radulov (2012)[a]
910–1014	727–866	430–482	94–121	22–27	11–13	12–15	10–11	0.8–0.9	n.a.	Lan et al. (2019)[b]

n.a., not available.

[a] *The data were obtained from five varieties of hempseed.*

[b] *The data were obtained from 10 varieties of hemp flours.*

comparison and analysis of these factors can be found in the previous review published by Farinon et al. (2020).

4.3 Hemp secondary metabolites

4.3.1 Phenolic compounds

Phenolic compounds are representative secondary metabolites in the hempseed and products. Attributed to their specific chemical structures, these compounds have an intrinsic antioxidant function to protect cell components against oxidative damage, reducing the risk of various degenerative diseases related to oxidative stress (Farinon et al., 2020). Coupling with the most considerable tocopherols aforementioned, phenolic compounds can provide the high oxidative stability of hempseed oil (Chen et al., 2012; Izzo et al., 2020; Smeriglio et al., 2016). Due to the variation in extraction and analytical methods and sources of raw materials, a striking range in terms of total phenolic content (TPC) among the collected data was observed (Table 4.7). Among them, Siano et al. (2019) characterized the biochemical properties of hemp fractions and reported that its oil contained the lowest TPC (0.02 mg gallic acid equivalent (GAE)/g) compared to seed (0.77 mg GAE/g) and flour (0.74 mg GAE/g). This phenomenon might be laboriously explained by that only a few polyphenols were isolated during oil extraction ascribed to their hydro-soluble feature, whereas most of them remain in the hempseed cake (Moccia et al., 2020). However, this low TPC in the hempseed oil was not in agreement with that determined in the cold-pressed hempseed oil from other studies being 0.44 mg GAE/g (Yu et al., 2005), 1.88 mg GAE/g (Teh & Birch, 2013), and 2.68 mg GAE/g (Smeriglio et al., 2016). Izzo et al. (2020) evaluated the TPC of 13 commercial hempseed oil from different origins at the Italian local market and found that the TPC reached a staggering amount of 22.1–160.8 mg GAE/g. Regarding the TPC of whole hempseed, Vonapartis et al. (2015) reported that the TPC of 10 hemp cultivars ranged from 13.68 to 51.60 mg GAE/g, which was notably higher than that (0.77–7.80 mg GAE/g) obtained from other studies (Table 4.7). Besides, several studies have investigated the TPC in the defatted kernel (3.9–15.6 mg GAE/g) (Chen et al., 2012) and hull (9.2–139.3 mg GAE/g) (Chen et al., 2012) and protein hydrolysates

TABLE 4.7 Total phenolic content (mg gallic acid equivalent (GAE)/g) of hempseed fractions.

Sample	TPC	References
Oil	0.02–160.8	Izzo et al. (2020), Kim and Lee (2011), Simopoulos (2008), Smeriglio et al. (2016), Yu, Zhou, and Parry (2005)
Seed	0.77–51.60	Chen et al. (2012), Frassinetti et al. (2018), Irakli et al. (2019), Mattila, Mäkinen, et al. (2018), Siano et al. (2019), Vonapartis et al. (2015)
Flour	0.74	Siano et al. (2019)
Defatted kernel	3.9–15.6	Chen et al. (2012)
Defatted hull	9.2–139.3	Chen et al. (2012)
Protein hydrolysates	0.42–0.48	Xu, Li, et al. (2021), Xu, Zhao, et al. (2021)

TPC, total phenolic content. Where more than one cultivar has been analyzed, the maximum and minimum values of the obtained data range are shown.

(0.42–0.48 mg GAE/g) of hempseed (Xu, Li, et al., 2021; Xu, Zhao, et al., 2021). The significant variation might be associated with the genotype of hempseed and the extraction pretreatment as well as determination methods used for total phenolic compounds.

It has been demonstrated that in whole hempseed, polyphenols are primarily located in the hull rather than in the kernel (Chen et al., 2012; Mattila, Mäkinen, et al., 2018). Therefore the interactions between dietary fiber and phenolic compounds should be considered. Dietary fiber can form chemical crosslinking with polyphenols via hydrophobic interactions, hydrogen bonding (polysaccharide's glycosidic chains provide oxygen atoms, while phenolic compounds offer hydroxyl groups), and covalent bonds (Saura-Calixto, 2011). The main phenolic compounds in hempseed products have been identified to be lignans also called phenylpropionamides, which belong to two main groups recognized as phenolic amides and lignanamides (Lesma et al., 2014; Sakakibara, Katsuhara, Ikeya, Hayashi, & Mitsuhashi, 1991). Phenolic amides are formed through the linkage between an amine moiety and a phenolic moiety, while lignanamides are originated from a random polymerization of phenolic amides (Farinon et al., 2020). The two compounds have been isolated from hempseed hull by Chen et al. (2012) using 60% ethanol with a macroporous resin absorption approach and they were qualitatively identified as *N-trans*-caffeoyltyramine and cannabisin B. In the work by Pojić et al. (2014), nine phenolic compounds of hemp meal fractions with different particle sizes were quantitatively characterized in terms of gallic acid (0.43–1.06 mg/kg), vanillic acid (0.35–0.55 mg/kg), protocatechuic acid (14.5–36.0 mg/kg), sinapic acid (17.3–66.8 mg/kg), ferulic acid (9.67–88.4 mg/kg), *p*-hydroxybenzoic acid (33.3–129 mg/kg), cannabisin B (4.27–153 mg/kg), *N-trans*-caffeoyltyramine (41.7–287 mg/kg), and catechin (107–744 mg/kg). In another study conducted by Irakli et al. (2019), where they reported that the major phenolic amide in seven hempseed extracts was *N-trans*-caffeoyltyramine (14.8–83.2 mg/100 g) and the predominant lignanamide was cannabisin A (51.1–159.1 mg/100 g). Apart from those, protocatechuic acid (0.4–1.6 mg/100 g), *p*-hydroxybenzoic acid (1.2–3.0 mg/100 g), and cinnamic acid (0.2–7.3 mg 100 g) have been identified as well. Moreover, the cannabisin F (30 mg/100 g) and grossamide (30 mg/100 g) in the hempseed have also been reported (Mattila, Mäkinen, et al., 2018). Based on these literature data, it can be found that most of phenylpropionamides are lignanamides, which is in accordance with the finding by Yan et al. (2015), where a total of 14 phenolic compounds, including four new lignanamides (cannabisin M, cannabisin N, cannabisin O, and 3,3′-demethyl-heliotropamide), were isolated and identified with 12 of them being lignanamides.

4.3.2 Bioactive peptides

Bioactive peptides, as functional compounds in the hempseed protein hydrolysates (HPH), have received considerable interest due to their versatile bioactivity (Rivero-Pino, Espejo-Carpio, & Guadix, 2020), such as antioxidant (Teh, Bekhit, Carne, & Birch, 2016), acetylcholinesterase-inhibitory (Malomo & Aluko, 2016), antihypertensive (Malomo, Onuh, Girgih, & Aluko, 2015), antiproliferative (Logarušić et al., 2019), anti-inflammatory (Samsamikor, Mackay, Mollard, & Aluko, 2020), hypocholesterolemic (Aiello, Lammi, Boschin, Zanoni, & Arnoldi, 2017), and neuroprotective (Rodriguez-Martin et al., 2019) properties. Herein, we will not extend the functionality further but focus on the types of obtained bioactive peptides in terms of molecular size and amino acid profile. Because hempseed proteins have the insufficient bioactive function, the hydrolysis process is inevitably needed to release these encrypted bioactive peptides from the virgin protein (Farinon et al., 2020; Wang & Xiong, 2019). This indicates that the characteristics of obtained peptides in the hydrolysates, including the molecular size and the amino acid profile, can be highly influenced by the hydrolysis and isolation conditions (Malomo & Aluko, 2015b, 2016; Tang, Wang, & Yang, 2009). All published information showed that most of the released bioactive peptides own a high hydrophobicity rate.

More specifically, Farinon et al. (2020) reported that compared to high-molecular-weight peptides, the low-molecular-weight ones have a strong resistance to enzymatic degradation in the gastrointestinal tract and interact with specific target sites. In this regard, Malomo et al. (2015) evaluated the structural properties of various forms of HPI and found that the molecular size of obtained peptides ranged from 0.3 to 10 kDa via size exclusion chromatography. Wang and Xiong (2019) concluded that small water-soluble peptides were more abundant (up to 63.4%) in HPH through measuring the zinc-binding behavior. Aiello et al. (2017) found that medium-size peptides containing 8–10 amino acid residues (characterized by a hydrophobic N-terminus and a negatively charged C-terminus) appeared to be particularly favorable for interacting with 3-hydroxy-3-methylglutaryl-coenzyme A reductase. In terms of amino acid profile, Orio et al. (2017) prepared the protein hydrolysates of defatted hempseed meal using extensive chemical hydrolysis under acid conditions and identified four short-chain peptides, including Glycine-Valine-Leucine-Tyrosine (GVLY), Leucine-Glycine-Valine (LGV), Arginine-Valine-Arginine (RVR), and Isoleucine-Glutamic Acid-Glutamic Acid (IEE). Among them, GVLY was the most active peptide to inhibit the angiotensin-converting enzyme, followed by LGV, RVR, and IEE. This finding was comparable to the work by Girgih et al. (2014), who simulated gastrointestinal tract digestion of hempseed proteins for HPH production and reported that among 23 short-chain ($5 \leq$ amino acids) peptides Tryptophan-Valine-Tyrosine-Tyrosine and Proline-Serine-Leucine-Proline-Alanine peptides are the most active antioxidant with strong metal chelation activity. In addition, Ren et al. (2016) identified two novel α-glucosidase inhibitory peptides in the *endo*-protease (Alcalase from Novozyme) treated HPH with their sequences of Leucine-Arginine (287.2 Da) and Proline-Leucine-Methionine-Leucine-Proline (568.4 Da).

4.3.3 Cannabinoids

Hempseed on its surface may contain a distinct class of terpenophenolic compounds which are actual products of Cannabis inflorescence—the cannabinoids (Adesina et al., 2020). THC is the principal psychoactive, intoxicant, and acute toxic cannabinoid component of the resin secreted by the flowering buds of the hemp plant, leading to the legislative regulations regarding THC use in the hempseed products produced for human consumption (Bosy & Cole, 2000; Cherney & Small, 2016), whereas cannabidiol (CBD), the main non-psychoactive cannabinoid component, has been proved to have a substantial medical potential (Adesina et al., 2020; Hanuš, Meyer, Muñoz, Taglialatela-Scafati, & Appendino, 2016). In the case of hempseed oil, Bosy and Cole (2000) reported that the THC concentrations in six commercially available hempseed oils ranged from 11.5 to 117.5 μg/g originating from the leaves and resin adhering to the seeds being processed into the oil. It was in accordance with the study by Holler et al. (2008), where the THC of 35 hempseed oils from different manufacturers were lower than 117.5 μg/g for all samples with several samples being less than 10 μg/g. In addition, Citti, Pacchetti, Vandelli, Forni, and Cannazza (2018) analyzed seven cannabinoids including cannabidiolic acid (CBDA), tetrahydrocannabinolic acid (THCA), CBD, THC, cannabinol (CBN), cannabigerol (CBG), and cannabidivarin (CBDV) in 13 commercial hempseed oils. Among them, CBG, CBN, THC, and THCA levels were lower than 1.4, 12.4, 1.4, and 9.5 μg/g, respectively; CBDV and CBD levels in 10 samples were lower than 8.8 and 8.7 μg/g, respectively. These low concentrations of cannabinoids were in agreement with their study (Citti et al., 2019), where apart from THC and CBD, additional 30 cannabinoids were identified in hempseed oil with all of their levels below 5.0 ppm.

4.4 Anti-nutritive compounds

Herein, the antinutritional factors present in the hempseed products are deleterious compounds such as phytic acid, trypsin inhibitors, cyanogenic glycosides, tannins, and saponins which can interfere with the absorption of biomolecules and impede their bioavailability (Farinon et al., 2020; Leonard et al., 2020). Regarding hempseed products, only a few studies have investigated the composition and levels of antinutritional compounds. In terms of hempseed, Mattila, Pihlava, et al. (2018) compared several antinutritional factors in commercial protein-rich plant products and found that the phytic acid contents in whole hempseed, fat-free hempseed, seed hull, and fat-free seed hull are 3.5, 5.3, 2.1, and 2.3 g/100 g, respectively, and the condensed tannins in whole hempseed and hempseed peel are 105 and 144 mg/100 g, respectively. Galasso et al. (2016) demonstrated the notable differences in phytic acid (43–75 g/kg) among genotypes. For hempseed flour, Russo (2013) reported the mean values of phytic acid (63.70 g/kg), condensed tannins (0.280 g/kg), trypsin inhibitors (105.96 U/mg of defatted flour), cyanogenic glycosides (0.017 g/kg), and saponins (161.63 mg/kg) in six hempseed flours. Russo and Reggiani (2015) characterized hempseed meal (after extraction of oil) from three dioecious and three monoecious varieties in terms of phytic acid (61.5–76.7 g/kg), condensed tannins (2.14–4.56 g/kg), trypsin inhibitors (10.8–27.7 U/mg of defatted flour), cyanogenic glycosides (0.05–0.17 g/kg), and saponins (0.47–0.70 mg/kg). Moreover, Pojić et al. (2014) identified the distribution of major antinutrients in different fractions of hempseed meal and concluded that antinutrients, including trypsin inhibitors (1.39–3.90 TIU/mg of protein), phytic acid (4.36–22.5 mg/g), glucosinolates (3.14–5.64 μmol/g), and condensed tannins (0.19–0.33 mg/g), were typically located in the cotyledon fractions. The use of hempseed products for human food and animal feed may be limited due to the presence of antinutritive compounds. For example, the presence of phytic acid may lead to mineral deficiencies (Russo, 2013).

4.5 Challenges and opportunities

Here, we only focus on the challenges and opportunities related to the nutritional and chemical composition of hempseed. The genotype, environment, and their interaction have been shown to primarily influence the chemical composition of hempseed and its products. The legislation on low THC hemp cultivation and processing is loosening, which will be beneficial to the acceleration of breeding objectives. Large amounts of attention have been attracted in terms of compositional and functional properties of hempseed protein, oil, and secondary metabolites, but diverse analytical methods and different sampling sources often lead to inconsistent results. In this regard, it might be reasonable and clear to express the component levels based on the original material to provide a convenient channel for data collection and comparison among literature. Hempseed contains a spectrum of complex macromolecules, particularly carbohydrates, dietary fibers, and anti-nutritive compounds, in addition to protein and oil. The utilization of carbohydrates, dietary fibers, and anti-nutritive compounds as platform materials could lead to the comprehensive valorization of hempseed. However, reports targeting their profiles based on hempseed are still limited, which could be an area of interest for future studies. In addition, multi-stream and sequential processing systems should be taken into consideration targeting specific compounds of interest as principal products or raw materials for subsequent valorization.

4.6 Conclusion

The chapter first summarized the proximate composition of hempseed in terms of protein (21%–32%), oil (25%–36%), and ash (4%–6%) content. Hempseed protein is characterized by its high edestin (up to 80%) and all essential amino acid profiles with specific acidic and basic subunits, while hempseed oil is known for its high percentages of polyunsaturated fatty acids, unsaponifiable matter, and the nutritional ω-6/ω-3 ratio of 2.8–5.5. However, carbohydrates and dietary fibers, accounting for around 30% of hempseed, have been insufficiently studied. The macro- (P, K, Mg, Ca, and Na) and micro-elements (Fe, Mn, Zn, Cu, and Cd) in hempseed or hemp flour were observed. Besides, the distribution, molecular size, and structural characteristics of phenolic compounds, bioactive peptides from hempseed protein hydrolysates, and cannabinoids that can be present on the outer surface of hempseed are illuminated. Finally, several representative antinutritive compounds, including phytic acid, trypsin inhibitors, cyanogenic glycosides, condensed tannins, and saponins, were discussed in terms of their composition in hempseed products.

References

Adesina, I., Bhowmik, A., Sharma, H., & Shahbazi, A. (2020). A review on the current state of knowledge of growing conditions, agronomic soil health practices and utilities of hemp in the United States. *Agriculture, 10*(4). https://doi.org/10.3390/agriculture10040129.

Aiello, G., Fasoli, E., Boschin, G., Lammi, C., Zanoni, C., Citterio, A., et al. (2016). Proteomic characterization of hempseed (*Cannabis sativa* L.). *Journal of Proteomics, 147*, 187–196. https://doi.org/10.1016/j.jprot.2016.05.033.

Aiello, G., Lammi, C., Boschin, G., Zanoni, C., & Arnoldi, A. (2017). Exploration of potentially bioactive peptides generated from the enzymatic hydrolysis of hempseed proteins. *Journal of Agricultural and Food Chemistry, 65*(47), 10174–10184. https://doi.org/10.1021/acs.jafc.7b03590.

Aladić, K., Jarni, K., Barbir, T., Vidović, S., Vladić, J., Bilić, M., et al. (2015). Supercritical CO_2 extraction of hemp (*Cannabis sativa* L.) seed oil. *Industrial Crops and Products, 76*, 472–478. https://doi.org/10.1016/j.indcrop.2015.07.016.

Anwar, F., Latif, S., & Ashraf, M. (2006). Analytical characterization of hemp (*Cannabis sativa*) seed oil from different agro-ecological zones of Pakistan. *Journal of the American Oil Chemists' Society, 83*(4), 323–329. https://doi.org/10.1007/s11746-006-1207-x.

Bosy, T. Z., & Cole, K. A. (2000). Consumption and quantitation of Δ^9-tetrahydrocannabinol in commercially available hemp seed oil products. *Journal of Analytical Toxicology, 24*(7), 562–566. https://doi.org/10.1093/jat/24.7.562.

Callaway, J. C. (2004). Hempseed as a nutritional resource: An overview. *Euphytica, 140*(1–2), 65–72. https://doi.org/10.1007/s10681-004-4811-6.

Chen, T., He, J., Zhang, J., Li, X., Zhang, H., Hao, J., et al. (2012). The isolation and identification of two compounds with predominant radical scavenging activity in hempseed (seed of *Cannabis sativa* L.). *Food Chemistry, 134*(2), 1030–1037. https://doi.org/10.1016/j.foodchem.2012.03.009.

Chen, T., He, J., Zhang, J., Zhang, H., Qian, P., Hao, J., et al. (2010). Analytical characterization of hempseed (seed of *Cannabis sativa* L.) oil from eight regions in China. *Journal of Dietary Supplements, 7*(2), 117–129. https://doi.org/10.3109/19390211003781669.

Cherney, J. H., & Small, E. (2016). Industrial hemp in North America: Production, politics and potential. *Agronomy, 6*(4). https://doi.org/10.3390/agronomy6040058.

Citti, C., Linciano, P., Panseri, S., Vezzalini, F., Forni, F., Vandelli, M. A., et al. (2019). Cannabinoid profiling of hemp seed oil by liquid chromatography coupled to high-resolution mass spectrometry. *Frontiers in Plant Science, 10*, 1–17. https://doi.org/10.3389/fpls.2019.00120.

Citti, C., Pacchetti, B., Vandelli, M. A., Forni, F., & Cannazza, G. (2018). Analysis of cannabinoids in commercial hemp seed oil and decarboxylation kinetics studies of cannabidiolic acid (CBDA). *Journal of Pharmaceutical and Biomedical Analysis, 149*, 532–540. https://doi.org/10.1016/j.jpba.2017.11.044.

Crescente, G., Piccolella, S., Esposito, A., Scognamiglio, M., Fiorentino, A., & Pacifico, S. (2018). Chemical composition and nutraceutical properties of hempseed: An ancient food with actual functional value. *Phytochemistry Reviews, 17*(4), 733–749. https://doi.org/10.1007/s11101-018-9556-2.

Crini, G., Lichtfouse, E., Chanet, G., & Morin-Crini, N. (2020). Applications of hemp in textiles, paper industry, insulation and building materials, horticulture, animal nutrition, food and beverages, nutraceuticals, cosmetics and hygiene, medicine, agrochemistry, energy production and environment: A review. *Environmental Chemistry Letters, 18*(5), 1451–1476. https://doi.org/10.1007/s10311-020-01029-2.

Da Porto, C., Decorti, D., & Natolino, A. (2015). Potential oil yield, fatty acid composition, and oxidation stability of the hempseed oil from four *Cannabis sativa* L. cultivars. *Journal of Dietary Supplements, 12*(1), 1–10. https://doi.org/10.3109/19390211.2014.887601.

Da Porto, C., Decorti, D., & Tubaro, F. (2012). Fatty acid composition and oxidation stability of hemp (*Cannabis sativa* L.) seed oil extracted by supercritical carbon dioxide. *Industrial Crops and Products, 36*(1), 401–404. https://doi.org/10.1016/j.indcrop.2011.09.015.

Da Porto, C., Natolino, A., & Decorti, D. (2015). Effect of ultrasound pre-treatment of hemp (*Cannabis sativa* L.) seed on supercritical CO_2 extraction of oil. *Journal of Food Science and Technology, 52*(3), 1748–1753. https://doi.org/10.1007/s13197-013-1143-3.

Dapčević-Hadnađev, T., Dizdar, M., Pojić, M., Krstonošić, V., Zychowski, L. M., & Hadnađev, M. (2019). Emulsifying properties of hemp proteins: Effect of isolation technique. *Food Hydrocolloids, 89*, 912–920. https://doi.org/10.1016/j.foodhyd.2018.12.002.

Dapčević-Hadnađev, T., Hadnađev, M., Lazaridou, A., Moschakis, T., & Biliaderis, C. G. (2018). Hempseed meal protein isolates prepared by different isolation techniques. Part II. Gelation properties at different ionic strengths. *Food Hydrocolloids*, *81*, 481–489. https://doi.org/10.1016/j.foodhyd.2018.03.022.

Devi, V., & Khanam, S. (2019). Comparative study of different extraction processes for hemp (*Cannabis sativa*) seed oil considering physical, chemical and industrial-scale economic aspects. *Journal of Cleaner Production*, *207*, 645–657. https://doi.org/10.1016/j.jclepro.2018.10.036.

Farinon, B., Molinari, R., Costantini, L., & Merendino, N. (2020). The seed of industrial hemp (*Cannabis sativa* L.): Nutritional quality and potential functionality for human health and nutrition. *Nutrients*, *12*(7), 1–60. https://doi.org/10.3390/nu12071935.

Faugno, S., Piccolella, S., Sannino, M., Principio, L., Crescente, G., Baldi, G. M., et al. (2019). Can agronomic practices and cold-pressing extraction parameters affect phenols and polyphenols content in hempseed oils? *Industrial Crops and Products*, *130*, 511–519. https://doi.org/10.1016/j.indcrop.2018.12.084.

Frassinetti, S., Moccia, E., Caltavuturo, L., Gabriele, M., Longo, V., Bellani, L., et al. (2018). Nutraceutical potential of hemp (*Cannabis sativa* L.) seeds and sprouts. *Food Chemistry*, *262*, 56–66. https://doi.org/10.1016/j.foodchem.2018.04.078.

Galasso, I., Russo, R., Mapelli, S., Ponzoni, E., Brambilla, I. M., Battelli, G., et al. (2016). Variability in seed traits in a collection of *Cannabis sativa* L. genotypes. *Frontiers in Plant Science*, *7*, 20–25. https://doi.org/10.3389/fpls.2016.00688.

Galves, C., Stone, A. K., Szarko, J., Liu, S., Shafer, K., Hargreaves, J., et al. (2019). Effect of pH and defatting on the functional attributes of safflower, sunflower, canola, and hemp protein concentrates. *Cereal Chemistry*, *96*(6), 1036–1047. https://doi.org/10.1002/cche.10209.

Girgih, A. T., He, R., Malomo, S., Offengenden, M., Wu, J., & Aluko, R. E. (2014). Structural and functional characterization of hemp seed (*Cannabis sativa* L.) protein-derived antioxidant and antihypertensive peptides. *Journal of Functional Foods*, *6*(1), 384–394. https://doi.org/10.1016/j.jff.2013.11.005.

Grijó, D. R., Piva, G. K., Osorio, I. V., & Cardozo-Filho, L. (2019). Hemp (*Cannabis sativa* L.) seed oil extraction with pressurized n-propane and supercritical carbon dioxide. *Journal of Supercritical Fluids*, *143*, 268–274. https://doi.org/10.1016/j.supflu.2018.09.004.

Hanuš, L. O., Meyer, S. M., Muñoz, E., Taglialatela-Scafati, O., & Appendino, G. (2016). Phytocannabinoids: A unified critical inventory. *Natural Product Reports*, *33*(12). https://doi.org/10.1039/c6np00074f.

Holler, J. M., Bosy, T. Z., Dunkley, C. S., Levine, B., Past, M. R., & Jacobs, A. (2008). Δ^9-Tetrahydrocannabinol content of commercially available hemp products. *Journal of Analytical Toxicology*, *32*(6), 428–432. https://doi.org/10.1093/jat/32.6.428.

House, J. D., Neufeld, J., & Leson, G. (2010). Evaluating the quality of protein from hemp seed (*Cannabis sativa* L.) products through the use of the protein digestibility-corrected amino acid score method. *Journal of Agricultural and Food Chemistry*, *58*(22), 11801–11807. https://doi.org/10.1021/jf102636b.

Irakli, M., Tsaliki, E., Kalivas, A., Kleisiaris, F., Sarrou, E., & Cook, C. M. (2019). Effect of genotype and growing year on the nutritional, phytochemical, and antioxidant properties of industrial hemp (*Cannabis sativa* L.) seeds. *Antioxidants*, *8*(10), 20–25. https://doi.org/10.3390/antiox8100491.

Izzo, L., Pacifico, S., Piccolella, S., Castaldo, L., Narváez, A., Grosso, M., et al. (2020). Chemical analysis of minor bioactive components and cannabidiolic acid in commercial hemp seed oil. *Molecules*, *25*(16). https://doi.org/10.3390/molecules25163710.

Jami, T., Karade, S. R., & Singh, L. P. (2019). A review of the properties of hemp concrete for green building applications. *Journal of Cleaner Production*, *239*. https://doi.org/10.1016/j.jclepro.2019.117852, 117852.

Kapoor, R., & Huang, Y.-S. (2006). Gamma linolenic acid: An antiinflammatory omega-6 fatty acid. *Current Pharmaceutical Biotechnology*, *7*(6), 531–534. https://doi.org/10.2174/138920106779116874.

Kim, J.-J., & Lee, M.-Y. (2011). Isolation and characterization of edestin from Cheungsam hempseed. *Journal of Applied Biological Chemistry*, *54*(2), 84–88. https://doi.org/10.3839/jabc.2011.015.

Kriese, U., Schumann, E., Weber, W. E., Beyer, M., Brühl, L., & Matthäus, B. (2004). Oil content, tocopherol composition and fatty acid patterns of the seeds of 51 *Cannabis sativa* L. genotypes. *Euphytica*, *137*(3), 339–351. https://doi.org/10.1023/B:EUPH.0000040473.23941.76.

Kritchevsky, D., & Chen, S. C. (2005). Phytosterols-health benefits and potential concerns: A review. *Nutrition Research*, *25*(5), 413–428. https://doi.org/10.1016/j.nutres.2005.02.003.

Lan, Y., Zha, F., Peckrul, A., Hanson, B., Johnson, B., Rao, J., et al. (2019). Genotype x environmental effects on yielding ability and seed chemical composition of industrial hemp (*Cannabis sativa* L.) varieties grown in North Dakota, USA. *Journal of the American Oil Chemists' Society*, *96*(12), 1417–1425. https://doi.org/10.1002/aocs.12291.

Leonard, W., Zhang, P., Ying, D., & Fang, Z. (2020). Hempseed in food industry: Nutritional value, health benefits, and industrial applications. *Comprehensive Reviews in Food Science and Food Safety, 19*(1), 282–308. https://doi.org/10.1111/1541-4337.12517.

Lesma, G., Consonni, R., Gambaro, V., Remuzzi, C., Roda, G., Silvani, A., et al. (2014). Cannabinoid-free *Cannabis sativa* L. Grown in the po valley: Evaluation of fatty acid profile, antioxidant capacity and metabolic content. *Natural Product Research, 28*(21), 1801–1807. https://doi.org/10.1080/14786419.2014.926354.

Logarušić, M., Slivac, I., Radošević, K., Bagović, M., Redovniković, I. R., & Srček, V. G. (2019). Hempseed protein hydrolysates' effects on the proliferation and induced oxidative stress in normal and cancer cell lines. *Molecular Biology Reports, 46*(6), 6079–6085. https://doi.org/10.1007/s11033-019-05043-8.

Malomo, S. A., & Aluko, R. E. (2015a). A comparative study of the structural and functional properties of isolated hemp seed (*Cannabis sativa* L.) albumin and globulin fractions. *Food Hydrocolloids, 43*, 743–752. https://doi.org/10.1016/j.foodhyd.2014.08.001.

Malomo, S. A., & Aluko, R. E. (2015b). Conversion of a low protein hemp seed meal into a functional protein concentrate through enzymatic digestion of fibre coupled with membrane ultrafiltration. *Innovative Food Science and Emerging Technologies, 31*, 151–159. https://doi.org/10.1016/j.ifset.2015.08.004.

Malomo, S. A., & Aluko, R. E. (2016). In vitro acetylcholinesterase-inhibitory properties of enzymatic hemp seed protein hydrolysates. *Journal of the American Oil Chemists' Society, 93*(3), 411–420. https://doi.org/10.1007/s11746-015-2779-0.

Malomo, S. A., He, R., & Aluko, R. E. (2014). Structural and functional properties of hemp seed protein products. *Journal of Food Science, 79*(8), 1512–1521. https://doi.org/10.1111/1750-3841.12537.

Malomo, S. A., Onuh, J. O., Girgih, A. T., & Aluko, R. E. (2015). Structural and antihypertensive properties of enzymatic hemp seed protein hydrolysates. *Nutrients, 7*(9), 7616–7632. https://doi.org/10.3390/nu7095358.

Mamone, G., Picariello, G., Ramondo, A., Nicolai, M. A., & Ferranti, P. (2019). Production, digestibility and allergenicity of hemp (*Cannabis sativa* L.) protein isolates. *Food Research International, 115*, 562–571. https://doi.org/10.1016/j.foodres.2018.09.017.

Matthäus, B., Schumann, E., Brühl, L., & Kriese, U. (2005). Hempseed oil-influence of the genotype on the composition in a two-year study. *Journal of Industrial Hemp, 10*(2), 45–65. https://doi.org/10.1300/J237v10n02_05.

Mattila, P., Mäkinen, S., Eurola, M., Jalava, T., Pihlava, J. M., Hellström, J., et al. (2018). Nutritional value of commercial protein-rich plant products. *Plant Foods for Human Nutrition, 73*(2), 108–115. https://doi.org/10.1007/s11130-018-0660-7.

Mattila, P. H., Pihlava, J. M., Hellström, J., Nurmi, M., Eurola, M., Mäkinen, S., et al. (2018). Contents of phytochemicals and antinutritional factors in commercial protein-rich plant products. *Food Quality and Safety, 2*(4), 213–219. https://doi.org/10.1093/fqsafe/fyy021.

Mihoc, M., Pop, G., Alexa, E., & Radulov, I. (2012). Nutritive quality of Romanian hemp varieties (*Cannabis sativa* L.) with special focus on oil and metal contents of seeds. *Chemistry Central Journal, 6*(1). https://doi.org/10.1186/1752-153X-6-122.

Moccia, S., Siano, F., Russo, G. L., Volpe, M. G., La Cara, F., Pacifico, S., et al. (2020). Antiproliferative and antioxidant effect of polar hemp extracts (*Cannabis sativa* L., Fedora cv.) in human colorectal cell lines. *International Journal of Food Sciences and Nutrition, 71*(4), 410–423. https://doi.org/10.1080/09637486.2019.1666804.

Montserrat-De La Paz, S., Marín-Aguilar, F., García-Giménez, M. D., & Fernández-Arche, M. A. (2014). Hemp (*Cannabis sativa* L.) seed oil: Analytical and phytochemical characterization of the unsaponifiable fraction. *Journal of Agricultural and Food Chemistry, 62*(5), 1105–1110. https://doi.org/10.1021/jf404278q.

Multari, S., Neacsu, M., Scobbie, L., Cantlay, L., Duncan, G., Vaughan, N., et al. (2016). Nutritional and phytochemical content of high-protein crops. *Journal of Agricultural and Food Chemistry, 64*(41), 7800–7811. https://doi.org/10.1021/acs.jafc.6b00926.

Orio, L. P., Boschin, G., Recca, T., Morelli, C. F., Ragona, L., Francescato, P., et al. (2017). New ACE-inhibitory peptides from hemp seed (*Cannabis sativa* L.) proteins. *Journal of Agricultural and Food Chemistry, 65*(48), 10482–10488. https://doi.org/10.1021/acs.jafc.7b04522.

Patel, S., Cudney, R., & McPherson, A. (1994). Crystallographic characterization and molecular symmetry of edestin, a legumin from hemp. *Journal of Molecular Biology, 235*(1), 361–363. https://doi.org/10.1016/S0022-2836(05)80040-3.

Pojić, M., Mišan, A., Sakač, M., Hadnađev, T. D., Šarić, B., Milovanović, I., et al. (2014). Characterization of byproducts originating from hemp oil processing. *Journal of Agricultural and Food Chemistry, 62*(51), 12346–12442. https://doi.org/10.1021/jf5044426.

Poo, H., Park, C., Kwak, M. S., Choi, D. Y., Hong, S. P., Lee, I. H., et al. (2010). New biological functions and applications of high-molecular-mass poly-γ-glutamic acid. *Chemistry and Biodiversity, 7*(6), 1555–1562. https://doi.org/10.1002/cbdv.200900283.

Potin, F., Lubbers, S., Husson, F., & Saurel, R. (2019). Hemp (*Cannabis sativa* L.) protein extraction conditions affect extraction yield and protein quality. *Journal of Food Science, 84*(12), 3682–3690. https://doi.org/10.1111/1750-3841.14850.

Raikos, V., Duthie, G., & Ranawana, V. (2015). Denaturation and oxidative stability of hemp seed (*Cannabis sativa* L.) protein isolate as affected by heat treatment. *Plant Foods for Human Nutrition, 70*(3), 304–309. https://doi.org/10.1007/s11130-015-0494-5.

Ren, Y., Liang, K., Jin, Y., Zhang, M., Chen, Y., Wu, H., et al. (2016). Identification and characterization of two novel α-glucosidase inhibitory oligopeptides from hemp (*Cannabis sativa* L.) seed protein. *Journal of Functional Foods, 26,* 439–450. https://doi.org/10.1016/j.jff.2016.07.024.

Rezvankhah, A., Emam-Djomeh, Z., Safari, M., Askari, G., & Salami, M. (2018). Investigation on the extraction yield, quality, and thermal properties of hempseed oil during ultrasound-assisted extraction: A comparative study. *Journal of Food Processing and Preservation, 42*(10), 1–11. https://doi.org/10.1111/jfpp.13766.

Rivero-Pino, F., Espejo-Carpio, F. J., & Guadix, E. M. (2020). Antidiabetic food-derived peptides for functional feeding: Production, functionality and in vivo evidences. *Foods, 9*(8). https://doi.org/10.3390/foods9080983.

Rodriguez-Martin, N. M., Toscano, R., Villanueva, A., Pedroche, J., Millan, F., Montserrat-De La Paz, S., et al. (2019). Neuroprotective protein hydrolysates from hemp (*Cannabis sativa* L.) seeds. *Food & Function, 10*(10), 6732–6739. https://doi.org/10.1039/c9fo01904a.

Russo, R. (2013). Variability in antinutritional compounds in hempseed meal of Italian and French varieties. *Plant, 1*(2), 25. https://doi.org/10.11648/j.plant.20130102.13.

Russo, R., & Reggiani, R. (2015). Evaluation of protein concentration, amino acid profile and antinutritional compounds in hempseed meal from dioecious and monoecious varieties. *American Journal of Plant Sciences, 06*(01), 14–22. https://doi.org/10.4236/ajps.2015.61003.

Sakakibara, I., Katsuhara, T., Ikeya, Y., Hayashi, K., & Mitsuhashi, H. (1991). Cannabisin A, an arylnaphthalene lignanamide from fruits of *Cannabis sativa*. *Phytochemistry, 30*(9), 3013–3016. https://doi.org/10.1016/S0031-9422(00)98242-6.

Samsamikor, M., Mackay, D., Mollard, R. C., & Aluko, R. E. (2020). A double-blind, randomized, crossover trial protocol of whole hemp seed protein and hemp seed protein hydrolysate consumption for hypertension. *Trials, 21*(1), 1–13. https://doi.org/10.1186/s13063-020-4164-z.

Saura-Calixto, F. (1998). Antioxidant dietary fiber product: A new concept and a potential food ingredient. *Journal of Agricultural and Food Chemistry, 46*(10), 4303–4306. https://doi.org/10.1021/jf9803841.

Saura-Calixto, F. (2011). Dietary fiber as a carrier of dietary antioxidants: An essential physiological function. *Journal of Agricultural and Food Chemistry, 59*(1), 43–49. https://doi.org/10.1021/jf1036596.

Schultz, C. J., Lim, W. L., Khor, S. F., Neumann, K. A., Schulz, J. M., Ansari, O., et al. (2020). Consumer and health-related traits of seed from selected commercial and breeding lines of industrial hemp, *Cannabis sativa* L. *Journal of Agriculture and Food Research, 2*(100025), 100025. https://doi.org/10.1016/j.jafr.2020.100025.

Shen, P., Gao, Z., Xu, M., Rao, J., & Chen, B. (2020). Physicochemical and structural properties of proteins extracted from dehulled industrial hempseeds: Role of defatting process and precipitation pH. *Food Hydrocolloids, 108*(106065). https://doi.org/10.1016/j.foodhyd.2020.106065, 106065.

Siano, F., Moccia, S., Picariello, G., Russo, G. L., Sorrentino, G., Di Stasio, M., et al. (2019). Comparative study of chemical, biochemical characteristic and ATR-FTIR analysis of seeds, oil and flour of the edible Fedora cultivar hemp (*Cannabis sativa* L.). *Molecules, 24*(1), 1–13. https://doi.org/10.3390/molecules24010083.

Simopoulos, A. P. (2008). The importance of the omega-6/omega-3 fatty acid ratio in cardiovascular disease and other chronic diseases. *Experimental Biology and Medicine, 233*(6), 674–688. https://doi.org/10.3181/0711-MR-311.

Smeriglio, A., Galati, E. M., Monforte, M. T., Lanuzza, F., D'Angelo, V., & Circosta, C. (2016). Polyphenolic compounds and antioxidant activity of cold-pressed seed oil from Finola cultivar of *Cannabis sativa* L. *Phytotherapy Research, 30,* 1298–1307. https://doi.org/10.1002/ptr.5623.

Sokoła-Wysoczańska, E., Wysoczański, T., Wagner, J., Czyż, K., Bodkowski, R., Lochyński, S., et al. (2018). Polyunsaturated fatty acids and their potential therapeutic role in cardiovascular system disorders—A review. *Nutrients, 10*(10), 1–21. https://doi.org/10.3390/nu10101561.

Subratti, A., Lalgee, L. J., & Jalsa, N. K. (2019). Liquified dimethyl ether (DME): A green solvent for the extraction of hemp (*Cannabis sativa* L.) seed oil. *Sustainable Chemistry and Pharmacy, 12*(100144). https://doi.org/10.1016/j.scp.2019.100144, 100144.

Tang, C. H., Ten, Z., Wang, X. S., & Yang, X. Q. (2006). Physicochemical and functional properties of hemp (*Cannabis sativa* L.) protein isolate. *Journal of Agricultural and Food Chemistry, 54*(23), 8945–8950. https://doi.org/10.1021/jf0619176.

Tang, C. H., Wang, X. S., & Yang, X. Q. (2009). Enzymatic hydrolysis of hemp (*Cannabis sativa* L.) protein isolate by various proteases and antioxidant properties of the resulting hydrolysates. *Food Chemistry, 114*(4), 1484–1490. https://doi.org/10.1016/j.foodchem.2008.11.049.

Teh, S. S., Bekhit, A. E. D., Carne, A., & Birch, J. (2014). Effect of the defatting process, acid and alkali extraction on the physicochemical and functional properties of hemp, flax and canola seed cake protein isolates. *Journal of Food Measurement and Characterization, 8*(2), 92–104. https://doi.org/10.1007/s11694-013-9168-x.

Teh, S. S., Bekhit, A. E. D. A., Carne, A., & Birch, J. (2016). Antioxidant and ACE-inhibitory activities of hemp (*Cannabis sativa* L.) protein hydrolysates produced by the proteases AFP, HT, Pro-G, actinidin and zingibain. *Food Chemistry, 203*, 199–206. https://doi.org/10.1016/j.foodchem.2016.02.057.

Teh, S. S., & Birch, J. (2013). Physicochemical and quality characteristics of cold-pressed hemp, flax and canola seed oils. *Journal of Food Composition and Analysis, 30*(1), 26–31. https://doi.org/10.1016/j.jfca.2013.01.004.

Vecka, M., Staňková, B., Kutová, S., Tomášová, P., Tvrzická, E., & Žák, A. (2019). Comprehensive sterol and fatty acid analysis in nineteen nuts, seeds, and kernel. *SN Applied Sciences, 1*(12), 1–12. https://doi.org/10.1007/s42452-019-1576-z.

Vonapartis, E., Aubin, M. P., Seguin, P., Mustafa, A. F., & Charron, J. B. (2015). Seed composition of ten industrial hemp cultivars approved for production in Canada. *Journal of Food Composition and Analysis, 39*, 8–12. https://doi.org/10.1016/j.jfca.2014.11.004.

Wang, Q., Jin, Y., & Xiong, Y. L. (2018). Heating-aided pH shifting modifies hemp seed protein structure, cross-linking, and emulsifying properties. *Journal of Agricultural and Food Chemistry, 66*(41), 10827–10834. https://doi.org/10.1021/acs.jafc.8b03901.

Wang, X. S., Tang, C. H., Yang, X. Q., & Gao, W. R. (2008). Characterization, amino acid composition and in vitro digestibility of hemp (*Cannabis sativa* L.) proteins. *Food Chemistry, 107*(1), 11–18. https://doi.org/10.1016/j.foodchem.2007.06.064.

Wang, Q., & Xiong, Y. L. (2019). Processing, nutrition, and functionality of hempseed protein: A review. *Comprehensive Reviews in Food Science and Food Safety, 18*. https://doi.org/10.1111/1541-4337.12450.

Wu, G., Bazer, F. W., Davis, T. A., Kim, S. W., Li, P., Marc Rhoads, J., et al. (2009). Arginine metabolism and nutrition in growth, health and disease. *Amino Acids, 37*(1), 153–168. https://doi.org/10.1007/s00726-008-0210-y.

Xu, Y., Li, J., Zhao, J., Wang, W., Griffin, J., Li, Y., et al. (2021). Hempseed as a nutritious and healthy human food or animal feed source: A review. *International Journal of Food Science and Technology, 56*(2), 530–543. https://doi.org/10.1111/ijfs.14755.

Xu, Y., Zhao, J., Hu, R., Wang, W., Griffin, J., Li, Y., et al. (2021). Effect of genotype on the physicochemical, nutritional, and antioxidant properties of hempseed. *Journal of Agriculture and Food Research, 3*(100119). https://doi.org/10.1016/j.jafr.2021.100119, 100119.

Yan, X., Tang, J., Dos Santos Passos, C., Nurisso, A., Simoes-Pires, C. A., Ji, M., et al. (2015). Characterization of lignanamides from Hemp (*Cannabis sativa* L.) seed and their antioxidant and acetylcholinesterase inhibitory activities. *Journal of Agricultural and Food Chemistry, 63*(49), 10611–10619. https://doi.org/10.1021/acs.jafc.5b05282.

Yin, S. W., Tang, C. H., Cao, J. S., Hu, E. K., Wen, Q. B., & Yang, X. Q. (2008). Effects of limited enzymatic hydrolysis with trypsin on the functional properties of hemp (*Cannabis sativa* L.) protein isolate. *Food Chemistry, 106*(3), 1004–1013. https://doi.org/10.1016/j.foodchem.2007.07.030.

Yu, L. L., Zhou, K. K., & Parry, J. (2005). Antioxidant properties of cold-pressed black caraway, carrot, cranberry, and hemp seed oils. *Food Chemistry, 91*(4), 723–729. https://doi.org/10.1016/j.foodchem.2004.06.044.

Zhao, J., Xu, Y., Wang, W., Griffin, J., Roozeboom, K., & Wang, D. (2020). Bioconversion of industrial hemp biomass for bioethanol production: A review. *Fuel, 281*(118725). https://doi.org/10.1016/j.fuel.2020.118725.

Zhou, Y., Wang, S., Lou, H., & Fan, P. (2018). Chemical constituents of hemp (*Cannabis sativa* L.) seed with potential anti-neuroinflammatory activity. *Phytochemistry Letters, 23*, 57–61. https://doi.org/10.1016/j.phytol.2017.11.013.

5

Industrial hempseed oil and lipids: Processing and properties

Biljana B. Rabrenović[a] and Vesna B. Vujasinović[b]

[a]Faculty of Agriculture, University of Belgrade, Belgrade-Zemun, Serbia [b]Faculty of Sciences, University of Novi Sad, Novi Sad, Serbia

Abbreviations

ADI	acceptable daily intake
ALA	alpha-linolenic acid
CAE	caffeic acid equivalent
CS	Carmagnola Selez
DHA	docosahexaenoic acid
EACP	enzyme-assisted cold pressing
EFA	essential fatty acid
EPA	eicosapentaenoic acid
FAO	Food and Agriculture Organization
GAE	gallic acid equivalent
GLA	γ-linolenic
LA	linoleic fatty acid
LNA	α-linolenic
LOAEL	lowest observed adverse effect level
MAE	microwave-assisted extraction
NOAEL	observable adverse effect level
NS	Novosadska
OSI	oxidative stability index
***p*-AnV**	*para*-anisidine value
PEF	pulsed electric field
PUFA	polyunsaturated fatty acids
PV	peroxide value
SDA	stearidonic acid
SFE	supercritical fluid extraction
THC	tetrahydrocannabinol
TPC	total phenolic content
TSz	Tiborszállási
UAE	ultrasound-assisted extraction
WHO	World Health Organization

5.1 Introduction to hempseed oil

Industrial hemp (*Cannabis sativa* L.), as a versatile crop, has contributed to human development like no other plant species. It has been used as a source of food, oil, medicines, and fiber since ancient times. However, due to the similarities between industrial hemp and the narcotic/medical type of Cannabis, industrial hemp production was prohibited in most countries for years (Rupasinghe, Davis, Kumar, Murray, & Zheljazkov, 2020; Siudem, Wawer, & Paradowska, 2019). But recently, Australia, Canada, United States, and some European countries have legalized the cultivation and consumption of hempseed and hempseed products at a low (< 0.3%) Δ^9-tetrahydrocannabinol-THC level. For that reason, nowadays, there is a growing interest in hemp cultivation for seed (Leonard, Zhang, Ying, & Fang, 2020). Nondrug industrial hemp with such a low THC content has no psychoactive effects.

The production of industrial hempseed has, so far, gained more interest due to the macronutrients and phytochemicals. However, among macro-compounds of seeds, peculiarly, the oil is a main and well-balanced health product. Moreover, hempseed oil has many bioactive components that can aid human health beyond that of basic nutrition. Accordingly, hempseed oil is valued primarily for its nutritional properties and the health benefits associated with it (Rupasinghe et al., 2020).

Hemp oil, valued not only because of its pleasant sensory properties such as taste and smell, is rich in major and minor lipid nutrients. It is especially important to emphasize that essential fatty acid (EFAs) are well represented in hempseed oil, such as linoleic fatty acid (LA, C18:2, ω-6) and alpha-linolenic acid (ALA, C18:3, ω-3) (Dimić, Romanić, & Vujasinović, 2009; Vogl, Mölleken, Lissek-Wolf, Surbök, & Kobert, 2004). Moreover, significant amounts of metabolic products of these fatty acids, such as γ-linolenic (GLA, C18:3, ω-6) and stearidonic acid (SDA, C18:4, ω-3), are found in hempseed oil too (Callaway & Pate, 2009). In particular, the γ-linolenic acid makes hempseed oil nutritionally superior to the other seed oils. It is a fact that most edible vegetable oils have some EFAs, but it is unusual to have all earlier mentioned fatty acids together in some oil. So, no other industrial crop, except hemp, can make this affirmation. Still, it must be noted that hempseed oil is very susceptible to oxidative rancidity because of the high ratio of unsaturated fatty acids.

The significant beneficial health effects of hempseed oil in humans, such as antihypertensive and cholesterol-lowering consequence, atopic dermatitis, wound healing etc., are well documented (Schwab et al., 2006). Furthermore, considering that hempseed oil is used for food preparation, health benefits for the population would be meaningful in preventing heart diseases.

Thus far, little efforts have been focused on investigating the unsaponifiable fraction of hempseed oil. However, the unsaponifiable fraction, which amounts to about 1.5%–2% of the oil, is an appreciable source of various minor compounds, such as tocopherols, phenols, sterols, vitamins etc. (Montserrat-de la Paz, Marín-Aguilar, García-Giménez, & Fernández-Arche, 2014).

Hempseed oil is the most expensive product accounting for about 16% of total hemp production usage. Almost exclusively, the oil is used for human food preparation, only 0.3% for cosmetics and 0.3% for animal feed (Karche & Singh, 2019).

In the following, this chapter will describe the industrial hempseed oil processing, major and minor constituents of hempseed oil, and its oxidative stability.

5.2 Conventional and novel hemp oil seed extraction

Several processes have been developed to extract hempseed oil based on two fundamental principles: cold pressing and solvent (n-hexane) extraction. Recently, various alternative/innovative possibilities of extraction of hempseed oil have been investigated, such as pressured n-propane extraction, ultrasound-assisted extraction, liquid CO_2 extraction, supercritical CO_2 extraction, liquefied methyl ether (DME) extraction, and similar, in order to achieve higher extracted oil yield which consequently means lower residual oil (less than 1%) in a cake. Devi and Khanam (2019) provided a comparative study of different extraction processes for hempseed oil considering physical, chemical, and industrial-scale economic aspects. Because of environmental concerns about solvents used for conventional oil extraction and consumers' preferences for healthier food, cold-pressed oils have received much attention since the late 1970s and are available in markets, especially in Europe, as well as around the world (Radočaj & Dimić, 2013).

The real cold-pressing process is without any heat treatment of raw material before pressing on the screw press. The temperature during the pressing is also very low, and the process deals milder to the beneficial natural components of oils. Cold-pressed oils retain higher amounts of natural antioxidants, providing additional health benefits to consumers in health promotion (Radočaj & Dimić, 2013; Siger, Nogala-Kałucka, & Lampart-Szczapa, 2008). So, mechanical press extraction concerning solvent extraction seems more acceptable for health, safety, and environmental reasons.

Regardless of which procedure is used for oil extraction, the characteristics of the oilseeds are very important, as they affect the extraction process itself. For that reason, the general characteristics and technical-technological characteristics of industrial hempseeds will be briefly discussed later.

5.2.1 General characteristics of industrial hempseeds

Hempseeds ripen gradually, and mature seeds fall from the plant, so the harvest period must be strictly observed. Hemp for seed should be harvested when most of the seeds from the upper (reproductive) part of the plant are mature. The fruit of hemp is not a true seed but an "achene," almost round in shape. The seed is wrapped in a hard shell (pericarp), which mechanically protects it. A reticulate sculpture is clearly visible on the seed coat (Deferne & Pate, 1996). The color of the seeds is not permanent but changes according to the degree of maturity of the seeds, hemp varieties, origin, etc. Immature seeds are greenish, and mature ones can be silvery gray or even dark brown (Dimić, 2005).

The ripened hempseed is an excellent source of dietary oil, approximately 35%, protein 25%, and fiber 27% (Dimić, 2005; House, Neufeld, & Leson, 2010). The oil content in the seeds is affected by hemp variety, agrotechnical measures during cultivation (use of fertilizers), production area (latitude), soil type, precipitation amounts, maturity of seeds, etc. The oil content in the seeds is increased by the cultivation of the hemp from the southern to the northern areas (Deferne & Pate, 1996; Karlović & Berenji, 1996).

Based on the database (Dimić, 2005; Schuster, 1993; Sharp et al., 1986) according to which certain varieties of industrial hemp give a yield of the stem of 14–18 t/ha and the seed of 0.8–1.6 t/ha (oil content 30%–35%), relative productivity of industrial hemp as an oilseed in relation to soybean is 0.7. Thus, compared to traditional oilseeds, industrial hemp lags far behind relative seed productivity and is similar to pumpkin seed. By the same calculation, relative productivity for rapeseed is 1.5, for sesame 1.7, sunflower 2.0, coconut and olive 3.5, and for oil palm, it is 10 (Dimić, 2005).

By Blade, Ampong-Nyarko, and Przybylski (2006), the average seed yield of industrial hemp cultivars grown in the high latitude region of Canada varied between 0.987 and 1.633 t/ha. Nevertheless, despite the unfavorable economic calculation for hemp as an oilseed, the fact is that recently in developed western countries, there is a growing interest in the use of hempseeds and derived oils. Hempseed products are particularly desired by special groups of consumers (vegans, vegetarians, or people suffering from immune deficiencies). Especially, the beneficial effects of hempseed oil need to be studied to improve the profitability of industrial hempseed worldwide.

5.2.2 Physical and technological characteristics of hempseeds

Physical and mechanical properties of hempseeds are significant for several reasons: from the aspect of seed storage, seed processing, the quality of oil, and the efficacy of the oil extraction process. Unfortunately, literature data on the basic physicomechanical and technological characteristics of industrial hempseeds are pretty lacking. Table 5.1 lists unpublished data (from the laboratory of the Department of edible oils—Faculty of Technology, University of Novi Sad, Serbia) of physical and technological properties for different hemp varieties cultivated under the same agroecological conditions on the experimental fields.

The mass of 1000 hempseeds varies significantly, leading to changes in the shape and size of the seed. The seeds are small since the weight of 1000 seeds is 9–26 g (average 16.5 g), by Schluttenhofer and Yuan (2017). The bulk density of seeds is about $0.5 \, kg/dm^3$, but the true density is approximately twice as high as the bulk density. The share of hulls is in the range of 35%–40% and the kernels of 60%–65%. Hempseed varies in shape (from almost spherical to somehow oblong) with a variable diameter of about 3–5 mm (Leonard et al., 2020).

5.2.3 Conventional hemp oil seed extraction—Cold pressing

During the past few decades, a new direction appeared in producing edible oils, favoring gourmet, aromatic, and spicy oils. These oils belong to the group of unrefined edible oils. The difference between unrefined and refined oils is significant, as reflected in the sensory attributes, chemical composition, nutritional value, and oxidative stability of oils. Unrefined edible oils are produced exclusively by mechanical pressing without the use of any subsequent refining step. In no case, the use of an organic solvent for the extraction of oil is not allowed. The idea of cold pressing is not a new one. It is originated with the previous use of hydraulic presses in the last centuries, whereby there was little or no heat generation in the pressing process (Dimić, 2005).

Nowadays, the extraction of oil from hempseed is carried out mainly by a cold-pressing process on small screw presses to preserve the natural quality of the oil. Unfortunately, the mentioned process does not allow an extraction yield equal to that of techniques employing solvents or high temperatures. Still, it has the advantage of minimizing degradative changes in the oil (Crimaldi, Faugnob, Sanninobc, & Arditoa, 2017; Dimić, 2005). Cold-pressed seed oils retain beneficial micro-components of the seeds, including natural antioxidants. As cold-pressed oils are free of any chemical treatments, they become a more interesting alternative

TABLE 5.1 Physical and technological properties of different industrial hempseeds.

Seed characteristics (mean values)	Cultivar/origin							
	Felina France	NS Serbia	TSz Hungary	Benico Poland	Secuieni Romania	CS Italy	Range[a]
Mass of 1000 seeds (g)	16.2	18.7	20.3	18.2	22.2	18.3	13–26
Bulk density (kg/dm^3)	0.57	0.53	0.57	0.54	0.56	0.52	0.49–0.55
True density (kg/dm^3)	1.09	1.07	0.94	0.98	1.01	0.92	–
Share of (%)							
– Hulls	35	40	37	39	35	38	35–40
– Kernels	66	60	63	61	65	62	60–65
Size (mm)							
– Length	4.5	4.1	–	–	–	–	3.3–5.0[b]
– Width	3.5	3.6					2.6–3.9
– Thickness	2.8	2.9					2.1–3.2

NS, Novosadska; TSz, Tiborszállási; CS, Carmagnola Selez.

[a] *Karlović and Berenji (1996).*

[b] *Oomah, Busson, Godfrey, and Drover (2002).*

for conventional practices due to the increased consumers' requirements for natural and safe food products. As novel dietary sources, cold-pressed oils are desired by consumers and food manufacturers to benefit human health through improved nutrition. That was the reason for establishing many small oil processing plants in Central and Southeast European Countries lately, where specialty oils are produced by employing cold-pressing process on small capacity screw presses. Hempseed is a suitable raw material for cold pressing in these plants because of chemical composition and mechanical characteristics of seed such as dimensions, the share of hull, hardness, and fracture characteristics (Berenji, Dimić, & Romanić, 2005; Dimić et al., 2009).

The production of the cold-pressed hempseed oil is a relatively simple process. However, many different factors are crucial to producing top quality oil. The process includes two phases as follows:

(a) preparation of raw material/seeds for oil extraction, and
(b) mechanical extraction of oil.

The oil extraction process has to be adjusted to the characteristics of the raw material to obtain high oil extraction efficacy. Anyway, the most important prerequisite for producing oils and meals (press residues) of premium quality is a high-quality raw material that has not been damaged during transportation, drying, storage, and processing. Fig. 5.1 presents a typical cold-press mill block diagram. The processing of hempseed by cold pressing results in oil and press cake with alimentary use.

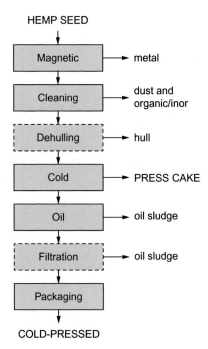

FIG. 5.1   Typical cold-pressed mill block diagram (Dimić & Vujasinović, 2011).

5.2.3.1 Pre-cleaning and storage of seed

The pre-cleaning process means cleaning the bulk material before entering the storage facility. The bulk of hempseed after harvesting contains impurities of organic and inorganic origin. Organic impurities, stalks and leaves, can be present in the mass of seeds in larger quantities. Their presence is especially problematic because the moisture content in organic impurities is always significantly higher than in the seed. Pre-cleaning is mostly performed using a coarse screen to remove organic impurities and stones that could damage the processing equipment.

Hempseed oil for human consumption is ideally produced from fresh, well-cleaned seeds that were air-dried at low temperatures (< 25°C) over several days or weeks. At the time of harvest, the hempseed moisture content is typically 15%–20% (Callaway & Pate, 2009).

Adequate storage is of great importance to preserve the seed quality. The crucial factor for sunflower seed's safe storage and oil quality is its moisture content (Dimić, Premović, Radočaj, Vujasinović, & Takači, 2018). For safe storage of hempseed, the moisture content should be below 10% and the temperature of bulk seed has to be less than 40°C. The seeds should be stored under dry and cool conditions. Furthermore, special care must be taken to ensure that the seed does not support mold growth until the drying (Callaway & Pate, 2009). If the moisture content in the seed is high, it is necessary to dry it immediately, to prevent the increase of acidity in the wet seed during storage. Namely, the limiting factor for the processing of hempseeds by cold pressing is the acidity of the oil in the seed itself. Also, seed damage generally contributes to the increase in oil acidity due to inadequate handling and/ or processing equipment.

Contrary to the oil's acidity, the hempseeds are not susceptible to oxidative deterioration during storage. In addition, the outer hull protects the oil in the seed for an extended period. Thus it is recommended to store the seeds and to process only sufficient amounts of hempseed oil manageable on the market (Matthäus & Brühl, 2008).

5.2.3.2 Seed cleaning

Before pressing, it is necessary to perform fine cleaning of hempseed. The fine cleaning is used to remove metal particles (using a magnet), dust, and other tiny organic particles of the hemp plant. Simple vibrating screens with variable aspiration are mostly used for fine cleaning.

Besides that, very careful cleaning of hempseed before pressing process is necessary for the following particular reason. Hempseeds contain virtually no Δ^9-tetrahydrocannabinol-THC, as the psychoactive substance is not synthesized in the seeds. However, THC contamination results from contact of the seeds with the resin secreted by the epidermal glands on the leaves and floral parts, and also by the failure to sift away all of the bracts (which have the highest concentration of THC of any parts of the plant) that cover the seeds. This result in small levels of THC appearing in hempseed oil made with cleaned seeds. Therefore it is important to underline that there exist special requirements concerning the purity of the industrial hempseed that means proper removing of impurities and other parts of the plant to avoid possible migration of THC to the processed products (Morar et al., 2010; Small & Marcus, 2002).

Petrović, Debeljak, Kezić, and Džidara (2015) concluded that the limit for THC (20 mg/kg) in hempseed oil would be acceptable, as there is no risk of exceeding No-Observed Effect Level (NOEL) and manufacturers would have no problem producing good quality oils by simply separating residues and immature fruits.

Cultivars of *C. sativa* employed for seed production with a low THC level generally contain a high concentration of cannabidiol (CBD), including cannabidiolic acid (CBDA). Citti, Pacchetti, Vandelli, Forni, and Cannazza (2018) developed and validated a highly sensitive and rapid HPLC-UV method for the qualitative and quantitative determination of the main cannabinoids present in commercial hempseed oils. This analysis can provide quantitative data of the cannabinoids content and qualitative information about the purity of the oil. According to these authors, CBDA/CBD ratio can be taken as a marker of cold pressing and good storage conditions.

5.2.3.3 Dehulling

Hempseed is usually processed by pressing as complete with their outer hull, i.e. without dehulling. However, there is a possibility of partial dehulling of the seeds to increase the protein content of the press cake or meal. The hull can be removed by an impact dehuller in conjunction with the aspiration system.

5.2.3.4 Cold pressing

The extraction of oil from hempseed is carried out using a conventional continuous screw press, sometimes called an expeller press. A mechanical screw press extracts oil by applying pressure to the seed by means of a decreasing volume Archimedean screw contained within a drained barrel. The screw elements are generally known as the "worm" and the "cage." The press is fed with a conveyor within the feeder unit. The feeder regulates the flow of seeds into the press without any heat treatment and thereby controls the loading on the main press motor. Oil released along the length of the cage is allowed to drain into the base of the press, where it is collected (Dimić, 2005) (Fig. 5.2).

The friction during the pressing process generates heat, reducing oil quality, but the temperature should be maintained as low as possible to produce so-called cold-pressed oil, or sometimes called "virgin oil." Therefore the screw presses used to extract the oil from the hempseeds must be of such a construction that there is no excessive rise in the temperature of the output oil during the pressing process. Although not specified by regulations, the temperature of the oil leaving the press should not exceed 50°C (Dimić, 2005; Panfilis, Toschi, & Lercker, 1998). That means that during the pressing process, the effective control of temperature has to be carried out. In addition, both the temperature and the pressing efficiency are highly dependent on the moisture content of the seed. Higher moisture content provides lower press temperatures and higher residual oil content in the press cake and vice versa. A permanent control of the seeds' moisture during pressing is of particular importance because the moisture directly affects both the capacity of the press and the content of residual oil in the press cake, as well as the quality of obtained oil. The seed dryer can be used to regulate the moisture content of the seed for pressing (Dimić, 2005; Fils, 2000). Parameters of hempseeds pressing of the variety Novosadska on the screw press Reinartz (capacity about 60 kg/h) (Berenji et al., 2005), variety Fedora 17 on the screw press SPU 20 (capacity 25 kg/h) (Jozinović et al., 2017), and Fedora cv. (Siano et al., 2018) are presented in Table 5.2.

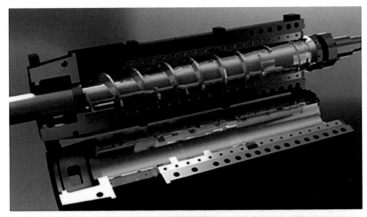

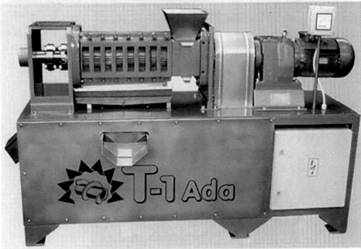

FIG. 5.2 The layout of worm assembly and typical small-scale screw press. *Courtesy of T-1 Ada Company, Serbia.*

The yield of hempseed oil from screw press is related to the characteristics of the press and input process characteristics. The yield is mostly influenced by screw rotational speed compared to extraction temperature and seed preheating. For example, by pressing of hempseed on screw press (capacity of about 20 kg/h) under the following conditions: nozzle size 8 mm, temperature of extraction 70°C, rotational speed 32 rpm, at no thermal treatment of seeds before pressing, Crimaldi et al. (2017) achieved oil recovery/yield of 73%. Morar et al. (2010) noticed that the pressing rate efficiency on small capacity screw press (25 kg hempseed/h) was dependent on the diameter of the press nozzle, the seed characteristics, and the moisture content. The highest capacity of hempseed oil was obtained using press nozzle diameters of 8 mm (moisture of seeds 6.32%) and 10 mm (moisture of seeds 8.18%) at fixed rotation speed of 60 rpm. In general, by cold-pressing process, 60%–80% of oil can be extracted from the seed, depending on the settings of the screw press (Matthäus & Brühl, 2008).

TABLE 5.2   Pressing efficiency parameters of hempseed on a screw press.

Parameter	Novosadska		Fedora 17		Fedora cv.	
	Seed	Cake	Seed	Cake	Seed	Cake
Moisture content (%)	9.73	10.75	6.61	7.57	7.3	7.9
Oil content (%)	29.30	11.28	31.69	6.72	24.5	13.6
Efficiency of the pressing (%)	61.5	n/a	78.7	n/a	55.5	n/a

At small capacities, high-pressure pressing, either full or double pressing, is a possible alternative for hempseed oil extraction. Depending on the seed characteristics and press performances, residual oil content in press cake of 12% and 5%, respectively, may be expected. Cold pressing of hempseed in milder conditions, i.e. lower pressures, may also be used, particularly when special quality oil is required. However, in that case, the oil extraction yield is significantly lower. Although hempseed oil from the cold press has a high quality, a high amount of residual oil remains in the cake. Today, there are many screw press manufacturers around the world that are divided into three categories: small local companies (operates in one country), bigger regional manufacturers (operates in several countries), and international companies that operate on a global scale such as De Smet, Krupp, Simon Rosedown, etc. All these manufacturers also produce presses of small or medium capacity that can be used for processing hempseed. Most contemporary producers of hempseed oil are either individual operators with a small press or small start-up enterprises with one, two, or several presses (Fils, 2000).

5.2.3.5 Oil clarification

Tiny solid particles (hull and endosperm fragments) and colloidally dissolved mucous substances (phospholipids, waxes, and minor amounts of other constituents) inevitably accompany the hempseed oil produced in a screw press. These matters have to be removed to get clear oil which better meets consumers' liking. Oil clarification usually involves one- or two-stage process, sedimentation/decantation and filtration/separation or only filtration. Decantation is cheaper, but more time-consuming process. The oil is allowed to settle for at least 1 or 2 weeks and is then decanted into smaller containers for retail sales.

Most solid particles can be removed simply by screening the oil over either a static or a vibratory screen. The screened oil is then finally clarified using a filter—usually a hermetically sealed filter with stainless steel leaves, in which the sludge from oil acts as a pre-coat through which the oil is filtered. Today, there are already filters of smaller or larger capacity. Since high-capacity presses are used for higher production, it is much more practical to perform fine filtration directly in bulk containers instead of waiting for gravity deposition (Callaway & Pate, 2009; Fils, 2000). There is another far more efficient way to separate mechanical impurities from oil by using a separator. In this case, the centrifugal force greatly accelerates the deposition rate, i.e. the speed of separation of solid particles and other impurities from the oil. Thus separators achieve the fastest and most efficient separation of sludge; however, due to their large capacity and high prices, they are not suitable for small capacity plants for cold-pressed oil production. On the other hand, the fine sediment, oil sludge, from freshly pressed hempseed oil is of high nutritive value, and it can be used as a nut-butter spread or in other human food product (Dimić, 2005; Pojić, Hadnađev, & Hadnađev-Dapčević, 2014).

5.2.3.6 Press cake—Nutritive value

After mechanical extraction of the oil from the seeds, the oil cake, sometimes called press cake, meal, or flour, remains. As far as the production of the cold-pressed hempseed oil is concerned, a certain amount of oil remains in the cake, which cannot be completely recovered by using mechanical pressing. By literature data, the residual oil content in the hemp cake after the screw-pressing process is approximately 7%–13% (Jozinović et al., 2017; Leonard et al., 2020; Siano et al., 2018). Solvent extraction of oil from the press cake using *n*-hexane is an effective method due to low residual oil (< 1%) in a meal. However, solvent-extracted oil, as such, is not suitable for human consumption and must be refined. Also, the defatted meal after solvent extraction is not suitable for human nutrition as residual solvent (mainly *n*-hexane) can contaminate the food product (Dimić, 2005).

Hemp press cake after cold pressing can be considered a good source of protein, dietary fiber, micronutrients (vitamins, phenolics, sterols), and minerals, with a higher relative content of certain macro- and micronutrients compared to seeds due to oil removal (Callaway & Pate, 2009; Jozinović et al., 2017; Radočaj, Dimić, & Tsao, 2014).

However, the relative content of some macro-compounds differs significantly in cake compared to seeds. After pressing, the relative enrichment of the cake with proteins, carbohydrates, and ash is obvious due to reducing the oil content (Table 5.3).

The most valuable compound of hempseed meal are proteins ranging from 30% to 50% in the dry matter. The protein content mostly depends on the variety of hemp, oil extraction method (cold pressing or solvent extraction), and process efficiency, i.e. processing parameters (Malomo, He, & Aluko, 2014).

TABLE 5.3 Proximate composition (%) of various hempseed press cake.

Content of	House et al. (2010)	Folegatti et al. (2014)	Pojić, Mišan, et al. (2014)	Siano et al. (2018)
Moisture	4.9	9.3	7.88	7.9
Oil	10.2	10.1	11.8	13.6
Crude protein (% N×6.25)	40.7	29.4	27.9	30.7
Total ash	6.7	7.1	6.74	6.2
Crude fiber	30.5	30.8	17.3	–
Total carbohydrates	37.5[a]	13.3[b]	1.49[c]	41.6

[a] *Calculated as the difference between all components.*
[b] *In the calculation is only fiber.*
[c] *Total sugar content.*

Hempseed proteins are of high biological value containing essential amino acids and highly digestible protein, whereby the removal of the hull fraction improves their digestibility. It is reported that the digestibility scores are higher for protein from dehulled hempseed (value of 61) than the protein from whole hempseed (value of 51) and hempseed meal (value of 48). Concerning the digestibility of protein, it must be noted that heat-damaged proteins have lower digestibility. This is one of the reasons that under the conditions of cold pressing special care must be taken not to allow an excessive increase in the temperature (House et al., 2010).

Hempseed meal contains respectable amounts of vitamins and minerals. Recently, Siano et al. (2018) published a comparative study of chemical and biochemical characteristics of seeds, oil, and flour of the Fedora genotype hemp cultivar, which is among the most popular hemp variety. The dominant macro-elements in flour were K (5064.45 mg/kg) and Mg (2310.54 mg/kg), followed by Ca (1907.20 mg/kg) and Na (90.55 mg/kg), which occurred at a lower concentration. Among the micro-elements, Fe (152.47 μg/kg) showed the highest concentrations followed by Mn (94.71 μg/kg), Zn (54.68 μg/kg), and Cu (11.94 μg/kg). Other micro-elements occurred in minor amounts. All these macro- and micro-elements had a higher concentration in flour than the seeds. Nevertheless, it should be said that these minerals in hempseeds and flour cover rather wide ranges (Mihoc, Pop, Alexa, & Radulov, 2012).

The phytochemical character of hempseed press cake, in terms of phytosterol profile, phenolic and tocopherol content, and antioxidant activity, is also of great interest. According to Siano et al. (2018), hemp flour contained 744 mg GAE/kg as total polyphenols. Among the phytosterols, β-sitosterol, campesterol, Δ^5-avenasterol, and stigmasterol were the most abundant. The DPPH free radical scavenging activity (expressed as % inhibition) of flour was 46.8% inhibition, and it was comparable to seed (51.5% inhibition), while it was much lower for oil (8.2% inhibition). Tocopherols, as the fraction of unsaponifiable, were also present in the oil of the flour. The most abundant isomer was γ-tocopherol, followed by α-tocopherol.

5.2.4 Novel hempseed oil extraction methods

The increase in the population and the need for larger quantities of food have led to the accelerated development of technological processes in all branches of industry, including the production of vegetable oil and fats. However, the increase in productivity and production volume has led to the inevitable negative impact on the environment due to the increased use of organic solvents to extract vegetable oils. However, the 21st century brought new aspirations and habits to consumers. Consumers are now opting more for products obtained with green technologies, which require minimal processing, less energy for production and transport. When it comes to oilseed processing and oil production, they include, in addition to cold pressing, some specific green extraction processes (Vavpot, Williams, & Williams, 2014).

In order to minimize the use of solvent and energy, recently, methods such as supercritical fluid extraction, ultrasound extraction, subcritical water extraction, controlled pressure drop process, pulsed electric field, and microwave extraction are employed for the extraction of hempseed oil.

5.2.4.1 Supercritical fluid extraction

The principle of supercritical fluid extraction (SFE) is the use of safe and capable solvents (fluids) in their critical state for efficient extraction. With good adjustment of working parameters, it is possible to extract more than 90% oil from the seeds by this method (Tomita et al., 2013).

Various fluids can be used for this purpose, such as ethene, water, methanol, carbon dioxide, nitrous oxide, sulfur hexafluoride, *n*-butene, and *n*-propane. Among them, carbon dioxide (CO_2) stands out as a non-toxic, non-flammable, relatively chemically inert, recyclable, and inexpensive solvent that is easy to remove from the extracts upon returning to room temperature and pressure conditions (Porto, Voinovich, Decorti, & Natolino, 2012).

Aladić et al. (2014) demonstrated the utilization of supercritical extraction with CO_2 as a subsequent process to cold pressing, allowing the complete extraction of the residual hemp oil from the pressed cake (only 0.39% of the oil was present after processing). In further research, Aladić et al. (2015) concluded that supercritical CO_2 extraction was more effective procedure for extraction of hempseed oil than Soxhlet extraction using *n*-hexane. In addition, it was shown that the SFE technique has many advantages over traditional methods, especially in the preservation of thermosensitive compounds using low extraction temperatures.

Grijó, Piva, Osorio, and Cardozo-Filho (2019) evaluated the various physicochemical properties of hempseed oil obtained using supercritical carbon dioxide (scCO_2), pressurized *n*-propane, and conventional technique (mixture of hexane and isopropanol at atmospheric conditions). The extraction yields and composition of hempseed oils obtained using supercritical CO_2 were similar to previously reported. However, the tocopherol content in oils obtained was higher than previously reported. The cost-effectiveness of using pressurized *n*-propane over supercritical CO_2 was demonstrated due to lower extraction pressures and lower solvent amounts (*n*-propane) required. Moreover, product obtained with pressurized *n*-propane contained higher concentrations of antioxidants and lower acidity and humidity indexes.

5.2.4.2 Ultrasound-assisted extraction

The application of ultrasound to increase the extraction yield began long ago, in the 1950s, in the laboratory. Today, however, ultrasound-assisted extraction (UAE) is a well-established technique commonly used to facilitate oilseed extraction. When ultrasound waves pass through a medium, they involve expansion and compression cycles. Expansion pulls molecules apart and compression pushes them together. These mechanisms create bubbles in a liquid which can grow and finally collapse. Such phenomenon is called cavitation. During cavitation, bubbles collapse near the surface of the cell wall and produce temperatures up to 5000 K and pressure up to 1000 atm. Cavitation contributes to the destruction of cell walls, which enables easier oil release or better mass transfer (Kate et al., 2016). Lin et al. (2012) demonstrated the advantages of ultrasonic extraction (power output: 200 W; exposure time: 25 min; acting on-off ratio: 20:20 (s/s); and solvent-to-solid ratio: 7:1 (v/w)) of hempseed oil over solvent extraction reflected in a shorter operating time, lower operating temperature, and less solvent consumption. These extraction conditions contributed to preserved bioactive components contained in the oil. Ultrasound without any solvent assistance was effectively applied as a hempseed pre-treatment prior to extraction with supercritical CO_2. The results showed that the best oil yield was achieved by pre-treatment of hempseeds with ultrasound for 10 min (Porto, Natolino, & Decorti, 2015).

In their research, Rezvankhah, Emam-Djomeh, Safari, Askari, and Salami (2018) compared the efficiency of UAE with Soxhlet extraction for hempseed oil. Better oil yield was obtained by Soxhlet extraction. Still, the processing time was shorter in UAE (the optimum point was at 91 W and 10 min compared to 8 h for Soxhlet extraction). The oil obtained by this process had better oxidative stability and stronger antioxidant activity. Esmaeilzadeh Kenari and Dehghan (2020) examined using RSM ultrasound-assisted solvent extraction of hemp (*C. sativa* L.) seed oil. They varied hexane and isopropanol ratio, time, and temperature of extraction. The optimal extraction conditions were obtained with a hexane-to-isopropanol ratio 3:2, a temperature of 40.26°C, and an ultrasonication time of 54.40 min.

5.2.4.3 Microwave-assisted extraction

The principle of microwave-assisted extraction (MAE) is based on the penetration of microwaves into the plant cell, their absorption and heat production, causing the expansion of the cells and their destruction, ensuring more effortless mass transfer.

Differences in fatty acid composition, physicochemical and thermal characteristics, and antioxidant activity of hempseed oil obtained by MAE and Soxhlet method were researched by Rezvankhah, Emam-Djomeh, Safari, Askari, and Salami (2019). As in the case of the UAE (Rezvankhah et al., 2018), which was examined by the same authors, better oil extraction was achieved by the Soxhlet method (37.93% w/w), which was the only advantage over the MAE which yielded 33.91% w/w of oil. The oil obtained by the MAE process, conducted at 450 W for 7.19 min, was characterized by higher content of tocopherols, better oxidative stability, and better thermal characteristics. The oil extraction process did not affect the fatty acid composition. Combining MAE and pulsed electric field (PEF), Teh, Niven, Bekhit, Carne, and Birch (2014) tried to obtain phenolic substances from the defatted hemp seed cake as much as possible. The applied methodology proved to be very successful, and results suggested that microwave processing and PEF can be integrated into processing defatted hemp seed cake to enhance polyphenol extraction (Teh et al., 2014).

5.2.4.4 Enzyme-assisted extraction

In order to improve oil yield by extraction, specific food-grade enzymes could be utilized, able to hydrolyze the structural polysaccharides from cell walls to enhance the release of the oil, which results in a higher oil yield. Their application is limited because they are very sensitive and specific requiring a tight control of pH and temperature.

Latif and Anwar (2009) examined the application of enzyme-assisted cold pressing (EACP) on the physicochemical characteristics of hempseed oil by utilization of single and multi-enzyme preparations comprising protease, cellulase, xylanase, beta glucanase, phytase, and alpha amylase. The results of this analysis, compared to the control, showed that the addition of enzymes during cold pressing of hempseed resulted in a significantly higher oil yield without a negative impact on oil quality. The enzymatic pretreatment influenced higher color intensity and tocopherol content, as well as improved oxidative stability and sensory property scores of the cold-pressed oils in comparison to non-treated seeds.

5.3 Major and minor constituents of hempseed oil

Hemp oil is produced mainly from whole hempseed and has a unique chemical composition. Like other vegetable oils, hempseed oil in general consists of major components, i.e. triacylglycerols (TAG) (approximately 98%) and non-TAG or unsaponifiable fraction (approximately up to 2%) (Leonard et al., 2020). Since hempseed oil is mainly produced by a cold-pressing process, in which neither high temperature nor chemical treatment is used, various minor seed components are extracted into pressed oil. Endogenous minor components of the unsaponifiable fraction present in hemp oil are phytosterols, polyphenols, tocopherols, pigments, vitamins, and dietary minerals. Some of these minor components have a protective role (e.g., tocopherols and polyphenols), protecting the oil from oxidation and providing health benefits to humans. On the contrary, some other minor components are undesirable as they are able to deteriorate the oil quality by affecting sensory attributes, nutritional value, or oxidative stability.

Lately, the potential health benefits of hempseed oil have become of particular interest. Hempseed oil is mostly used fresh, without any thermal treatment, as a salad oil rather than cooking or frying oil due to its low smoke point of 165°C (Xu et al., 2021). From the aspect of direct consumption of this type of oil, the sensory quality of the oil is of great importance for consumers.

5.3.1 Sensory quality of hempseed oil

The sensory quality of hempseed oil, which includes color, brightness, smell, flavor, and aroma, is influenced by several factors, primarily the production process (cold-pressed or refined oil) and the age of the oil. In this sense, hempseed oils' sensory attributes are as follows (Table 5.4) (Callaway & Pate, 2009).

Cold-pressed hempseed oil is greenish, sometimes even extremely dark green in the case of a higher content of chlorophyll (Teh & Birch, 2013). According to Spano et al. (2020), some commercial cold-pressed hempseed oils were very dark, but most of the investigated commercial samples were very light in color. Concerning the reflectance profiles of other commercial oils, hempseed oils showed a peculiar curve trend that could represent a fingerprint for this product.

Some producers roast hempseeds before pressing in order to achieve a special aroma of oil and meet the consumers' preferences. However, as far as "hot pressing on hydraulic presses" is used for oil extraction, this method can provide some special oil flavor, but also contributes to the formation of some toxic compounds such as benzo-a-pyrene and 3-MCPD (3-monochloropropanediol) in oil (Xu et al., 2021).

TABLE 5.4 Sensory attributes of hempseed oil.

Type of oil	Sensory attributes
– Fresh cold-pressed oil from the seed of excellent quality	Clear, bright, dark green in color; fresh, nutty taste and smell; a delicious combination of citrus, mint, and pepper flavors
– Old (degraded) cold-pressed oil	Clear, olive-green to yellow in color; fishy, paint smell; without delicate flavor; bitter aftertaste; unwanted taste and smell that are reminiscent of jute rope or burlap sacks
– Refined oil	Clear, colorless to light yellow in color; odorless to paint smell

TABLE 5.5　Fatty acid composition (%) of different[a] hempseed oils.

Fatty acid	1	2	3	4	5	6	7	8
C 16:0	7.32	6.63	5	7.15	5–7	6.21	6.1	5.37
C 18:0	2.97	3.30	2	2.73	1–2	2.40	2.6	1.56
C 18:1	14.55	13.0	9	12.75	8–13	10.64	12.5	11.51
C 18:2, ω-6	55.50	54.8	56	56.08	52–62	55.21	55.2	59.16
C 18:3, ω-3	14.35	18.4	22	14.89	12–23	19.08	18.9	17.96
C 18:3, ω-6	0.80	2.76	4	3.03	3–4	4.62	2.2	3.48
C 20:0	0.87	1.10	–	0.89	0.39–0.79	0.93	0.7	0.18
C 20:1	–	–	–	0.26	0.51	0.92	0.4	0.80
C 20:2	–	–	–	1.03	0.00	–	–	–
C 22:0	–	–	–	0.20	–	0.08	–	–
Σ SAFA	12.18	11.0	–	10.97	–	10.01	9.9	7.74
Σ MUFA	14.55	13.0	–	13.01	–	11.96	13.1	12.31
Σ PUFA	70.67	76.0	84	75.03	–	78.91	77.0	80.60
ω-6/ω-3	3.87	3.13	2.50	3.97	–	3.13	3.04	3.29

[a] 1—Novosadska cv (Dimić et al., 2009); 2—Helena cv (Pojić, Mišan, et al., 2014); 3—Finola cv (Callaway & Pate, 2009); 4—Fedora cv grown in Italy (Siano et al., 2018); 5—Fedora 19 cv grown in Canada (Leizer, Ribnicky, Poulev, Dushenkov, & Raskin, 2000); 6—commercial hemp oil found in the EU market (Radočaj & Dimić, 2013); 7—two oils from a local grocery store and two from manufacturers of cold-pressed oils in Poland (Prescha, Grajzer, Dedyk, & Grajeta, 2014); 8—hempseed oil extracted by supercritical carbon dioxide (Xu et al., 2021). SAFA, saturated fatty acids; MUFA, monounsaturated fatty acids, PUFA, polyunsaturated fatty acids.

5.3.2　Major constituents of hempseed oil—Fatty acids

The largest share in the composition of edible vegetable oils consists of fatty acids. The hempseed oil contains a variety of fatty acids (Table 5.5), among which the essential unsaturated linoleic (LA 18:2, cis-9,12, ω-6) and α-linolenic (LNA 18:3, cis-9,12,15, ω-3) acids are dominant. On the other hand, the oil is low in saturated fatty acids, mainly palmitic and stearic acid.

There is no doubt that the quality of oil is determined by its fatty acid composition. Small and Marcus (2002) emphasized that hempseed oil is of high nutritional quality since it contains high amounts of unsaturated fatty acids, primarily oleic acid (10%–16%), linoleic acid (50%–60%), α-linolenic acid (20%–25%), and γ-linolenic acid (2%–5%) (GLA C18:3, ω-6). Essential fatty acids serve as building materials for cellular structures (i.e., cellular membrane) and precursors for biosynthesis of the body's regulatory eicosanoids. Eicosanoids are short-lived, hormone-like substances that play an important role in controlling vital processes (Matthäus & Brühl, 2008). These compounds ultimately become prostaglandins and thromboxane, which affect blood clotting, inflammation response, and immunoregulation (Abedi & Sahari, 2014). As the human body cannot synthesize essential fatty acids, therefore they must be supplied by dietary sources.

Linoleic acid is a ω-6 fatty acid, and it means that the first of its two double bonds are positioned at the sixth carbon atom counted from the methyl end of the fatty acid molecule. Once in the body, various desaturase and elongase enzymes set to work, creating downstream metabolites: γ-linolenic, dihomo-γ-linolenic, and arachidonic acid, the major members of the ω-6 family. Alpha-linolenic acid is a ω-3 fatty acid with a double bond positioned at the third carbon atom from the methyl end. It is the precursor for two additional ω-3 long-chain polyunsaturated fatty acids: eicosapentaenoic acid (EPA) (C20:5, ω-3) and docosahexaenoic acid (DHA) (C22:6, ω-3) (Fig. 5.3) (Dimić, 2005; Kapoor & Nair, 2005; Leizer et al., 2000; Newton, 1996; Watkins, 2004).

The presence of certain amounts of γ-linolenic acid (C18:3, ω-6) and stearidonic acid (C18:4, ω-3) makes hemp oil almost unique. Both fatty acids in the human body are precursors for synthesizing very-long-chain fatty acids and for eicosanoids. According to literature data, γ-linolenic acid in hempseed oil varies in wide range, from 0.34% to 6.8%, whereas stearidonic acid from 0.3% to 2.5%. It should be noted that there is considerable variance regarding these fatty acids between northern and southern varieties of hempseed. Origin of the seed and

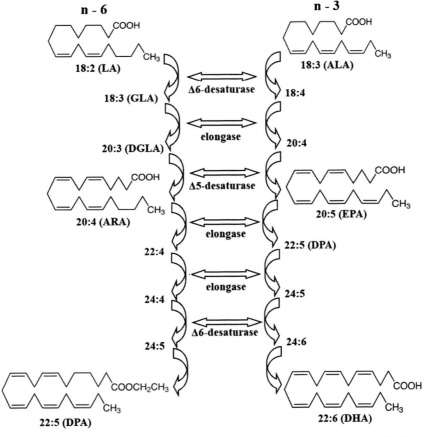

FIG. 5.3 The metabolic pathway for linoleic and α-linolenic acid (Rodriguez-Leyva & Pierce, 2010).

maturity of hemp fruits also influence the fatty acid composition (Callaway & Pate, 2009; Mölleken & Theimer, 1997).

The ratio of linoleic to α-linolenic acids in the oil, i.e. ω-6:ω-3 ratio, is important as it is the key factor for the balanced synthesis of eicosanoids in the body. The ratio of ω-6:ω-3 essential fatty acids in hempseed oil is almost an ideal value approaching 4:1. Although the essential fatty acids are considered the basic nutritional components of daily nutrition, their efficacy is dependent upon the ratio of ω-6:ω-3 and the condition being treated. It has been established that only lower (ω-6:ω-3) ratios between 2.5:1 and 5:1 are beneficial, while a daily intake of 2.5:1 has been proven to act beneficially in cases of colorectal cancer, 2–3:1 on rheumatoid arthritis, and 5:1 on asthma. In addition to their reputed beneficial effects, ω-3 fatty acids have one notable adverse side effect; they stimulate oxidation which necessitates the regular use of an antioxidant, commonly vitamin E (Gogus & Smith, 2010). Radočaj and Dimić (2013) reported the ratio of essential fatty acids ω-6:ω-3 in the following oils: rapeseed 2.2:1, hemp 2.9:1, and walnut 4.5:1, being in the range recommended by the nutritional guidelines (Dubois, Breton, Linder, Fanni, & Parmentier, 2007).

In Table 5.6, the average content of fatty acids determined in cold-pressed hempseed oil samples is compared with the literature data for a fatty acid profile of different unrefined oils. According to the results, hempseed oil has an ideal fatty acid composition suitable for nutrition claims.

A number of studies have well documented the importance of fatty acid composition, particularly polyunsaturated fatty acids in oils, and their nutritional value for human health. Several nutritional disorders resulting from low essential fatty acid levels in the human diet are treated with dietary supplements containing LA, LNA, and/or GLA. Diseases such as tuberculosis and AIDS-related wasting syndromes, skin conditions (psoriasis, atopic eczema and mastalgia), stress, hypertension, and diabetes can be associated with low GLA levels

TABLE 5.6　Fatty acid composition (%) of different unrefined edible oils.

Oil	Essential fatty acids content	LA	LNA	GLA	Oleic fatty acid	Saturated fatty acids, C 16:0 and C 18:0	Ratio ω-6:ω-3
Hempseed oil[a]	68.20	54.25	13.95	1.64	15.07	10.18	3.9:1
Hemp oil[b]	80	55–60	18–23	< 2	12	8	3–4:1
Flax[b]	72	14	58	–	19	9	1:4
Sunflower[b]	65	65	Traces	–	23	12	65:1
Canola[b]	37	30	7	–	54	7	4.3:1
Olive[b]	9	8.5	0.5	–	75	16	18:1
Evening primrose[b]	70.83	70.83	–	8.83	12.55	7.89	8.02:1

[a] *The average values of analyzed oil samples (Dimić, Romanić, Berenji, & Bodroža-Solarov, 2006; Vujanić, 2005).*
[b] *By literature data (Dimić, Romanić, Tešanović, & Dimić, 2005; Mölleken, Mothes, & Dudek, 2000).*

TABLE 5.7 Tocopherol contents (mg/100 g) of hempseed oils.

| Tested sample | Tocopherols | | | | | References |
	α	β	γ	δ	Total (%-a ratio of γ)	
n-Hexane extracted oil from cultivar Fasamo	3.4	0.6	73.3	2.5	79.7 (91.6)	Oomah et al. (2002)
Commercial cold-pressed oil	3.8–5.2	1.0–2.2	73.7–79.1	1.6–2.5	81.9–86.8 (89.9–91.1)	Oomah et al. (2002)
Average values of oils from 51 genotypes (2000 and 2001 data pooled)	1.82	0.16	21.68	1.2	24.86 (87.2)	Kriese et al. (2004)
Commercial cold-pressed oil from EU market	7.87	–	81.74 (β + γ)	–	91.2	Radočaj and Dimić (2013)
Refined oil	3.02	0.81	73.38	2.87	80.28 (91.4)	Montserrat-de la Paz et al. (2014)
Different commercial cold-pressed oils	–	–	–	–	3.47–13.25	Izzo et al. (2020)

and can be treated with EFAs. GLA acts on the circulatory and nervous systems and assists in the repair of damaged myelin sheath tissue surrounding nerves. Its beneficial action in the treatment of diabetes nerve deadening has also been proven, which, if left untreated, results in gangrene of the fingers and toes. A wide range of EFAs and GLA experimental uses include the treatment of cardiovascular, psychiatric, and immunological disorders (Deferne & Pate, 1996; Rodriguez-Leyva & Pierce, 2010; Small & Marcus, 2002; Yu, Zhou, & Parry, 2005), and it is effective against rheumatoid arthritis (Deluca, Rothman, & Zurier, 1995), some cancers, acute respiratory distress syndrome, premenstrual syndrome (Kapoor & Nair, 2005), etc. Numerous studies are still being conducted on animals or in carefully controlled clinical trials on human subjects in order to examine the health benefit effects of these fatty acids.

On the other hand, oil with high content of dominant linoleic fatty acid is susceptible to oxidation, leading to an increased risk of atherosclerosis, tumor development, and an essential role in aging. Thus, nowadays, it is recommended to increase the uptake of monounsaturated oleic acid. Therefore hempseed oil should be consumed moderately in the diet (Matthäus & Brühl, 2008). In an ideal diet, the daily consumption of three to five tablespoon of the hempseed oil is recommended (Leizer et al., 2000).

5.3.3 Minor constituents of hempseed oil

Many well-known minor bioactive compounds are present in hempseed oil, including tocopherols, polyphenols, phytosterols, and carotenoids that contribute to the nutritional value of the oil. Lately, these natural compounds of edible oils are increasingly valued by consumers.

5.3.3.1 Tocopherols

Tocopherols are very important minor compounds of edible oils because they act as natural antioxidants and protect the oil from oxidative deterioration, slowing down lipid peroxidation. In addition, they also act as vitamin E in human nutrition. Four different isomers, known as α-, β-, γ-, and δ-, of tocopherols are found in natural vegetable oils. The biological, i.e. vitamin E, in vivo activity of tocopherols decreases in the order from α- to δ-isomers, while their antioxidant activity in vitro is opposite (Bjelica, Dimić, & Vujasinović, 2018; Mag, Mag, & Reichert, 2002).

Regarding hempseed oil, according to the published data (Table 5.7), the content of total tocopherols, as well as the share of individual tocopherol isomers, differs somewhat. Total tocopherols content is in the range of about 3.5 mg/100 g to 91.2 mg/100 g. Still, all the data indicate that the major tocopherol in hempseed oil is the γ-isomer which amounts to about 90% of the total tocopherols.

The ratio of the tocopherol isomers α: β: γ: δ in hempseed oil by Oomah et al. (2002) was 5:2:90:3, while the ratio of 7:1:87:5 was found by Kriese et al. (2004). The high share of γ-tocopherol in the oil protects unsaturated fatty acids from oxidation in vitro, positively affecting the oil storage. Elmadfa and Park (1999) reported that the combination of high levels of γ- and α-tocopherol isomers might be associated with the protection against DNA damage and reduce the risk of cancer in vivo. The wide range of tocopherol content in hempseed oil is undoubtedly influenced by several seed-related factors (varieties, agronomic conditions, extraction methods, and storage conditions of seeds) (Liang, Aachary, & Thiyam-Holländer, 2015; Xu et al., 2021). However, loss of tocopherols content in the oil is also attributed to the storage time, oxygen exposure, and oil temperature.

5.3.3.2 Phenolic compounds

Research on the total phenolic content (TPC) and phenolic profile of hempseed oil is rare, probably due to its lower concentration than that in whole seed, hull, or cake. However, the author's opinion is still unique that the polyphenol levels in hempseed fractions are in the following order: hull > cake > oil. Accordingly, in the study with Fedora cultivar, Siano et al. (2018) found total phenolic content in whole hempseed, flour, and oil at 767 ± 41 mg gallic acid equivalent (GAE)/kg, 744 ± 29 mg GAE/kg, and 21 ± 5 mg GAE/kg, respectively. When

TABLE 5.8   Main phytosterol contents (mg/kg) of hempseed oil.

Main phytosterols	Montserrat-de la Paz et al. (2014)[a]	Matthäus and Brühl (2008)[b]	
	Average content	Average content	Range
Campesterol	505	709	257–1001
Stigmasterol	100	133	97–181
β-Sitosterol	1905	3191	2704–4434
Δ5-Avenasterol	142	336	209–572
Total sterols	2793	4724	3922–6719

[a] *Refined hempseed oil.*
[b] *Virgin hempseed oil from 10 different cultivars harvested in 2000.*

measured as caffeic acid equivalents (CAEs), hempseed oil showed 2.45 mg CAEs/100 g total phenolic content. This value, as with pumpkin oil (2.46 mg CAE/100 g), was the highest compared to other types of oil (soybean, sunflower, rapeseed, corn, grapeseed, flax, and rice bran) (Siger et al., 2008). The content of total phenols as bioactive components in different commercial hempseed oils was in a wide range of 22.1–160.8 mg GAE/g (Izzo et al., 2020).

Recently, Gaca, Kludská, Hradecký, Hajslova, and Jeleń (2021) confirmed a TPC of 51.41 ± 7.07 mg/kg in commercial cold-pressed hempseed oil (based on the phloroglucinol absorbance value). However, after 10 days of storage at 60°C, the TPC slightly decreased and amounted to 49.06 ± 4.16 mg/kg.

Teh and Birch (2013) reported that hempseed oil had the highest total phenolic acids (natural hydrophilic antioxidants) content of 188.23 mg GAE/kg compared to flaxseed and canola oil. Also, hempseed oil had the highest flavonoid (secondary metabolites) content of 19.50 mg/100 g (as luteolin equivalent) among these three oils.

The nutritional value of oil is not only affected by the TPC but also by the composition of phenolic compounds. By examining the composition of phenolic compounds, Siger et al. (2008) found that the dominant phenolic acids in hempseed oil were *p*-hydroxybenzoic acid (6.0 mg/100 g), followed by sinapic acid (3.0 mg/100 g), vanillic acid (2.0 mg/100 g), *p*-coumaric acid (2.0 mg/100 g), and ferulic acid (1.0 mg/100 g). The phenolic profile of hempseed oil detected by Siger et al. (2008) was mainly in accordance with that of Pojić, Mišan, et al. (2014) found in hemp meal fractions, although in much lower concentrations in the oil. As for the phenolic profile, Finola hempseed oil contained a significant amount of phenolic compound, of which 2780.4 mg of total flavonoids expressed as quercetin equivalent/100 g of fresh weight (Smeriglio et al., 2016).

5.3.3.3 Phytosterols

Phytosterols are the most abundant group of unsaponifiable lipids fraction in vegetable oils, and they occur as free alcohols or esters of fatty acids. Investigations about phytosterols revealed that hempseed oil contains from 2.8 g/kg (Montserrat-de la Paz et al., 2014) to 3.9 and 6.7 g/kg of total phytosterols (Matthäus & Brühl, 2008). The main component of phytosterol composition is β-sitosterol, corresponding to about 70% of the total phytosterol content (Table 5.8). Brassicasterol was not detected in any published investigations.

TABLE 5.9 Initial peroxide value and induction period of cold-pressed hempseed oils.

	Novosadska	Secuieni	Beniko	Felina 34	Futura 75	Tiborszállási	Carmagnola S.
PV—initial (meqO$_2$/kg)	12.84	12.86	9.02	8.52	8.48	8.20	10.40
Induction period[a] (h)	6.8	6.4	6.4	7.0	7.6	6.8	6.6
Schaal oven test PV[b]	73.46	80.78	64.02	59.38	59.8	60.48	70.5

[a] *Rancimat test at 100°C, airflow 20 L/h.*
[b] *PV (meqO$_2$/kg) at 60°C after 96 h.*

Siano et al. (2018) found also β-sitosterol, campesterol, Δ5-avenasterol, and stigmasterol as main phytosterols in hempseed, oil, and flour products of Fedora cultivar. Still, their contents in unsaponifiable fraction were considerably lower. Hempseed oil does not contain considerable amounts of Δ7-phytosterols. Their content is only about 2% of total phytosterols (Matthäus & Brühl, 2008).

5.4 Oxidative stability of hempseed oil

Cold-pressed hempseed oil is a rich source of unsaturated fatty acids, prone to oxidation processes, i.e. rancidity. The nutritional value and sensory characteristics of vegetable oils are compromised by rancidity development, which is very common spoilage. The extreme sensitivity of hempseed oil to oxidative deterioration can be explained by its high degree of unsaturation. About 80% of polyunsaturated fatty acids (Radočaj & Dimić, 2013) are highly prone to atmospheric oxygen. Fatty acid oxidation occurs in two steps: firstly, unsaturated fatty acids react with oxygen-producing peroxides, whereby the degree of this oxidation process is evaluated by peroxide value (PV); secondly, peroxides are decomposed into aldehydes, ketones, alcohols, esters, and many others, responsible for the different deteriorative changes, e.g. unpleasant rancid taste of oils. The *para*-anisidine value (*p*-AnV) provides the assessment of the secondary oxidation step. Conjugated diene and triene contents (such as K232, K270, and ΔK), peroxide, acid, and *p*-AnV values, as well as radical scavenging activity (DPPH assay), are commonly used to measure oxidative state and stability in cold-pressed oils (Choe & Min, 2006; Prescha et al., 2014; Spano et al., 2020; Vujasinović, Djilas, Dimić, Romanić, & Takači, 2010).

However, as previously stated, hempseed oil contains many other natural constituents like polyphenols, tocopherols, and carotenoids, which act as antioxidants and could protect the oil against lipid oxidation processes. In contrast, the natural chlorophylls in hempseed oil are a photosensitizer (potent prooxidant). Therefore they can induce and accelerate the oxidation of PUFA, associated with a reduction of the shelf life of oil (Izzo et al., 2020). Several authors have analyzed commercial hempseed oils to provide information about the quality of the oils present on the markets. The peroxide value (PV) (the content of lipid hydroperoxides) is still the most common empirical chemical analysis of measuring the oxidative deterioration of oils. PV in examined oil samples showed great variability, but in most of the samples, PV was lower than the recommended maximum limit for cold-pressed oils of 15 meqO$_2$/kg by Codex Alimentarius Commission (Izzo et al., 2020; Prescha et al., 2014; Teh & Birch, 2013). However, PV above 20 meqO$_2$/kg in oil was also found (Spano et al., 2020).

The PV of commercial hempseed oils in clear glass packaging inevitably increases during storage. Prescha et al. (2014) reported a mean increase of PV in oil of 168%, after 6 months of storage at 20°C (in their original glass bottles, 12/12h light/dark regime), while Gaca et al. (2021) found an increase of PV from 14.47 meqO$_2$/kg up to 1543.98 meqO$_2$/kg after 10 days storage of oil in an oven at 60°C. Reported *p*-AnV in hempseed oils had low values, less than 3.5, reflecting good quality of oil from the aspect of secondary oxidation products (Oomah et al., 2002; Spano et al., 2020).

The Rancimat test and Schaal oven test based on the accelerated oxidation of oil at elevated temperature (even 100°C and above) have been proved to be a good predictor of the oxidative stability of oils. Anyway, the oxidative stability of the oil depends on the initial degree of oxidation, i.e. initial PV. Table 5.9 presents the data for oxidative stability of cold-pressed oils of different hempseed varieties based on initial peroxide values and the induction period by Schaal oven test (Vujanić, 2005) and Rancimat test (Dimić et al., 2009). Cold-pressed hempseed oil was obtained from seven hemp varieties cultivated under the same agroecological conditions on the experimental field. The induction period by Rancimat test ranging between 6.4 and 7.6 h at 100°C indicated poor oxidative stability of hempseed oil at high temperatures. The same authors correlated the decrease in oil's oxidative stability with unsaturation ($R^2 = 0.8586$). A significant increase of PV (5.7–7.3 times) according to the Schaal oven test also indicated poor oxidative stability of oil samples at 60°C.

Matthäus and Brühl (2008) confirmed significantly shorter stability of virgin hempseed oil in the Rancimat test at 120°C in comparison to rapeseed and olive virgin oils. Virgin rapeseed oil and olive oil have oxidative stability of 4 and 6 h, respectively, but hempseed oil less than 1 h. The oxidative stability index (OSI) at 110°C and airflow of 20 L/h for cold-pressed hempseed oil was found to be 2.4 h (initial PV 0.6 meqO$_2$/kg) (Radočaj & Dimić, 2013). Although hempseed oil is very susceptible to oxidation, the roasting process of hempseeds before pressing could protect PUFA in seeds from oxidation due to the simultaneous increase of antioxidative activities of phenolics and tocopherols. Hempseed oil from seeds roasted at 120°C for 30 min had the best oxidation stability. Higher roasting temperature (180°C) and longer roasting time (60 min) resulted in a shorter induction period and higher PV of oils from roasted hempseeds (Bryś et al., 2019). Similar results have also been published with pumpkin seed oil (Vujasinović, Djilas, Dimić, Basić, & Radočaj, 2012).

Oxidation is the main problem for the storage of polyunsaturated oils, so special care should be taken to maintain an inert atmosphere throughout the processing until the oil is bottled and capped. In general, hempseed is rarely pressed under an inert atmosphere. Because of that, bottles of hempseed oils prepared for sale should be stored at the lowest possible temperatures and should be protected from light. Light protection is especially important due to high chlorophyll content, which was found to be 98.6 mg/kg in cold-pressed hempseed oil. Similarly, the total chlorophyll content in cold-pressed hempseed oil, expressed as pheophytin, was 75.21 mg/kg and was higher than that of cold-pressed canola (0.86 mg/kg) and flaxseed oil (6.78 mg/kg) (Teh & Birch, 2013). In addition, chlorophylls can capture light energy and accelerate the oxidation of the oil. Therefore exposure of oil to light must be avoided. For that reason, bottles for consumers should be made of dark glass that will prevent the effect of light.

Ultrasound-assisted bleaching with various clays is a new approach for reducing the chlorophyll content in hempseed oil. During the bleaching process, varying the process conditions can influence the intensities of oil color, all depending on the consumer's habits. The color can be from light green to light yellow to colorless when it comes to hemp oil. Colorless hempseed oil would be suitable for a topographic application, which may be the subject of further research. Aachary, Liang, Hydamaka, Eskin, and Thiyam-Holländer (2016) reported that 47%–99% of chlorophylls in cold-pressed hempseed oil could be removed by a new ultrasound-assisted clay adsorption bleaching technique. The content of removed chlorophylls varied depending on clay types (activated/non-activated) and ultrasound treatment conditions (ultrasound power, time). The authors concluded that the bleaching efficiency increased significantly as the ultrasound power (20%–60% amplitude) and concentration of bleaching clay (20–40 g/kg) increased. However, the negative side of such treatment is that clay also reduces phenolics contents in oil by 27%–35%.

5.4.1 Safety of hempseed oil

The results of auto-oxidation and photo-oxidation are noticeable by varnish-like smell, which reminds of linoleum or putty. Matthäus and Brühl (2008) considered that virgin hempseed oil stored for more than 2 months in opened bottles should not be used for human consumption. Moreover, hempseed oil shouldn't be used for frying or cooking practice because of very weak oxidative stability, which needs temperatures above 120°C (Oomah et al., 2002). The best way to use cold-pressed hempseed oil is as a salad oil or as an ingredient in salad dressing, dips and sandwich spreads, or as a substitute for butter/margarine on cooked foods (bread, pasta, vegetables) (Callaway & Pate, 2009; Radočaj, Dimić, Diosady, & Vujasinović, 2011).

The issue of THC in cold-pressed hemp oil is of particular importance. To be legally classified as hemp plant, it is not allowed to contain more than 0.2% or 0.3% (on dry bases) of the Δ^9-tetrahydrocannabinol (THC-psychoactive substance) in Europe and North American countries, respectively. This pretty low level of THC is insufficient to induce intoxication in the human organism. A precautionary guidance value for THC in the hempseed oil of 5000 μg/kg was defined in the year 2000 by the Federal Institute for Risk Assessment, Germany (Matthäus & Brühl, 2008). This limit is based on protecting consumers from psychoactive effects. Almost the same values were published by Grotenhermen, Leson, and Pless (2001), stating that 5 μg/g for hemp oil should be considered by regulatory agencies as a conservative and enforceable choice of THC limits in hempseed derivatives.

In general, THC should be no problem for cold-pressed hempseed oil, but it is not always so, as it can be transferred to hempseed oil during the crushing process. If THC is found in the oil, it may predominantly result from the external contact of the seed hulls with cannabinoid-containing resins in the bracts and leaves of the plant. For that reason, for the producers of cold-pressed hempseed oil, it is highly recommended to carefully control the seed cleaning (Callaway & Pate, 2009; Matthäus & Brühl, 2008).

5.5 Challenges and opportunities

Virgin hempseed oil is not widespread on the market, although it is characterized by an interesting fatty acid composition with a high content of polyunsaturated fatty acids.

To enable further development of hempseed oil industry, the lack of a specific regulation for hempseed oils and quality parameters which exist for olive oils must be overcome. Thus the preservation of consumer awareness and safety related to this valorized product will be achieved.

Although the price of hempseed oil varies widely, it is generally an expensive food item, which opens up the possibility of its adulteration with other lower price vegetable oils. To supervise the potential hempseed oil adulteration, different spectral methods, such as nuclear magnetic resonance (NMR) and Fourier-transformed infrared spectroscopy (FTIR), have proved to be suitable for a fast screening of oil quality, the origin of the oil, and evaluating the ratio of ω-6/ω-3 fatty acids (Siudem et al., 2019). On the other hand, the growing demand for exclusive vegetable oils requires an additional extensive characterization of hempseed oil to the innovative use of its bioactive components.

Some minor components of the unsaponifiable fraction of hempseed oil have not been sufficiently investigated yet. There are indications that hempseed oil, among natural antioxidants, contains plastohromanol-8 (P-8) (Kriese et al., 2004). Even though the content of P-8 was found very low, it is more effective in protecting unsaturated fatty acids from oxidation than tocopherols in hempseed.

Another interesting fraction of the unsaponifiable matters of the hempseed oil is the aliphatic alcohols. Montserrat-de la Paz et al. (2014) announced that phytol (167.59 ± 1.81 mg/kg) and geranylgeraniol (26.06 ± 0.08 mg/kg) were predominant in the aliphatic alcohol fraction of hempseed oil. As a constituent of chlorophyll, phytol is mainly found in human food, such as raw green vegetables, and has anti-cancer and antioxidant actions (Vetter, Schröder, & Lehnert, 2012). The secondary metabolites from the unsaponifiable fraction of hempseed oil are also potential complementary products derived from industrial hempseed. Accordingly, further analytical and phytochemical characterization of the unsaponifiable fraction of hempseed oil would be a significant challenge. It should be kept in mind that, as Leizer et al. (2000) stated, β-sitosterol and methyl salicylate also complement the nutritious value of hempseed oil and increases its effectiveness as a functional food. Hence, they need to be examined further for additional beneficial characterization of oil.

Considering the variability in composition and price of oil, distinctive quality parameters need to be established to preserve consumer awareness and safety, as valorized products. There is an obvious need for a disciplinary booklet that would define agronomic and post-harvest management conditions for achieving a valuable food item because of the diversity in bioactive compounds in hempseed oils.

5.6 Conclusion

The results of our considerations confirm the great nutritional features of hempseed oil in terms of macro- and micronutrients. It also highlighted the nutritional value and possibility of utilizing the cake and flour (waste after oil extraction from seed) to prepare functional food products and nutraceuticals.

The nutritional values and health benefits of hempseed oil are attributed to its unique fatty acid composition and minor bioactive components. It should be kept in mind that hempseed oil has an optimal PUFA balance, and the ratio of linoleic (ω-6) to alpha-linolenic (ω-3) acids approaches 4:1, indicating a desired nutritional profile of all hemp products, according to a recommended nutritional consensus. Furthermore, hempseed oil is characterized by the simultaneous appearance of γ-linolenic acid and stearidonic acid, rare in vegetable oils. Because of a specific fatty acid profile, hempseed oil is an interesting supplement for humans with a lack of enzyme system for the production of eicosanoids. Furthermore, the minor biocomponents, such as tocopherols, phenolic compounds, phytosterols, etc., prevent the oil from oxidative deterioration and contribute to humans' health benefits.

Many advances have been made in recent years in the technologies for the extraction of hempseed oil. Some innovative extraction methods, such as supercritical fluid extraction, ultrasonic extraction, microwave extraction, enzyme extraction, etc., have given very good results in laboratory research, but time will show how realistic their implementation on industrial scale. However, there is still no industry consensus on the best oil extraction method because it largely depends on the planned production capacity and the end use of oil and by-products.

Nowadays, oil extraction from hempseed is mostly carried out by a pressing process to produce cold-pressed oil. By applying the cold-pressing process on small capacity screw presses, the oil yield is much lower, but it has the advantage of minimizing degradative changes in the oil. Moreover, as cold-pressed hempseed oil is free of chemical treatments, it is a much more interesting alternative for conventional practices because consumers desire natural and safe food products. Regardless, it would be of great interest to standardize procedures of hempseed oil production. That would guarantee a good quality foodstuff, consumer safety, and the further expansion of the hemp food industry.

Acknowledgments

The authors are grateful to the Ministry of Education, Science and Technological Development, Republic of Serbia, Grant no. 451-03-9/2021-14/200116 (University of Belgrade, Faculty of Agriculture) and specially to Prof Dr. Etelka Dimić for her significant help with the chapter preparation.

References

Aachary, A., Liang, J., Hydamaka, A., Eskin, N., & Thiyam-Holländer, U. (2016). A new ultrasound-assisted bleaching technique for impacting chlorophyll content of cold-pressed hemp seed oil. *LWT—Food Science and Technology, 72*, 439–446.

Abedi, E., & Sahari, M. (2014). Long-chain polyunsaturated fatty acid sources and evaluation of their nutritional and functional properties. *Food Science and Nutrition, 2*, 443–463.

Aladić, K., Jarni, K., Barbir, T., Vidovic, S., Vladić, J., Bilić, M., et al. (2015). Supercritical CO_2 extraction of hemp (*Cannabis sativa* L.) seed oil. *Industrial Crops and Products, 76*, 472–478.

Aladić, K., Jokić, S., Moslavac, T., Tomas, S., Vidovic, S., Vladić, J., et al. (2014). Cold pressing and supercritical CO_2 extraction of hemp (*Cannabis sativa*) seed oil. *Chemical and Biochemical Engineering Quarterly, 28*, 481–490.

Berenji, J., Dimić, E., & Romanić, R. (2005). Hemp—The potential raw material for cold-pressed oil. In *Proceedings of 46th conference: Production and processing of oilseeds with international participation* (pp. 127–137). Petrovac na moru, Serbia and Montenegro.

Bjelica, M., Dimić, S., & Vujasinović, V. (2018). Vitamin E activity of cold-pressed grapeseed oils. In *Proceedings of 59th conference: Production and processing of oilseeds with international participation* (pp. 175–184). Herceg Novi, Montenegro.

Blade, S., Ampong-Nyarko, K., & Przybylski, R. (2006). Fatty acid and tocopherol profiles of industrial hemp cultivars grown in the high latitude Prairie Region of Canada. *Journal of Industrial Hemp, 10*, 33–43.

Bryś, A., Bryś, J., Mellado, Á. F., Glowacki, S., Tulej, W., Ostrowska-Ligęza, E., et al. (2019). Characterization of oil from roasted hemp seeds using the PDSC and FTIR techniques. *Journal of Thermal Analysis and Calorimetry, 138*, 2781–2786.

Callaway, J. C., & Pate, D. W. (2009). Hemp seed oil. In R. A. Moreau, & A. Kamal-Eldin (Eds.), *Gourmet and health-promoting specialty oils* (pp. 185–213). Urbana: AOCS Press.

Choe, E., & Min, D. B. (2006). Mechanisms and factors for edible oil oxidation. *Comprehensive Reviews in Food Science and Food Safety, 5*, 169–186.

Citti, C., Pacchetti, B., Vandelli, M. A., Forni, F., & Cannazza, G. (2018). Analysis of cannabinoids in commercial hemp seed oil and decarboxylation kinetics studies of cannabidiolic acid (CBDA). *Journal of Pharmaceutical and Biomedical Analysis, 149*, 532–540.

Crimaldi, M., Faugnob, S., Sanninobc, M., & Arditoa, L. (2017). Optimization of hemp seeds (*Canapa sativa* L.) oil mechanical extraction. *Chemical Engineering Transactions, 58*, 373–378.

Deferne, J., & Pate, D. (1996). Hemp seed oil: A source of valuable essential fatty acids. *Journal of the International Hemp Association, 3*(1), 1. 4–7.

Deluca, P., Rothman, D., & Zurier, R. (1995). Marine and botanical lipids as immunomodulatory and therapeutic agents in the treatment of rheumatoid arthritis. *Rheumatic Diseases Clinics of North America, 21*, 759–777.

Devi, V., & Khanam, S. (2019). Comparative study of different extraction processes for hemp (*Cannabis sativa*) seed oil considering physical, chemical and industrial-scale economic aspects. *Journal of Cleaner Production, 207*, 645–657.

Dimić, E. (2005). *Hladno ceđena ulja (Cold-pressed oils)*. Novi Sad: University of Novi Sad, Faculty of Technology.

Dimić, E., Premović, T., Radočaj, O., Vujasinović, V., & Takači, A. (2018). Influence of seed quality and storage time on the characteristics of cold-pressed sunflower oil: Impact on bioactive compounds and color. *Rivista Italiana Delle Sostanze Grasse, 95*(2–3), 23–36.

Dimić, E., Romanić, R., Berenji, J., & Bodroža-Solarov, M. (2006). Essential fatty acids and nutritive value of cold pressed hempseed oil, *Cannabis sativa* L. In *Proceedings of 4th conference on medicinal and aromatic plats of South-East European countries* (pp. 408–414). Iaşi—România. 28–31 May.

Dimić, E., Romanić, R., Tešanović, D., & Dimić, V. (2005). Essential fatty acids and tocopherols of evening primrose oil (*Oenothera biennis* L.). In *Proceedings of 10th symposium: Vitamine und Zusatzstoffe in der Ernährung von Mensch und Tier* (pp. 236–239). Jena/Thüringen, Germany.

Dimić, E., Romanić, R., & Vujasinović, V. (2009). Essential fatty acids, nutritive value and oxidative stability of cold pressed hempseed (*Cannabis sativa* L.) oil from different varieties. *Acta Alimentaria, 38*, 229–236.

Dimić, E., & Vujasinović, V. (2011). Extraction technology, quality and biologically active components of pumpkin oil. In J. Berenji (Ed.), *Uljana tikva (Oil pumpkin) Cucurbita pepo L* (pp. 257–288). Novi Sad: The Institute of Field and Vegetable Crops.

Dubois, V., Breton, S., Linder, M., Fanni, J., & Parmentier, M. (2007). Fatty acid profiles of 80 vegetable oils with regard to their nutritional potential. *European Journal of Lipid Science and Technology, 109*, 710–732.

Elmadfa, I., & Park, E. (1999). Impact of diets with corn oil or olive/sunflower oils on DNA damage in healthy young men. *European Journal of Nutrition, 38*, 286–292.

Esmaeilzadeh Kenari, R., & Dehghan, B. (2020). Optimization of ultrasound-assisted solvent extraction of hemp (*Cannabis sativa* L.) seed oil using RSM: Evaluation of oxidative stability and physicochemical properties of oil. *Food Science and Nutrition, 8*, 4976–4986.

Fils, J. M. (2000). The production of oils. In W. Hamm, & R. J. Hamilton (Eds.), *Edible oil processing* (pp. 47–78). England: Sheffield Academic Press Ltd.

Folegatti, L., Rovellini, P., Baglio, D., De Cearei, S., Fusari, P., Venturini, S., et al. (2014). Caratterizzazione chimica della farina ottenuta dopo la spremitura a freddo dei semi di *Cannabis sativa* L. *Rivista Italiana Delle Sostanze Grasse, 91*, 3–16.

Gaca, A., Kludská, E., Hradecký, J., Hajslova, J., & Jeleń, H. (2021). Changes in volatile compound profiles in cold-pressed oils obtained from various seeds during accelerated storage. *Molecules, 26*(2). https://doi.org/10.3390/molecules26020285, 285.

Gogus, U., & Smith, C. J. (2010). N-3 Omega fatty acids: A review of current knowledge. *International Journal of Food Science and Technology, 45*, 417–436.

Grijó, D. R., Piva, G. K., Osorio, I. V., & Cardozo-Filho, L. (2019). Hemp (*Cannabis sativa* L.) seed oil extraction with pressurized n-propane and supercritical carbon dioxide. *Journal of Supercritical Fluids, 143*, 268–274.

Grotenhermen, F., Leson, G., & Pless, P. (2001). *Assessment of exposure to and human health risk from THC and other cannabinoids in hemp food*. HempFoodRiskAss.doc 10/11/2001 http://www.nova-institut.de/pdf/hempFoodRiskAss.PDF.

House, J., Neufeld, J., & Leson, G. (2010). Evaluating the quality of protein from hemp seed (*Cannabis sativa* L.) products through the use of the protein digestibility-corrected amino acid score method. *Journal of Agricultural and Food Chemistry, 58*, 11801–11807.

Izzo, L., Pacifico, S., Piccolella, S., Castaldo, L., Narváez, A., Grosso, M., et al. (2020). Chemical analysis of minor bioactive components and cannabidiolic acid in commercial hemp seed oil. *Molecules, 25*(16). https://doi.org/10.3390/molecules25163710, 3710.

Jozinović, A., Ačkar, Đ., Jokić, S., Babić, J., Balentić, J. P., Banožić, M., et al. (2017). Optimization of extrusion variables for the production of corn snack products enriched with defatted hemp cake. *Czech Journal of Food Sciences, 35*, 507–516.

Kapoor, R., & Nair, H. (2005). Gamma linolenic acid oils. In F. Shahidi (Ed.), *Edible oil and fat products: Specialty oils and oil products: Vol. 3. Bailey's industrial oil and fat products* (pp. 67–119). New Jersey: J. Wiley & Sons, Inc.

Karche, T., & Singh, R. (2019). The application of hemp (*Cannabis sativa L.*) for a green economy: A review. *Turkish Journal of Botany, 43*, 710–723.

Karlović, D., & Berenji, J. (1996). Ulje zrna konoplje: *Pro et contra. Zbornik radova Naučnog instituta za ratarstvo i povrtarstvo, Novi Sad, 26*, 131–136.

Kate, A., Singh, A., Shahi, N., Jp, E., Prakash, O., & Singh, T. P. (2016). Novel eco-friendly techniques for extraction of food based lipophilic compounds from biological materials. *Natural Products Chemistry and Research, 4*, 1–7.

Kriese, U., Schumann, E., Weber, W., Beyer, M., Brühl, L., & Matthäus. (2004). Oil content, tocopherol composition and fatty acid patterns of the seeds of 51 *Cannabis sativa* L. genotypes. *Euphytica, 137*, 339–351.

Latif, S., & Anwar, F. (2009). Physicochemical studies of hemp (*Cannabis sativa*) seed oil using enzyme-assisted cold-pressing. *European Journal of Lipid Science and Technology, 111*, 1042–1048.

Leizer, C., Ribnicky, D., Poulev, A., Dushenkov, S., & Raskin, I. (2000). The composition of hemp seed oil and its potential as an important source of nutrition. *Journal of Nutraceuticals, Functional and Medical Foods, 2*, 35–53.

Leonard, W., Zhang, P., Ying, D., & Fang, Z. (2020). Hempseed in food industry: Nutritional value, health benefits, and industrial applications. *Comprehensive Reviews in Food Science and Food Safety, 19*, 282–308.

Liang, J., Aachary, A., & Thiyam-Holländer, U. (2015). Hemp seed oil: Minor components and oil quality. *Lipid Technology, 27*, 231–233.

Lin, J., Zeng, Q., An, Q., Zeng, Q., Jian, L., & Zhu, Z. (2012). Ultrasonic extraction of hempseed oil. *Journal of Food Process Engineering, 35*, 76–90.

Mag, T. K., Mag, T., & Reichert, R. D. (2002). A new recommended calculation of vitamin E activity: Implications for the vegetable oil industry. *Inform, 13*, 836–839.

Malomo, S., He, R., & Aluko, R. (2014). Structural and functional properties of hemp seed protein products. *Journal of Food Science, 79*, C1512–C1521.

Matthäus, B., & Brühl, L. (2008). Virgin hemp seed oil: An interesting niche product. *European Journal of Lipid Science and Technology, 110*, 655–661.

Mihoc, M., Pop, G., Alexa, E., & Radulov, I. (2012). Nutritive quality of romanian hemp varieties (*Cannabis sativa* L.) with special focus on oil and metal contents of seeds. *Chemistry Central Journal, 6*, 122.

Mölleken, H., Mothes, R., & Dudek, S. (2000). Quality of hemp fruits and hemp oil relation to the maturity of the fruits. In *Proceedings of symposium: Bioresource hemp* (pp. 1–7). Wolfsburg, Germany http://www.nova-institute.de.

Mölleken, H., & Theimer, R. R. (1997). Survey of minor fatty acids in *Cannabis sativa* L. fruits of various origins. *Journal of the International Hemp Association, 4*(1), 13–17.

Montserrat-de la Paz, S., Marín-Aguilar, F., García-Giménez, M. D., & Fernández-Arche, M.Á. (2014). Hemp (*Cannabis sativa* L.) seed oil: Analytical and phytochemical characterization of the unsaponifiable fraction. *Journal of Agricultural and Food Chemistry, 62*, 1105–1110.

Morar, M., Dragan, K., Bele, C., Matea, C., Tarta, I., Suharovschi, R., et al. (2010). Researches regarding the processing of the hemp seed by cold pressing. *Bulletin of University of Agricultural Sciences and Veterinary Medicine Cluj-Napoca, 67*, 284–290.

Newton, I. S. (1996). Food enrichment with long chain n-3 PUFA. *Inform, 7*, 169–177.

Oomah, B., Busson, M., Godfrey, D., & Drover, J. (2002). Characteristics of hemp (*Cannabis sativa* L.) seed oil. *Food Chemistry, 76*, 33–43.

Panfilis, F. D., Toschi, T. G., & Lercker, G. (1998). Quality control for cold-pressed oils. *Inform, 9*, 212–221.

Petrović, M., Debeljak, Ž., Kezić, N., & Džidara, P. (2015). Relationship between cannabinoids content and composition of fatty acids in hempseed oils. *Food Chemistry, 170*, 218–225.

Pojić, M., Hadnađev, M., & Hadnađev-Dapčević, T. (2014). Reutilization of hemp seed oil sludge in creating value-added functional spreads. In *COST conference on food waste in the European food supply chain: Challenges and opportunities. Book of abstracts* (p. 14). Athens, Greece, 12–13 May.

Pojić, M., Mišan, A., Sakač, M., Dapčević Hadnadjev, T., Šarić, B., Milovanović, I., et al. (2014). Characterisation of by-products originating from hemp oil processing. *Journal of Agricultural and Food Chemistry, 62*, 12436–12442.

Porto, C., Natolino, A., & Decorti, D. (2015). Effect of ultrasound pre-treatment of hemp (Cannabis sativa L.) seed on supercritical CO_2 extraction of oil. *Journal of Food Science and Technology, 52*(3), 1748–1753.

Porto, C., Voinovich, D., Decorti, D., & Natolino, A. (2012). Response surface optimization of hemp seed (*Cannabis sativa* L.) oil yield and oxidation stability by supercritical carbon dioxide extraction. *Journal of Supercritical Fluids, 68*, 45–51.

Prescha, A., Grajzer, M., Dedyk, M., & Grajeta, H. (2014). The antioxidant activity and oxidative stability of cold-pressed oils. *Journal of the American Oil Chemists' Society, 91*, 1291–1301.

Radočaj, O., & Dimić, E. (2013). Physico-chemical and nutritive characteristics of selected cold-pressed oils found in the European market. *Rivista Italiana Delle Sostanze Grasse, 90*(4), 219–228.

Radočaj, O., Dimić, E., Diosady, L. L., & Vujasinović, V. (2011). Optimization of the texture of fat-based spread containing hull-less pumpkin (*Cucurbita pepo* L.) seed press-cake. *Journal of Texture Studies, 42*, 394–403.

Radočaj, O., Dimić, E., & Tsao, R. (2014). Effects of hemp (*Cannabis sativa* L.) seed oil press-cake and decaffeinated green tea leaves (*Camellia sinensis*) on functional characteristics of gluten-free crackers. *Journal of Food Science, 79*, C318–C325.

Rezvankhah, A., Emam-Djomeh, Z., Safari, M., Askari, G., & Salami, M. (2018). Investigation on the extraction yield, quality, and thermal properties of hempseed oil during ultrasound-assisted extraction: A comparative study. *Journal of Food Processing and Preservation, 42*(10), e13766.

Rezvankhah, A., Emam-Djomeh, Z., Safari, M., Askari, G., & Salami, M. (2019). Microwave-assisted extraction of hempseed oil: Studying and comparing of fatty acid composition, antioxidant activity, physiochemical and thermal properties with Soxhlet extraction. *Journal of Food Science and Technology, 56*, 4198–4210.

Rodriguez-Leyva, D., & Pierce, G. N. (2010). The cardiac and haemostatic effects of dietary hemp seed. *Nutrition and Metabolism, 7*, 32–42.

Rupasinghe, H., Davis, A., Kumar, S. K., Murray, B., & Zheljazkov, V. (2020). Industrial hemp (*Cannabis sativa* subsp. sativa) as an emerging source for value-added functional food ingredients and nutraceuticals. *Molecules, 25*(18). https://doi.org/10.3390/molecules25184078, 4078.

Schluttenhofer, C. M., & Yuan, L. (2017). Challenges towards revitalizing hemp: A multifaceted crop. *Trends in Plant Science, 22*, 917–929.

Schuster, W. (1993). *Ölpflanzen in Europe*. Frankfurt am Main: DLG-Verlag.

Schwab, U., Callaway, J. C., Erkkilä, A., Gynther, J., Uusitupa, M., & Järvinen, T. (2006). Effects of hempseed and flaxseed oils on the profile of serum lipids, serum total and lipoprotein lipid concentrations and haemostatic factors. *European Journal of Nutrition, 45*, 470–477.

Sharp, W. R., Whitaker, R. J., Sondahl, M. R., Evans, D. A., Bravo, J. E., Marsden, J. F., et al. (1986). Opportunities for biotechnology in the development of new edible vegetable oil products. *Journal of the American Oil Chemists' Society, 63*(5), 594–600.

Siano, F., Moccia, S., Picariello, G., Russo, G., Sorrentino, G., Stasio, M. D., et al. (2018). Comparative study of chemical, biochemical characteristic and ATR-FTIR analysis of seeds, oil and flour of the edible fedora cultivar hemp (*Cannabis sativa* L.). *Molecules, 24*(1). https://doi.org/10.3390/molecules24010083, 83.

Siger, A., Nogala-Kałucka, M., & Lampart-Szczapa, E. (2008). The content and antioxidant activity of phenolic compounds in cold-pressed plant oils. *Journal of Food Lipids, 15*, 137–149.

Siudem, P., Wawer, I., & Paradowska, K. (2019). Rapid evaluation of edible hemp oil quality using NMR and FT-IR spectroscopy. *Journal of Molecular Structure, 1177*, 204–208.

Small, E., & Marcus, D. (2002). Hemp: A new crop with new uses for North America. In J. Janick, & A. Whipkey (Eds.), *Trends in new crops and new uses* (pp. 284–326). Alexandria: ASHS Press.

Smeriglio, A., Galati, E. M., Monforte, M. T., Lanuzza, F., D'Angelo, V., & Circosta, C. (2016). Polyphenolic compounds and antioxidant activity of cold-pressed seed oil from Finola cultivar of *Cannabis sativa* L. *Phytotherapy Research, 30*, 1298–1307.

Spano, M., Matteo, G. D., Rapa, M., Ciano, S., Ingallina, C., Cesa, S., et al. (2020). Commercial hemp seed oils: A multimethodological characterization. *Applied Sciences, 10*, 6933.

Teh, S., & Birch, J. (2013). Physicochemical and quality characteristics of cold-pressed hemp, flax and canola seed oils. *Journal of Food Composition and Analysis, 30*, 26–31.

Teh, S., Niven, B., Bekhit, A., Carne, A., & Birch, E. J. (2014). The use of microwave and pulsed electric field as a pretreatment step in ultrasonic extraction of polyphenols from defatted hemp seed cake (*Cannabis sativa*) using response surface methodology. *Food and Bioprocess Technology, 7*, 3064–3076.

Tomita, K., Machmudah, S., Quitain, A., Sasaki, M., Fukuzato, R., & Goto, M. (2013). Extraction and solubility evaluation of functional seed oil in supercritical carbon dioxide. *Journal of Supercritical Fluids, 79*, 109–113.

Vavpot, V. J., Williams, R. J., & Williams, M. A. (2014). Extrusion/Expeller® pressing as a means of processing green oils and meals. In W. E. Farr, & A. Proctor (Eds.), *Green vegetable oils processing* (pp. 1–54). Urbana: AOCS Press.

Vetter, W., Schröder, M., & Lehnert, K. (2012). Differentiation of refined and virgin edible oils by means of the trans- and cis-phytol isomer distribution. *Journal of Agricultural and Food Chemistry, 60*, 6103–6107.

Vogl, C. R., Mölleken, H., Lissek-Wolf, G., Surbök, A., & Kobert, J. (2004). Hemp (*Cannabis sativa* L.) as a resource for green cosmetics: Yield of seed and fatty acid compositions of 20 varieties under the growing conditions of organic farming in Austria. *Journal of Industrial Hemp, 9*, 51–68.

Vujanić, S. (2005). *Potential of hemp (Cannabis sativa L.) as oilseed* (B.Sci. thesis). Novi Sad, Serbia: University of Novi Sad, Faculty of Technology.

Vujasinović, V., Djilas, S., Dimić, E., Basić, Z., & Radočaj, O. (2012). The effect of roasting on the chemical composition and oxidative stability of pumpkin oil. *European Journal of Lipid Science and Technology, 114*, 568–574.

Vujasinović, V., Djilas, S., Dimić, E., Romanić, R., & Takači, A. (2010). Shelf life of cold-pressed pumpkin (*Cucurbita pepo* L.) seed oil obtained with a screw press. *Journal of the American Oil Chemists' Society, 87*, 1497–1505.

Watkins, C. (2004). Fundamental fats. *Inform*, *15*, 638–640.

Xu, Y., Li, J., Zhao, J., Wang, W., Griffin, J., Li, Y., et al. (2021). Hempseed as a nutritious and healthy human food or animal feed source: A review. *International Journal of Food Science and Technology*, *56*, 530–543.

Yu, L., Zhou, K., & Parry, J. (2005). Antioxidant properties of cold-pressed black caraway, carrot, cranberry, and hemp seed oils. *Food Chemistry*, *91*, 723–729.

6

Industrial hemp proteins: Processing and properties

Anne Pihlanto, Markus Nurmi, and Sari Mäkinen

Natural Resources Institute Finland, Jokioinen, Finland

6.1 Introduction

Cereals, legumes, and oilseeds are the main sources of plant protein. It is estimated that only 35% of the total plant protein produced by the agricultural sector is used for human consumption, while the rest is used for livestock feed, nonfood applications, or simply treated as a waste. Conventional sources of protein, such as animal and cereal sources, will no longer be sufficient to meet the global demand for protein in the coming years. Finding alternative sources of protein is strongly recommended (Sari, Mulder, Sanders, & Bruins, 2015). Hemp provides an ecologically and economically interesting source of food. Hemp (*Cannabis sativa* L.) is an herbaceous annual belonging to the family Cannabaceae and has been cultivated for food, fiber, and medicine for more than six millennia (Salentijn, Zhang, Amaducci, Yang, & Trindade, 2015; Kerckhoffs, Kavas, Millner, Anderson, & Kawana-Brown, 2015; O'Brien & Arathi, 2019). To be legally classified as food, hempseeds may contain no more than 0.2% or 0.3% (by dry weight) of the intoxicating compound Δ^9-tetrahydrocannabinol (THC) in Europe and North American countries, respectively (Leonard, Zhang, Ying, & Fang, 2020).

Hemp has several advantages in agriculture as it has strong and rapid growth and can be harvested 4 months after seeding and produces a significant amount of biomass (Ash, 1948). It does not need phytosanitary products because it can eradicate weeds and is not affected by any pests or diseases. It is one of the few agricultural products to be grown without phytosanitary products on nonorganic farms (Carus & Sarmento, 2016). It is used in rotation culture and increases the yield of the next culture by 10%–15%. It can remove large quantities of heavy metals from the soil, and its long root system helps to prevent soil erosion. It also has low water and input requirements (Ranalli & Venturi, 2004). Recent interest in industrial hemp as a multipurpose crop has grown rapidly worldwide. Since 2015 for the first time, it has exceeded 20,000 ha as a dual-purpose crop for both the seeds and the fiber (Carus & Sarmento, 2016). The plant genotype and growing environment have significant effects on seed production. Although a renewed interest in hemp cultivation for the combination of seeds and fiber is apparent, genetic and agronomic information to support the dual-purpose hemp cultivation is scare (Amaducci et al., 2015; Cherney & Small, 2016).

Hempseeds are an excellent source of nutrients. Whole hempseeds contain 25% to 35% oil, 20% to 25% protein, 20% to 30% carbohydrates, 10% to 15% insoluble fibers, and vitamins and minerals such as phosphorus, potassium, magnesium, sulfur, calcium, iron, and zinc (Callaway, 2004; House, Neufeld, & Leson, 2010). Hempseed proteins have an excellent digestibility and a desirable essential amino acid composition (Tang, Ten, Wang, & Yang, 2006; Wang, Tang, Yang, & Gao, 2008). A recent proteomic characterization of hempseed concluded that hempseed is an underexploited nonlegume, protein-rich seed (Aiello et al., 2016). Because of the high nutritional value, hempseed protein has drawn increasing attention in scientific research (Fig. 6.1). Despite the nutritive potential, the application of hempseed appears to be limited due to its poor performance on some functional properties such as poor protein solubility, emulsifying capacity, and stability.

This chapter provides a review of the available research on the structure and chemical composition of hempseed proteins, functional properties, and recent advances to improve the functional properties as well as their potential use in food processes. Moreover, potential hempseed protein-derived bioactive peptides are discussed with their potential health benefits.

6.2 Hemp protein concentrates and isolates

6.2.1 Composition and properties of hemp proteins

A total of 181 proteins have been identified in hempseed, and the two major storage proteins are the legumin-type globulin edestin (67%–75%) and globular-type albumin (25%–37%) (House et al., 2010; Aiello et al., 2016).

6.2.1.1 Edestin

Globular edestin is located inside the aleurone grains as large crystalloid substructures. The crystallographic characterization of edestin has revealed its six identical subunits and each subunit consists of acidic (34.0 kDa) and basic subunits (20.0 and 18.0 kDa) linked by one disulfide bond (Patel, Cudney, & McPherson, 1994; Wang et al., 2008). The molecular weight (MW) of edestin is estimated to be approximately 300 kDa and consists of mainly the 11S and 7S protein types, which can be separated based on pH shifts (Tang et al., 2006; Wang et al., 2008). Wang et al. (2008) showed that the main component in hemp 11S is the basic subunit of edestin, while a subunit of about 4.5 kDa makes up the 7S hemp protein type. Kim and Lee (2011) isolated and characterized the edestin protein from a Korean variety. The first seven and six amino acid residues of the acid subunit had a sequence of Ile-SerArg-Ser-Ala-Val-Tyr in the N-terminus, while two constituents of the basic subunit had an identical N-terminus of Gly-Leu-Glu-Glu-Thr-Phe. Edestin genes were identified and arranged into two groups (type1 and type2) based on differences in their primary structures (Docimo, Caruso, Ponzoni, Mattana, & Galasso, 2014) and later a type 3 of edestin with two isoforms (CsEde3A and CsEde3B) was isolated and identified (Ponzoni, Brambilla, & Galasso, 2018). All edestin types are rich in arginine and glutamic acid. Types 2 and 3 edestin are particularly rich in

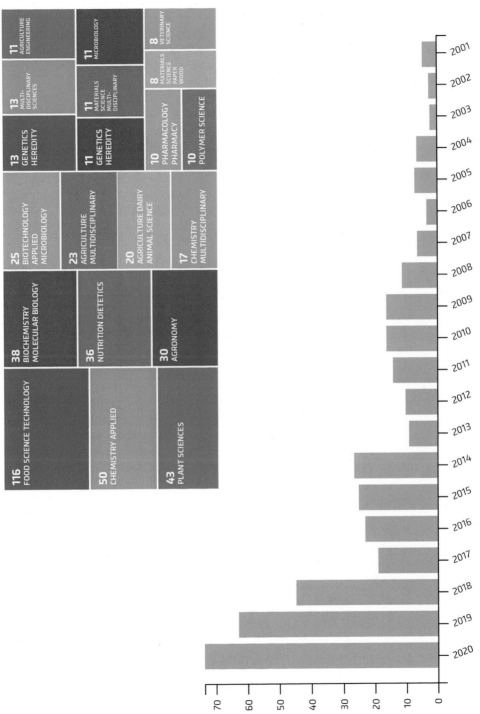

FIG. 6.1   Number of publications in Web of Science containing hemp protein in the title, abstract, and keywords within the years 2001 to 2020.

methionine (2.36% and 2.98%, respectively), exceeding the methionine content in soybean glycinins. Methionine-rich proteins are very important because the limited Met content reduces the nutritional value of crop plants. Therefore hempseeds, containing 20%–25% of the total protein per dry seed mass, of which approximately 80% is edestin and 13% is sulfur-rich albumin (Tang et al., 2006; Wang et al., 2008), hold good potential as a new source of high-value protein for both human and animal nutrition.

6.2.1.2 Albumin

The albumin fraction constitutes about 25% of the hempseed storage protein and mainly consists of two polypeptide chains with 27 and 61 amino acid residues, which includes 18% wt of sulfur-containing amino acids (methionine and cysteine). The albumin fraction with a few disulfide-bonded proteins exhibits a less compact structure with greater flexibility than the globulin fraction. This has been further confirmed by intrinsic fluorescence and circular dichroism analyses which have illustrated greater exposures of tyrosine residues when compared with globulin (Malomo & Aluko, 2015a). Albumins had a highly ordered secondary structure and very little tertiary conformation at pH 3.0 but the tertiary conformation increased at higher pH values. The high degree of flexibility and ordered secondary structure are probably structural factors that contribute to the high solubility and foaming capacity of albumin in comparison to the more compact or aggregated globulin.

6.2.1.3 Other proteins

A methionine- and cysteine-rich seed protein (10 kDa protein, 2S albumin) has been isolated from hempseed. The protein consists of two polypeptide chains (small and large) with 27 and 61 amino acid residues, respectively (Odani & Odani, 1998). The two polypeptide chains contain 18% by weight of sulfur-containing amino acids (cysteine and methionine) and are held together by two disulfide bonds. This protein has no trypsin inhibitory activity and could serve as a rich thiol source to improve the nutritional quality of plant-based foods since various plant food proteins, especially legumin proteins from soybeans, peas, and beans, are deficient in sulfur-containing amino acids. The gene families encoding the precursor polypeptides of 2S albumin have recently been identified by Ponzoni et al. (2018), and two genomic isoforms for 2S albumin were obtained, namely, Cs2S-1 and Cs2S-2. The alignment of the deduced gene with the mature 2S protein sequence published in the literature (Odani & Odani, 1998) showed that Cs2S is 97% identical to the mature 2S protein. There is not much information available concerning protease inhibitors in hempseeds. Pojić et al. (2014) and Mattila et al. (2018) were able to measure trypsin inhibitory activity from hempseed meal (HSM), seed hulls, and whole hempseeds.

6.2.2 Concentrates and isolates

Due to its superior essential amino acid content (House et al., 2010; Mattila et al., 2018) research on hempseed protein has focused on extraction, isolation, and functional properties (Hadnađev et al., 2018; Malomo, He, & Aluko, 2014; Yin, Tang, Wen, & Yang, 2009). Among these, concentrates and isolates have been isolated from hempseeds (Fig. 6.2). Protein composition and functionality are influenced by the isolation and concentration methods used to prepare the concentrates or isolates. Oil removal is needed to extract plant protein

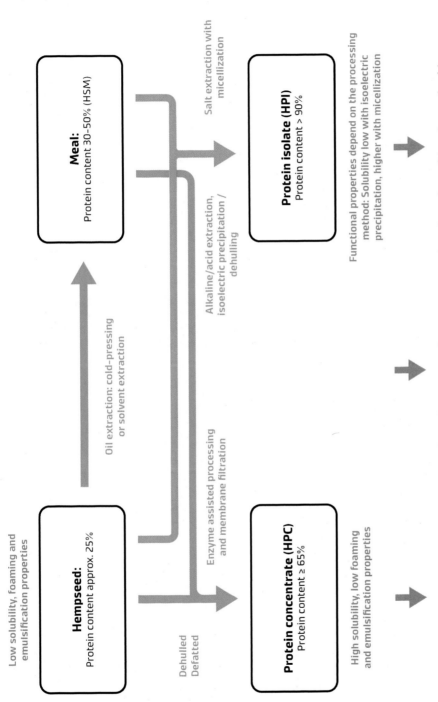

FIG. 6.2 Processing options to produce hempseed protein concentrates in isolates.

Hempseed:
Protein content approx. 25%

Low solubility, foaming and emulsification properties

Oil extraction: cold-pressing or solvent extraction

Meal:
Protein content 30-50% (HSM)

Dehulled
Defatted

Enzyme assisted processing and membrane filtration

Salt extraction with micellization

Alkaline/acid extraction, isoelectric precipitation / dehulling

Protein concentrate (HPC)
Protein content ≥ 65%

High solubility, low foaming and emulsification properties

Protein isolate (HPI)
Protein content > 90%

Functional properties depend on the processing method: Solubility low with isoelectric precipitation, higher with micellization

Possible to enhance the functional properties with enzymatic treatment and conjugation with polysaccharides

- Dairy alternatives; emulsifying properties are important, as well as colour, mouthfeel, flavour and solubility characteristics

- Meat analogues; limiting functional properties are water holding capacity, emulsifying capacity, and emulsion stability (ES), fat binding and thermal stability

- Enzymatic treatment can release bioactive peptides with potential health benefits (e.g antioxidative, antihypertensive) since lipid-protein complexes hinder the recovery of protein, especially alkaline extraction (Manamperi, Wiesenborn, Chang, & Pryor, 2011). High fiber and phytate contents, as well as the traditional extraction methods to prepare hempseed protein concentrates and isolates, reduce the functionality of the preparations, which limits their use in food applications (Tang et al., 2006). Studies have indicated that whole hempseed, dehulled hempseed, hempseed meal, and hempseed hulls possess different profiles of macronutrient composition and amino acid composition (House et al., 2010). Seed dehulling is a food processing and seed treatment which partially removes the hull/husk from the seeds. It has been widely used to reduce or eliminate the antinutrients that affect protein utilization (Pal et al., 2016).

The oil extraction by-product of crushed hempseeds is commonly referred to as hempseed meal or cake (HSM). The protein content in HSM ranges from 30% to 50% in the dry matter depending on the used hempseed variety and the oil extraction method (cold pressing or solvent extraction) and efficiency (Malomo et al., 2014). Hemp protein concentrate (HPC) is prepared from dehulled and defatted hempseed or HSM by removing most of the water-soluble nonprotein constituents. HPC contains at least 65% protein ($N \times 6.25$) on a dry weight basis. The most purified and enriched form of the commercial protein product, hemp protein isolate (HPI, > 90% protein), is prepared to meet food processing needs that entail minimal influence of unwanted nonprotein components. Alkaline extraction followed by isoelectric precipitation is the most common method to prepare HPI (Malomo et al., 2014; Wang, Jiang, & Xiong, 2018; Wang, Jin, & Xiong, 2018). Depending on specific extraction conditions (e.g. pH, temperature, time), a purity of up to 94% can be obtained. With the alkaline extraction, pH is generally 9–10, since the native hempseed proteins are tightly compacted and may be integrated with other components, like phenolic compounds and phytic acid.

HPI has been extracted exclusively from HSM obtained after processing the nondehulled hempseeds into edible hemp oil (Hadnađev et al., 2018; Malomo et al., 2014; Teh, Bekhit, Carne, & Birch, 2014; Yin et al., 2009). This, however, neglects the impact of the dehulling process on the composition and properties of the obtained hempseed protein. The dehulling process could greatly increase the protein content of the raw materials, as well as influencing the extractability, physicochemical properties, and in vitro digestion of protein from plant sources (Joshi & Varma, 2016; Pal et al., 2016; Rommi et al., 2014). Shen et al. (2020), Shen, Gao, Xu, Rao, and Chen (2020) found that the dehulling process significantly increased the extraction and protein recovery yields of HPI and achieved a higher protein purity and Arg/Lys ratio with improved color. The aromatic profile suggested that dehulled HPI accumulated more terpenes and less lipid oxidation volatiles than nondehulled HPI. These findings indicate that the dehulling process greatly enhances the extraction yield with a higher protein content and more characteristic aroma of HPI. In addition, the dehulling process has been shown to increase the yield of protein extracted from rapeseed press cakes (Rommi et al., 2014). Lentil protein isolate prepared with a dehulling process has shown a higher surface hydrophobicity than those with hulls (Aryee & Nickerson, 2012). Moreover, improved protein digestion has also been achieved in dehulled white bean flour (*Phaseolus vulgaris* L.) as evaluated by an in vitro infant digestion model (Romano, Giosafatto, Masi, & Mariniello, 2015).

Pojić et al. (2014) separated HSM using different particle sizes (> 350, > 250, > 180, and <180 μm) into four fractions. The highest yield (39.4%) was with particle size in the range of 180–250 μm, which was composed of ground cotyledon particles, and sieving yielded a 33.0% coarse fraction consisting of hulls. The finest fraction (< 180 μm) was lighter than those with higher particle sizes. The protein content of the fractions varied between 10.62% for the coarsest fraction and 44.36% for the finest cotyledon fraction. Similar results have been shown for chickpeas (Maaroufi et al., 2000). Malomo and Aluko (2015a) treated HSM with carbohydrase and phytases coupled with membrane ultrafiltration that enriched the protein content up to 70%. The protein digestibility of the obtained freeze dried retentate was significantly higher than that of HSM and a traditional isoelectric protein isolate.

Teh et al. (2014) used acid extraction to prepare HPI. The yield of protein extracted at an acidic pH was lower than that extracted at an alkaline pH. In addition, a method, known as "salt extraction with micellization," has been described for HPI (Dapčević-Hadnađev, Hadnađev, Lazaridou, Moschakis, & Biliaderis, 2018). HPI obtained by this method has a very high purity (98.9% protein, on a dry basis). Adverse chemical reactions such as the formation of lysinoalanine compounds from cysteine and serine residues can occur in highly alkaline conditions during heating. For example, HPI extracted at pH 10 and room temperature had a low level of lysinoalanine (0.8 mg/100 g protein), but at pH 12 at 40 °C for 5 min, the lysinoalanine content increased to 4 mg/100 g protein (Wang, Jin, & Xiong, 2018).

Enzyme-assisted extraction of proteins is generally considered a mild extraction with lower environmental impacts in comparison to acid- and alkaline-assisted extraction (Sari et al., 2015; Sari, Bruins, & Sanders, 2013). Rommi et al. (2015) demonstrated the effects of pectinolytic enzymes on the extraction yield of proteins from rapeseed press cakes. The treatment resulted in a 1.7-fold higher protein yield in comparison without enzymes. Tirgar, Silcock, Carne, and Birch (2017) showed that enzymatic and enzymatic-solvent extractions yielded higher protein contents from flaxseed meal in comparison to alkali extraction and with better emulsifying properties. Pap et al. (2020) studied the effect of FlavorPRO 750 MDP and Promod 439 L on protein yield, phytochemicals, and organoleptic quality of HSM and defatted HSM. The results showed that proteins and phytochemicals were mainly found in the sediment fractions, which were also regarded as tasty and lighter in color.

6.2.3 Nutritional properties of hempseed protein

Hemp protein contains all nine essential amino acids required by humans. Its amino acid profile is characterized by very high levels of arginine and glutamic acid, with a moderate quantity of sulfur-containing amino acids (Mattila, Mäkinen, et al., 2018). Generally, the amino acid profile of hempseed protein is comparable to that of egg white and soybeans, with a high concentration of arginine, glycine, and histidine (Callaway, 2004). Protein isolated from dehulled hempseed had higher protein digestibility-corrected amino acid scores (61) than that from whole wheat (40), pinto beans (57), or lentils (52); however, significantly lower than that from egg whites (100) or beef (92) (House et al., 2010). This research also reported that the digestibility scores were higher for the protein from dehulled hempseed (61) than the protein from whole hempseed (51) and hempseed meal (48). This suggests that the high concentration of antinutritional factors such as phytic acid and fiber may impede protein absorption. Amino acid scores for lysine, leucine, and tryptophan lied between 0.5 and 1.0, and suggested that hempseed alone was insufficient to meet the minimum daily intake of this amino acid as recommended by the FAO/WHO (House et al., 2010). Additionally, food processing conditions may further exacerbate this loss, as the ε-amino group of lysine is vulnerable to the Maillard reaction. Thus food manufacturers may need to further supplement hempseed-containing products with lysine in order to achieve good amino acid composition as required for human intake.

6.3 Hemp protein hydrolysates and bioactive peptides

Bioactive peptides are released during hydrolysis. Smaller peptides are better since they can avoid degradation by intestinal enzymes, enter the blood stream intact, and interact more easily with their target sites. The formation of the peptides heavily depends on the enzymes used and the hydrolysis time. Additionally, heat treatment before hydrolysis can alter the protein availability for the enzyme degradation. Overall, hempseed proteins are susceptible to proteolysis. Hydrophobicity is an important feature for bioactive peptides since it enhances the permeability through the gastrointestinal barrier and cell membranes (Aiello, Lammi, Boschin, Zanoni, & Arnoldi, 2017) (Table 6.1).

6.3.1 Antioxidative effect

Antioxidative activity is beneficial for prolonged food shelf life and increased quality. It can also have health benefits by protecting cells against reactive oxygen species-mediated damage. Several reports indicate that hempseed protein hydrolysates have good antioxidative

TABLE 6.1   Reported bioactivities of peptides derived from hempseed protein.

Processes	Sequence	Bioactivity	Reference
Fermentation (LAB)	ALASIGKATR	Antioxidative	Pontonio et al. (2020)
	IGQSHPQALMYPLLVACKSISNLR		
	AQVSVGGGR		
	AIENGAVSVSEPEEK		
	DLQIIAPSR		
Pepsin + pancreatin	WVYY	Antioxidative/hypotensive	Girgih et al. (2014)
	PSLPS		
	WYT	Hypotensive	Girgih et al. (2014)
	SVYT		
	IPAGV		
Alcalase	LR	α-Glucosidase inhibitor	Ren et al. (2016)
	PLMLP		
Acid hydrolysis	GVLY	ACE inhibitor	Orio et al. (2017)
	IEE		
	LGV		
	RVR		

activity. Hydrolysates by several proteases (Alcalase, Flavourzyme, Neutrase, Protamex, pepsin, trypsin) showed antioxidant properties with DPPH radical scavenging and Fe^{2+} chelating abilities. The properties showed strong correlation with trichloroacetic (TCA) soluble peptide content and hydrophobicity. Alcalase hydrolysis of HPI and further separation with macroporous adsorption resin improved DPPH radical scavenging activity. The active fraction was further separated by gel filtration and RP-HPLC to obtain two purified peptides, namely NHAV and HVRETALV (Lu et al., 2010). Girgih, Udenigwe, and Aluko (2011) subjected HPI to the sequential action of pepsin and pancreatin and applied membrane ultrafiltration to separate the protein hydrolysate into different molecular weight (MW) fractions. Three peptide fractions with MWs of 1 to 3, 3 to 5, and 5 to 10 kDa exhibited significantly greater hydroxyl radical scavenging activity but weaker metal chelation activity than the whole protein hydrolysate.

Reverse-phase HPLC purification resulted to eight less heterogeneous peptide fractions based on the elution time (Girgih, Udenigwe, & Aluko, 2011). Some of the peptide fractions demonstrated excellent antioxidant activity with higher oxygen radical absorbance capacity as well as the ability to scavenge superoxide anion and hydroxyl radicals than HPH. Several amino acids, including Lys, Leu, and Pro, appeared to contribute to the observed radical scavenging activity. Totally 23 short-chain peptides were isolated and sequenced from the pepsin pancreatin digests and Trp-Val-Tyr-Tyr and Pro-Ser-Leu-Pro-Ala were found to be the most potential antioxidant peptides (Girgih et al., 2014). Another technique used to produce bioactive peptides is fermentation. Pontonio et al. (2020) found that whole hempseed flour fermentation with lactic acid bacteria (LAB), *Lactiplantibacillus plantarum* 18S9 and *Leuconostoc mesenteroides* 12MM1, produced even stronger antioxidative capabilities than hydrolysis with commercial enzymes xylanase (Depol761P) and protease (VeronPS). The fermented hemp was fractionated through RP-FPLC and five peptides (ALASIGKATR, IGQSHPQALMYPLLVACKSISNLR, AQVSVGGGR, AIENGAVSVSEPEEK, and DLQIIAPSR) were identified from the most active fraction (Pontonio et al., 2020). A protective effect was further shown on human keratinocytes predisposed to hydrogen peroxide. Hydrolysates, especially when LAB fermented, significantly reduced cytotoxicity even when compared to when treated using α-tocopherol. On the other hand, the fractions contained also phenolic compounds, cannabisin C and D, *N-trans*-caffeoyltyramine, and *N-trans*-feruloyltyramine, which also increased during bioprocessing, and they are known for their powerful DPPH and ABTS radical scavenging activity (Pontonio et al., 2020).

6.3.2 Antihypertensive effect

Angiotensin-converting enzyme (ACE) and renin are two enzymes regulating human blood pressure levels. The renin-angiotensin aldosterone system is the primary pathway for the blood pressure (BP) regulation. Renin cleaves angiotensinogen that is further cleaved by ACE producing angiotensin-II (AT-II). AT-II then narrows the blood vessels causing a BP increase.

Different forms of enzymatic hydrolysis (pepsin, Alcalase, papain, and pepsin + pancreatin) have been used to produce antihypertensive protein hydrolysates and tested for in vitro inhibition of renin and ACE. The hempseed protein hydrolysate (HPH) treated by the sequential action of pepsin and pancreatin led to significantly high in vitro inhibition of the activities of ACE and renin (Girgih, Udenigwe, Li, Adebiyi, & Aluko, 2011). Teh, Bekhit, Carne, and Birch (2016) found that the type of HPI extract (alkali/acid soluble protein isolate), protease (AFP4000, HT proteolytic concentrate, Protease G, actinidin, and zingipain), and hydrolysis time had a significant effect on the resulting ACE inhibitory activity. Among enzymes HT generated the highest ACE inhibitory activity at 2h hydrolysis time, but reduced upon further hydrolysis to 4h. Generally, HPI produced by alkali treatment showed higher inhibitory potencies than the ones with acid treatment (Teh et al., 2016). In addition to enzymes, extensive chemical hydrolysis (6M HCl) released four peptides (GVLY, IEE, LGV, and RVR) with ACE inhibitory activity. GVLY possessed the highest ACE inhibitory activity, while IEE was almost inactive in inhibiting ACE (Orio et al., 2017).

The overall results indicated 1% Alcalase HPH was the most effective systolic blood pressure-reducing agent (-32.5 ± 0.7 mmHg after 4h), whereas the pepsin HPHs produced longer-lasting effects (-23.0 ± 1.4 mmHg after 24h). Therefore a combination of Alcalase treatment with pepsin can produce a long-lasting and effective BP-lowering effect (Malomo, Onuh, Girgih, & Aluko, 2015). Short-chain peptides (≤ 5 amino acids), such as WVYY and PSLPA, showed maximum systolic blood pressure reduction in spontaneously hypertensive rats by 34 (2h) and 40mmHg (4h), respectively, after oral administration of 30mg kg^{-1} body weight dose. The same peptides have shown also antioxidant properties (see earlier).

6.3.3 Antiproliferative effect

Hempseed protein hydrolysate has been proposed to have anticancer properties (Wei et al., 2021). Seed hydrolysate produced with papain and neutral proteases were reported to cause apoptosis of liver cancer cells (Hep3B) but have no effect on normal cells (L02). An antiproliferative effect is caused via modulated activity of the Akt/GSK/β-catenin signaling pathway. Additionally, bioactive peptides have been found to induce increased ROS levels in cells and in the mitochondria, which leads to cell death. However, no responsible peptide has been identified (Wei et al., 2021). The hydrolyzation of hempseed protein with gastrointestinal enzymes probably releases peptides that can inhibit the growth of human colon cancer cells (HCT-116, HT-21). It is important to note that heat pretreatment stops the antiproliferative activity (Lin, Pangloli, Meng, & Dia, 2020). Thus nonheated hempseed protein could be beneficial for colon cancer avoidance. However, no effect of the hydrolysate on normal colon cells has been studied and the peptide or active compound responsible for the effect has not been reported. Protein hydrolysate digested with Alcalase, Neutrase, and Protamex has been shown to be able to inhibit the growth of keratinocytes with carcinoma (HeLa) in a dose-dependent manner, whereas no cytotoxic effect has been seen on the normal keratinocyte cells (HaCaT) (Logarušić et al., 2019). The active peptide has not been identified. These studies show that hempseed protein hydrolysates can contain potential antiproliferative agents, thus more research is needed.

6.3.4 Hypocholesterolemic effect

The key enzyme in cholesterol metabolism is 3-hydroxy-3-methyl-glutaryl-coenzyme A (HMGCoAR). HMGCoAR inhibition leads to lowered blood cholesterol levels and it is the target of statins. HMGCoAR activity is regulated by phosphorylation via the AMPK pathway.

In vitro measurements have revealed bioactive peptides from hempseed hydrolysate to be an effective inhibitor of HMGCoAR. Effective peptides have been found to be released with pepsin, trypsin, pancreatin, and with a combination of these three enzymes (mimicking gastrointestinal digestion). In tested conditions, tryptic hydrolysis led to the highest HMGCoAR inhibition whereas pancreatin hydrolysate was the least effective (Aiello et al., 2017). Zanoni, Aiello, Arnoldi, and Lammi (2017) found similar results by using pepsin hydrolysate. Further, pepsin hydrolysate has been able to increase the low-density lipoprotein (LDL) intake from the extracellular matrix in HepG2 cells by increasing the level of low-density lipoprotein receptor (LDLR) responsible for LDL intake. Bioactive peptides have been shown to be able to cause the phosphorylation of AMPK which in turn leads to the phosphorylation of HMGCoAR and thus inactivation. Furthermore, the quantity of regulatory element binding protein 2 (SREBP2), as well as the transcription factor affecting the HMGCoAR and LDLR mRNA, was increased by the bioactive peptides. Since the authors were able to detect 90 peptides from the hydrolysate, there are a large number of possible candidates which could be the key peptide for the hypocholesterolemic effect (Zanoni et al., 2017). Further investigation is needed to show if these peptides are able to pass into blood circulation from the gastrointestinal track, and thus be beneficial to health. In addition to the earlier mentioned functionalities, hempseed protein hydrolysates have been shown to have antiinflammatory, neuroprotective, and α-glucosidase inhibitory properties (Malomo & Aluko, 2016, 2019; Ren et al., 2016; Rodriguez-Martin et al., 2019, 2020). Hempseed is a promising source of bioactive peptides and future research is needed to identify the functional peptides and further study the effect on in vivo.

6.4 Techno-functional properties of hempseed proteins

In addition to the nutritional quality, protein ingredients need to have certain techno-functional properties to be applicable in food products. Behavior and performance of the ingredients in food systems during preparation, processing, storage, and consumption is largely determined by the techno-functional properties. Since many food products exist as beverages, emulsions, foams, or solids, the solubility of protein ingredients and their ability to bind water or fat, form gels, films, foams, and emulsions are essential to the quality of the final food products. These properties are largely determined by the nature and extent of interactions between protein molecules and other components (for example, water and oil) in the food systems. Depending on the processing methods applied for producing the proteinaceous ingredients, proteins can undergo some level of denaturation which can adversely affect the techno-functionality. Information on the functionality of hempseed proteins is still rather limited; however, scientific interest has grown significantly in the recent years. Hereafter, the present knowledge on the techno-functional properties of hempseed protein ingredients and the effect of processing methods on the functionality are summarized.

6.4.1 Solubility

Solubility and dispersion stability of proteins are important factors that affect various other functional properties. Proteins are generally most soluble at the native state, whereas denaturation by heat, shear, or chemicals may change the surface properties of proteins in the unfolded state and thereby promote aggregation. Solubility of the hempseed proteins is determined mainly by the salt-soluble globulins, edestin, and water-soluble albumin, which comprise ~ 75% and ~ 25% of the total protein in hempseed. Solubility of proteins is greatly affected by the pH value. Solubility of hempseed protein is generally low at neutral pH values; solubility values from 8 to 38% have been reported depending on the solubilization method and centrifugation force. However, hempseed protein solubility can increase from 65% to 90% at pH values above 8.0 (Hadnađev et al., 2018; Tang et al., 2006). The alkaline soluble character of hempseed protein has been affiliated to the dissociation of edestin (Goring & Johnson, 1955). In comparison with soy protein, hempseed protein is more soluble at pH values below pH 8.0 whereas at pH values greater than 8.0, the solubility is similar (Tang et al., 2006).

Regarding the processes for producing the protein extracts, hempseed meal proteins as such are rather insoluble due to crosslinking with phytate and high content of insoluble fiber which hinders protein-water interaction. Protein extraction and concentration with membrane filtration techniques have been used to produce hempseed protein concentrates with high solubility values, up to 74% at pH 4.0–5.0 where the solubility of hempseed protein is typically low (Malomo & Aluko, 2015a). On the other hand, if the protein concentrates are purified further, it may affect adversely the solubility. For example, isoelectric precipitation is known to cause protein denaturation which results in the formation of insoluble aggregates. Micellization method has been reported to produce protein isolates with higher solubility values, which have been related to the higher presence of proteins in native state.

6.4.2 Water-holding capacity

The interactions between protein and water, and the ability to entrap water voluminously, are highly important for many food product systems. Water-holding capacity (WHC) is a commonly used term to describe the amount of water entrapped in the protein matrix and the capacity of proteins to retain water against gravity. WHC is dependent on the protein concentration and conformation, amino acid profile, level of hydrophobicity, pH, ionic strength, and temperature (reviewed by Moure, Sineiro, Domínguez, & Parajó, 2006). Hempseed flour has been shown to possess higher WHC in comparison to, for example, green pea and fava bean, buckwheat and wheat (Raikos, Neacsu, Russell, & Duthie, 2014). This was suggested to be due to the high protein content of hempseed flour. Also, hempseed protein isolates have shown superior WHC values in comparison to the flaxseed and canola protein isolates, both in alkaline and acidic extractions (Teh et al., 2014). Alkaline extraction of hempseed proteins results in higher WHC values than both acidic extraction (Teh et al., 2014) and micellization technique (Hadnađev et al., 2018). However, enzymatic treatment to digest fiber to enhance the release of proteins followed by membrane ultrafiltration has been shown to produce hempseed protein concentrates with remarkably higher WHC values than hempseed protein isolates or hempseed meal (Malomo & Aluko, 2015a). This was related to the loosened protein structure by the enzyme treatment which may promote the protein-water interactions and also due to the removal of nonprotein compounds with low molecular weight, e.g. phytate, by the membrane filtration (Malomo & Aluko, 2015a).

6.4.3 Emulsifying properties

Emulsifying properties are important functional characteristics in the development of new plant protein ingredients for foods. In most food emulsion, proteins are the dominating components. Emulsions consist of an oil phase containing hydrophobic compounds and an aqueous phase containing water-soluble compounds. One of the phases is dispersed into the other defined as oil-in-water (O/W) where lipid droplets are in aqueous media, or water-in-oil (W/O) where aqueous solution droplets are in a continuous lipid phase. Amphiphilic components such as proteins can decrease the interfacial tension at the water-oil interface and form films, and thereby affect greatly on the structure, thickness, mechanical properties, and stability of the emulsion systems. Emulsifying properties of proteins are related to many factors, for example, the rate of protein adsorption at the oil–water interface, the conformation of the proteins at the interface, the extent of interfacial tension reduction, and the rheology of the cohesive film. Even if proteins are not fully soluble, they can participate in the emulsion forming and stabilization. To describe the emulsifying properties of proteins in food emulsion systems, terms such as emulsion capacity (EC), emulsion stability (ES), and emulsifying activity (EA) are generally used. Among these, EA is the most commonly used term; it describes the ability of the protein to participate in emulsion formation and to stabilize the newly created emulsion.

Reports on the EA of hempseed protein products are highly variable. One probable reason for the variability is the different protein extraction procedures causing variation in the edestin and albumin ratio and configuration of the proteins. Malomo and Aluko (2015b) used carbohydrase treatment and consequent membrane filtration (10 kDa) for producing protein concentrate from the industrial hempseed meal. The resulting hempseed protein concentrate (HPC) formed emulsions with rather large oil droplets; the emulsion forming properties were weaker in comparison to the hempseed protein isolate prepared with traditional isoelectric precipitation. However, the emulsions prepared with the HPC were very stable. According to Tang et al. (2006), emulsifying activity of hempseed protein isolate produced with isoelectric precipitation (HPI) is remarkably lower in comparison to soy protein isolate.

According to Tang et al. (2006), emulsifying activity of HPI produced with isoelectric precipitation is remarkably lower in comparison to soy protein isolate. Wang, Jin, and Xiong (2018) studied the possibility to enhance the functional properties of HPI by alkaline pH shift and heat treatments. Significant improvement in emulsifying activity of HPI was obtained when pH 12 was applied with heat treatment at 30–60°C. The change was attributed to the dissociation of the protein subunits and induced level of cross-linked polymers, resulting in an overall increase in protein solubility and surface hydrophobicity. Heating alone didn't change the emulsifying activity of HPI and showed only a minor effect on protein solubility. Wang, Jiang, and Xiong (2018) applied high-pressure homogenization and pH shift treatment to produce a nonthermally processed, physically and oxidative stable hemp milk. Both, the pH shift treatment process and high-pressure homogenization process showed stabilizing effect on hemp milk emulsion and their combination exhibited the maximum benefit. The effects of the pH shift treatment and consequent high-pressure homogenization were attributed to the formation of large protein clusters and aggregates that stabilized the emulsion. The milk was also good in terms of microbiological quality and level of lipid oxidation. Hence, the combination of pH shift and high-pressure homogenization seems an interesting opportunity for producing hemp protein ingredients for food emulsion systems.

6.4.4 Foaming properties

The property of proteins to form stable foams is important for various food products. Foam is a two-phase system consisting of air bubbles separated by a thin continuous liquid layer. Foams with equal distribution and small size of air bubbles form body, smoothness, and lightness to the food. Proteins in foams contribute to the uniform distribution of fine air bubbles in the structure of foods. Food foams also participate in the volatilization of flavors and thereby, taste of the food products. Foaming properties of protein ingredients are determined mainly by protein concentration, solubility, and the level of hydrophobicity (Malomo et al., 2014). These properties determine the ability of proteins to reduce surface tension. Just as pH affects solubility it also affects foaming ability of hempseed proteins. Foaming capacity of hempseed meal is known to improve when pH values are increased from 3.0 to 9.0, which is related to the increased protein unfolding and interaction with water at neutral and alkaline pH values (Malomo & Aluko, 2015a). However, foaming properties of hempseed meal are reported to be much lower in comparison to the more purified hempseed proteins ingredients. This is most probably due to the high level of nonprotein compounds in hempseed meal, which inhere the ability of hempseed meal to form and stabilize foams. Hempseed protein isolates have shown better foaming capacity at pH 3.0 in comparison to pH 5.0, 7.0, and 9.0 (Malomo et al., 2014). When comparing albumin and globulin fractions of hempseed, albumin fraction has shown significantly higher foaming capacity than globulin fraction at pH range from 3.0 to 9.0 (Malomo & Aluko, 2015a). This is suggested to be due to the higher solubility of albumin which increases the capacity of protein molecules to encapsulate air particles (Sai-Ut, Ketnawa, Chaiwut, & Rawdkuen, 2009).

6.4.5 Effects of processing methods on the techno-functional properties of the hempseed proteins

As described before, hempseed protein ingredients have varying properties depending on the processing methods used for the production. Often, the functionality of hempseed protein ingredients is lower in comparison to, for example, milk or soy proteins. Therefore research activities have been put developing methods for improving the functionalities of hemp proteins to increase their application potential in different food products. For example, pH shift and heat treatments have demonstrated promising efficacies.

6.4.5.1 Defatting

Hempseeds are rich in lipids and hempseed meal after industrial oil removal still contains typically around 10% (of fresh weight) lipids (e.g. House et al., 2010). Although these lipids are high in nutritional quality, they may adversely affect the sensory quality and functional properties of hempseed protein ingredients. Thus different methods are needed for removing the residual oil prior to protein extraction. Defatting conditions, namely the solvent nature, temperature, and pressure, influence the functional properties of further extracted proteins by inducing changes in their conformational state (Espinosa-Pardo, Savoire, Subra-Paternault, & Harscoat-Schiavo, 2020). Shen, Gao, Xu, Rao, and Chen (2020) tested the effect of different solvents used for defatting on the chemical composition, structure, and functional properties of hemp protein isolates. According to the results, hexane extraction and Folch defatting process can increase the protein recovery yields, result in higher protein purity, and increase thermal stability of hempseed protein isolates. However, it needs to be noted that Folch defatting includes toxic solvents, chloroform and methanol. Supercritical carbon dioxide is recognized as an adequate solvent for extracting lipids from plant sources as it is nonexplosive, nontoxic, selective, and easily separated from the final extract, what preserves its biological properties. Pihlanto, Nurmi, Pap, Mäkinen, and Mäkinen (2021) applied the supercritical carbon dioxide extraction (SFE) for defatting hempseed, peeled hempseed, geminated hempseed, and hempseed meal prior to protein extraction. SFE increased protein recovery yields but didn't show effect on the emulsification properties of the protein extracts.

6.4.5.2 Dehulling

Dehulling has been recently studied for increasing protein recovery levels and to enhance the sensory quality and functional attributes of hempseed protein isolates (Pihlanto et al., 2021; Shen, Gao, Xu, Ohm, et al., 2020; Shen, Gao, Xu, Rao, & Chen, 2020). Dehulling as a pretreatment prior to protein extraction increased protein yields and sensory properties of the resulting protein extracts in both studies. However, dehulling didn't affect emulsifying properties (Pihlanto et al., 2021) and reduction in thermal stability was observed (Shen, Gao, Xu, Rao, & Chen, 2020).

6.4.5.3 The pH treatments

Alkaline pH 12 shift treatment has been successful for increasing hempseed protein solubility and emulsifying properties (Wang, Jiang, & Xiong, 2018; Wang, Jin, & Xiong, 2018) as discussed previously. The pH shift in combination with heat treatment enhanced solubility and emulsifying properties of hempseed protein isolates (Wang, Jin, & Xiong, 2018). When the pH shift was applied prior to high-pressure homogenization of crude hempseed protein extract, a physically and oxidative stable hempseed milk was obtained (Wang, Jiang, & Xiong, 2018). In traditional isoelectric precipitation process, the protein precipitation pH has also shown effects on the functional attributes of the protein isolates (Shen, Gao, Xu, Ohm, et al., 2020; Shen, Gao, Xu, Rao, & Chen, 2020). Adjusting the precipitation pH to 6.0 significantly increased the protein recovery yield compared to pH 5.0. Precipitation at pH 6.0 together with Folch defatting process greatly enhanced the protein purity meanwhile reducing the level of lipids and carbohydrates. Adjusting pH to 6.0 during precipitation process also increased the thermal stability of hemp protein. The results suggest that applying precipitation pH to 6.0 instead of commonly used 5.0 can increase the hempseed protein extraction yield with higher protein purity and improved functional properties. The effect of pH on the properties of hempseed protein extracts has also been studied recently by Potin, Lubbers, Husson, and Saurel (2019). At acidic pH (2 to 7) the extraction yields were low corresponding mainly to solubilization of hempseed albumins. The extraction of globulins increased significantly above pH 8, with protein extraction yields varying from 8% at pH 8 to 67% at pH 12.

Heat treatment has been applied to modify hempseed protein solubility and functional properties. High temperature has been shown to cause mainly adverse effects on hempseed protein solubility (Raikos et al., 2014; Yin et al., 2008). However, heat treatments have been shown to increase emulsifying activity (Wang, Jiang, & Xiong, 2018; Wang, Jin, & Xiong, 2018), water-holding capacity, and foaming stability (Yin et al., 2008) of hempseed proteins.

6.4.5.4 Other methods

Mild enzymatic treatments, conjugation with polysaccharides and extrusion, are examples of other processes applied for enhancing the functional properties of hempseed proteins. Proteases, such as trypsin, have been applied to increase hempseed protein solubility, water-holding and fat-absorption capacity. Solubility was enhanced by the slightly increased hydrolyzation degree of proteins; however, water- and fat-holding capacity was decreased (Yin et al., 2008). Conjugation with polysaccharides is suggested to be a suitable option for improving functional properties of plant proteins. This approach could be suitable especially at low and neutral pH values (reviewed by (Akhtar & Ding, 2017). Recently also extrusion has been applied successfully for producing high moisture meat analogues with excellent functional properties (Zahari et al., 2020).

6.5 Hempseed protein allergenicity

Hempseeds have been considered a "mild" allergenic product, especially when compared to other seeds. Severe cases of allergenic reactions have been reported when consuming roots, leaves, or flowers of the plant. This is possible because the allergens in hemp seem to be photosynthesis-related proteins (Nayak et al., 2013). There are also reports of anaphylaxis after the ingestion of hempseeds (Bortolin et al., 2016). Since the consumption of hempseeds is increasing, more allergy cases will most probably arise. Hemp allergies are studied mostly using other parts of the plant than the seeds, and thus information on hempseed causative food allergies is limited.

Decuyper et al. (2015) listed five known allergenic proteins in hemp from the literature. These are nsLTP/PR-14 (Can s3), profilin, oxygen-evolving enhancer protein, TLP/PR-5, and ribulose-1,5-biphosphate carboxylase/oxygenase (Rubisco), and Rubisco is the only protein found in seeds (Mamone, Picariello, Ramondo, Nicolai, & Ferranti, 2019). It is important to note that nsLTP type of proteins are widely spread in the plant kingdom and the use of cannabis leaves and flowering parts can sensitize people to other vegetables and fruits via cross-reactivity (Decuyper et al., 2015).

Mamone et al. (2019) used two hempseed products (hemp flour and hemp protein isolate) to study the survival of possible allergenic peptides during intestinal digestion using in vitro models. Gastrointestinal digestion was completed with brush border enzymes. In this study, hempseed proteins were found to be efficiently digested and the peptides survived only from 12 proteins of which none were known allergenic proteins. However, six peptides detected after the digestion belonged to Z-serpins, a group of protease inhibitors. Z-serpins in wheat and barley can be potential triggers for IgE-mediated food allergies. They also showed that simple precipitation-based processing to produce hempseed protein isolates from defatted meal would be sufficient to eliminate all known allergenic proteins from the product.

6.6 Challenges and opportunities in hemp protein utilization

For the use of hempseed proteins, multiple steps and aspects are needed to evaluate carefully before the process can be scaled from lab or pilot scale to industrial relevant level.

The research on processes using different plant sources and side streams is increasing, however, economic, environmental evaluation needs to be considered thoroughly in the future.

Multiresponse optimization has become an important issue in complex industrial processes, particularly in situations where more than one correlated response must be assessed simultaneously. Several publications have presented approaches using principal component analysis (PCA) for addressing the correlation among multiple quality characteristics to obtain uncorrelated components (De Paiva, Gomes, Peruchi, Leme, & Balestrassi, 2014; Salmasnia, Baradaran Kazemzadeh, & Niaki, 2012). Response surface methodology (RSM) is frequently used. This technique provides comprehensive and informative insight into the system, which leads to faster process optimization. PCA-based multiresponse optimization mostly assumes that only the first principal component (PC) or few PCs are dominant to represent the original responses, however, this is not always enough to explain most of the variance-covariance structure in the data set. Some authors have proposed a method called weighted principal components (WPC) or weighted multivariate index (WMI). It is a strategy that aggregates all the significant PC scores weighted by their respective eigen values (De Paiva et al., 2014; Salmasnia et al., 2012). Purkayastha, Dutta, Barthakur, and Mahanta (2015) presented systematic optimization technique, using WMI, RSM, and desirability function, and applied on a multiresponse process to extract light-colored rapeseed protein having reduced phytate level. This combined approach was used to optimize many multiple correlated quality characteristics, including color parameters, protein yield, and phytate level. The advantage of the methodology is that the optimal factor combination reflects a compromise between the partially quality characteristics. To prepare a good quality HPI, antinutrients, functional properties and color of the end products should be included in addition to protein yield.

The superior amino acid content makes hempseed a very interesting source for food manufactures to develop a wide range of products. The low allergenicity of hemp proteins permits it as a substitute for other proteins in some food products. The major challenges are related to the physicochemical and sensory qualities of hempseed protein fractions. Studies by Mikulec et al. (2019), Korus, Gumul, Krystyjan, Juszczak, and Korus (2017), and Radočaj, Dimić, and Tsao (2014) observed a negative correlation between the level of hempseed inclusion and sensory evaluation scores for bread and biscuits. Inclusion of 10% HPC or hempseed flour improved the tested sensory properties, except structure and porosity, in gluten-free bread (Korus, Witczak, Ziobro, & Juszczak, 2017). Therefore structure-modifying technologies must be vigorously explored through scientific research to convert hemp protein into a more soluble and diversely functional protein. Additional research is also needed to investigate the health benefits using molecular, cellular, and animal models. Such basic and applied research is essential to the development of this valuable protein source and broadening its market potential in the food industry.

6.7 Conclusion

Hempseed is an excellent nutrition source with good fatty acid and amino acid profiles. Nutritional value and health benefits of hempseed can be attributed with rich source of beneficial amino acids and the encrypted bioactive peptides within the proteins. Research related to the understanding of chemical composition, nutritional value, health benefits, and processing impact, as well as functional properties of hempseed proteins, has been progressed in the recent years; however, much remains unknown and requires further research. Especially, the chemical component interaction in food and its variety–geography–composition–nutrition relationship need further investigations. Breeding is required to develop low THC varieties with suitability to different geographical areas, high grain yield and quality, superior amino acid and fatty acid profile, and other nutritional compounds. Technologies, especially mild processing, are needed to increase the sensory and functional properties of hempseed proteins for development of attractive and acceptable food products.

References

Aiello, G., Fasoli, E., Boschin, G., Lammi, C., Zanoni, C., Citterio, A., et al. (2016). Proteomic characterization of hempseed (C. sativa L.). *Journal of Proteomics, 147*, 187–196. https://doi.org/10.1016/j.jprot.2016.05.033.

Aiello, G., Lammi, C., Boschin, G., Zanoni, C., & Arnoldi, A. (2017). Exploration of potentially bioactive peptides generated from the enzymatic hydrolysis of hempseed proteins. *Journal of Agricultural and Food Chemistry, 65*(47), 10174–10184. https://doi.org/10.1021/acs.jafc.7b03590.

Akhtar, M., & Ding, R. (2017). Covalently cross-linked proteins & polysaccharides: Formation, characterization and potential applications. *Current Opinion in Colloid and Interface Science, 28*, 31–36. https://doi.org/10.1016/j.cocis.2017.01.002.

Amaducci, S., Scordia, D., Liu, F. H., Zhang, Q., Guo, H., Testa, G., et al. (2015). Key cultivation techniques for hemp in Europe and China. *Industrial Crops and Products, 68*, 2–16. https://doi.org/10.1016/j.indcrop.2014.06.041.

Aryee, F. N. A., & Nickerson, M. T. (2012). Formation of electrostatic complexes involving mixtures of lentil protein isolates and gum Arabic polysaccharides. *Food Research International, 48*(2), 520–527. https://doi.org/10.1016/j.foodres.2012.05.012.

Ash, A. L. (1948). Hemp-production and utilization. *Economic Botany, 2*(2), 158–169. https://doi.org/10.1007/BF02858999.

Bortolin, K., Ben-Shoshan, M., Kalicinsky, C., Lavine, E., Lejtenyi, C., Warrington, R. J., et al. (2016). Case series of 5 patients with anaphylaxis to hemp seed ingestion. *Journal of Allergy and Clinical Immunology*. https://doi.org/10.1016/j.jaci.2015.12.969, AB239.

Callaway, J. C. (2004). Hempseed as a nutritional resource: An overview. *Euphytica, 140*(1–2), 65–72. https://doi.org/10.1007/s10681-004-4811-6.

Carus, M., & Sarmento, L. (2016). *The European hemp industry: Cultivation, processing and applications for fibers, shivs, seeds and flowers* (pp. 1–9). European Industrial Hemp Association.

Cherney, J. H., & Small, E. (2016). Industrial hemp in North America: Production, politics and potential. *Agronomy, 6*(4), 58. https://doi.org/10.3390/agronomy6040058.

Dapčević-Hadnađev, T., Hadnađev, M., Lazaridou, A., Moschakis, T., & Biliaderis, C. G. (2018). Hempseed meal protein isolates prepared by different isolation techniques. Part II. Gelation properties at different ionic strengths. *Food Hydrocolloids, 81*, 481–489. https://doi.org/10.1016/j.foodhyd.2018.03.022.

De Paiva, A., Gomes, J., Peruchi, R., Leme, R., & Balestrassi, P. (2014). A multivariate robust parameter optimization approach based on principal component analysis with combined arrays. *Computers & Industrial Engineering, 74*, 186–198.

Decuyper, I., Ryckebosch, H., Van Gasse, A. L., Sabato, V., Faber, M., Bridts, C. H., et al. (2015). Cannabis allergy: What do we know anno 2015. *Archivum Immunologiae et Therapiae Experimentalis, 63*(5), 327–332. https://doi.org/10.1007/s00005-015-0352-z.

Docimo, T., Caruso, I., Ponzoni, E., Mattana, M., & Galasso, I. (2014). Molecular characterization of edestin gene family in C. sativa L. *Plant Physiology and Biochemistry, 84*, 142–148. https://doi.org/10.1016/j.plaphy.2014.09.011.

Espinosa-Pardo, F. A., Savoire, R., Subra-Paternault, P., & Harscoat-Schiavo, C. (2020). Oil and protein recovery from corn germ: Extraction yield, composition and protein functionality. *Food and Bioproducts Processing, 120*, 131–142. https://doi.org/10.1016/j.fbp.2020.01.002.

Girgih, A. T., He, R., Malomo, S., Offengenden, M., Wu, J., & Aluko, R. E. (2014). Structural and functional characterization of hemp seed (*C. sativa* L.) protein-derived antioxidant and antihypertensive peptides. *Journal of Functional Foods*, 6(1), 384–394. https://doi.org/10.1016/j.jff.2013.11.005.

Girgih, A. T., Udenigwe, C. C., & Aluko, R. E. (2011). In vitro antioxidant properties of hemp seed (*C. sativa* L.) protein hydrolysate fractions. *Journal of the American Oil Chemists' Society*, 88(3), 381–389. https://doi.org/10.1007/s11746-010-1686-7.

Girgih, A. T., Udenigwe, C. C., Li, H., Adebiyi, A. P., & Aluko, R. E. (2011). Kinetics of enzyme inhibition and antihypertensive effects of hemp seed (*C. sativa* L.) protein hydrolysates. *Journal of the American Oil Chemists' Society*, 88(11), 1767–1774. https://doi.org/10.1007/s11746-011-1841-9.

Goring, D. A. I., & Johnson, P. (1955). The preparation and stability of ultracentrifugally monodisperse edestin. *Archives of Biochemistry and Biophysics*, 56(2), 448–458. https://doi.org/10.1016/0003-9861(55)90265-4.

Hadnađev, M., Dapčević-Hadnađev, T., Lazaridou, A., Moschakis, T., Michaelidou, A. M., Popović, S., et al. (2018). Hempseed meal protein isolates prepared by different isolation techniques. Part I. physicochemical properties. *Food Hydrocolloids*, 79, 526–533. https://doi.org/10.1016/j.foodhyd.2017.12.015.

House, J. D., Neufeld, J., & Leson, G. (2010). Evaluating the quality of protein from hemp seed (*C. sativa* L.) products through the use of the protein digestibility-corrected amino acid score method. *Journal of Agricultural and Food Chemistry*, 58(22), 11,801–11,807. https://doi.org/10.1021/jf102636b.

Joshi, P., & Varma, K. (2016). Effect of germination and dehulling on the nutritive value of soybean. *Nutrition and Food Science*, 46(4), 595–603. https://doi.org/10.1108/NFS-10-2015-0123.

Kerckhoffs, H., Kavas, Y., Millner, J., Anderson, C., & Kawana-Brown, E. (2015). Industrial hemp in New Zealand—potential for cash cropping for a better environment in the Taranaki region. In *Proceedings of the 17th ASA conference*.

Kim, J.-J., & Lee, M.-Y. (2011). Isolation and characterization of Edestin from Cheungsam hempseed. *Journal of Applied Biological Chemistry*, 84–88. https://doi.org/10.3839/jabc.2011.015.

Korus, A., Gumul, D., Krystyjan, M., Juszczak, L., & Korus, J. (2017). Evaluation of the quality, nutritional value and antioxidant activity of gluten-free biscuits made from corn-acorn flour or corn-hemp flour composites. *European Food Research and Technology*, 243(8), 1429–1438.

Korus, J., Witczak, M., Ziobro, R., & Juszczak, L. (2017). Hemp (*C. sativa* subsp. sativa) flour and protein preparation as natural nutrients and structure forming agents in starch based gluten-free bread. *LWT Food Science and Technology*, 84, 143–150. https://doi.org/10.1016/j.lwt.2017.05.046.

Leonard, W., Zhang, P., Ying, D., & Fang, Z. (2020). Hempseed in food industry: Nutritional value, health benefits, and industrial applications. *Comprehensive Reviews in Food Science and Food Safety*, 19(1), 282–308. https://doi.org/10.1111/1541-4337.12517.

Lin, Y., Pangloli, P., Meng, X., & Dia, V. P. (2020). Effect of heating on the digestibility of isolated hempseed (*C. sativa* L.) protein and bioactivity of its pepsin-pancreatin digests. *Food Chemistry*, 314. https://doi.org/10.1016/j.foodchem.2020.126198.

Logarušić, M., Slivac, I., Radošević, K., Bagović, M., Redovniković, I. R., & Srček, V. G. (2019). Hempseed protein hydrolysates' effects on the proliferation and induced oxidative stress in normal and cancer cell lines. *Molecular Biology Reports*, 46(6), 6079–6085. https://doi.org/10.1007/s11033-019-05043-8.

Lu, R. R., Qian, P., Sun, Z., Zhou, X. H., Chen, T. P., He, J. F., et al. (2010). Hempseed protein derived antioxidative peptides: Purification, identification and protection from hydrogen peroxide-induced apoptosis in PC12 cells. *Food Chemistry*, 123(4), 1210–1218. https://doi.org/10.1016/j.foodchem.2010.05.089.

Maaroufi, C., Melcion, J. P., De Monredon, F., Giboulot, B., Guibert, D., & Le Guen, M. P. (2000). Fractionation of pea flour with pilot scale sieving. I. Physical and chemical characteristics of pea seed fractions. *Animal Feed Science and Technology*, 85(1–2), 61–78. https://doi.org/10.1016/S0377-8401(00)00127-9.

Malomo, S. A., & Aluko, R. E. (2015a). A comparative study of the structural and functional properties of isolated hemp seed (*C. sativa* L.) albumin and globulin fractions. *Food Hydrocolloids*, 43, 743–752. https://doi.org/10.1016/j.foodhyd.2014.08.001.

Malomo, S. A., & Aluko, R. E. (2015b). Conversion of a low protein hemp seed meal into a functional protein concentrate through enzymatic digestion of fiber coupled with membrane ultrafiltration. *Innovative Food Science and Emerging Technologies*, 31, 151–159. https://doi.org/10.1016/j.ifset.2015.08.004.

Malomo, S. A., & Aluko, R. E. (2016). In vitro acetylcholinesterase-inhibitory properties of enzymatic hemp seed protein hydrolysates. *Journal of the American Oil Chemists' Society*, 93(3), 411–420.

Malomo, S. A., & Aluko, R. E. (2019). Kinetics of acetylcholinesterase inhibition by hemp seed protein-derived peptides. *Journal of Food Biochemistry*, 43(7). https://doi.org/10.1111/jfbc.12897.

Malomo, S. A., He, R., & Aluko, R. E. (2014). Structural and functional properties of hemp seed protein products. *Journal of Food Science, 79*(8), C1512–C1521. https://doi.org/10.1111/1750-3841.12537.

Malomo, S. A., Onuh, J. O., Girgih, A. T., & Aluko, R. E. (2015). Structural and antihypertensive properties of enzymatic hemp seed protein hydrolysates. *Nutrients, 7*(9), 7616–7632. https://doi.org/10.3390/nu7095358.

Mamone, G., Picariello, G., Ramondo, A., Nicolai, M. A., & Ferranti, P. (2019). Production, digestibility and allergenicity of hemp (*C. sativa* L.) protein isolates. *Food Research International, 115*, 562–571. https://doi.org/10.1016/j.foodres.2018.09.017.

Manamperi, W. A. R., Wiesenborn, D. P., Chang, S. K. C., & Pryor, S. W. (2011). Effects of protein separation conditions on the functional and thermal properties of canola protein isolates. *Journal of Food Science, 76*(3), E266–E273. https://doi.org/10.1111/j.1750-3841.2011.02087.x.

Mattila, P., Mäkinen, S., Eurola, M., Jalava, T., Pihlava, J. M., Hellström, J., et al. (2018). Nutritional value of commercial protein-rich plant products. *Plant Foods for Human Nutrition, 73*(2), 108–115. https://doi.org/10.1007/s11130-018-0660-7.

Mattila, P. H., Pihlava, J. M., Hellström, J., Nurmi, M., Eurola, M., Mäkinen, S., et al. (2018). Contents of phytochemicals and antinutritional factors in commercial protein-rich plant products. *Food Quality and Safety, 2*(4), 213–219. https://doi.org/10.1093/fqsafe/fyy021.

Mikulec, A., Kowalski, S., Sabat, R., Skoczylas, Ł., Tabaszewska, M., & Wywrocka-Gurgul, A. (2019). Hemp flour as a valuable component for enriching physicochemical and antioxidant properties of wheat bread. *LWT, 102*, 164–172. https://doi.org/10.1016/j.lwt.2018.12.028.

Moure, A., Sineiro, J., Domínguez, H., & Parajó, J. C. (2006). Functionality of oilseed protein products: A review. *Food Research International, 39*(9), 945–963. https://doi.org/10.1016/j.foodres.2006.07.002.

Nayak, A. P., Green, B. J., Sussman, G., Berlin, N., Lata, H., Chandra, S., et al. (2013). Characterization of *C. sativa* allergens. *Annals of Allergy, Asthma & Immunology, 111*(1), 32–e4. https://doi.org/10.1016/j.anai.2013.04.018.

O'Brien, C., & Arathi, H. S. (2019). Bee diversity and abundance on flowers of industrial hemp (*C. sativa* L.). *Biomass and Bioenergy, 122*, 331–335. https://doi.org/10.1016/j.biombioe.2019.01.015.

Odani, S., & Odani, S. (1998). Isolation and primary structure of a methionine- and cystine-rich seed protein of *cannabis sativa*. *Bioscience, Biotechnology, and Biochemistry, 62*(4), 650–654. https://doi.org/10.1271/bbb.62.650.

Orio, L. P., Boschin, G., Recca, T., Morelli, C. F., Ragona, L., Francescato, P., et al. (2017). New ACE-inhibitory peptides from hemp seed (*C. sativa* L.) proteins. *Journal of Agricultural and Food Chemistry, 65*(48), 10,482–10,488. https://doi.org/10.1021/acs.jafc.7b04522.

Pal, R. S., Bhartiya, A., ArunKumar, R., Kant, L., Aditya, J. P., & Bisht, J. K. (2016). Impact of dehulling and germination on nutrients, antinutrients, and antioxidant properties in horsegram. *Journal of Food Science and Technology, 53*(1), 337–347. https://doi.org/10.1007/s13197-015-2037-3.

Pap, N., Hamberg, L., Pihlava, J. M., Hellström, J., Mattila, P., Eurola, M., et al. (2020). Impact of enzymatic hydrolysis on the nutrients, phytochemicals and sensory properties of oil hemp seed cake (*C. sativa* L. FINOLA variety). *Food Chemistry, 320*. https://doi.org/10.1016/j.foodchem.2020.126530, 126530.

Patel, S., Cudney, R., & McPherson, A. (1994). Crystallographic characterization and molecular symmetry of edestin, a legumin from hemp. *Journal of Molecular Biology, 235*(1), 361–363. https://doi.org/10.1016/S0022-2836(05)80040-3.

Pihlanto, A., Nurmi, M., Pap, N., Mäkinen, J., & Mäkinen, S. (2021). The effect of processing of hempseed on protein recovery and emulsification properties. *International Journal of Food Science, 2021*, 1–12. https://doi.org/10.1155/2021/8814724.

Pojić, M., Mišan, A., Sakač, M., Dapčević Hadnađev, T., Bojana, Š., Milovanović, I., et al. (2014). Characterization of byproducts originating from hemp oil processing. *Journal of Agricultural and Food Chemistry, 62*, 12436–12442. https://doi.org/10.1021/jf5044426.

Pontonio, E., Verni, M., Dingeo, C., Diaz-De-cerio, E., Pinto, D., & Rizzello, C. G. (2020). Impact of enzymatic and microbial bioprocessing on antioxidant properties of hemp (*C. sativa* L.). *Antioxidants, 9*(12), 1–26. https://doi.org/10.3390/antiox9121258.

Ponzoni, E., Brambilla, I. M., & Galasso, I. (2018). Genome-wide identification and organization of seed storage protein genes of *C. sativa*. *Biologia Plantarum, 62*(4), 693–702. https://doi.org/10.1007/s10535-018-0810-7.

Potin, F., Lubbers, S., Husson, F., & Saurel, R. (2019). Hemp (*C. sativa* L.) protein extraction conditions affect extraction yield and protein quality. *Journal of Food Science, 84*(12), 3682–3690. https://doi.org/10.1111/1750-3841.14850.

Purkayastha, M. D., Dutta, G., Barthakur, A., & Mahanta, C. L. (2015). Tackling correlated responses during process optimisation of rapeseed meal protein extraction. *Food Chemistry, 170*, 62–73.

Radočaj, O., Dimić, E., & Tsao, R. (2014). Effects of hemp (*C. sativa* L.) seed oil press-cake and decaffeinated green tea leaves (*Camellia sinensis*) on functional characteristics of gluten-free crackers. *Journal of Food Science, 79*(3), C318–C325. https://doi.org/10.1111/1750-3841.12370.

Raikos, V., Neacsu, M., Russell, W., & Duthie, G. (2014). Comparative study of the functional properties of lupin, green pea, fava bean, hemp, and buckwheat flours as affected by pH. *Food Science & Nutrition, 2*(6), 802–810. https://doi.org/10.1002/fsn3.143.

Ranalli, P., & Venturi, G. (2004). Hemp as a raw material for industrial applications. *Euphytica, 140*(1–2), 1–6. https://doi.org/10.1007/s10681-004-4749-8.

Ren, Y., Liang, K., Jin, Y., Zhang, M., Chen, Y., Wu, H., et al. (2016). Identification and characterization of two novel α-glucosidase inhibitory oligopeptides from hemp (*C. sativa* L.) seed protein. *Journal of Functional Foods, 26*, 439–450. https://doi.org/10.1016/j.jff.2016.07.024.

Rodriguez-Martin, N. M., Montserrat-De la Paz, S., Toscano, R., Grao-Cruces, E., Villanueva, A., Pedroche, J., et al. (2020). Hemp (*C. sativa* l.) protein hydrolysates promote anti-inflammatory response in primary human monocytes. *Biomolecules, 10*(5). https://doi.org/10.3390/biom10050803.

Rodriguez-Martin, N. M., Toscano, R., Villanueva, A., Pedroche, J., Millan, F., Montserrat-De La Paz, S., et al. (2019). Neuroprotective protein hydrolysates from hemp (*C. sativa* L.) seeds. *Food & Function, 10*(10), 6732–6739. https://doi.org/10.1039/c9fo01904a.

Romano, A., Giosafatto, C. V. L., Masi, P., & Mariniello, L. (2015). Impact of dehulling on the physico-chemical properties and in vitro protein digestion of common beans (*Phaseolus vulgaris* L.). *Food & Function, 6*(4), 1345–1351. https://doi.org/10.1039/c5fo00021a.

Rommi, K., Hakala, T. K., Holopainen, U., Nordlund, E., Poutanen, K., & Lantto, R. (2014). Effect of enzyme-aided cell wall disintegration on protein extractability from intact and dehulled rapeseed (*Brassica rapa* L. and *Brassica napus* L.) press cakes. *Journal of Agricultural and Food Chemistry, 62*(32), 7989–7997. https://doi.org/10.1021/jf501802e.

Rommi, K., Holopainen, U., Pohjola, S., Hakala, T. K., Lantto, R., Poutanen, K., et al. (2015). Impact of particle size reduction and carbohydrate-Hydrolyzing enzyme treatment on protein recovery from rapeseed (*B. rapa* L.) press cake. *Food and Bioprocess Technology, 8*(12), 2392–2399. https://doi.org/10.1007/s11947-015-1587-8.

Sai-Ut, S., Ketnawa, S., Chaiwut, P., & Rawdkuen, S. (2009). Biochemical and functional properties of proteins from red kidney, navy and adzuki beans. *Asian Journal of Food and Agro-Industry, 2*(4), 493–504.

Salentijn, E. M. J., Zhang, Q., Amaducci, S., Yang, M., & Trindade, L. M. (2015). New developments in fiber hemp (*C. sativa* L.) breeding. *Industrial Crops and Products, 68*, 32–41. https://doi.org/10.1016/j.indcrop.2014.08.011.

Salmasnia, A., Baradaran Kazemzadeh, R., & Niaki, N. (2012). An approach to optimize correlated multiple responses using principal component analysis and desirability function. *The International Journal of Advanced Manufacturing Technology, 62*(5), 835–846.

Sari, Y. W., Bruins, M. E., & Sanders, J. P. M. (2013). Enzyme assisted protein extraction from rapeseed, soybean, and microalgae meals. *Industrial Crops and Products, 43*(1), 78–83. https://doi.org/10.1016/j.indcrop.2012.07.014.

Sari, Y. W., Mulder, W. J., Sanders, J. P. M., & Bruins, M. E. (2015). Towards plant protein refinery: Review on protein extraction using alkali and potential enzymatic assistance. *Biotechnology Journal, 10*(8), 1138–1157. https://doi.org/10.1002/biot.201400569.

Shen, P., Gao, Z., Xu, M., Ohm, J. B., Rao, J., & Chen, B. (2020). The impact of hempseed dehulling on chemical composition, structure properties and aromatic profile of hemp protein isolate. *Food Hydrocolloids, 106*. https://doi.org/10.1016/j.foodhyd.2020.105889, 105889.

Shen, P., Gao, Z., Xu, M., Rao, J., & Chen, B. (2020). Physicochemical and structural properties of proteins extracted from dehulled industrial hempseeds: Role of defatting process and precipitation pH. *Food Hydrocolloids, 108*. https://doi.org/10.1016/j.foodhyd.2020.106065.

Tang, C. H., Ten, Z., Wang, X. S., & Yang, X. Q. (2006). Physicochemical and functional properties of hemp (*C. sativa* L.) protein isolate. *Journal of Agricultural and Food Chemistry, 54*(23), 8945–8950. https://doi.org/10.1021/jf0619176.

Teh, S. S., Bekhit, A. E. D., Carne, A., & Birch, J. (2014). Effect of the defatting process, acid and alkali extraction on the physicochemical and functional properties of hemp, flax and canola seed cake protein isolates. *Journal of Food Measurement and Characterization, 8*(2), 92–104.

Teh, S. S., Bekhit, A. E. D. A., Carne, A., & Birch, J. (2016). Antioxidant and ACE-inhibitory activities of hemp (*C. sativa* L.) protein hydrolysates produced by the proteases AFP, HT, pro-G, actinidin and zingibain. *Food Chemistry, 203*, 199–206. https://doi.org/10.1016/j.foodchem.2016.02.057.

Tirgar, M., Silcock, P., Carne, A., & Birch, E. J. (2017). Effect of extraction method on functional properties of flaxseed protein concentrates. *Food Chemistry, 215*, 417–424. https://doi.org/10.1016/j.foodchem.2016.08.002.

Wang, Q., Jiang, J., & Xiong, Y. L. (2018). High pressure homogenization combined with pH shift treatment: A process to produce physically and oxidatively stable hemp milk. *Food Research International, 106,* 487–494. https://doi.org/10.1016/j.foodres.2018.01.021.

Wang, Q., Jin, Y., & Xiong, Y. L. (2018). Heating-aided pH shifting modifies hemp seed protein structure, cross-linking, and emulsifying properties. *Journal of Agricultural and Food Chemistry, 66*(41), 10827–10834. https://doi.org/10.1021/acs.jafc.8b03901.

Wang, X. S., Tang, C. H., Yang, X. Q., & Gao, W. R. (2008). Characterization, amino acid composition and in vitro digestibility of hemp (*C. sativa* L.) proteins. *Food Chemistry, 107*(1), 11–18. https://doi.org/10.1016/j.foodchem.2007.06.064.

Wei, L. H., Dong, Y., Sun, Y. F., Mei, X. S., Ma, X. S., Shi, J., et al. (2021). Anticancer property of hemp bioactive peptides in Hep3B liver cancer cells through Akt/GSK3β/β-catenin signaling pathway. *Food Science & Nutrition, 9*(4), 1833–1841. https://doi.org/10.1002/fsn3.1976.

Yin, S. W., Tang, C. H., Cao, J. S., Hu, E. K., Wen, Q. B., & Yang, X. Q. (2008). Effects of limited enzymatic hydrolysis with trypsin on the functional properties of hemp (*C. sativa* L.) protein isolate. *Food Chemistry, 106*(3), 1004–1013. https://doi.org/10.1016/j.foodchem.2007.07.030.

Yin, S. W., Tang, C. H., Wen, Q. B., & Yang, X. Q. (2009). Functional and structural properties and in vitro digestibility of acylated hemp (*C. sativa* L.) protein isolates. *International Journal of Food Science and Technology, 44*(12), 2653–2661. https://doi.org/10.1111/j.1365-2621.2009.02098.x.

Zahari, I., Ferawati, F., Helstad, A., Ahlström, C., Östbring, K., Rayner, M., et al. (2020). Development of high-moisture meat analogues with hemp and soy protein using extrusion cooking. *Foods, 9*(6). https://doi.org/10.3390/foods9060772, 772.

Zanoni, C., Aiello, G., Arnoldi, A., & Lammi, C. (2017). Hempseed peptides exert hypocholesterolemic effects with a statin-like mechanism. *Journal of Agricultural and Food Chemistry, 65*(40), 8829–8838.

The significance of industrial hemp knowledge management

Anamarija Koren[a], Milica Pojić[b], and Vladimir Sikora[a]

[a]Institute of Field and Vegetable Crops, National Institute of the Republic of Serbia, Novi Sad, Serbia [b]Institute of Food Technology (FINS), University of Novi Sad, Novi Sad, Serbia

7.1 Introduction

Since the time of hunter-gatherers, hemp has been one of the plants that accompanies humanity. Different parts of the plant had their daily practical use. Grain was used as a source of food and oil, fiber from the stalk for making cloth and binders, and herb (leaves and flowers) for medical and ritual purposes. Targeted hemp cultivation in the vicinity of human habitats is thought to have begun about 10,000 years ago. With the migration of the population from Central Asia, hemp spreads over time to most of the world (Clarke & Merlin, 2013).

7.2 A historical overview of the knowledge development of industrial hemp

The first knowledge about the best way of growing hemp was orally spread through accidental or targeted transmission of hempseeds. The earliest notes about the hemp plant and its cultivation come from its Asian origin. Information of this kind can be found in translations of ancient Chinese records (Liu, Hu, Du, Deng, & Yang, 2017). The first written traces of hemp in the Western world date from the second half of the 18th century and relate to the description and general considerations of cultivation, processing, and use of the plant (Hamel [Duhamel], 1747; Marcandier, 1764).

During the 19th century, along with flax hemp became a traditional plant grown for fiber. In Europe, the first publications appear that discuss in more detail the process of production and processing with information based on practical experience (Delamer, 1854). The plant itself has already been approached more studiously, primarily in terms of more detailed descriptions of anatomical characteristics (Briosi & Tognini, 1894). By the end of the century, hemp found its stable place in the plant production system of some US states (Dodge, 1896), where information based mainly on European information and experience was published (Dodge, 1898).

From the end of the 19th century until the beginning of the WWII, the publication of data related to biological characteristics, the technological process of growing not only for fiber, but also for seeds, increased rapidly. In addition to popular European texts that present information on regional production (Neppi, 1920), there are more complex publications which, in addition to being partially based on exact scientific results (primarily in terms of chemical composition of the plant), look at the complete issue in relative detail. Processed data from the aspect of production itself can already be found, where information on the needs of hemp plants for soil quality, climatic conditions, and irrigation is systematized. There are also detailed descriptions of the machinery used for harvesting, as well as the process of processing and obtaining fibers (Boyce, 1900).

At that time, mostly hemp populations were grown that were created in nature without human influence as a result of spontaneous crossing, natural mutations, and natural selection. Populations were named after the region they came from, and the seeds for sowing were taken from previous year's commercial crop. Man's creative work on the planned evaluation of hemp varieties with desired properties started at the beginning of the 20th century. The earliest planned breeding of hemp was recorded in the United States, where (before the beginning of such activities in Europe) around 1910, the creation of the first hemp varieties began (Dewey, 1928). Dewey is considered to be the first hemp breeder who created several exceptional late-maturing varieties ("Kymington," "Chington," and "Arlington") originating from Chinese populations. In addition, he is the author of the "Ferramington" variety, which is a selection from the Italian "Ferrara" population. Given that the production of hemp was abandoned relatively quickly in the United States, which was later even banned, this pioneering work on hemp breeding has almost completely fallen into oblivion and the created assortment has been lost. Hemp breeding has long been neglected in Europe. After the WWI, due to the increased demand for fiber and disruptions in the supply of seeds from Italy, hemp breeding and organized seed production began in several countries (Fruhvirth, 1922).

Hemp was primarily grown as a fiber plant, while the seeds, in addition to reproduction, were used partly in bird nutrition and for the production of technical oil. The seed material used included dioecious populations, in which male plants mature significantly earlier than female ones. In nature, in addition to purely male and purely female, hermaphroditic plants with different proportions of male and female sex also appear. Research related to the natural occurrence and isolation of various sexual forms was initiated in Eastern Europe in the first half of the 19th century (Breslavec & Zaurov, 1937). The identification of a certain percentage of monoecious and intersex types in dioecious hemp populations becomes a significant moment in the further development of hemp breeding and scientific approach to the plant. Hermaphroditic plants were the initial material for the selection of first monoecious varieties. Already in the 30s of the XX century, the first monoecious and at the same time maturing dioecious varieties were selected. Simultaneous maturation of all plants in the population was essentially the goal, since in this way the problem of harvesting was solved. This assortment belongs to the southern type or hybrid offspring from a cross between Central Russian and southern hemp (Grishko, 1935).

The expression and determination of sex in hemp is one of the first areas that intrigued scientists. The first studies related to the expression of sex in hemp (Schaffner, 1927), based on the influence of different environmental conditions, neglected the role of sex chromosomes, preferring the epigenetic mechanism of determination. In his cytogenetic studies, Hirata (1927) found that the haploid number of chromosomes is 10, and the sex chromosomes are described within the karyotype. Based on the defined sex chromosomes, it was assumed that the XY system was responsible for the determination of the hemp sex. Since there are different transitional types of hermaphroditic plants in hemp populations, Hoffmann (1952) later assumed that a monofactorial mechanism of inheritance participates in the expression of monoeciousness, according to which factors with a tendency to feminization are present on the X chromosome and factors with masculinization tendency on autosomes.

7.2.1 Knowledge of industrial hemp breeding and cultivation in the 20th century

Agronomic studies during the first half of the 20th century were focused on optimizing the factors that determine stalk yield and fiber quality. Nitrogen fertilization, determining the optimum harvest time, and achieving the ideal growth density are defined as the most important factors of yield and quality (Zatta, Monti, & Venturi, 2012). During this period, hemp was grown exclusively as a fiber plant and breeding took place in that direction. In order to speed up and optimize the process of defining and selecting plants with the best characteristics, Bredemann (1942) developed the non-destructive method for relatively fast determination of the fiber content in the stalk. This method is often used in contemporary breeding programs.

The second part of the 20th century was marked by the legislative elimination of industrial hemp from the plant production system in some regions where it had been traditionally grown until then. As a result, research and breeding work has been reduced mainly to European countries where the cultivation of industrial hemp has not been banned and discontinued. As a basis for the evaluation of assortments at the national level, the starting material for dioecious varieties was the populations grown in a particular region. The application of individual selection of male and female plants based on in vivo determination of the fiber content in the stalk led to a significant increase in the content and total fiber yield (Bredemann, Garber, Huhnke, & Sengbusch, 1961). In addition to selective breeding, at that time work was done on examining the possibility of exploitation of heterosis (Bócsa, 1954). In crosses of selected male plants of one population with selected female plants of another population (variety), a significant hybrid vigor was recorded for the components of yield and quality of quantitative traits to which the selection is directed. Significant progress in the breeding of monoecious hemp was achieved by Neuer and Sengbusch (1943), who fixed monoeciousness in the variety "Fibrimon," from which several modern monoecious varieties originate. In the research of all aspects of monoeciousness in hemp, the Ukrainian program (Marynchenko & Chunjing, 2018) is especially emphasized, which, in addition to the development of cultivation technology and assortment, also includes research related to inheritance and maintenance of monoeciousness. After monoecious varieties found their place in the hemp production systems, further research was conducted with this material which led to the possibility of exploitation of unisexuality (Bócsa, 1967). Offspring from crossing a female plant as a maternal component with a purely monoecious pollinator contain 70%–85% of female, 10%–15% of monoecious, and 1%–2% of male plants. A variety of this constitution can serve as a maternal component in the wide production of hybrid seeds. During this period, a significant assortment, partially relevant even today, was created in certain regional improvement programs. From this period come the well-known Italian dioecious ("Carmagnola Selezionata," "Eletta Campana," "Fibranova"), French monoecious ("Felina 34," "Futura 77"), Polish monoecious ("Beniko," "Bialobrzeskia"), Romanian dioecious ("Fibramulta 151," "Lovrin 110"), and Ukrainian monoecious varieties ("USO 11," "USO 13," "USO 31"). Of the other industrial hemp programs during this period, the former Yugoslavia and the former USSR should also be mentioned. One of the most important breeding programs of that period was Hungarian, which looked at a broader aspect of biology, genetics, and use of hemp. The first registered hybrid dioecious variety "Kompolty Hybrid TC," yellow stalk dioecious variety "Kompolty Sargaszáru," and the first unisexual female variety "Uniko B" originated from this program.

7.2.2 Knowledge of industrial hemp processing in the 20th century

Along with the creation of varieties, during the second half of the 20th century, intensive research was conducted related to the optimization of the production process and processing of hemp stalks. Although during this period there are not many published separate papers, the results of the research are usually presented through thematic monographs (Bócsa, 1969; Pasković, 1966; Senchenko & Timonin, 1978; Váša, 1965). The main feature of the research is that the agronomic parameters of hemp production from the sowing time and densities, through the agrotechnical operations, fertilization, abiotic and biotic influences, all the way to the harvest are defined in detail. Although most of the information refers to fiber hemp, the data that define the hempseed production are also extensive.

7.3 Modern knowledge of industrial hemp

Due to the expansion of global trends in healthy food and natural raw materials, hemp as a multipurpose crop at the end of the 20th and at the beginning of the 21st century experienced its renaissance. The indulgence of legal restrictions in a significant part of the world also contributed to increasing interest in growing and using hemp. With the development of the hemp products market, a significant worldwide expansion of research concerning various aspects of hemp among its controversies regarding taxonomy (Koren et al., 2020), genome, or GGE interactions has been observed.

As a result of increased breeding and research efforts, more than 70 varieties of hemp for various purposes have been registered on the EU's "Plant Variety Database of Agricultural Plant Species" so far. In addition to conventional methods that were previously used (selection, recombination, inbreeding, and hybridization), more advanced methods have begun to be used in hemp breeding. Boháč (1990) indicated ploidization and mutation as methods that can find some application. Using a 0.1% solution of colchicine, about 11% of autotetraploids were obtained ($2n = 40$). Compared to diploid plants, tetraploid plants are taller, mature 10–15 days later, and have larger seeds. It is of special importance that male and female plants of autotetraploid hemp mature almost simultaneously and are characterized by increased fiber content. Spontaneous mutations are important in hemp breeding, of which a spontaneous mutation for the yellow color of the hemp stalk should certainly be mentioned. The formation of monoecious from dioecious forms can also be caused by mutagenic agents

(Boháč, 1990). Feeney and Punja (2003) achieved a successful genetic transformation of hemp using *Agrobacterium tumefaciens* as a vector. Transgenic varieties of hemp are not found anywhere in production, but it is important to take into account that, if necessary, there are methods of creating genetically modified hemp.

7.3.1 Knowledge of hemp genome

The application of modern biotechnological methods has opened the possibility of genome sequencing, i.e., determining the entire DNA sequence in the hemp genome (Van Bakel et al., 2011). Genetic polymorphism is used to examine the diversity of available germplasm at the gene level. This has largely enabled the explanation of the cause-and-effect relationships between the genetic basis and phenotypic expression of traits. Defined genetic maps with markers associated with a trait of practical breeding importance can help in the selection of complex traits. Although biotechnological studies are becoming more extensive and significant (Chandra, Lata, & ElSohly, 2017), hemp has its own specifics in this regard. The wide genetic variability and high level of heterozygosity (most alleles occur in populations at a frequency < 0.30) in nature is caused by free pollination and the dioecious character of the plant. When differentiating individual populations, MAS has little discriminant potential and cannot clearly define the difference between marijuana and industrial hemp. Due to the high degree of variability and the common gene pool of the genus *Cannabis* with limited segregation between populations, the practical use of genetic maps and molecular markers in hemp is limited.

7.3.2 Knowledge of multiple uses of hemp

Inventions of new ways of utilization of all hemp plant parts: fiber in the textile industry, hemp biomass as an alternative energy source (Prade, 2011), hemp hurd in the automotive industry and construction (Karche & Singh, 2019), grain in the food industry (Farinon, Molinari, Costantini, & Merendino, 2020), and aerial parts as a source of biomolecules for pharmaceuticals and cosmetics (Andre, Hausman, & Guerriero, 2016), led hemp research beyond the framework of biology and agronomy.

Along with the development of knowledge related to the production and use value of the plant, significant progress has recently been made in understanding the environmental benefits that the inclusion of hemp in the global system of plant production brings. The environmentally friendly status of hemp arises from its impact on climate change, eutrophication, acidification, and terrestrial ecotoxicity, as well as energy consumption for plant growth and development (11.4 GJ/ha), which is lower compared to some other crops (Van der Werf, 2004). Biological properties, such as the rapid growth of a large amount of biomass with a relatively deep and rich root system and a solid stalk, make hemp one of the most favorable plants for phytoremediation (environmental cleansing). Phytoremedial capacity is also affected by specific microbial activity in the rhizosphere of hemp plants (Citterio et al., 2005). The bioconcentration factor of pollutants, including heavy metals and radionuclides in leaves and stems, has no effect on fiber and hurd quality and is significantly higher compared to other phytoremediators (Linger, Mussig, Fischer, & Kobert, 2002; Vandenhove & Van Hees, 2005). In addition to heavy metals and radionuclides, hemp is also tolerant to soils contaminated with organic pollutants (Campbell, Paquin, Awaya, & Li, 2002).

In the context with the trend of healthy food, hempseeds are increasingly finding their place in the human diet due to their nutritional value. The value of hempseeds as food is given by the amino acid composition of the protein and the fatty acid composition of the oil. In that sense, the biofortification ability, i.e., the accumulation of microelements in the edible parts of the plant, comes to the fore. The increase in the content of microelements occurs through phytoremediation, cultivation on soils that are naturally rich in microelements, or through fertilization (Stonehouse, 2019). Owning to a content of bioactive molecules, aerial parts of industrial hemp are more and more finding their place in pharmaceutical, nutraceutical, and cosmeceutical sector (Drinić et al., 2018, 2020, 2021).

7.4 The role of different knowledge creators and knowledge providers in hemp breeding, agriculture, and (food) processing

Despite very extensive and lively research on different aspect of hemp breeding, cultivation, processing, and utilization in recent years, a still existing knowledge gap of hemp is perceived as most concerning (Dingha et al., 2019). Due to its multi-purpose nature, industrial hemp can be designated as more knowledge-intensive crop than others. According to a survey conducted among North Carolina organic farmers in the United States, inadequate knowledge and information about industrial hemp, as well as questionable social attitude of the community, was associated with the inhibiting acceptance of industrial hemp cultivation. It was shown that (lack of) knowledge of industrial hemp is directly related to respondents' openness to its cultivation. What is encouraging is the fact that respondents exhibited high enthusiasm and willingness to learning more about the hemp production practices and adapted cultivars (87.9% of respondents) (Dingha et al., 2019).

The knowledge-based economy provides a framework for knowledge creation, its acquisition, transmission, and effective utilization by industry and business sector, research organizations, policy makers, citizens, and general public to enable economic growth, wealth creation, and employment (Skrodzka, 2016). Knowledge creators and providers contribute to creation of unique knowledge bases and to break through key technological bottlenecks (Ge & Liu, 2021). Several forms of knowledge creation and knowledge provision will be listed.

7.4.1 Relevant projects that have addressed the industrial hemp

Projects are suitable means for bringing together experts, including farmers, researchers, advisers, and industry professionals, to gain experience and knowledge and enable the progression of the industrial hemp sector. Thus the progression of the sector is enabled through the implementation of different types of projects: (i) coordination and support projects—aimed at compiling and disseminating the existing knowledge and raising awareness of stakeholders; (ii) demonstration projects—aimed at practical information and best practice sharing and demonstration of business models for farmers and industry; and (iii) research and innovation projects—aimed at exploring the potentials to increase yields, obtain high product quality and safety and low dependence on inputs, to produce value-added products (food, feed, pharmaceuticals, nutraceuticals, cosmeceuticals, bio-based chemicals and materials, and bio-energies), find alternative applications and processing methods, to enable the progression at higher technology readiness levels (TRLs), etc. Funding of listed types of the projects is of crucial importance since they are offering a sustainable and long-term recovery of the economy. The project activities could be designed to enable the strong participation of targeted group(s) of stakeholders—farmers, researchers, industry, and policy makers.

The projects that addressed the industrial hemp that have been completed by to date and ongoing, mainly in Europe and United States, are listed in Table 7.1.

TABLE 7.1   Projects related to industrial hemp.

MULTIHEMP (2012–17)
Multipurpose hemp for industrial bioproducts and biomass
http://multihemp.eu/
EU FP7 program
Description: Development of an integrated hemp-based biorefinery to obtain fiber, oil, construction materials, fine chemicals, and biofuels from all components of the harvested hemp biomass. The project reported the improved hemp target traits, optimized hemp cultivation, harvesting and processing, assessed end use applications for hemp raw material (e.g., fiber for blow-in insulation and building materials, seed oil for cosmetic applications)

LIGNOFOOD (2013–15)
Ingredients for food and beverage industry from a lignocellulosic source
http://www.lignofood.eu/
EU FP7 program
Description: Development of ingredients for food and beverage industry from a lignocellulosic hemp waste. Health-enhancing dietary additives—xylooligosaccharides (XOS) with the prebiotic potential and xylitol for gut health improvement were obtained

4F CROPS (2008–10)
Future crops for food, feed, fiber, and fuel
http://www.cres.gr/4fcrops/pdf/poznan/alexopoulou.pdf
EU FP7 program
Description: Survey and analysis of all the parameters that contribute to successful non-food cropping systems in the agriculture of the EU, by considering 15 selected crops—hemp was analyzed for oil, solid biomass, and fibers. Economic viability of non-food crops in different European countries and climatic zones was assessed, whereby industrial hemp exhibited attractive financial potential. The status of current genomic research in hemp crop was evaluated and different marker systems were developed for hemp's germplasm fingerprinting

CROPS2INDUSTRY (2009–12)
Non-food crops-to-industry schemes in EU27
http://www.cres.gr/crops2/Project.htm
EU FP7 program
Description: Exploration of the potential of non-food crops which can be domestically grown in EU for selected industrial applications (oils, fibers, resins, and other specialty products) to support sustainable, economic viable, and competitive bio-based industry and agriculture. Hemp exhibited the highest bio-physical biomass production potentials on EU27 croplands

FLHEA (2013–15)
Flax and hemp advanced fiber based composites
EU FP7 program
Description: The upscaling and modification of micro and nano reinforcements based on hemp (and flax), and its subsequent processing (by sheet extrusion and thermoforming) to obtain composites with improved properties for food packaging trays. Biodegradable materials for food packaging purposes were developed contributing to economical production and reduction of the environmental costs of packaging

Continued

TABLE 7.1   Projects related to industrial hemp—cont'd

GRACE (2017–22)

Growing advanced industrial crops on marginal lands for biorefineries

https://www.grace-bbi.eu/

EU H2020 program

Description: Demonstration of the upscaling of crop production of hemp genotypes matched to end use and their suitability for marginal, contaminated, and unused land. Upscaling demonstration of the most promising biomass valorization chains with tailored genotypes

---

DanuBioValNet (2017–19)

Cross-clustering partnership for boosting eco-innovation by developing a joint bio-based value-added network for the Danube Region

http://www.interreg-danube.eu/approved-projects/danubiovalnet

Interreg Danube Transnational Programme

Description: Development of new methods, strategies, and tools to connect SMEs, farmers, universities, and research institutes from Danube region within three bio-based value chains (phytopharma, eco-construction, bio-based packing), with special emphasis on hemp industry due to the significance of hemp as a raw material for all three value chains. Links between existing clusters and a bio-based economy sector were created and strengthened as a main prerequisite for bringing sustainable and environmentally responsible products to the market

---

Cultivation of industrial hemp (*Cannabis sativa* L.) in Slovenia (2016–19)

http://www.ihps.si/en/plant-soil-environment/v4-1611-cultivation-industrial-hemp-cannabis-sativa-l-slovenia/

Slovenian Research Agency, Ministry of Agriculture, Forestry and Food RS

Description: Finding suitable varieties for production in Slovenia for seed and fiber production, definition of suitable agrotechnical measures to ensure high quality, stable production, and secure yield, creating conditions for independent production of hemp varieties. The adequacy of European hemp varieties for cultivation in Slovene agricultural areas was assessed, nutritional value of hempseed oil and quality of hemp fiber from different varieties were determined, the economic potential of production of industrial hemp harvesting machine was assessed

---

Industrial hemp production, processing, and marketing in the United States (2018–23)

https://www.nimss.org/projects/view/mrp/outline/17716

Description: Determination of the effects of agronomic practices on grain, fiber, and dual-purpose productivity; hemp quality determination in relation to fiber, grain, and cannabinoid; identification of genes for advanced traits of interest; assessment of crop value grown for different uses and in different cropping systems

---

SCARABEO (2017–19)

Hemp residues—reuse for food and energy recovery with oils

https://www.psrscarabeo.it/

Rural development 2014–20 for Operational Groups, Italy

Description: Applying innovative approaches to diversify hemp by-products and to valorize the hemp stems waste for polyphenols extraction for food or pharmaceutical end uses, among others. Stems were mechanically pretreated before maceration which reduced the amount of water needed, maceration times, and the amount of waste produced, affecting positively the environment, among the other outcomes

---

CATERPILLAR (2020–22)

Fiber hemp industrial chain for the production of functional food and animal feed additives

https://www.gocaterpillar.it/

Rural development 2014–20 for Operational Groups, Italy

Description: Enhancing the maximum potential of organically cultivated monoecious hemp, by the generation of new high value-added products and by-products—oil and flour and fiber and shives; developing new functional and gluten-free food formulations and the production of feed industry additives

TABLE 7.1   Projects related to industrial hemp—cont'd

MAGIC (2017–21)
Marginal lands for growing industrial crops: Turning a burden into an opportunity
http://magic-h2020.eu/
EU H2020 program
Description: Promotion of the sustainable development through exploration of resource-efficient and economically profitable industrial crops grown on marginal lands, which can be used for high value-added products and bioenergy. Among others, hemp has been selected as promising industrial crop to be grown on marginal lands facing natural constraints

Production and enhancement of plant bioproducts for a total and characterizing use of new nutritional formulations rich in functional components (2020–22)
Rural development 2014–20 for Operational Groups, Italy
Description: Promotion of hemp cultivation (among other crops) in eco-sustainable and biological systems and validation of the cultivation practices to enhance the productivity and the quality of the functional components for new food formulations

The role of industrial hemp in adaptation to climate change and in protection of agricultural resources (2020–23)
Rural development 2014–20 for Operational Groups, Slovenia
Description: Evaluation of the role of hemp for adaptation to climate change and for the protection of agricultural resources. Dissemination of technological know-how between farmers and experts and provision of new knowledge and data on the usefulness of hemp for soil improvement, water protection, carbon sequestration, and climate change mitigation

Multi-purpose applications to resume the hemp supply chain (2020–22)
Rural development 2014–20 for Operational Groups, Italy
Description: Creation of the technical, economic, and market conditions to support the development of a multi-purpose hemp production chain in Emilia Romagna

CANAPRO (2019–22)
Enhancement of the hemp supply chain through product and process innovation
Rural development 2014–20 for Operational Groups, Italy
Description: Identification of the most suitable varieties to improve agronomic and environmental sustainability, development of growth models for open field and greenhouse cultivation, evaluation of yield and quality of extra-seasonal hemp production through greenhouse cultivation, identification of the varieties with the highest oil yield, determination of quality of different hemp varieties, enhancement of the hemp by-products in zootechnical applications

Preservation of the original, natural CBD content of the hemp plant for long-term storage by evaluation and optimization of various drying methods (2020–23)
Rural development 2014–20 for Operational Groups, Germany
Description: Processing of three hemp varieties by innovative drying to preserve high content of CBD and product quality, determination of optimal storage conditions to achieve high CBD stability and quality. Thus the contribution to high-priced values of the harvested product, stabilization and diversification of Bavarian farms through new sources of income, and an increase of the cultivation of hemp are expected

ORGANIC FARMING: Evaluation of biopesticides obtained from hemp's by-products and evaluation of toxicity for the operator (2019–22)
https://scienzaelode.unicam.it//ricerca/agricoltura-biologica-grazie-alla-canapa?fbclid=IwAR2sfpYONX0-LeQT9zD4Fvvhy3H-TLJFMgJ61ZJTbSuKR1yShdVQtODx Q6U
Rural development 2014–20 for Operational Groups, Italy
Description: Extraction of essential oil from hemp waste products and development of a product with insecticidal and/or fungicidal effects to be commercialized and used in organic agriculture

Continued

TABLE 7.1   Projects related to industrial hemp—cont'd

CANAPA IN FILIERA_Produce hemp in the food and agroindustrial chain (2019–22)
Rural development 2014–20 for Operational Groups, Italy
Description: Testing different cultivation techniques, organizing a producers' network to ensure high efficiency in production, logistics, and business expenses; experimental tests with zootechnical transformation of hemp-based products for verification of nutritive composition; reduction of the fertilizers, the elimination of chemical herbicides, and a significant increase in the absorption of CO_2

Cultivation, harvesting and processing of hemp straw and hemp seeds (2019–23)
Rural development 2014–20 for Operational Groups, Germany
Description: Establishment of hemp cultivation in the Werra-Meißner district with associated value chains. Creating conditions for the partial processing of hemp plant on site and the optimization of hemp cultivation and breeding of plants adapted to the local conditions

---

Innovative solutions in treatment and processing of industrial hemp (2019–22)
Rural development 2014–20 for Operational Groups, Latvia
Description: Evaluation and improvement of hemp valuation chain; examination of agrotechnical measures for hemp productivity, quantitative and qualitative parameters, output, suitability for production of high value-added products; development of recommendations for selection of optimal hemp growing and harvesting technologies in Latvia

---

Development of new product group of cold-pressed organic bars (2015–16)
Rural development 2014–20 for Operational Groups, Estonia
Description: Development of organic bars from hemp cake and traditional fruit raw materials from Estonia (pear, apple, buckwheat, chokeberry, and apple juice pomace)

COBRAF (2019–21)
By-products for biorefineries
https://www.cobraf.it/
Rural development 2014–20 for Operational Groups, Italy
Description: Utilization of by-products of four oilseed crops (hemp among others) to develop a biorefinery system for a maximum exploitation of the oil crops' biomass that can be used in rotation. Biorefinery system is based on the utilization of different parts of biomass and by-products for production of bio-products for at least six sectors of the Tuscan industry: food, cosmetics, pharmaceutical, construction, wood, and automotive

---

From ecological intercropping to fine fiber (2016–19)
Rural development 2014–20 for Operational Groups, Germany
Description: Solving the operational intercropping problem; optimization and evaluation of winter hemp quality; enhancement of sustainability and economic and environmental effects for farms by launching marketable products; transfer of the procedures into practices

---

Effects of hemp seed meal on physiologic, nutritional, and residue chemistry endpoints in feedlot cattle (2020–24)
https://www.ars.usda.gov/research/project/?accnNo=438497
USDA
Description: Examination of physiologic effects of dietary hemp and its by-products as utilized as value-added animal feed and description of the fate and/or clearance of pharmacologically active phytochemicals (CBD, THC)

Sustainable Incorporation of Hemp into American Agricultural Cropping Systems (2020–23)
https://www.ars.usda.gov/research/project/?accnNo=438331
USDA
Description: Development of science-based approaches for sustainable incorporation of hemp into American agricultural systems; determination of the optimal adaptation of hemp essential oil variety types and genetics across the United States; determination of hemp crop water use across diverse production regions; determination of the optimal uses and safety of hemp and hemp by-products fed to livestock; determination of protocols for optimal sampling of hemp fields and laboratory analyses for accurate analysis of THC content at harvest

---

TABLE 7.1   Projects related to industrial hemp—cont'd

Economic Potential of Pacific Northwest Hemp Production Systems (2020–22)
https://www.ars.usda.gov/research/project/?accnNo=438346
USDA
Description: Development of publicly available economic information needed for producers and the industry to evaluate the potential for commercial adoption of hemp by crop production farms in Oregon

---

Fostering a well-rounded and sustainable hemp industry in the United States (2020–22)
https://www.ars.usda.gov/research/project/?accnNo=438474
USDA
Description: Identification of the best management practices for grain, fiber, and floral hemp production in a transition zone environment; evaluation of the economic potential of hemp grown for fiber, grain, and floral products; evaluation of the potential of hemp foliage, grain, and by-products for use as animal feed

---

Developing decision support tools to incorporate hemp into existing farming enterprises (2020–24)
https://www.ars.usda.gov/research/project/?accnNo=438146
USDA
Description: Gathering and integration of current hemp-related data and collection of new data for establishment and improvement of hemp production models; development of modeling techniques to serve as decision support tools to help farmers integrate hemp production into existing farm systems

---

Determining the predominant industrial hemp diseases to assist breeding and management efforts (2021–23)
https://www.ars.usda.gov/research/project/?accnNo=440965
USDA
Description: Determination of the predominate diseases affecting hemp in different stages of crop phenological development in Oregon and Washington and the frequency of their occurrence; determination of the effects of different growing environments and plant growth stages on disease outbreaks; help farmers to identify hemp diseases and make related information available in handbooks and open access peer review articles

---

The agronomy of hemp and its uses in forage-animal agriculture (2021–22)
https://www.ars.usda.gov/research/project/?accnNo=440911
USDA
Description: Gathering and integration of current hemp-related data and collection of new data to support hemp production model systems; support of the use of hemp and hemp residual biomass as a livestock feed; exploration of possible benefits of hemp-based compounds in animal production

---

Development of new value-added processes and products from advancing oilseed crops (2020–25)
https://www.ars.usda.gov/research/project/?accnNo=436964
USDA
Description: Development of analysis of new crop germplasm and agronomic traits of oilseed crops (among others hemp); development of processes for the commercial production of oils, meal, gums, waxes, and value-added products; conversion of oils and gums into marketable new value-added bio-based products

---

Improved processes and technologies for comprehensive utilization of speciality grains in functional food production for digestive health and food waste reduction (2020–25)
https://www.ars.usda.gov/research/project/?accnNo=438210
USDA
Description: Improvement of the properties of underutilized crops (among others hemp) and their by-products by innovative processing (thermomechanical, chemical/enzymatic processing, and their combination); determination of the digestive health effects of different compounds typical of crops (e.g., hemp); development of different food and non-food products on the basis of different bio-product ingredients

---

Continued

TABLE 7.1   Projects related to industrial hemp—cont'd

Development of effective analytical methods for evaluation of hemp materials and processing products for phytochemical composition, incl cannabinoids (2020–21)
https://www.ars.usda.gov/research/project/?accnNo=439216
USDA
Description: Formation of standardized hemp samples to develop methodology for compositional analyses of hemp (HPLC-UV) for use in analytical laboratories

---

Integration of hemp production into US farming systems (2020–23)
https://www.ars.usda.gov/research/project/?accnNo=438043
USDA
Description: Provision of the agronomic information (e.g., irrigation, fertilizer use, soil properties, breeding, disease and pest management, crop rotations, interactions with other crops) needed to support the incorporation of hemp into farming operations within the United States

---

Strengthening hemp research and educational capacity at Alabama A&M University by morphological evaluation of diverse hemp germplasm
https://www.ars.usda.gov/research/project/?accnNo=440866
USDA
Description: Establishment of the USDA-ARS hemp germplasm collection in Geneva, New York; strengthening hemp research and educational capacity at Alabama A&M University

---

Conservation and utilization of hemp genetic resources and associated information (2021–26)
https://www.ars.usda.gov/research/project/?accnNo=439751
USDA
Description: Acquisition of safe, health, and viable hemp genetic resources; development of effective and regulatory compliant genetic resource maintenance, evaluation, testing, and characterization methods and their application to priority hemp genetic resources; development, update, document, and implementation of the best management practices for priority hemp genetic resources; development and application of research tools, knowledge of hemp genetics, the genetic control of priority traits, and genetic resources for hemp research, breeding, and crop improvement

---

Documentation, evaluation and characterization of industrial hemp germplasm (2019–24)
https://www.ars.usda.gov/research/project/?accnNo=435200
USDA
Description: Exploration of the status of germplasm research, breeding, and conservation of industrial hemp; establishment of the capacity and infrastructure for documentation, evaluation, and characterization of industrial hemp; development of genomic resources for industrial hemp

---

7.4.2 Extension services

Dissemination of knowledge, provision of inputs for agriculture modernization, development of perceptions and attitudes on agricultural innovations, and promotion of rural development are highly dependent on extension trainings (Dingha et al., 2019; Kassem, Alotaibi, Muddassir, & Herab, 2021). Agricultural extension services are defined as "the entire set of organizations that facilitate and support people engaged in agricultural activities to solve problems and to obtain information, skills, and technologies to improve their livelihoods and well-being" (Davis, Babu, & Ragasa, 2020). Several authors have explored the satisfaction of farmers with the quality of agricultural extension services. It was shown that farmers' satisfaction is dependent not only on availability, accessibility, diversity, relevance, and effectiveness of services, but also on farm size, diversity of farming activities, annual income, and participation in extension services (Kassem et al., 2021). The extension services specialized in industrial hemp are listed in Table 7.2.

TABLE 7.2 Extension services.

CLEMSON EXTENSION, South Carolina, United States
https://www.clemson.edu/extension/
Serves hemp producers of varying levels of experience by sharing information, providing in-field consultations, and diagnosing plant and pest problems

VIRGINIA COOPERATIVE EXTENSION, Virginia, United States
https://www.vaes.vt.edu/about/faq-hemp.html
Fields of expertise: Variety trials, planting date assessments, best planting practices, successful pest management

TEXAS A&M AGRILIFE EXTENSION, Texas, United States
https://agrilifeextension.tamu.edu/browse/hemp/

THE UNIVERSITY OF VERMONT, UVM EXTENSIONS, Vermont, United States
https://www.uvm.edu/extension/nwcrops/industrial-hemp
Offers industrial hemp training program—for agricultural/technical service providers

UNIVERSITY OF WYOMING, UW EXTENSIONS, Wyoming, United States
http://www.uwyo.edu/ipm/ag-ipm/hemp-ipm.html

OREGON STATE UNIVERSITY, OSU EXTENSION SERVICE
https://extension.oregonstate.edu/crop-production/hemp

ICANNA—International Institute for Cannabinoids, Slovenia, EU
https://www.institut-icanna.com/en/
Education and training services

In recent years, US state legislatures have taken steps to establish a state-licensed industrial hemp program and promote hemp as an agricultural commodity. The Farms Act 2018 removes hemp from the Controlled Substances Act and considers hemp as an agricultural product under certain restrictions (the concentration of delta-9 tetrahydrocannabinol (THC) in the dry plant mass must not exceed 0.3%). The US Department of Agriculture (USDA) is the responsible federal regulatory agency that oversees hemp production in the United States. The Farms Act 2018 allows states and tribes to regulate hemp production in their territory after submitting a plan and applying to the primary regulatory body. The Farms Act 2018 allows states and tribes, after approval of their application and submitted, to regulate the production of hemp on their territory. Hemp production or hemp research has so far been legally regulated in at least 47 states in the United States (Table 7.3).

Government bodies and agencies are also important actors in the industrial hemp knowledge creation and management which, by analyzing the data of the hemp industry, legally regulate the hemp value chain (Table 7.4).

Colleges and universities in Canada are starting a number of programs designed to help students to find jobs in fast-growing cannabis and industrial hemp industry (Table 7.5).

7.4.3 Hemp associations and centers of excellences

The recent hemp renaissance has led to the creation of multiple organizations and associations gathering hemp farmers, researchers, medical expert, representatives of hemp industry, representatives of the public working groups, and other stakeholders with a common aim to contribute to the development of the industrial hemp industry (Table 7.6).

TABLE 7.3 Industrial hemp state research programs in the United States.

State/country	Program's focus
Alabama	Agricultural program overseen by the Alabama Department of Agriculture and Industries to study hemp **for research purposes**. Grower or processor must be approved to participate in the program http://agi.alabama.gov/divisions/plant-protection/industrial-hemp/program-applications
Alaska	Alaska Department of Natural Resources, through the Division of Agriculture, designed an Industrial Hemp Pilot Program to **research the production and marketing of industrial hemp**. The purpose of the pilot program is to increase the knowledge of how industrial hemp cultivation may increase agricultural production in the state https://plants.alaska.gov/industrialhemp.htm
Arizona	**Research** of industrial hemp production and marketing https://agriculture.az.gov/plantsproduce/industrial-hemp-program
Arkansas	The Program **licenses** hemp producers, processors, and handlers to conduct research and generate the industry data https://www.agriculture.arkansas.gov/plant-industries/feed-and-fertilizer-section/hemp-home/industrial-hemp-research-pilot-program-overview/
Colorado	• A **seed certification** program • A grant program for state institutions of higher education **to develop new hemp cultivars for grain production** • https://ag.colorado.gov/plants/industrial-hemp
Connecticut	**Research** of industrial hemp **and licensing** of producers/processors
Delaware	**Research** purposes
Florida	Assessment of the crop, cropping systems, and their impacts on economy and ecology https://programs.ifas.ufl.edu/hemp/
Georgia	Grower **licensing**, processor permits http://agr.georgia.gov/georgia-hemp-program.aspx
Hawaii	Hemp **research** https://hdoa.hawaii.gov/hemp/
Illinois	Educational and research https://www2.illinois.gov/sites/agr/Plants/Pages/Industrial-Hemp.aspx
Indiana	Hemp licensing and crop management https://www.oisc.purdue.edu/hemp/index.html
Kentucky	University of Kentucky Agricultural Experiment Station oversees a 5-year hemp research program https://www.uky.edu/ccd/production/crop-resources/gffof/hemp
Louisiana	Provides for responsibilities of the government agencies to develop an criteria for seed approval, licensing, testing, and other necessary rules and regulations https://www.ldaf.state.la.us/industrial-hemp/
Michigan	https://www.canr.msu.edu/hemp/
Missouri	Industrial hemp agricultural pilot program implemented by the Missouri Department of Agriculture (MDA) to study the cultivation, processing, feeding, and marketing https://agriculture.mo.gov/plants/industrial-hemp/
Oregon	Global Hemp Innovation Center, Overseen by College of agricultural sciences, Oregon State University Research of innovative approaches to hemp as food, feed, and fiber https://agsci.oregonstate.edu/hemp

TABLE 7.3 Industrial hemp state research programs in the United States—cont'd

State/country	Program's focus
South Carolina	• Farming program • https://agriculture.sc.gov/divisions/consumer-protection/hemp/ • Field trials, Clemson University • https://www.clemson.edu/extension/industrial-hemp/
Utah	Optimization of hemp growing practices for botanical medicines https://caas.usu.edu/cultivate/spring19/hemp-research
Virginia	Virginia Tech College of Agriculture and Life Sciences (CALS)
Washington	Center for Cannabis Policy, Research, and Outreach, Washington State University https://research.wsu.edu/cannabis/
West Virginia	Studying what environmental factors lead to THC accumulation in industrial hemp plants, West Virginia University

TABLE 7.4 Governmental bodies and agencies involved in industrial hemp knowledge management.

INDUSTRIAL HEMP TASKFORCE

https://agriculture.vic.gov.au/crops-and-horticulture/cannabis-in-victoria/industrial-hemp-taskforce#

Secretariat support: Agriculture Victoria, Australia

Description: A working group formed to explore the industrial hemp industry and to examine the challenges and opportunities that industry is facing

Findings: The working group published its findings highlighting a number of potential uses of hemp in the construction, food, beverage, cosmetics, and pet feed industries. Challenges to address include lack of processing capacity, regulatory barriers, gaps in knowledge of crop cultivars and agricultural crop management, as well as data gaps to support business sustainability

PIRSA—THE DEPARTMENT OF PRIMARY INDUSTRIES AND REGIONS IN SOUTH AUSTRALIA

https://www.pir.sa.gov.au/primary_industry/industrial_hemp

Support: Government of South Australia

Description: The Department is responsible for issuing licenses to authorize the possession, cultivation, processing, and supply of industrial hemp

Deliverables: In cooperation with the South Australian Research and Development Institute (SARDI), PIRSA conducted research trials in the South East of the state, the region with the greatest concentration of hemp cultivation licenses. The trials included six hemp cultivars with objective of optimization of cultivation practices. The cultivars were assessed as suitable for cultivation in the two regions for production of grains and fibers, and the information collected will provide hemp producers the best advice when considering industrial hemp as a potential crop for their farming enterprise

DPIRD—THE DEPARTMENT OF PRIMARY INDUSTRIES AND REGIONS

https://www.agric.wa.gov.au/hemp/industrial-hemp-western-australia-0

Support: Government of Western Australia

Main responsibilities:

• Industry development and general enquiries
• Agronomy and hemp research trials: hemp variety trials and hemp fed livestock trials
• Industrial hemp licensing, Industrial Hemp Registrar

Deliverables:

• **Manjimup hemp processing prefeasibility study** (2017–18) commissioned by the Manjimup Agriculture Expansion Project, jointly steered by Shire of Manjimup, DPIRD, South West Development Commission and the Southern Forests Food Council. The report considers production of hemp fiber and seed, supply chain, and markets for hemp products
• **New opportunities in new and emerging agricultural industries in Australia** (2017)—report of opportunities for Australian new and emerging agricultural industries, listing hemp among 20–30 priority opportunities for new rural industry development

Continued

TABLE 7.4   Governmental bodies and agencies involved in industrial hemp knowledge management—cont'd

MANITOBA MINISTRY OF AGRICULTURE AND RESOURCE DEVELOPMENT, CANADA
Agriculture and Resource Development
https://www.gov.mb.ca/agriculture/crops/crop-management/hemp-production.html
Support: Government of Manitoba, Canada
Overall responsibilities: The ministry promotes an integrated approach to the development of sustainable agriculture and natural resources of the Province
Deliverables: A comprehensive and detailed manual for hemp crop production and management

UNITED STATES DEPARTMENT OF AGRICULTURE, AGRICULTURAL MARKETING SERVICE
Deliverables:
• Establishment of a Domestic Hemp Production Program—A Rule

TABLE 7.5   Colleges and universities with Cannabis programs in Canada.

Institution	Name of the course	Description, focus
Niagara College	Commercial Cannabis Production	Training in cannabis production
Durham College	Cannabis Industry Specialization program	The program is part of a partnership between the school and GrowWise Health Limited
Kwantlen Polytechnic University	Retail Cannabis Consultant certificate program	"Focus on best practices highlights compliance, customer service and competence in a complex and evolving industry"
McGill University	Medical Cannabis production and quality control	Workshops
Community College of New Brunswick	N/A	Training program for medical cannabis cultivation technicians
Loyalist College	N/A	Cannabis applied science program
University of Ottawa	N/A	Focused on cannabis law
St. Francis Xavier University	N/A	Partnership with THC Dispensaries
Ryerson University	The Business of Cannabis	Legislation, financing, and regulations for cannabis businesses
College of the Rockies	Online course	The retail aspects of cannabis
Olds College	N/A	Combination of both online courses and a 2-week job placement

Centers of Excellence (CoEs) are intended to support innovation activities, conduct research and development in breakthrough areas, and are characterized with unique technical, intellectual, and human resources (Yakovlev, Kostikov, Kozyreva, & Martyushev, 2015). Although there is no generally accepted definition of centers of excellence and no standardized metrics have been established for their designation and implementation in international context, the common thing is that they are considered leaders in one or more areas of science and technology (Li, Burson, Clapp, & Fleisher, 2020; Yakovlev et al., 2015) (Table 7.7).

TABLE 7.6 Hemp associations.

Name	Mission	URL	Working groups	Head office
European Industrial Hemp Association (EIHA)	Steering and promotion of hemp farming, processing, and trading in the EU; development of a single and safe common market of high-quality hemp products on social, environmental, and economic sustainability principles	https://eiha.org	− Fiber and textile − Food and food supplements − Agriculture and environment	Brussels, Belgium
Global Hemp Association (GHA)	Connection of farmers, manufacturers, distributors, and policy makers to advance hemp for textiles, food, biofuels, construction, education, and sustainability	https://globalhempassociation.org	− Building/construction − Textiles − Carbon sequestration and sustainability − Grain and feed	Riverton, Utah, United States
Latin American Industrial Hemp Association (LAIHA)	Promotion of good practices and high quality standards in hemp industry; support of scientific research on hemp production and processing; promotion of the regulatory framework development in Latin America; development of sustainable supply chains for local industry players	https://laiha.org	− Hemp fiber − Hemp flowers and biomass − Hemp grains	Sao Paulo, Brasil
Australian Hemp Council (2020)	Addressing the issues impacting the hemp industry; engagement with policy makers and community; driving the policies to foster productivity of the Australian Hemp Industry; improving the industry's capacity to compete at a global level; ensuring long-term sustainability of a profitable industry	https://australianhempcouncil.org.au	N/A	Boambee, New South Wales, Australia
British Hemp Alliance (BHA) (2019)	Lobbing for change and barriers removal to enable UK hemp industry growth; support progressive changes in hemp legislation; promotion of UK's hemp industry	https://britishhempalliance.co.uk	N/A	London, United Kingdom
Canadian Hemp Trade Alliance (CHTA) (2003)	Promotion of Canadian hemp and hemp products globally; dissemination of information, promotion of the use of nutritional and industrial hemp products; development of standards and coordination of research	https://www.hemptrade.ca	− Farming − Food − Feed − Fiber − Fractions	Calgary, Canada

Continued

TABLE 7.6   Hemp associations—cont'd

Name	Mission	URL	Working groups	Head office
Hokkaido Industrial Hemp Association (HIHA) (2012)	Dissemination of hemp information; Lobbying to Government and assembly for deregulation; conducting research—farming, processing, products, and marketing; support getting licenses for cultivation/research; networking with overseas hemp industries; raising public awareness	https://hokkaido-hemp.net	N/A	Hokkaido, Japan
National Hemp Association (NHA) (2013)	Networking of farmers, processors, manufactures, researchers, investors, and policy makers to accelerate the growth of hemp industry; dissemination of information about the health, environmental, and economic benefits of hemp; building a community of individuals, businesses, and organizations to facilitate the growth of hemp in the industry; implementation of industrial hemp standards, certifications, and regulations	https://nationalhempassociation.org/	– Standing committee of hemp organizations – Standing committee for social equity	Washington, United States
New Zealand Hemp Industries Association (NZHIA) (1997)	Promotion of the economic, environmental, health, and social benefits of a New Zealand hemp industry	https://nzhia.com	– Research committee	Auckland, New Zealand
United States Hemp Growers Association (USHGA)	Education of hemp growers on the best agricultural practices; provision and sharing of research, legal, and regulatory information; facilitation of the access to certified seed and/or stable genetics	https://ushempga.org	N/A	Lenexa, United States
Hemp Industries Association (HIA)	Connection of farmers and industry experts to share knowledge; consumers education on hemp's history and potential; support of science-based regulations; promotion of hemp row-crop adoption	https://thehia.org		

TABLE 7.7   Hemp related centers of excellence.

HEMP SCIENCE CENTER OF EXCELLENCE
Partners: Front Range Biosciences, Shimadzu Scientific Instruments
https://www.frontrangebio.com/
https://www.ssi.shimadzu.com/
Location: Boulder, Colorado, United States
The Center fosters research collaborations from both public and private sectors that include academic, governmental, and non-profit agencies
Focus: Biobanking, genetics, and breeding analysis utilizing Shimadzu instruments
Objective: Development of new hemp cultivars for the production of Natural Biocompounds (NBCs) for medical and wellness (cannabinoids and terpenoids), food and cosmetics (protein, lipid, and wax ingredients), and industrial (fibers) applications

COLORADO HEMP CENTER OF EXCELLENCE
Partners: Center for Research and Education Addressing Cannabis and Health (CU REACH), University of Colorado Boulder, Colorado Department of Agriculture (CDA)
https://ag.colorado.gov/plants/industrial-hemp
Location: Colorado, United States
Description: Established by a bill approved in 2018 by the Colorado Legislature, it is intended to help guide research, promotion, and educational efforts approved by the US Department of Agriculture
Challenges: The law stipulates that the launch of the center and early strategic development are managed by a third party—selected by a board established by the CDA. The committee selected the Marijuana Policy Group (now MPG Consulting). Although MPG was at the forefront of political work on drafting state-legal legislation on marijuana and THC, it still lacks the experience with industrial hemp regulations

INDUSTRIAL HEMP CENTER OF EXCELLENCE
Partners: Academic (Lambert Center for the Study of Medicinal Cannabis & Hemp, Thomas Jefferson University), government and industry
https://www.jefferson.edu/academics/colleges-schools-institutes/health-professions/emerging-health-professions/centers/lambert-center/research/hemp-research.html
Location: Philadelphia, Pennsylvania, United States
Objective: Support and design of several new materials, processes, products, and business models necessary to develop and sustain industry of industrial hemp
Description: Interdisciplinary academic research program that has been developed to address key points of hemp production and manufacturing with targeted public and private partners
Results: Five "product-by-process" patents for hemp-based materials in various stages of commercialization

THE VIRTUAL CANNABIS CENTER OF EXCELLENCE, INC. (CCOE)
https://www.cannacenterofexcellence.org/
Type: Non-profit organization
Partners: Citizens, academy, policy makers, and all other stakeholders in the cannabis industry
Objective: Engagement of community into cannabis scientific research, projects, and education
Goals: To provide citizens with access to documented science-based cannabis data and research through a virtual platform developed to promote data-driven recommendations, best practices, and meaningful restorative justice in the industry

CANNABIS CENTER OF EXCELLENCE (CoE)
Partners: Lake Superior State University, Agilent Technologies Inc.
https://www.lssu.edu/news/opening-of-the-cannabis-center-of-excellence-for-the-first-cannabis-program-in-the-united-states/
Location: SAULT STE. MARIE, Michigan, United States
Focus: Training of undergraduate students for quantitative chemists to use innovative instrumentation and protocols of cannabis analytics industry
Objective: Approaching education and research among the sciences related to the fast-growing cannabis industry

7.4.4 Dedicated journals

Research interest in industrial hemp as a raw material with health-promoting properties has been considerably expanding since the 2000s. In Web of Science core collection, by using "industrial hemp, agriculture" keywords, 419 research papers; "industrial hemp, food" keywords, 287 research papers; "industrial hemp, oil", 241 research papers; "industrial hemp, nutritional supplement," 6 research papers; "industrial hemp, feed," 40 research papers, and "industrial hemp, CBD" keywords, 94 research papers were listed. They are scattered throughout different journals dealing with agricultural topics (agricultural engineering, agronomy, economics and policy, and multidisciplinary agriculture), biochemistry, botany, chemistry (analytical, applied, multidisciplinary, clinical, and medicinal), food science and technology, etc. However, certain publishers have introduced journals dedicated to the scientific exploration of hemp from the standpoint of various scientific disciplines, as listed in Table 7.8.

TABLE 7.8 Hemp-related scientific journals.

Name	Scope	URL	Publisher	Journal metrics
Journal of Industrial Hemp (1900–2008)	Different aspects of hemp except the medicinal uses: agronomy, taxonomy, breeding, crop physiology, modeling, crop yield and quality, control of diseases, harvesting technologies, utilization of hemp fiber, seeds, and oil *Incorporated in the Journal of Natural Fibers (2004–current)*	https://www.tandfonline.com/loi/wjih20	Taylor & Francis Note: An official journal of the International Hemp Association	N/A
Journal of Cannabis Therapeutics (2001–2004)	History of cannabis; clinical applications; essential fatty acids and essential oils; biochemical and pharmacological functions of cannabinoids in humans and animals; legislative/legal issues	http://www.cannabis-med.org/membersonly/mo.php	International Association for Cannabinoid Medicines	N/A
Journal of Cannabis Research	History, ethnobotany, and domestication; genetics, genomics, breeding, and synthetic biology; plant biology; chemistry of hemp; CBD: medical effects, production and regulation; hemp production and management; fiber hemp; hempseed and oil; new materials developed from hemp; plant-microbe interactions; current status: regulation and policy	https://jcannabisresearch.biomedcentral.com/	BioMed Central, Springer Nature Note: An official journal of the Institute of Cannabis Research	IF 3.0 (2020)
Cannabis and Cannabinoid Research	Clinical cannabis, cannabinoids, and the biochemical mechanisms of endocannabinoids; human and animal studies; behavioral, social, and epidemiological issues; ethical, legal, and regulatory issues	https://home.liebertpub.com/publications/cannabis-and-cannabinoid-research/633/	Mary Ann Liebert, Inc. publishers	IF 5.764 (2020)

TABLE 7.8 Hemp-related scientific journals—cont'd

Name	Scope	URL	Publisher	Journal metrics
Molecules, special issue "Industrial Hemp Chemistry and Nutraceutical Perspectives" (IF)	Hemp chemistry; bioactivity and effectiveness of hemp phytochemicals; food usages of hemp; innovative products in food and nutraceutical sectors	https://www.mdpi.com/journal/molecules/special_issues/Hemp_Chemistry_Nutraceutical	MDPI	IF 4.411 (2020)

7.4.5 Industrial hemp dedicated events

With renewed interest in industrial hemp as a sustainable raw material, a growing number of events dedicated to industrial hemp have sprung up around the world providing policy makers, industry experts, researchers, and innovators with plenty of networking and education opportunities (Table 7.9). Connecting businesses with the science and investors is the primary objective of such events, allowing for growth and development of the industrial hemp industry.

TABLE 7.9 Workshops, field days, symposiums, conferences, expos, and fairs on different aspects of industrial hemp industry.

INDUSTRIAL HEMP VIRTUAL WORKSHOP
https://programs.ifas.ufl.edu/hemp/news/events/
Organizer: University of Florida's Institute of Food and Agricultural Sciences (UF/IFAS); Industrial hemp pilot project
Topics:
- Research outcomes (hemp physiology and management, project overview, variety trials, invasion risk, pests and diseases)
- On-farm trial updates
- Extension

INDUSTRIAL HEMP CONFERENCE
https://campus.extension.org/course/view.php?id=1686
Organizer: University of Vermont (UVM) Extension Northwest Crops and Soils Program
Support: Vermont Agency of Agriculture, Food and Markets
Topics:
- Cannabis taxonomy and chemistry
- CBD hemp: propagation, selection, and breeding
- Grain and fiber hemp crop production
- Hemp grain: harvest, processing, storage, products, and markets
- Hemp markets
- Integrated pest management in hemp production
- Hemp irrigation and soil fertilization

2021 TN HEMP MARKET OUTLOOK WEBINAR
https://arec.tennessee.edu/hemp-workshops-and-webinars/
Organizer: Department of Agricultural and Resource Economics, The University of Tennessee Institute of Agriculture
Topics: Hemp production in Tennessee, outlook of hemp market and production, hemp certification

Continued

TABLE 7.9 Workshops, field days, symposiums, conferences, expos, and fairs on different aspects of industrial hemp industry—cont'd

---

HEMP RESEARCH FIELD WALK

https://extension.psu.edu/hemp-research-field-walk

Organizer: PennState Extension, College of Agricultural Sciences, The Pennsylvania State University

Topics:

- CBD hemp: autoflower and photosensitive varieties trials, genetics, no-till cropping system, weed control and management

---

- **INDUSTRIAL HEMP FIELD DAY**
- **MISSOURI HEMP FORUM**
- **MISSOURI HEMP FARMER FORUM**
- https://bluetigerportal.lincolnu.edu/web/hemp-institute/home

Organizer: Lincoln University Hemp Institute

Topics: Hempseed genetics, processing, hemp production, extension

---

NATIONAL HEMP SYMPOSIUM

https://nationalhempsymposium.org/

Organizer: National Academies of Sciences Board on Agriculture and Natural Resources (BANR) and Oregon State University's Global Hemp Innovation Center (GHIC)

Topics: State and the future of the industry of industrial hemp

---

HEMP SYMPOSIA

https://ag.colorado.gov/plants/industrial-hemp/general-information/hemp-symposia

Organizer: Industrial hemp program, Colorado Department of Agriculture, Colorado, United States

Topics:

- HOP—Hemp online portal
- USDA Final Report—pros, cons, and CDA suggestions
- Hemp cultivation in Colorado in relation to legal water supply and soil health

---

EIHA CONFERENCES

18th Conference: Hemp for Europe—Emerging opportunities for the Green Recovery (2021)

https://eiha-conference.org/

Organizer: European Industrial Hemp Association

Topics:

- Hemp fiber: processing, importance for sustainable fashion industry
- Hemp value chain
- Whole hemp plant utilization
- Hemp food and food supplements
- Hemp extracts
- Hemp market and regulations

AUSTRALIAN INDUSTRIAL HEMP CONFERENCE

https://www.agrifutures.com.au/product/australian-industrial-hemp-conference-2020/

Organizer: AgriFutures Australia

Topics: Industrial hemp production and marketing

HEMP BUSINESS SUMMIT

https://hempindustrial.com/our-events/2021-hemp-business-summit/

Organizer: The National Industrial Hemp Council

Topics:

- Banking and farm crediting
- Grain and fiber hemp breeding
- International trade and hemp markets

TABLE 7.9 Workshops, field days, symposiums, conferences, expos, and fairs on different aspects of industrial hemp industry—cont'd

NEW ZEALAND iHEMP 2021 SUMMIT AND EXPO
https://nzhia.com/events/hemp-summit-nz/
Organizer: New Zealand Hemp Industries Association—NZHIA
Where: Rotorua, NZ
Topics: Food, fiber, health

ANNUAL INDUSTRIAL HEMP SUMMIT
https://www.industrialhempsummit.info/
Organizer: Southeast Hemp Association
Topics: Technology, business and science of hemp food, fiber, and flower

SOUTHERN HEMP EXPO 2021
https://www.southernhempexpo.com/
Organizer: Hemp feed coalition
Topics: Genetics, soil amendments, processing and extraction methods, packaging solutions, hemp foods, hemp pet products

HEMP PRODUCTS & ANIMAL HEALTH VIRTUAL CONFERENCE
Organizer: The University of Tennessee Institute of Agriculture
Topics: Hempseed as animal feed, impacts of CBD to animal health, government policy

FLORIDA INDUSTRIAL HEMP CONFERENCE AND EXHIBITION
https://floridahempconference.org/
Topics: Development of the industrial hemp industry in Florida, hemp components

CULTIVA HANFEXPO
https://www.cultiva.at/page.asp/lang=en/index.htm
Topics: Hemp applications, law regulations, cultivation, products, CBD products for pets

EarthxHemp
https://earthx.org/about/hemp/
Topics: Environmental benefits of hemp-based products such as food, household items, and personal products

7.5 Challenges and opportunities

Industrial hemp as a multipurpose crop is increasingly finding a place in the wide market of products from construction to the pharmaceutical industry, as a result of which the expansion of its cultivation in the plant production systems can be expected. Due to its multipurpose nature, industrial hemp can be designated as more knowledge-intensive crop than others. Taking into account the fact that the hemp cultivation has been revived worldwide and the fact that technological process of industrial hemp production/processing has generally mastered to a sufficient extent to make production/processing economically viable in different agro-climatic conditions, it is still considered an underutilized crop requiring efficient knowledge creation, absorption, acquisition, and dissemination. Future efforts should be directed toward the development of an assortment specialized for specific purposes (fiber, hurds, oil, proteins, specific biomolecules). Due to climate change and expansion of the cultivation area in the conditions of pronounced crop-environment-management ($G \times E \times M$) interactions, attention should be focused on defining the adequate assortment with the appropriate technology for a particular region. Nonetheless, given its overall multipurpose potential, a top priority should be set to the promotion of currently underutilized industrial hemp and its prospects to improve environmental sustainability and reduce the carbon footprint. Due to a number of knowledge creators in the field, more harmonized protocols for their provision and dissemination should be developed globally.

Acknowledgment

M. Pojić would like to acknowledge the financial support of the Ministry of Education, Science and Technological Development of the Republic of Serbia (Contract No. 451-03-68/2022-14/ 200222).

References

Andre, C. M., Hausman, J.-F., & Guerriero, G. (2016). *Cannabis sativa*: The plant of the thousand and one molecules. *Frontiers in Plant Science, 7*, 19. https://doi.org/10.3389/fpls.2016.00019.
Bócsa, I. (1954). A kender heterózis-nemesítésének eredményei [Results of hemp heterosis breeding]. *Növenytermelés, 3*(4), 301–316 (in Hungarian).
Bócsa, I. (1967). Kender fajtahibrid előállításához szükséges unisexuális (hímmentes) anyafajta nemesítése [Breeding of a unisexual female variety required for the production of a hemp hybrid]. *Rostnövények, 1*, 3–8 (in Hungarian).
Bócsa, I. (1969). *Kender [Hemp]*. Budapest: Magyar növénynemesítés, Akadémiai Kiadó (in Hungarian).
Boháč, J. (1990). *Šľachtenie rastlin [Plant breeding]*. Bratislava: Priroda (in Slovak).
Boyce, S. S. (1900). *Hemp (Cannabis sativa). A practical treatise on the culture of hemp for seed and fiber with a sketch of the history and nature of the hemp plant*. New York: Orange Judd Company.
Bredemann, G. (1942). Die Bestimmung des Fasergehaltes bei Massenunter-suchungen von Hanf, Flachs, Fasernesseln und anderen Bastfaserpflanzen [The determination of the fiber content in mass examinations of hemp, flax, fiber acorns and other bast fiber plants]. *Faserforschung, 16*, 14–39 (in German).
Bredemann, G., Garber, K., Huhnke, W., & Sengbusch, R. (1961). Die Züchtung von monözischen, fasereitragreichen Hanfsorten Fibrimon und Fibridia [The breeding of monoean, fiber-rich hemp varieties Fibrimon and Fibridia]. *Zeitschrift für Pflanzenzüchtung, 46*, 235–245 (in German).
Breslavec, L. P., & Zaurov, E. (1937). Исследование конопли-гермафродита, обнаруженной в полевых культурах [A study of hermaphrodite hemp found in field crops]. *Doklady Akademii Nauk SSSR, 16*, 285–288 (in Russian).
Briosi, G., & Tognini, F. (1894). *Intorno alla anatomia della canapa (Cannabis sativa L.) [Around the anatomy of hemp (Cannabis sativa L.)]. Vol. 3*. Atti dell'Istituto Botanico, University of Pavia. II. (in Italian).
Campbell, S., Paquin, D., Awaya, J. D., & Li, Q. X. (2002). Remediation of benzo[a]pyrene and chrysene-contaminated soil with industrial hemp (*Cannabis sativa*). *International Journal of Phytoremediation, 4*(2), 157–168. https://doi.org/10.1080/15226510208500080.
Chandra, S., Lata, H., & ElSohly, M. A. (2017). *Cannabis sativa L.—Botany and biotechnology*. Springer International Publishing.
Citterio, S., Prato, N., Fumagalli, P., Aina, R., Massa, N., Santagostino, A., et al. (2005). The arbuscular mycorrhizal fungus *Glomus mosseae* induces growth and metal accumulation changes in *Cannabis sativa* L. *Chemosphere, 59*(1), 21–29. https://doi.org/10.1016/j.chemosphere.2004.10.009.
Clarke, R. C., & Merlin, M. D. (2013). *Cannabis: Evolution and ethnobotany*. University of California Press.
Davis, K. E., Babu, S. C., & Ragasa, C. (2020). *Agricultural extension: Global status and performance in selected countries*. Washington, DC: International Food Policy Research Institute.
Delamer, E. S. (1854). *Flax and hemp, culture and manipulation*. London: G. Routledge and Co.
Dewey, L. H. (1928). Hemp varieties of improved type are result of selection. In *USDA Yearbook 1927* (pp. 358–361). Washington, DC: United States Department of Agriculture.
Dingha, B., Sandler, L., Bhowmik, A., Akotsen-Mensah, C., Jackai, L., Gibson, K., et al. (2019). Industrial hemp knowledge and interest among North Carolina organic farmers in the United States. *Sustainability, 11*(9), 2691. https://doi.org/10.3390/su11092691.
Dodge, C. R. (1896). *A report on the culture of hemp and jute in the United States*. U.S.D.A. Office of Fiber Investigations, Report 8. Washington, DC: United States Government Printing Office.
Dodge, C. R. (1898). *A report on the culture of hemp in Europe*. U.S.D.A. Office of Fiber Investigations, Report 11. Washington, DC: United States Government Printing Office.

Drinić, Z., Vidović, S., Vladić, J., Koren, A., Kiprovski, B., & Sikora, V. (2018). Effect of extraction solvent on total polyphenols content and antioxidant activity of industrial hemp (Cannabis sativa L.). *Lekovite Sirovine, 38*, 17–21. https://doi.org/10.5937/leksir1838017DDu.

Drinić, Z., Vladić, J., Koren, A., Zeremski, T., Stojanov, N., Kiprovski, B., et al. (2020). Microwave-assisted extraction of cannabinoids and antioxidants from *Cannabis sativa* aerial parts and process modeling. *Journal of Chemical Technology and Biotechnology, 95*(3), 831–839. https://doi.org/10.1002/jctb.6273.

Drinić, Z., Vladić, J., Koren, A., Zeremski, T., Stojanov, N., Tomić, M., et al. (2021). Application of conventional and high-pressure extractions for the isolation of bioactive compounds from the aerial parts of hemp (*Cannabis sativa* L.) assortment Helena. *Industrial Crops and Products, 171*. https://doi.org/10.1016/j.indcrop.2021.113908, 113908.

Farinon, B., Molinari, R., Costantini, L., & Merendino, N. (2020). The seed of industrial hemp (Cannabis sativa L.): Nutritional quality and potential functionality for human health and nutrition. *Nutrients, 12*, 1935. https://doi.org/10.3390/nu12071935.

Feeney, M., & Punja, Z. K. (2003). Tissue culture and agrobacterium-mediated transformation of hemp (*Cannabis sativa* L.). *In Vitro Cellular and Developmental Biology, 39*(6), 578–585. https://doi.org/10.1079/IVP2003454.

Fruhvirth, C. (1922). Zur Hanfzüchtung [Hemp breeding]. *Zeitschrift fur Pflanzenzüchtung.* H. 4, B. VIII, Berlin. (in German).

Ge, S., & Liu, X. (2021). The role of knowledge creation, absorption and acquisition in determining national competitive advantage. *Technovation, 102396.* https://doi.org/10.1016/j.technovation.2021.102396.

Grishko, N. N. (1935). Die Züchtung des einhäusigen Hanfes und von Hanf mit gleichzeitig ausreifenden männlichen und weiblichen Pflanzen. In *Bericht. der Allruss. Akad. der Wiss. F. Ldw. Ser 3* (in German).

Hamel [Duhamel], H. L. (1747). *Traite de la Fabrique de Manoeuvres pour les Vaisseaux ou l'Art de la Corderie Perfectionné.* [publisher unknown] Paris; cited by Wissett, R. 1808 [1804] A treatise on hemp London: J. Harding.

Hirata, K. (1927). Sex determination in hemp (*Cannabis sativa* L.). *Journal of Genetics, 19,* I.

Hoffmann, W. (1952). Die vererbung der geschlechtsformen des hanfes (*Canabis sativa* L.) [Inheritance of the sex forms of hemp (*Canabis sativa* L.)] II. *Theoretical and Applied Genetics, 22,* 147–158.

Karche, T., & Singh, M. R. (2019). The application of hemp (*Cannabis sativa* L.) for a green economy: A review. *Turkish Journal of Botany, 43,* 710–723. https://doi.org/10.3906/bot-1907-15.

Kassem, H. S., Alotaibi, B. A., Muddassir, M., & Herab, A. (2021). Factors influencing farmers' satisfaction with the quality of agricultural extension services. *Evaluation and Program Planning, 85.* https://doi.org/10.1016/j.evalprogplan.2021.101912, 101912.

Koren, A., Sikora, V., Kiprovski, B., Brdar-Jokanović, M., Aćimović, M., Konstantinović, B., et al. (2020). Controversial taxonomy of hemp. *Genetika, 52*(1), 1–13. https://doi.org/10.2298/GENSR2001001K.

Li, J., Burson, R. C., Clapp, J. T., & Fleisher, L. A. (2020). Centers of excellence: Are there standards? *Healthcare, 8*(1), 20. 100388 https://doi.org/10.1016/j.hjdsi.2019.100388.

Linger, P., Mussig, J., Fischer, H., & Kobert, J. (2002). Industrial hemp (Cannabis sativa L.) growing on heavy metal contaminated soil: Fibre quality and phytoremediation potential. *Industrial Crops and Products, 16,* 33–42. https://doi.org/10.1016/S0926-6690(02)00005-5.

Liu, F. H., Hu, H. R., Du, G. H., Deng, G., & Yang, Y. (2017). Ethnobotanical research on origin, cultivation, distribution and utilization of hemp (*Cannabis sativa* L.) in China. *Indian Journal of Traditional Knowledge, 16*(2), 235–242. https://doi.org/10.1111/gcbb.12451.

Marcandier, M. (1764 (1758)). *Traité du Chanvre cited by Wissett, R. 1808 [1804] A treatise on hemp.* London: J. Harding. Wissett.

Marynchenko, I., & Chunjing, G. (2018). Коноплярство: наукови здобутки і перспективи *[Hemp growing: scientific achievements and perspectives].* Суми: ФОП Щербина І.В (in Russian).

Neppi, C. (1920). *Per la canapicoltura in Italia [Hempculture in Italy].* Ferrara. (in Italian).

Neuer, H., & Sengbusch, R. (1943). Die Geschlechtsvererbung bei Hanf und die Züchtung eines monözischen Hanfed [The gender inheritance in hemp and the breeding of a monoecious hemped]. *Der Züchter, 15,* 49–62 (in German).

Pasković, F. (1966). *Predivo bilje. Konoplja, lan i pamuk [Fiber plants. Hemp, flax and coton].* Zagreb: Nakladni zavod znanje (in Croatian).

Prade, T. (2011). *Industrial Hemp (Cannabis sativa L.)—A high-yielding energy crop.* Doctoral Thesis Uppsala: Swedish University of Agricultural Sciences.

Schaffner, I. H. (1927). Sex and sex-determination in the light of observations and experiments on dioecious plants. *The American Naturalist, 61*(675), 319–332.

Senchenko, G. I., & Timonin, M. A. (1978). Конопля *[Hemp].* Moskow: Колос (in Russian).

Skrodzka, I. (2016). Knowledge-based economy in the European Union—Cross-country analysis. *Statistics in Transition, 17*(2), 281–294.

Stonehouse, G. C. (2019). *Phytoremediation and biofortification potential of Cannabis sativa L.* Doctoral Thesis Fort Collins: Colorado State University, Colorado.

Van Bakel, H., Stout, J. M., Cote, A. G., Tallon, C. M., Sharpe, A. G., Hughes, T. R., et al. (2011). The draft genome and transcriptome of *Cannabis sativa. Genome Biology, 12*(10), R102. https://doi.org/10.1186/gb-2011-12-10-r102.

Van der Werf, H. M. (2004). Life cycle analysis of field production of fibre hemp, the effect of production practices on environmental impacts. *Euphytica, 140*(1), 13–23. https://doi.org/10.1007/s10681-004-4750-2.

Vandenhove, H., & Van Hees, M. (2005). Fibre crops as alternative land use for radioactively contaminated arable land. *Journal of Environmental Radioactivity, 81*, 131–141. https://doi.org/10.1016/j.jenvrad.2005.01.002.

Váša, F. (1965). *Pradne rostliny [Fiber plants].* Praha: SZN (in Czech).

Yakovlev, N. A., Kostikov, S. K., Kozyreva, N. I., & Martyushev, V. N. (2015). From high technologies to the technological superiority. *Procedia—Social and Behavioral Sciences, 166*, 232–234. https://doi.org/10.1016/j.sbspro.2014.12.516.

Zatta, A., Monti, A., & Venturi, G. (2012). Eighty years of studies on industrial hemp in the Po Valley (1930–2010). *Journal of Natural Fibers, 9*, 180–196. https://doi.org/10.1080/15440478.2012.706439.

8

Nutraceutical potential of industrial hemp

Viviana di Giacomo*, Claudio Ferrante*, Luigi Menghini*, and Giustino Orlando*

Department of Pharmacy, Medicinal Plant Unit (MPU), Botanic Garden "Giardino dei Semplici", "G. d'Annunzio" University of Chieti-Pescara, Chieti, Italy

8.1 Introduction

Industrial hemp has been extensively used as a valuable source of nutrients and fibers from its fruit and stem, rather than solely for cannabinoids. Hemp has long been cultivated, and the economic and pharmaceutical effects of hemp are currently increasing exponentially throughout the world, with a global market for low THC varieties ranging from $100 million to $2000 million per year (Montserrat-de la Paz, Marín-Aguilar, García-Giménez, & Fernández-Arche, 2014). Throughout Europe, including Italy, governments promote the cultivation of industrial hemp within agricultural policies aimed at implementing environmentally friendly crops. Specifically, this policy is sensitively evident in the middle Italy region where industrial hemp is the object of renewed interest, as indicated by increased land use dedicated to its cultivation. National and supranational interests toward this crop have led to the development of a huge number of new varieties that, besides being certified for low THC content, are sources of many hemp seed-derived foods. Hemp seed oil and hemp flour, obtained from seeds after the oil extraction process, are present as ingredients in many foods, which have recently been gaining in popularity. Hemp seeds and flour display remarkable nutritional value, which is due, albeit partially, to their content in primary metabolites such as minerals, vitamins (mostly A, C, and E complexes), lipids, proteins, and carbohydrates. In this regard, hemp seed lipid fraction is particularly rich (almost 80%) in essential fatty acids. They mostly consist of a large amount of linoleic (ω-6) and α-linolenic (ω-3) acids, often in a 3:1 ratio, ideal for human nutrition and for contrasting the onset of different pathological and chronic conditions, such as cancer and cardiovascular, degenerative,

---

* All authors equally contributed to the manuscript writing.

173

and inflammatory diseases (Kiralan, Gül, & Metin Kara, 2010). Interestingly, the ω-6 and ω-3 acids are precursors of eicosanoids, a class of compounds deeply involved in homeostatic processes such as inflammation, immunity, and vascular tone. In *Cannabis (C.) sativa*, THC content typically varies from 3% to 15%, but industrial hemp cultivars are bred to synthesize it only in traces (≤ 0.3% w/w) (De Backer, Maebe, Verstraete, & Charlier, 2012). Currently, only varieties published by EU (Regulation (EC) N° 1251/99 and subsequent amendments) are approved for planting in Europe, and some European countries, such as Nederland, Lithuania, France, and Romania, have recently become primary worldwide hemp producers after Canada. The selected varieties cultivated in these countries are eligible for cultivation only after verification of their THC content, which must be less than 0.2% w/w (Regulation EC N°. 1124/2008–12 November 2008) (Da Porto, Decorti, & Natolino, 2014). However, recent findings suggest the presence of other terpenophenolics potentially endowed with THC-like effects (Linciano et al., 2020). On January 14, 2017, Italian regulation N°172/2017 was published, allowing and regulating hemp production, commerce, and therapeutic use in Italy. Different genotypes have been selected and registered through time, as well as cultivation methods, to avoid uncontrolled birth of new hybrids. Currently, more than 60 varieties are registered in Europe (Farinon, Molinari, Costantini, & Merendino, 2020). Among others, *Eletta campana*, *Futura 75*, *Carmagnola selezionata*, and *Kc Virtus* have been recently studied to unravel their pharmacological and health-promoting properties against inflammatory and infectious diseases (di Giacomo et al., 2021; Ferrante et al., 2019; Menghini et al., 2021; Orlando et al., 2020, 2021).

Industrial hemp has long been considered a multiuse crop with promising applications in different fields, including textile, agrifood, and pharmaceuticals. The fiber isolated from the stalk is used to produce rope, paper, construction materials, clothing, and as a reinforcement in manufacturing composite parts, for example, in thermal and acoustic insulation (Vonapartis, Aubin, Seguin, Mustafa, & Charron, 2015). The quantity and quality of hemp fiber production is determined by different factors that should be controlled to provide suitable products. Harvesting time is one of the most important factors to be considered (Amaducci et al., 2015). Although fiber and seed are hemp's main products, there is growing interest in the valorization of a plethora of hemp's secondary metabolites, including terpenes, terpenophenolics, and flavonoids, mostly accumulated into the inflorescences, which display potential pharmacological effects (Amaducci et al., 2015). In this regard, hemp's essential oil from inflorescences is reported to have intriguing antimicrobic activity (Menghini et al., 2021; Orlando et al., 2021; Zengin et al., 2018), whereas the whole decocted plant is used at very low doses against migraine, as a pain reliever, and to prevent cognitive decline (Nissen et al., 2010).

8.2 Phenolic and polyphenolic compounds from hemp seeds

Phenolic compounds are plant-derived secondary metabolites biosynthesized during biotic or abiotic stresses, naturally linked to sugars (glycosides), which makes them more soluble in water and therefore mostly found in vacuoles. It is a wide class of chemicals, with more than 9000 identified compounds. Natural phenolic antioxidants are commonly used as food additives to protect food products against oxidation/rancidity, but they also exhibit a multitude of physiological activities (Vonapartis et al., 2015). In humans, phenolic compounds are known to have protective and promoting health effects. In this regard, multiple mechanisms have been ascribed to these secondary metabolites, including intrinsic scavenging/reducing properties, which make them efficacious in reducing the levels of reactive oxygen species (ROS) and the capability to act as inducers of cell antioxidant defense systems (Ferrante et al., 2020; Vasileva et al., 2020). Consequently, foods containing high concentrations of phenolic compounds have recently obtained great attention in the market of food supplements and nutraceuticals. Specifically, nutraceuticals are natural compounds extracted from medicinal and edible plants and formulated as high concentration ingredients of functional foods. Nutraceuticals show an improved pharmacological profile in terms of both efficacy and pharmacokinetics, also occurring at concentrations much higher than those present in the plant of origin. However, in the case of phenolic compounds, the isolation of pure phytochemicals from the plant of origin could lead to a paradoxical reduction in bioavailability, especially when the phenolic compounds are stored in the plant's vacuoles. In those cases, the use of unpurified extracts could be more promising in terms of bioavailability. This apparent contradiction could be explained by the fact that the bioavailability of phenolic compounds, in vivo, could be influenced by other phytoconstituents that exert synergistic effects in pharmacokinetic interactions (Gómez-Juaristi, Martínez-López, Sarria, Bravo, & Mateos, 2018; Lin, Teo, Leong, & Zhou, 2019; Shukla et al., 2017).

Given that industrial hemp is characterized by the presence of low concentrations of psychoactive compounds, the phenolic fraction represents one of the most significant and varied chemical classes possible to extract from commercial varieties. Indeed, industrial hemp is rich in flavonoids and lignanamides characteristic of the plant. The phenolic compounds could be extracted from different parts of the plant, even though the seeds are the elective source for phenolic compounds of health-promoting interest, given their extensive use in many foods. Specifically, flavonoids, lignanamides, and tocopherols are present in hemp seed oil (Irakli et al., 2019), and its valuable content in phenolic compounds is also responsible for its good oxidative stability. In this regard, phenolics may function as antioxidants in cold-pressed seed oil, thus preventing oil rancidity (DeMan, 2000). Flavonoids, including naringenin, kaempferol, and epicatechin, appear to be prominent in seed oil compared to phenolic acids, and these compounds can also be present as both free aglycones and/or glycoside derivatives (Smeriglio et al., 2016). However, other plant materials, including the female inflorescences, have gained much attention in recent years (di Giacomo et al., 2021; Ferrante et al., 2019; Izzo et al., 2020; Orlando et al., 2020). These recent findings examined the fingerprint of polar extracts from inflorescences of different industrial hemp varieties, leading to the identification of phenolics, including phenolic acids (caffeic acid), lignanamides, and phenolic amides (N-trans-caffeoyltyramine). However, due to variable concentrations, Izzo et al. (2020) suggested that both biotic and abiotic stresses could be responsible for such variability in phenolic compound concentration, thus suggesting a minor importance of these phytochemicals as quality markers of industrial hemp. On the other hand, the same authors analyzed the flavonoid profile of industrial hemp inflorescences and found that the flavones cannflavin A and cannflavin B were most frequently detected in different varieties. Indeed, previous studies also suggested these compounds as *Cannabis* specific, and therefore responsible, at least partially, for the observed anti-inflammatory effects (Rea et al., 2019).

8.3 Cannabinoids and other hemp components with pharmacological effects

Cannabinoids are hemp's characteristic secondary metabolites that show numerous pharmacological effects, through multitarget mechanisms, displaying potential efficacy and applicability in the management of a wide spectrum of chronic diseases. Among cannabinoids, Δ^9-THC has long been considered the sole psychotropic molecule, but a novel phytocannabinoid, the Δ^9-tetrahydrocannabiphorol, which is even more potent than THC, was recently identified in the *C. sativa* phytocomplex (Linciano et al., 2020). THC seems to have a promising dermatological role in the treatment of melanoma, itch, allergic contact dermatitis, and most of the inflammatory skin diseases, thanks to cannabinoid's anti-inflammatory properties (Mounessa, Siegel, Dunnick, & Dellavalle, 2017). Several studies have also been performed for exploring THC's use as an analgesic for chronic pain, as an appetite enhancer for anorexia, and as a treatment for nausea and emesis associated with chemotherapy. Other cannabinoids present in the phytocomplex, such as cannabigerol (CBG), cannabinol (CBN), and tetrahydrocannabivarin (THCV), are known to possess biological activities. Interestingly, the therapeutic effects of these hemp phytoconstituents may be enhanced, and the unwanted side effects mitigated, by other cannabinoids (Aizpurua-Olaizola et al., 2014). In the last three decades, the study of the pharmacological properties of these phytocompounds have led to the identification and characterization of endocannabinoid signaling, consisting of arachinonic acid-deriving molecules including anandamide and 2-acylglycerole, and the related metabotropic receptors, namely cannabinoid type 1 (CB1), expressed in prevalence at the presynaptic level, and CB2, which is present especially but not exclusively at the peripheral level (Starowicz & Di Marzo, 2013). The neuroprotective effects exerted by endocannabinoid are mostly due to CB1 and CB2 receptor activation, whereas the neuroprotection resulting from phytocannabinoid administration could be better explained by a multitarget mechanism, mainly involving peroxisome proliferator-activated receptors (PPARs) and transient receptor potential (TRP) channels (Aso & Ferrer, 2014; Marchalant et al., 2009; Muller, Morales, & Reggio, 2019). In addition, the CB1 and CB2 receptors were considered promising targets for the development of antiobesity drugs (Bi, Galaj, He, & Xi, 2020) to the point that rimonabant, the prototype of the CB1 blockers, was clinically employed for its anorexigenic effects. Nevertheless, rimonabant was soon after retired from the market because of an increased frequency in psychiatric disorders observed in obese patients undergoing the treatment (Jager & Witkamp, 2014). The therapeutic failure was ascribed, at least partially, to the specific mechanism of action of the drug, whose orthosteric inverse agonism on CB1 could lead to supraphysiological receptor alterations (Wootten, Christopoulos, & Sexton, 2013).

Among cannabinoids, cannabidiol (CBD) has attracted the attention of the scientific community for its promising role as a protective agent in terms of anti-inflammatory and neuromodulatory properties. CBD could indeed represent an innovative pharmacological approach; although it is described as a CB1 ligand, CBD showed low affinity for the CB1 orthosteric site (Ibeas Bih et al., 2015). Basically, this could be considered as one of the main factors influencing the lack of psychiatric symptoms following CBD administration (Lee, Bertoglio, Guimarães, & Stevenson, 2017). Additionally, Laprairie, Bagher, Kelly, and Denovan-Wright (2015) suggested that CBD could act as a negative allosteric modulator rather than an orthosteric ligand. CB1 negative allosteric modulators are molecules devoid of intrinsic receptor activity, which depends solely on the presence of the endogenous ligands, i.e., the endocannabinoids (Ross,

2007). Considering that the brain level of endocannabinoids is upregulated in obese mice (Di Marzo et al., 2001), CBD could reduce, without annulling, the endocannabinoid-stimulated CB1 signaling, thus restoring the physiological activity of the endocannabinoid system (Lee et al., 2017). Another possible antiobesity target is the CB2 receptor. Ignatowska-Jankowska, Jankowski, and Swiergiel (2011) demonstrated that the anorexigenic effect following CBD administration could depend on the activation of the hypothalamic CB2 pool. Therefore, the anorexigenic effect exerted by CBD could be the result of a multitarget mechanism involving the whole endocannabinoid receptor system, particularly in the hypothalamus, where peripheral afferents are transduced by neuropeptides and neurotransmitters, including serotonin, in appetite and satiety signals (Valassi, Scacchi, & Cavagnini, 2008). In this regard, the increased serotonin synthesis in the hypothalamus seems to be one of the putative mechanisms underlying the anorexigenic effect of CBD (di Giacomo et al., 2020).

In addition to those previous described, despite having a chemical structure very close to that of THC, CBD was reported to have more than 60 receptor proteins putatively related to its pharmacological effects. However, the most probable target-proteins mediating CBD's observed pharmacological effects include the CB1 receptor, the transient receptor potential vanilloid-1 (TRPV1), the TRP cation channel subfamily M member 8 (TRPM8), the peroxisome proliferator-activated receptor γ (PPARγ), and the 5-hydroxytryptamine receptor 1A (5-HT1A).

The multitarget mechanism of CBD could be at the foundation of the potential use of *C. sativa* in the management of epilepsy; the use of this plant as antiepileptic remedy had been reported since the late 19th century (Fasinu, Phillips, ElSohly, & Walker, 2016). Epilepsy is a neurological disorder characterized by abnormal brain electrical activity and spontaneous recurrent seizures. About 4% to 10% of children suffering from epilepsy manifest seizure symptoms within adolescence; 30% of them have intractable seizures that cannot be controlled by first-order antiepileptic drugs (Schmidt & Schachter, 2014). Even though there are discrepancies in the scientific literature about the use of *Cannabis* in treating the epileptic crisis, with some studies reporting proconvulsant effects (Devinsky et al., 2014), epilepsy remains one of the most promising clinical applications of CBD (Fasinu et al., 2016). In a clinical study carried out by Devinsky et al. (2014), CBD administration was found to be safe and effective in childhood-onset treatment-resistant epilepsy. Although the involvement of 5-HT in the epilepsy pathogenesis is still debated (Ibeas Bih et al., 2015), one of the putative mechanisms of CBD could be the stimulation of 5-HT1A receptors. Their activation could lead to improved brain antioxidant defense with subsequent reduction of neuron membrane lipid peroxidation, especially after exposure to pro-oxidant stimuli (Atalay, Jarocka-Karpowicz, & Skrzydlewska, 2019). In this regard, it is sensible to highlight the tight relationships between epilepsy and neuroinflammation (Vezzani, Balosso, & Ravizza, 2019), whereas CBD was reported to downregulate proinflammatory pathways underlying epileptogenesis (Cheung, Peiris, Wallace, Holland, & Mitchell, 2019). A list of putative receptors and enzymes directly influenced by CBD is reported in Table 8.1. Specifically, the target proteins were predicted by the bioinformatics platform SwissTargetPrediction, which delivered a list of proteins that could directly interact with CBD.

CBG, another *C. sativa* terpenophenol that shares a similar mechanism of action with CBD, was also reported to act as a neuroprotective agent and bind the 5-HT1A receptor (Cascio, Gauson, Stevenson, Ross, & Pertwee, 2010). In a recent study, the treatment of isolated rat

TABLE 8.1 Putative target proteins interacting with cannabidiol.

Target	Common name	Uniprot ID	ChEMBL ID	Target class
G-protein coupled receptor 55	GPR55	Q9Y2T6	CHEMBL1075322	Family A G protein-coupled receptor
Cannabinoid receptor 1	CNR1	P21554	CHEMBL218	Family A G protein-coupled receptor
Cannabinoid receptor 2	CNR2	P34972	CHEMBL253	Family A G protein-coupled receptor
Arachidonate 5-lipoxygenase	ALOX5	P09917	CHEMBL215	Oxidoreductase
Cystic fibrosis transmembrane conductance regulator	CFTR	P13569	CHEMBL4051	Other ion channel
NAD-dependent deacetylase sirtuin 2	SIRT2	Q8IXJ6	CHEMBL4462	Eraser
Calcium-sensing receptor	CASR	P41180	CHEMBL1878	Family C G protein-coupled receptor
Neurokinin 3 receptor	TACR3	P29371	CHEMBL4429	Family A G protein-coupled receptor
Arachidonate 15-lipoxygenase	ALOX15	P16050	CHEMBL2903	Enzyme
Receptor protein-tyrosine kinase erbB-2	ERBB2	P04626	CHEMBL1824	Kinase
DNA polymerase beta (by homology)	POLB	P06746	CHEMBL2392	Enzyme
N-arachidonyl glycine receptor	GPR18	Q14330	CHEMBL2384898	Family A G protein-coupled receptor
Glycine receptor subunit alpha-1	GLRA1	P23415	CHEMBL5845	Ligand-gated ion channel
Cyclooxygenase-2	PTGS2	P35354	CHEMBL230	Oxidoreductase
Cathepsin D	CTSD	P07339	CHEMBL2581	Protease
Beta-secretase 1	BACE1	P56817	CHEMBL4822	Protease
Voltage-gated potassium channel subunit Kv1.5	KCNA5	P22460	CHEMBL4306	Voltage-gated ion channel
Corticotropin releasing factor receptor 1	CRHR1	P34998	CHEMBL1800	Family B G protein-coupled receptor
L-Lactate dehydrogenase B chain	LDHB	P07195	CHEMBL4940	Enzyme
Vascular endothelial growth factor receptor 2	KDR	P35968	CHEMBL279	Kinase
Cholesteryl ester transfer protein	CETP	P11597	CHEMBL3572	Other ion channel
Bile acid receptor FXR	NR1H4	Q96RI1	CHEMBL2047	Nuclear receptor
Prenyl protein specific protease	RCE1	Q9Y256	CHEMBL3411	Protease

Target name	Gene	UniProt	ChEMBL	Target class
11-beta-hydroxysteroid dehydrogenase 1	HSD11B1	P28845	CHEMBL4235	Enzyme
Carnitine O-palmitoyltransferase 1, liver isoform	CPT1A	P50416	CHEMBL1293194	Enzyme
Cholecystokinin A receptor	CCKAR	P32238	CHEMBL1901	Family A G protein-coupled receptor
Phosphodiesterase 4B	PDE4B	Q07343	CHEMBL275	Phosphodiesterase
Thymidylate synthase	TYMS	P04818	CHEMBL1952	Transferase
Serotonin 2b (5-HT2b) receptor	HTR2B	P41595	CHEMBL1833	Family A G protein-coupled receptor
Interleukin-8 receptor A	CXCR1	P25024	CHEMBL4029	Family A G protein-coupled receptor
Metabotropic glutamate receptor 2	GRM2	Q14416	CHEMBL5137	Family C G protein-coupled receptor
p53-binding protein Mdm-2	MDM2	Q00987	CHEMBL5023	Other nuclear protein
ADAMTS5	ADAMTS5	Q9UNA0	CHEMBL2285	Protease
6-phosphofructo-2-kinase/fructose-2,6-bisphosphatase 3	PFKFB3	Q16875	CHEMBL2331053	Enzyme
Isocitrate dehydrogenase [NADP] cytoplasmic	IDH1	O75874	CHEMBL2007625	Enzyme
Neuropeptide Y receptor type 5	NPY5R	Q15761	CHEMBL4561	Family A G protein-coupled receptor
G-protein coupled bile acid receptor 1	GPBAR1	Q8TDU6	CHEMBL5409	Family A G protein-coupled receptor
Cholecystokinin B receptor (by homology)	CCKBR	P32239	CHEMBL298	Family A G protein-coupled receptor
Histone deacetylase 6	HDAC6	Q9UBN7	CHEMBL1865	Eraser
Histone deacetylase 8	HDAC8	Q9BY41	CHEMBL3192	Eraser
Mitogen-activated protein kinase kinase kinase 8	MAP3K8	P41279	CHEMBL4899	Kinase
Protein farnesyltransferase	FNTA FNTB	P49354 P49356	CHEMBL2094108	Enzyme
L-lactate dehydrogenase A chain	LDHA	P00338	CHEMBL4835	Enzyme
MAP kinase p38 alpha	MAPK14	Q16539	CHEMBL260	Kinase
Steryl-sulfatase	STS	P08842	CHEMBL3559	Enzyme
ATP-binding cassette sub-family G member 2	ABCG2	Q9UNQ0	CHEMBL5393	Primary active transporter
Translocator protein (by homology)	TSPO	P30536	CHEMBL5742	Membrane receptor

Bioinformatics prediction was conducted on the platform SwissTargetPrediction.

cortex with CBD resulted in a decreased 5-HT turnover, and a similar trend was found for CBG (di Giacomo et al., 2020). The results of this study also suggested different putative mechanisms of action for the two terpenophenols. On one side, the efficacy of CBD was partly related to its neuromodulatory action on serotonin receptors. On the other side, the protective effects induced by CBG in preventing the reduction of serotonin pool in an isolated cortex exposed to a neurotoxic stimulus was mostly due to its intrinsic antiradical effect rather than to discrete receptor mechanisms. However, further studies need to confirm these observations.

Another interesting application of cannabinoids in the nervous system is represented by the treatment of anxiety disorders and depression. Accumulating evidence suggests that CBD can have anxiolytic properties, which have been highlighted by many preclinical studies. The anxiolytic effects of CBD were found mostly related to the activation of 5-HT1A receptors (Melas, Scherma, Fratta, Cifani, & Fadda, 2021), while TRPV1 receptors were found to be involved in just one study (Campos & Guimarães, 2009). Interestingly, in a murine model of chronic stress, CBD was found to exert an anxiolytic effect by inducing hippocampal neurogenesis (Campos et al., 2013), while Crippa et al. (2011) used the SPECT technique to demonstrate that CBD reduces cerebral flow in (para)limbic areas (i.e., hippocampus, para-hippocampal and inferior temporal gyrus) in subjects with social anxiety, suggesting that the anxiolytic effects of CBD may be related to the capacity of this compound to modify brain activity in specific areas.

The activation of 5-HT1A receptors and consequent increase of 5-HT and glutamate levels were also found to be involved in the antidepressant effects of CBD in several animal models of depression (Melas et al., 2021). Another factor targeted by CBD in mood disorders is the Brain-Derived Neurotrophic Factor (BDNF), whose serum level is usually found to be low in depressant patients and is increased by antidepressant drugs (Sen, Duman, & Sanacora, 2008). The acute antidepressant effects of CBD in mice were accompanied by increased BDNF levels in the hippocampus and medial prefrontal cortex (Sales et al., 2019). A study from El-Alfy et al. (2010) confirmed the antidepressant-like actions in animal models of behavioral despair and, at the same time, found no antidepressant properties ascribable to CBG. These findings agree with other studies (di Giacomo, Chiavaroli, Orlando, et al., 2020; di Giacomo, Chiavaroli, Recinella, et al., 2020) demonstrating a lower neuromodulatory activity of CBG when compared to CBD.

The effects of CBD and CBG were explored in the digestive system as well. Pagano et al. (2016) reported the efficacy of CBD as an anti-inflammatory agent in murine models of ulcerative colitis. As for CBG, the same research group also described anti-inflammatory and anticancer properties, in experimental preclinical models of ulcerative colitis and colon cancer (Borrelli et al., 2013, 2014). These results seem to be related to the pharmacological antagonism toward the TRPM8 receptor as already suggested by previous in vitro studies (De Petrocellis et al., 2008). Although the bioinformatics platform SwissTargetPrediction did not predict putative interactions between CBG, CBD, and this receptor (Tables 8.1 and 8.2), the docking runs suggested the ability of CBG to bind to TRPM8 with a micromolar affinity (Fig. 8.1).

According to the cannabinoid distribution found in colon tissue (30–40 ng/mg tissue) after oral administration (Pagano et al., 2016), the in silico and in vitro studies further support the inhibition of TRPM8 as one of the main mechanisms underlying the efficacy of CBG as a protective agent.

TABLE 8.2 Putative target proteins interacting with cannabigerol.

Target	Common name	Uniprot ID	ChEMBL ID	Target class
Cannabinoid receptor 1	CNR1	P21554	CHEMBL218	Family A G protein-coupled receptor
Cannabinoid receptor 2	CNR2	P34972	CHEMBL253	Family A G protein-coupled receptor
G-protein coupled receptor 55	GPR55	Q9Y2T6	CHEMBL1075322	Family A G protein-coupled receptor
Arachidonate 5-lipoxygenase	ALOX5	P09917	CHEMBL215	Oxidoreductase
DNA polymerase beta (by homology)	POLB	P06746	CHEMBL2392	Enzyme
Peroxisome proliferator-activated receptor gamma	PPARG	P37231	CHEMBL235	Nuclear receptor
Peroxisome proliferator-activated receptor alpha	PPARA	Q07869	CHEMBL239	Nuclear receptor
Arachidonate 15-lipoxygenase	ALOX15	P16050	CHEMBL2903	Enzyme
ADAMTS5	ADAMTS5	Q9UNA0	CHEMBL2285	Protease
Corticotropin releasing factor receptor 1	CRHR1	P34998	CHEMBL1800	Family B G protein-coupled receptor
Cyclooxygenase-2	PTGS2	P35354	CHEMBL230	Oxidoreductase
Adenosine A2a receptor	ADORA2A	P29274	CHEMBL251	Family A G protein-coupled receptor
Interleukin-8 receptor B	CXCR2	P25025	CHEMBL2434	Family A G protein-coupled receptor
Interleukin-8 receptor A	CXCR1	P25024	CHEMBL4029	Family A G protein-coupled receptor
Adenosine A1 receptor	ADORA1	P30542	CHEMBL226	Family A G protein-coupled receptor
Tyrosine-protein kinase LCK	LCK	P06239	CHEMBL258	Kinase
Neurokinin 3 receptor	TACR3	P29371	CHEMBL4429	Family A G protein-coupled receptor
HERG	KCNH2	Q12809	CHEMBL240	Voltage-gated ion channel
Muscle glycogen phosphorylase	PYGM	P11217	CHEMBL3526	Enzyme
Voltage-gated potassium channel subunit Kv1.5	KCNA5	P22460	CHEMBL4306	Voltage-gated ion channel
Gamma-secretase	PSEN2 PSENEN NCSTN APH1A PSEN1 APH1B	P49810 Q9NZ42 Q92542 Q96BI3 P49768 Q8WW43	CHEMBL2094135	Protease
L-Lactate dehydrogenase A chain	LDHA	P00338	CHEMBL4835	Enzyme

Continued

TABLE 8.2   Putative target proteins interacting with cannabigerol—cont'd

Target	Common name	Uniprot ID	ChEMBL ID	Target class
Cholecystokinin A receptor	CCKAR	P32238	CHEMBL1901	Family A G protein-coupled receptor
Muscarinic acetylcholine receptor M4	CHRM4	P08173	CHEMBL1821	Family A G protein-coupled receptor
Muscarinic acetylcholine receptor M5	CHRM5	P08912	CHEMBL2035	Family A G protein-coupled receptor

Bioinformatics prediction was conducted on the platform SwissTargetPrediction.

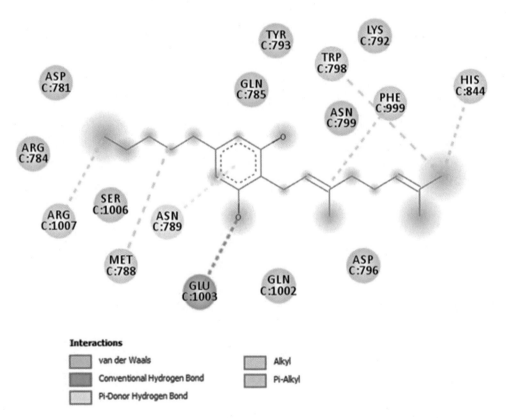

FIG. 8.1   Putative interactions between cannabigerol and TRPM8 receptor (PDB: 6NR3). Free binding energy and affinity are − 6.6 Kcal/mol and 14.7 μM, respectively.

8.4 The role of hemp products in disease prevention

Despite being characterized by a high nutritional value, hemp seeds were initially considered a by-product of hemp chain production, mainly focused on the fiber fraction. Therefore, seeds were only employed to feed animals (Della Rocca & Di Salvo, 2020). However, interest in hemp seed has greatly increased in recent years as a result of new agricultural policies for the introduction of environmentally friendly crops. Nevertheless, although scientific studies clearly describe the nutrient compositions of hemp seeds, there is still lack of scientific research assessing its health-promoting effects with particular focus on the prevention of human diseases (Farinon et al., 2020). Hemp seeds have been suggested to be a complete food with a valuable content in primary metabolites including fatty acids (20%–35%), proteins (20%–25%), and carbohydrates (20%–30%).

Regarding the fatty acid fraction, about 70% to 80% is constituted by polyunsaturated fatty acids (PUFAs) and monounsaturated fatty acids (MUFAs). Oleic acid and linoleic acid are the prominent MUFA and PUFA, respectively, with percentages in the range of 18% to 20%. Among PUFAs, it is worth mentioning the presence of α-linolenic acid that, with linolenic acid, acts as a precursor of n-3 PUFAs in mammals. The latter are known to exert numerous health-promoting effects including hypotriglyceridemic and anti-inflammatory effects. Numerous studied also suggest their efficacy as antihypertensive and antiproliferative agents. These findings are consistent, albeit partially, with the capability of n-3 PUFAs to reduce the synthesis of proinflammatory prostaglandin-E_2 and the expression of NFkB (nuclear factor kappa-light-chain-enhancer of activated B cells) (Siriwardhana, Kalupahana, & Moustaid-Moussa, 2012).

Whether the content in primary metabolites could represent a cornerstone for considering hemp seed-derived products as active ingredients in food (van den Driessche, Plat, & Mensink, 2018), thus reflecting improvement of the whole hemp chain with regard to this plant material, it is equally important to consider other parts of industrial hemp, such as inflorescences, up to now considered only as waste materials. Recent studies have pointed to the analysis of the phytocomplex of different polar extracts and essential oils from inflorescences, identifying a wide plethora of secondary metabolites including terpenes, terpenophenolics, and phenolics (di Giacomo et al., 2021; Ferrante et al., 2019; Menghini et al., 2021; Orlando et al., 2021; Smeriglio et al., 2016). Although the essential oils of the inflorescences proved to be particularly rich in the sesquiterpene caryophyllene (Nagy, Cianfaglione, Maggi, Sut, & Dall'Acqua, 2019), prior research found, in different hemp varieties, the presence of CBD and its acidic form, the cannabidiolic acid (CBDA), as prominent terpenophenolics (di Giacomo et al., 2021; Menghini et al., 2021; Nagy et al., 2019; Orlando et al., 2021). Besides showing antimicrobial and antiparasistic activities, the essential oil was described as an efficacious agent in downregulating the gene expression of angiotensin-converting enzyme 2 (ACE2) and transmembrane protease serine 2 (TMPRSS2) (Orlando et al., 2021), two proteins deeply involved in the SARS-CoV-2 virus entry in the human host (Sungnak et al., 2020; Turner, Hiscox, & Hooper, 2004). The efficacy against ACE2 and TMPRSS2 was also demonstrated by hemp polar extracts (Anil et al., 2021; Wang et al., 2020), thus further suggesting the involvement of terpenophenolics as active phytocompounds with promising anti-COVID-19 properties. Indeed, the presence of terpenophenolics was demonstrated in both polar extracts and essential oils by different research groups, with CBD and CBDA as the most representative of this class of compounds. In addition, the polar extracts of industrial hemp are also rich in flavonoids, and a study demonstrated the presence of phenolic compounds in the essential oil (Zengin et al., 2018). Although only one study investigated the presence of these compounds in the essential oil, the observed biopharmacological effects could still be ascribed to the antimicrobial and anti-COVID-19 properties of both lipophilic and hydrophilic fractions of the inflorescences. In this context, the whole hemp chain production should take these findings into consideration to formulate nutraceutical and cosmeceutical products, as well as protection devices, functioning as physical barriers against infectious diseases.

8.5 Conclusions

The use of commercial hemp cultivar extracts and essential oils, formulated in food supplements and/or cosmetic preparations, could provide significant improvements to health. It is sensible to note that the polar fraction of industrial hemp was revealed to be particularly effective as an antioxidant/anti-inflammatory agent, consistently with valuable levels of phenolic compounds, especially flavonoids. On the other hand, the lipophilic fraction, i.e., the essential oil, showed moderate to high efficacy as antimicrobial agents (Nissen et al., 2010) as demonstrated by multiple studies pointing to inhibitory effects against leishmanial, bacterial, and fungal strains involved in skin damage. Additionally, essential oils and extracts were also efficacious in inhibiting the gene expression of ACE2 and TMPRSS proteins playing key roles in COVID-19 entry in the human host. Collectively, recent studies conducted on industrial hemp support an improvement in its cultivation and valorization.

Nevertheless, considering the limits of the preclinical studies currently available, clinical trials are required to confirm both the efficacy and biocompatibility of industrial hemp-derived extracts and essential oils.

8.6 Challenges and opportunities

Today's renewed interest in hemp cultivation is oriented toward the development of sustainable crops able to generate innovative products within a productive chain in the context of a circular economy. The plant itself is suitable for multipurpose cultivation that includes the use of seeds in the food industry; the fibers from stem for fabrics, building materials, and bioplastics; and the inflorescences as a source of high-quality bioactive metabolites (di Giacomo et al., 2021; Izzo et al., 2020; Smeriglio et al., 2016). To correctly orient cultivation for a specific target, the selection of the variety is critical. Multiple products could derive from the same process, such as seed oil, characterized by the valuable content in fatty acids, and flour, the latter obtained by fine grinding of solid residue after oil extraction. However, the cultivation plan is more frequently defined for a selected target product. In this context, results of relevant interest derive from the experimental approach to the valorization of botanical chain by-products that usually represent a relevant part of the total plant biomass. The possibility of activating multitarget cultivations and parallel productive chains should be a strategy to support hemp farming, a traditional crop with a modern and innovative development perspective for economy, employment, environment, as well as for human health. It also represents one of the main actors in the challenge for a virtuous green economy process oriented toward the ambitions of the EU 2030 climate & energy framework.

Innovative and promising economic perspectives for hemp cultivation have been recently focused on industrial hemp inflorescences. After having been considered a waste product for centuries, the female flowers, as valuable sources of protective secondary metabolites, can completely support the economic sustainability of the crop in its current context, where the fiber productive chain is no longer available . On the other hand, the achievement of economic sustainability of the inflorescence production induces a deep revision of the variety-environment previously optimized for fiber production with particular emphasis on the selective production of terpenophenolics with no psychotropic effects. Because the number of admitted varieties is limited, agronomical factors, such as the selection of monoecious/dioecious variety and climate interactions, can play a pivotal role in the achievement of selective secondary metabolism activation. Regarding the perspectives for inflorescence valorization, literature studies point to the antimicrobial activity of the essential oil and the anti-inflammatory effects of the polar extracts, both prepared from the female inflorescences (Nissen et al., 2010). The raw plant material, as well as purified extracts and isolated compounds, are today considered relevant natural products with high potential and a plethora of therapeutic benefits that can support its use in pharmaceutical, nutraceutical, cosmetic, food, and beverage applications. CBD and other minor cannabinoids, such as CBN, CBG, cannabichromene (CBC), THCV, cannabidivarin (CBDV), cannabinodiol (CBND), and cannabinidiol (CBDL), are widely investigated for their activity as analgesic, anti-inflammatory, antipyretic, antiepileptic, anticancer, and antiemetic agents with potential applications in the managing of chronic pain, regulation of the nervous system, treatment of nausea and vomiting, cardiovascular disease, or appetite loss. An increasing request for medical and nutritional cannabinoids, mainly CBD, is evident and mainly focused on food products (Salami et al., 2020). Also, the steep growth of CBD formulations in the pharmaceutical sector is evident, but very few approved pharmaceutical products are available due to limited clinical data, safety data, and efficacy testing. Furthermore, well-defined and shared regulatory issues on proper classification and approval of CBD-based therapies, as well as for food uses with defined dosing range for each indication, are still missing. In addition, low bioavailability coupled with low water solubility represents one of the most impacting factors on the clinical efficacy of CBD and related cannabinoids, which require a technological formulation to express pharmacological potential and, at the same time, to limit the classical homemade practice of herbal preparations, such as tea and infusion. On the other hand, an increased level of education and awareness of consumers of the wide potential health benefits of hemp-based products in general and CBD products in particular demand new, high-quality products that combine healthy efficacy with the safety related to the THC-free grade. This approach will help to overcome classical prejudices related to hemp products.

The interest in the cultivation has deeply increased in recent years, and both commercial and academic interests are converging toward joint projects that could lead to the creation of new synergies and working opportunities, at both local and international levels. Middle Italy, where increasing land use dedicated to the cultivation of industrial hemp is paralleled by both public calls and private investments, are consistent with the requalification of uncultivated lands with an environmentally friendly crop. This is surely one of the stimuli driving the development of a wide plethora of new varieties (more than 60 throughout Europe) that, besides being certified for low THC content, are sources of many hemp seed-derived foods. This context of promising interest results in the production of plant material for extraction of CBD including the challenge to reduce the effects of genetic and environmental factors affecting the qualitative and quantitative composition of extracts from natural sources to a standardization of the active ingredient production.

Acknowledgments

The present chapter is part of the third mission activities of the Botanic Garden "Giardino dei Semplici" planned for the 20th anniversary of the establishment. The writing of the chapter was also conducted within the joint project between Department of Pharmacy of "G. d'Annunzio" University and Veridia S.r.l. aimed at the valorization of industrial hemp cultivation in Middle Italy, Abruzzo Region (Research Program 2020-2023). Project coordinators are Giustino Orlando, Luigi Menghini, and Claudio Ferrante.

Conflict of interest

The authors declare no conflict of interest.

References

Aizpurua-Olaizola, O., Omar, J., Navarro, P., Olivares, M., Etxebarria, N., & Usobiaga, A. (2014). Identification and quantification of cannabinoids in Cannabis sativa L. plants by high performance liquid chromatography-mass spectrometry. *Analytical and Bioanalytical Chemistry, 406*(29), 7549–7560. https://doi.org/10.1007/s00216-014-8177-x.

Amaducci, S., Scordia, D., Liu, F. H., Zhang, Q., Guo, H., Testa, G., & Cosentino, S. L. (2015). Key cultivation techniques for hemp in Europe and China. *Industrial Crops and Products, 68*, 2–16.

Anil, S. M., Shalev, N., Vinayaka, A. C., Nadarajan, S., Namdar, D., Belausov, E., … Koltai, H. (2021). Cannabis compounds exhibit anti-inflammatory activity in vitro in COVID-19-related inflammation in lung epithelial cells and pro-inflammatory activity in macrophages. *Scientific Reports, 11*(1), 1462. https://doi.org/10.1038/s41598-021-81049-2.

Aso, E., & Ferrer, I. (2014). Cannabinoids for treatment of Alzheimer's disease: Moving toward the clinic. *Frontiers in Pharmacology, 5*, 37. https://doi.org/10.3389/fphar.2014.00037.

Atalay, S., Jarocka-Karpowicz, I., & Skrzydlewska, E. (2019). Antioxidative and anti-inflammatory properties of cannabidiol. *Antioxidants, 9*(1), 21. https://doi.org/10.3390/antiox9010021.

Bi, G. H., Galaj, E., He, Y., & Xi, Z. X. (2020). Cannabidiol inhibits sucrose self-administration by CB1 and CB2 receptor mechanisms in rodents. *Addiction Biology, 25*(4). https://doi.org/10.1111/adb.12783, e12783.

Borrelli, F., Fasolino, I., Romano, B., Capasso, R., Maiello, F., Coppola, D., … Izzo, A. A. (2013). Beneficial effect of the non-psychotropic plant cannabinoid cannabigerol on experimental inflammatory bowel disease. *Biochemical Pharmacology, 85*(9), 1306–1316. https://doi.org/10.1016/j.bcp.2013.01.017.

Borrelli, F., Pagano, E., Romano, B., Panzera, S., Maiello, F., Coppola, D., … Izzo, A. A. (2014). Colon carcinogenesis is inhibited by the TRPM8 antagonist cannabigerol, a Cannabis-derived non-psychotropic cannabinoid. *Carcinogenesis, 35*(12), 2787–2797. https://doi.org/10.1093/carcin/bgu205.

Campos, A. C., & Guimarães, F. S. (2009). Evidence for a potential role for TRPV1 receptors in the dorsolateral periaqueductal gray in the attenuation of the anxiolytic effects of cannabinoids. *Progress in Neuro-Psychopharmacology & Biological Psychiatry, 33*(8), 1517–1521. https://doi.org/10.1016/j.pnpbp.2009.08.017.

Campos, A. C., Ortega, Z., Palazuelos, J., Fogaça, M. V., Aguiar, D. C., Díaz-Alonso, J., … Guimarães, F. S. (2013). The anxiolytic effect of cannabidiol on chronically stressed mice depends on hippocampal neurogenesis: Involvement of the endocannabinoid system. *The International Journal of Neuropsychopharmacology, 16*(6), 1407–1419. https://doi.org/10.1017/S1461145712001502.

Cascio, M. G., Gauson, L. A., Stevenson, L. A., Ross, R. A., & Pertwee, R. G. (2010). Evidence that the plant cannabinoid cannabigerol is a highly potent alpha2-adrenoceptor agonist and moderately potent 5HT1A receptor antagonist. *British Journal of Pharmacology, 159*(1), 129–141. https://doi.org/10.1111/j.1476-5381.2009.00515.x.

Cheung, K., Peiris, H., Wallace, G., Holland, O. J., & Mitchell, M. D. (2019). The interplay between the endocannabinoid system, epilepsy and cannabinoids. *International Journal of Molecular Sciences, 20*(23), 6079. https://doi.org/10.3390/ijms20236079.

Crippa, J. A., Derenusson, G. N., Ferrari, T. B., Wichert-Ana, L., Duran, F. L., Martin-Santos, R., ... Hallak, J. E. (2011). Neural basis of anxiolytic effects of cannabidiol (CBD) in generalized social anxiety disorder: A preliminary report. *Journal of Psychopharmacology, 25*(1), 121–130. https://doi.org/10.1177/0269881110379283.

Da Porto, C., Decorti, D., & Natolino, A. (2014). Separation of aroma compounds from industrial hemp inflorescences (*Cannabis sativa* L.) by supercritical CO_2 extraction and on-line fractionation. *Industrial Crops and Products, 58*, 99–103.

De Backer, B., Maebe, K., Verstraete, A. G., & Charlier, C. (2012). Evolution of the content of THC and other major cannabinoids in drug-type cannabis cuttings and seedlings during growth of plants. *Journal of Forensic Sciences, 57*(4), 918–922. https://doi.org/10.1111/j.1556-4029.2012.02068.x.

De Petrocellis, L., Vellani, V., Schiano-Moriello, A., Marini, P., Magherini, P. C., Orlando, P., & Di Marzo, V. (2008). Plant-derived cannabinoids modulate the activity of transient receptor potential channels of ankyrin type-1 and melastatin type-8. *The Journal of Pharmacology and Experimental Therapeutics, 325*(3), 1007–1015. https://doi.org/10.1124/jpet.107.134809.

Della Rocca, G., & Di Salvo, A. (2020). Hemp in veterinary medicine: From feed to drug. *Frontiers in Veterinary Science, 7*, 387. https://doi.org/10.3389/fvets.2020.00387.

DeMan, J. M. (2000). Chemical and physical properties of fatty acids. In C. Ching Kuang (Ed.), *Fatty acids in foods and their health implications* (2nd ed., pp. 17–46). Basel: Marcel Dekker, Inc.

Devinsky, O., Cilio, M. R., Cross, H., Fernandez-Ruiz, J., French, J., Hill, C., ... Friedman, D. (2014). Cannabidiol: Pharmacology and potential therapeutic role in epilepsy and other neuropsychiatric disorders. *Epilepsia, 55*(6), 791–802. https://doi.org/10.1111/epi.12631. and other neuropsychiatric disorders.

di Giacomo, V., Chiavaroli, A., Orlando, G., Cataldi, A., Rapino, M., Di Valerio, V., ... Ferrante, C. (2020). Neuroprotective and neuromodulatory effects induced by cannabidiol and cannabigerol in rat hypo-E22 cells and isolated hypothalamus. *Antioxidants, 9*(1), 71. https://doi.org/10.3390/antiox9010071.

di Giacomo, V., Chiavaroli, A., Recinella, L., Orlando, G., Cataldi, A., Rapino, M., ... Ferrante, C. (2020). Antioxidant and neuroprotective effects induced by cannabidiol and cannabigerol in rat CTX-TNA2 astrocytes and isolated cortexes. *International Journal of Molecular Sciences, 21*(10), 3575. https://doi.org/10.3390/ijms21103575.

di Giacomo, V., Recinella, L., Chiavaroli, A., Orlando, G., Cataldi, A., Rapino, M., ... Ferrante, C. (2021). Metabolomic profile and antioxidant/anti-inflammatory effects of industrial hemp water extract in fibroblasts, keratinocytes and isolated mouse skin specimens. *Antioxidants, 10*(1), 44. https://doi.org/10.3390/antiox10010044.

Di Marzo, V., Goparaju, S. K., Wang, L., Liu, J., Bátkai, S., Járai, Z., ... Kunos, G. (2001). Leptin-regulated endocannabinoids are involved in maintaining food intake. *Nature, 410*(6830), 822–825. https://doi.org/10.1038/35071088.

El-Alfy, A. T., Ivey, K., Robinson, K., Ahmed, S., Radwan, M., Slade, D., ... Ross, S. (2010). Antidepressant-like effect of delta9-tetrahydrocannabinol and other cannabinoids isolated from *Cannabis sativa* L. *Pharmacology, Biochemistry, and Behavior, 95*(4), 434–442. https://doi.org/10.1016/j.pbb.2010.03.004.

Farinon, B., Molinari, R., Costantini, L., & Merendino, N. (2020). The seed of industrial hemp (*Cannabis sativa* L.): Nutritional quality and potential functionality for human health and nutrition. *Nutrients, 12*(7), 1935. https://doi.org/10.3390/nu12071935.

Fasinu, P. S., Phillips, S., ElSohly, M. A., & Walker, L. A. (2016). Current status and prospects for cannabidiol preparations as new therapeutic agents. *Pharmacotherapy, 36*(7), 781–796. https://doi.org/10.1002/phar.1780.

Ferrante, C., Chiavaroli, A., Angelini, P., Venanzoni, R., Angeles Flores, G., Brunetti, L., ... Orlando, G. (2020). Phenolic content and antimicrobial and anti-inflammatory effects of Solidago virga-aurea, *Phyllanthus niruri, Epilobium angustifolium, Peumus boldus,* and *Ononis spinosa* extracts. *Antibiotics, 9*(11), 783. https://doi.org/10.3390/antibiotics9110783.

Ferrante, C., Recinella, L., Ronci, M., Menghini, L., Brunetti, L., Chiavaroli, A., ... Orlando, G. (2019). Multiple pharmacognostic characterization on hemp commercial cultivars: Focus on inflorescence water extract activity. *Food and Chemical Toxicology: An International Journal Published for the British Industrial Biological Research Association, 125*, 452–461. https://doi.org/10.1016/j.fct.2019.01.035.

Gómez-Juaristi, M., Martínez-López, S., Sarria, B., Bravo, L., & Mateos, R. (2018). Bioavailability of hydroxycinnamates in an instant green/roasted coffee blend in humans. Identification of novel colonic metabolites. *Food & Function, 9*(1), 331–343. https://doi.org/10.1039/c7fo01553d.

Ibeas Bih, C., Chen, T., Nunn, A. V., Bazelot, M., Dallas, M., & Whalley, B. J. (2015). Molecular targets of cannabidiol in neurological disorders. *Neurotherapeutics, 12*(4), 699–730. https://doi.org/10.1007/s13311-015-0377-3.

Ignatowska-Jankowska, B., Jankowski, M. M., & Swiergiel, A. H. (2011). Cannabidiol decreases body weight gain in rats: Involvement of CB2 receptors. *Neuroscience Letters*, *490*(1), 82–84. https://doi.org/10.1016/j.neulet.2010.12.031.

Irakli, M., Tsaliki, E., Kalivas, A., Kleisiaris, F., Sarrou, E., & Cook, C. M. (2019). Effect of genotype and growing year on the nutritional, phytochemical, and antioxidant properties of industrial hemp (*Cannabis sativa* L.) seeds. *Antioxidants*, *8*(10), 491. https://doi.org/10.3390/antiox8100491.

Izzo, L., Castaldo, L., Narváez, A., Graziani, G., Gaspari, A., Rodríguez-Carrasco, Y., & Ritieni, A. (2020). Analysis of phenolic compounds in commercial *Cannabis sativa* L. inflorescences using UHPLC-Q-Orbitrap HRMS. *Molecules*, *25*(3), 631. https://doi.org/10.3390/molecules25030631.

Jager, G., & Witkamp, R. F. (2014). The endocannabinoid system and appetite: Relevance for food reward. *Nutrition Research Reviews*, *27*(1), 172–185. https://doi.org/10.1017/S0954422414000080.

Kiralan, M., Gül, V., & Metin Kara, S. (2010). Fatty acid composition of hemp seed oils from different locations in Turkey. *Spanish Journal of Agricultural Research*, *8*(2), 385–390.

Laprairie, R. B., Bagher, A. M., Kelly, M. E., & Denovan-Wright, E. M. (2015). Cannabidiol is a negative allosteric modulator of the cannabinoid CB1 receptor. *British Journal of Pharmacology*, *172*(20), 4790–4805. https://doi.org/10.1111/bph.13250.

Lee, J., Bertoglio, L. J., Guimarães, F. S., & Stevenson, C. W. (2017). Cannabidiol regulation of emotion and emotional memory processing: Relevance for treating anxiety-related and substance abuse disorders. *British Journal of Pharmacology*, *174*(19), 3242–3256. https://doi.org/10.1111/bph.13724.

Lin, J., Teo, L. M., Leong, L. P., & Zhou, W. (2019). In vitro bioaccessibility and bioavailability of quercetin from the quercetin-fortified bread products with reduced glycemic potential. *Food Chemistry*, *286*, 629–635. https://doi.org/10.1016/j.foodchem.2019.01.199.

Linciano, P., Citti, C., Russo, F., Tolomeo, F., Laganà, A., Capriotti, A. L., … Cannazza, G. (2020). Identification of a new cannabidiol n-hexyl homolog in a medicinal cannabis variety with an antinociceptive activity in mice: Cannabidihexol. *Scientific Reports*, *10*(1), 22019. https://doi.org/10.1038/s41598-020-79042-2.

Marchalant, Y., Brothers, H. M., Norman, G. J., Karelina, K., DeVries, A. C., & Wenk, G. L. (2009). Cannabinoids attenuate the effects of aging upon neuroinflammation and neurogenesis. *Neurobiology of Disease*, *34*(2), 300–307. https://doi.org/10.1016/j.nbd.2009.01.014.

Melas, P. A., Scherma, M., Fratta, W., Cifani, C., & Fadda, P. (2021). Cannabidiol as a potential treatment for anxiety and mood disorders: Molecular targets and epigenetic insights from preclinical research. *International Journal of Molecular Sciences*, *22*(4), 1863. https://doi.org/10.3390/ijms22041863.

Menghini, L., Ferrante, C., Carradori, S., D'Antonio, M., Orlando, G., Cairone, F., … Iqbal, K. (2021). Chemical and bioinformatics analyses of the anti-leishmanial and anti-oxidant activities of hemp essential oil. *Biomolecules*, *11*(2), 272. https://doi.org/10.3390/biom11020272.

Montserrat-de la Paz, S., Marín-Aguilar, F., García-Giménez, M. D., & Fernández-Arche, M. A. (2014). Hemp (*Cannabis sativa* L.) seed oil: Analytical and phytochemical characterization of the unsaponifiable fraction. *Journal of Agricultural and Food Chemistry*, *62*(5), 1105–1110. https://doi.org/10.1021/jf404278q.

Mounessa, J. S., Siegel, J. A., Dunnick, C. A., & Dellavalle, R. P. (2017). The role of cannabinoids in dermatology. *Journal of the American Academy of Dermatology*, *77*(1), 188–190. https://doi.org/10.1016/j.jaad.2017.02.056.

Muller, C., Morales, P., & Reggio, P. H. (2019). Cannabinoid ligands targeting TRP channels. *Frontiers in Molecular Neuroscience*, *11*, 487. https://doi.org/10.3389/fnmol.2018.00487.

Nagy, D. U., Cianfaglione, K., Maggi, F., Sut, S., & Dall'Acqua, S. (2019). Chemical characterization of leaves, male and female flowers from spontaneous Cannabis (*Cannabis sativa* L.) growing in Hungary. *Chemistry & Biodiversity*, *16*(3). https://doi.org/10.1002/cbdv.201800562, e1800562.

Nissen, L., Zatta, A., Stefanini, I., Grandi, S., Sgorbati, B., Biavati, B., & Monti, A. (2010). Characterization and antimicrobial activity of essential oils of industrial hemp varieties (*Cannabis sativa* L.). *Fitoterapia*, *81*(5), 413–419. https://doi.org/10.1016/j.fitote.2009.11.010.

Orlando, G., Adorisio, S., Delfino, D., Chiavaroli, A., Brunetti, L., Recinella, L., … Ferrante, C. (2021). Comparative investigation of composition, antifungal, and anti-inflammatory effects of the essential oil from three industrial hemp varieties from Italian cultivation. *Antibiotics*, *10*(3), 334. https://doi.org/10.3390/antibiotics10030334.

Orlando, G., Recinella, L., Chiavaroli, A., Brunetti, L., Leone, S., Carradori, S., Di Simone, S., Ciferri, M. C., Zengin, G., Ak, G., Abdullah, H. H., Cordisco, E., Sortino, M., Svetaz, L., Politi, M., Angelini, P., Covino, S., Venanzoni, R., Cesa, S., … Ferrante, C. (2020). Water extract from inflorescences of industrial hemp Futura 75 variety as a source of anti-inflammatory, anti-proliferative and antimycotic agents: Results from in silico, in vitro and ex vivo studies. *Antioxidants*, *9*(5), 437. https://doi.org/10.3390/antiox9050437.

Pagano, E., Capasso, R., Piscitelli, F., Romano, B., Parisi, O. A., Finizio, S., ... Borrelli, F. (2016). An orally active cannabis extract with high content in Cannabidiol attenuates chemically-induced intestinal inflammation and hypermotility in the mouse. *Frontiers in Pharmacology, 7*, 341. https://doi.org/10.3389/fphar.2016.00341.

Rea, K. A., Casaretto, J. A., Al-Abdul-Wahid, M. S., Sukumaran, A., Geddes-McAlister, J., Rothstein, S. J., & Akhtar, T. A. (2019). Biosynthesis of cannflavins A and B from *Cannabis sativa* L. *Phytochemistry, 164*, 162–171. https://doi.org/10.1016/j.phytochem.2019.05.009.

Ross, R. A. (2007). Allosterism and cannabinoid CB(1) receptors: The shape of things to come. *Trends in Pharmacological Sciences, 28*(11), 567–572. https://doi.org/10.1016/j.tips.2007.10.006.

Salami, S. A., Martinelli, F., Giovino, A., Bachari, A., Arad, N., & Mantri, N. (2020). It is our turn to get cannabis high: Put cannabinoids in food and health baskets. *Molecules, 25*, 4036. https://doi.org/10.3390/molecules25184036.

Sales, A. J., Fogaça, M. V., Sartim, A. G., Pereira, V. S., Wegener, G., Guimarães, F. S., & Joca, S. (2019). Cannabidiol induces rapid and sustained antidepressant-like effects through increased BDNF signaling and synaptogenesis in the prefrontal cortex. *Molecular Neurobiology, 56*(2), 1070–1081. https://doi.org/10.1007/s12035-018-1143-4.

Schmidt, D., & Schachter, S. C. (2014). Drug treatment of epilepsy in adults. *BMJ, 348*. https://doi.org/10.1136/bmj.g254.

Sen, S., Duman, R., & Sanacora, G. (2008). Serum brain-derived neurotrophic factor, depression, and antidepressant medications: Meta-analyses and implications. *Biological Psychiatry, 64*(6), 527–532. https://doi.org/10.1016/j.biopsych.2008.05.005.

Shukla, M., Jaiswal, S., Sharma, A., Srivastava, P. K., Arya, A., Dwivedi, A. K., & Lal, J. (2017). A combination of complexation and self-nanoemulsifying drug delivery system for enhancing oral bioavailability and anticancer efficacy of curcumin. *Drug Development and Industrial Pharmacy, 43*(5), 847–861. https://doi.org/10.1080/03639045.2016.1239732.

Siriwardhana, N., Kalupahana, N. S., & Moustaid-Moussa, N. (2012). Health benefits of n-3 polyunsaturated fatty acids: Eicosapentaenoic acid and docosahexaenoic acid. *Advances in Food and Nutrition Research, 65*, 211–222. https://doi.org/10.1016/B978-0-12-416003-3.00013-5.

Smeriglio, A., Galati, E. M., Monforte, M. T., Lanuzza, F., D'Angelo, V., & Circosta, C. (2016). Polyphenolic compounds and antioxidant activity of cold-pressed seed oil from Finola cultivar of *Cannabis sativa* L. *Phytotherapy Research, 30*(8), 1298–1307. https://doi.org/10.1002/ptr.5623.

Starowicz, K., & Di Marzo, V. (2013). Non-psychotropic analgesic drugs from the endocannabinoid system: "magic bullet" or "multiple-target" strategies? *European Journal of Pharmacology, 716*(1–3), 41–53. https://doi.org/10.1016/j.ejphar.2013.01.075.

Sungnak, W., Huang, N., Bécavin, C., Berg, M., Queen, R., Litvinukova, M., ... Lung Biological Network, H. C. A. (2020). SARS-CoV-2 entry factors are highly expressed in nasal epithelial cells together with innate immune genes. *Nature Medicine, 26*(5), 681–687. https://doi.org/10.1038/s41591-020-0868-6.

Turner, A. J., Hiscox, J. A., & Hooper, N. M. (2004). ACE2: From vasopeptidase to SARS virus receptor. *Trends in Pharmacological Sciences, 25*(6), 291–294. https://doi.org/10.1016/j.tips.2004.04.001.

Valassi, E., Scacchi, M., & Cavagnini, F. (2008). Neuroendocrine control of food intake. *Nutrition, Metabolism, and Cardiovascular Diseases: NMCD, 18*(2), 158–168. https://doi.org/10.1016/j.numecd.2007.06.004.

van den Driessche, J. J., Plat, J., & Mensink, R. P. (2018). Effects of superfoods on risk factors of metabolic syndrome: A systematic review of human intervention trials. *Food & Function, 9*(4), 1944–1966. https://doi.org/10.1039/C7FO01792H.

Vasileva, L. V., Savova, M. S., Amirova, K. M., Balcheva-Sivenova, Z., Ferrante, C., Orlando, G., ... Georgiev, M. I. (2020). Caffeic and chlorogenic acids synergistically activate Browning program in human adipocytes: Implications of AMPK- and PPAR-mediated pathways. *International Journal of Molecular Sciences, 21*(24), 9740. https://doi.org/10.3390/ijms21249740.

Vezzani, A., Balosso, S., & Ravizza, T. (2019). Neuroinflammatory pathways as treatment targets and biomarkers in epilepsy. *Nature Reviews. Neurology, 15*(8), 459–472. https://doi.org/10.1038/s41582-019-0217-x.

Vonapartis, E., Aubin, M. P., Seguin, P., Mustafa, A. F., & Charron, J. B. (2015). Seed composition of ten industrial hemp cultivars approved for production in Canada. *Journal of Food Composition and Analysis, 39*, 8–12.

Wang, B., Kovalchuk, A., Li, D., Rodriguez-Juarez, R., Ilnytskyy, Y., Kovalchuk, I., & Kovalchuk, O. (2020). In search of preventive strategies: Novel high-CBD Cannabis sativa extracts modulate ACE2 expression in COVID-19 gateway tissues. *Aging, 12*(22), 22425–22444. https://doi.org/10.18632/aging.202225.

Wootten, D., Christopoulos, A., & Sexton, P. M. (2013). Emerging paradigms in GPCR allostery: Implications for drug discovery. *Nature Reviews. Drug Discovery, 12*(8), 630–644. https://doi.org/10.1038/nrd4052.

Zengin, G., Menghini, L., Di Sotto, A., Mancinelli, R., Sisto, F., Carradori, S., ... Grande, R. (2018). Chromatographic analyses, in vitro biological activities, and cytotoxicity of *Cannabis sativa* L. essential oil: A multidisciplinary study. *Molecules, 23*(12), 3266. https://doi.org/10.3390/molecules23123266.

Further reading

Girgih, A. T., Alashi, A., He, R., Malomo, S., & Aluko, R. E. (2014). Preventive and treatment effects of a hemp seed (*Cannabis sativa* L.) meal protein hydrolysate against high blood pressure in spontaneously hypertensive rats. *European Journal of Nutrition, 53*(5), 1237–1246. https://doi.org/10.1007/s00394-013-0625-4.

Girgih, A. T., Alashi, A. M., He, R., Malomo, S. A., Raj, P., Netticadan, T., & Aluko, R. E. (2014). A novel hemp seed meal protein hydrolysate reduces oxidative stress factors in spontaneously hypertensive rats. *Nutrients, 6*(12), 5652–5666. https://doi.org/10.3390/nu6125652.

Girgih, A. T., He, R., & Aluko, R. E. (2014). Kinetics and molecular docking studies of the inhibitions of angiotensin converting enzyme and renin activities by hemp seed (*Cannabis sativa* L.) peptides. *Journal of Agricultural and Food Chemistry, 62*(18), 4135–4144. https://doi.org/10.1021/jf5002606.

Khan, M. K., Huma, Z. E., & Dangles, O. (2014). A comprehensive review on flavanones, the major citrus polyphenols. *Journal of Food Composition and Analysis, 33*, 85–104.

Malinowska, B., Baranowska-Kuczko, M., Kicman, A., & Schlicker, E. (2021). Opportunities, challenges and pitfalls of using cannabidiol as an adjuvant drug in COVID-19. *International Journal of Molecular Sciences, 22*(4), 1986. https://doi.org/10.3390/ijms22041986.

Nusse, R., & Clevers, H. (2017). Wnt/β-catenin signaling, disease, and emerging therapeutic modalities. *Cell, 169*(6), 985–999. https://doi.org/10.1016/j.cell.2017.05.016.

Rodriguez-Martin, N. M., Toscano, R., Villanueva, A., Pedroche, J., Millan, F., Montserrat-de la Paz, S., & Millan-Linares, M. C. (2019). Neuroprotective protein hydrolysates from hemp (*Cannabis sativa* L.) seeds. *Food & Function, 10*(10), 6732–6739. https://doi.org/10.1039/c9fo01904a.

Wei, L. H., Dong, Y., Sun, Y. F., Mei, X. S., Ma, X. S., Shi, J., ... Song, S. M. (2021). Anticancer property of hemp bioactive peptides in Hep3B liver cancer cells through Akt/GSK3β/β-catenin signaling pathway. *Food Science & Nutrition, 9*(4), 1833–1841. https://doi.org/10.1002/fsn3.1976.

Industrial hemp nutraceutical processing and technology

Saša Đurović[a], *Rubén Domínguez*[b], *Mirian Pateiro*[b], *Nemanja Teslić*[c], *José M. Lorenzo*[b,d], *and Branimir Pavlić*[e]

[a]Institute of General and Physical Chemistry, Belgrade, Serbia [b]Centro Tecnológico de la Carne de Galicia, Parque Tecnológico de Galicia, Ourense, Spain [c]Institute of Food Technology, University of Novi Sad, Novi Sad, Serbia [d]Área de Tecnología de los Alimentos, Facultad de Ciencias de Ourense, Universidad de Vigo, Ourense, Spain [e]Faculty of Technology, University of Novi Sad, Novi Sad, Serbia

9.1 Introduction to hemp nutraceutical processing and technology

Isolation of the nutraceuticals form the plant material may be done with different extraction techniques. Some of them are conventional solid-liquid techniques such as maceration, Soxhlet extraction, and hydrodistillation. The Soxhlet extraction is well known and widely used technique for extraction of lipids and other compounds using different organic solvents (hexane, methylene chloride, alcohols, etc.). However, listed techniques have several disadvantages which limit their applications, so they are usually changed with newer techniques which are developed to overcome disadvantages of the classical methods. These novel extraction techniques are developed within the concept of "green" extraction. In that sense, sub- and supercritical carbon dioxide, subcritical water, microwave- and ultrasound-assisted extraction, and pressurized solvent extraction represent promising alternative methods, which provide high quality products, with complete fitting into the concept of "green" and eco-friendly process. Each technique is based on different physical and/or chemical phenomenon, which will be described in the following section. All of these techniques have been successfully applied for isolation of the biomolecules from the plant material using environmentally friendly solvents.

"Green" process is mainly oriented toward the discovery and design of new extraction process with non-hazardous solvents, less time consuming, with low energy requirements, low cost, renewable natural products, and high-quality extracts with bioactive molecules. "Green" extraction of biomolecules is based on use of alternative solvents which are non-toxic,

191

non-flammable, and without toxic residue in obtained herbal extracts. Among green solvents, bio-solvents, e.g., water, ethanol, methyl esters of fatty acids, carbon dioxide, etc., play an important role in the substitution of organic solvents. In the "green" extraction process, the use of renewable resources or plant cultivation is important instead of uncontrolled harvesting of natural resources. Safety, economic, and environmental aspects, but also consumers' requirements are forcing the industry to turn to greener alternatives (Chemat, Vian, & Cravotto, 2012). Therefore the minimization of energy consumption through the reduction of number of unit operations and optimization of existing processes is of interest for industrial production of natural extracts. Also, the concept of "green" extraction is commonly based on the valorization of by- or co-products and bio-waste to different products such as biofuels, building materials, packaging, and products for food and cosmetic industry.

There are several techniques which are falling in the concept of "green" or novel extraction techniques: ultrasound-assisted, microwave-assisted, subcritical water, and supercritical fluid extraction.

One of many plants which is drawing a significant attention of the scientific community and whose extracts are widely investigated is hemp. Such significance this plant owns to its chemical profile, containing the cannabinoids as the most significant class of the compounds. The most important is cannabidiol (CBD), having numerous benefits to human health (Andre, Hausman, & Guerriero, 2016). Besides cannabinoids, hemp contains terpenoids, polyphenols, and other bioactive compounds with proven biological activity (Zheljazkov et al., 2020).

Due to the increased attention paid to hemp bioactive extracts, this chapter has been compiled in an effort to evaluate the contribution of novel alternative extraction process in the isolation of valuable biomolecules from hemp as a natural source. Since "green" extraction has been gaining the increasing interest of researchers over 50 years, an overview of novel extraction techniques is based on the main characteristics of individual process, operating parameters, equipment, and practical application of these technologies.

Techniques for the extraction of bioactive compounds are generally divided into two major groups:

- Conventional extraction techniques—maceration, Soxhlet extraction, hydrodistillation.
- Novel extraction techniques—ultrasound-assisted, microwave-assisted, supercritical fluid extraction, and subcritical water extraction.

Fig. 9.1 summarizes the most notable advantages and disadvantages of conventional and emerging extraction techniques.

9.2 Conventional extraction techniques

Conventional techniques are based on the application of the solvents (usually organic solvents) combined with the heating and/or agitation.

9.2.1 Maceration

Maceration is a convenient, simple inexpensive, and favorable technique, especially in the case of small-scale extraction, such as that at laboratory scale. It is based on the induction of

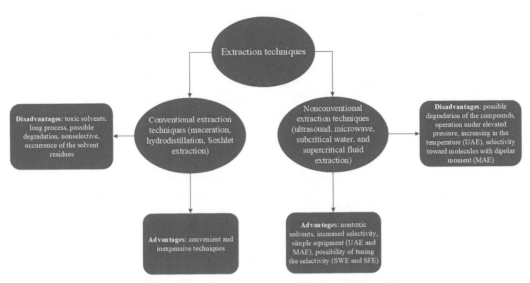

FIG. 9.1   Advantages and disadvantages of conventional and novel extraction techniques.

the mass transfer with shaking until the solid/liquid equilibrium is reached. However, this technique usually requires second step for the concentration of the extract. Maceration is also a well-known technique used for the preparation of the tonics. It is carried out in closed vessels with the occasional mixing and involves several consecutive steps: grinding of the plant material, immersion of grinded material into solvent, removing the solvent, and pressing of the sample in order to recover crude extract. Those steps are performed for (Azmir et al., 2013):

- Increasing diffusion of desired compounds from plant material to solvent.
- Removing concentrated solution from the surface of plant material which actually increases extraction yield.

Despite its simplicity, maceration possesses many disadvantages. They acquire solvents for the extraction which are usually toxic and environmentally non-friendly. Extraction processes are usually long with the possibility of degradation of the compounds of interest due to their thermolability and/or due to the oxidation. The technique is insufficiently selective and requires additional steps for separation and/or purification of the compounds of interest. Final extracts usually contain solvents' residues whose presence is not desirable due to health and product's safety issues.

9.2.2 Hydrodistillation

Hydrodistillation is another classical extraction technique, usually applied for isolation of essential oils from the spices and aromatic plants. Vapor withdraws the volatile compounds from the plant material which are further condensing together in the cooling condenser. Volatile compounds are further collected in suitable organic solvent (e.g., *n*-hexane), while condensed water returns to the round flask.

9.2.3 Soxhlet extraction

Soxhlet extraction technique has a long history in a laboratory practice. This technique is still widely applied and represents a reference technique for the evaluation of the other techniques' performances. Although this technique is superior over the other techniques regarding the exploration of the plant material, it is also limited with the thermal instability of the compounds extracted from the plant material. The apparatus for Soxhlet extraction on either laboratory or production scale is composed of the following parts: round flask filled with the solvent, Soxhlet extractor with a shell filled with the plant material, and condenser. Evaporated solvent condenses and fills the thimble-solder. When it fills the thimble-solder to the overflow level, it returns to the round flask through the syphon. This is one cycle of the extraction process, which usually continues in 15 or more cycles until the exhausting of the plant material (Wang & Weller, 2006).

To overcome listed drawbacks, novel extraction techniques emerged that are widely applied by the scientific community for versatile applications.

9.3 Novel extraction techniques

Novel extraction techniques are based on the application of ultrasounds and microwaves combined with the use green solvent such as ethanol or water, or supercritical fluid and subcritical water extraction which are based on the application of the fluids in super or subcritical state, are proved to be a promising substitute for conventional extraction techniques. Recent application of novel extraction techniques on isolation of hemp bioactives with optimized conditions and achieved yields is summarized in Table 9.1.

9.3.1 Ultrasound-assisted extraction

Ultrasound-assisted extraction technique (UAE) is based on the application of the ultrasonic waves with the frequencies in the range of 20 kHz to 100 MHz. This range may be divided into diagnostic ultrasound and power ultrasound. The diagnostic ultrasound is characterized by the low power and high frequency (1–10 MHz). On the other hand, power ultrasound is characterized by the high power and low frequency (20 kHz to 1 MHz) and is applied in the chemistry for producing the chemical and physical effects in the medium (Meullemiestre, Breil, Abert-Vian, & Chemat, 2015). Physical effects of the ultrasound are major at the lower frequencies (20–100 kHz), while the chemical effects dominate at the higher frequencies (200–500 kHz) (Tiwari, 2015). This extraction technique offers many advantages, such as low costs, simplified manipulation, low solvent, and energy consumption. Besides mentioned advantages, the disadvantages which may occur include possible changes in the chemical composition and the degradation as a consequence of the creation of the free radicals inside the gas bubbles (Paniwnyk, Beaufoy, Lorimer, & Mason, 2001).

The mechanism of the UAE is based on the compression and expansion. This phenomenon is well known as a cavitation and occurs when ultrasonic waves pass through the medium. As a consequence of the compression and expansion, bubbles are formed, expand,

TABLE 9.1 Applied novel extraction techniques on recovery of various hemp bioactives.

Plant organ	Separation technique	Process parameters[a]	Total extraction yield[b]	Major compounds	References
Inflorescence	SFE[c] with decarboxylation pretreatment	$S = CO_2$ $p = 380\,bar$ $T = 60°C$	14% dw[b]	≈ 50% Cannabidiol	Marzorati, Friscione, Picchi, and Verotta (2020)
	MAC[d] with decarboxylation pretreatment	$S =$ methanol $t = 2h$ $T =$ room temperature $S/P = 10\,mL/g$	22% dw	≈ 45.2% Cannabidiol	
Inflorescence	SFE with three separators	$S = CO_2$ $p = 340\,bar$ $T = 55°C$	18.5% dw	$Δ^9$-Tetrahydrocannabinol, $Δ^9$-tetrahydrocannabinolic acid and cannabidiol	Rovetto and Aieta (2017)
Seeds	SFE with 10% co-solvent	$S = CO_2$ $p = 350\,bar$ $T = 40°C$ $q = 15\,g/min$	36.26% dw	Fatty acids (ω-6 linoleic acid, ω-3 α-linolenic acid, palmitic acid, oleic acid, and stearic acid)	Devi and Khanam (2019)
	SOX[e]	$S = n$-hexane $t = 24h$ $T =$ boiling temperature $S/P = 10\,mL/g$	36.30% dw		
	PER[f]	$S = n$-hexane $t = 24h$ $T =$ room temperature $S/P = 10\,mL/g$	30.78% dw		
	UAE[g]	$S = n$-hexane $t = 2700s$ Amplitude = 60% Duty cycle = 0.2 $S/P = 10\,mL/g$	35.72% dw		
Seeds	SFE	$S = CO_2$ $p = 300\,bar$ $T = 40°C$ $d = 0.71\,mm$	21.5% dw	Fatty acids (ω-6 linoleic acid, ω-3 α-linolenic acid, palmitic acid, oleic acid, and stearic acid)	Da Porto, Voinovich, Decorti, and Natolino (2012)

TABLE 9.1 Applied novel extraction techniques on recovery of various hemp bioactives—cont'd

Plant organ	Separation technique	Process parameters[a]	Total extraction yield[b]	Major compounds	References
Leaves	SFE	$S = CO_2$ $p = 85\,bar$ $T = 45°C$	0.039% dw	Volatile terpenoids (caryophyllene, humulene, *trans*-α-bergamotene, *cis*-β-farnesene, and δ-limonene)	Naz, Hanif, Bhatti, and Ansari (2017)
	SD[h]	$T = 130°C$	0.032% dw		
	HD[i]	$T = 110°C$	0.035% dw		
Threshing residue	SFE	$S = CO_2$ $p = 465\,bar$ $T = 70°C$ $t = 120\,min$	8.3% dw	Cannabidiolic acid and cannabidiol	Kitrytė, Bagdonaitė, and Rimantas Venskutonis (2018)
	PLE[j]	$S = 80\%$ ethanol $T = 100°C$ $t = 45\,min$	18.9% dw	Polyphenols (mostly flavonoids)	
Inflorescence	UAE	$S = $ ethanol $t = 15\,min$ $S/P = 5.25\,mL/g$ $P = 180\,W$ $F = 40\,kHz$	0.71% dw	Luteolin and rosmarinic acid	Palmieri, Pellegrini, Ricci, Compagnone, and Lo Sterzo (2020)
	SOX	$S = $ ethanol $t = 6\,h$ $T = $ boiling temperature $S/P = 5\,mL/g$	10.00% dw	Luteolin, gallic acid, and rosmarinic acid	
	MAC	$S = $ ethanol $t = 30$ days $T = $ room temperature $S/P = 5.5\,mL/g$	0.95% dw	Rosmarinic acid and luteolin	
	RSLDE[k]	$S = $ ethanol $t = 2\,h$ $S/P = 5\,mL/g$	6.00% dw	Luteolin and rosmarinic acid	

Inflorescence	UAE	S = methanol t = 2h T = 40°C S/P = 12 mL/g P = 500 W F = 50 Hz	~14.00% dw	Cannabidiolic acid and cannabidiol	Rožanc et al. (2021)
	SOX	S = methanol t = 6h T = boiling temperature S/P = 12 mL/g	20.50% dw	Cannabidiol and cannabidiolic acid	
	DMAC[l]	S = methanol t = 4h T = room temperature S/P = 12 mL/g	16.30% dw	Cannabidiol and cannabidiolic acid	
	SFE	S = CO_2 p = 100 bar t = 3h T = 40°C q = 2 mL/min	~9.80% dw	Cannabidiolic acid and cannabidiol	
Seeds	MAE[m]	S = n-hexane t = 7.19 min S/P = 10 mL/g P = 450 W	33.91% w/w[n]	Linoleic acid and α-linolenic acid	Rezvankhah, Emam-Djomeh, Safari, Askari, and Salami (2019)
	SOX	S = n-hexane t = 8h T = boiling temperature	37.93% w/w		

Continued

TABLE 9.1 Applied novel extraction techniques on recovery of various hemp bioactives—cont'd

Plant organ	Separation technique	Process parameters[a]	Total extraction yield[b]	Major compounds	References
Inflorescence	MAE	S = ethanol t = 5 min T = 60°C S/P = 40 mL/g	~26.5 mg/g[o]	Cannabidiolic acid and cannabidiol	Brighenti, Pellati, Steinbach, Maran, and Benvenuti (2017)
	UAE	S = ethanol t = 15 min T = 40°C S/P = 40 mL/g	~22.5 mg/g		
	DMAC	S = ethanol t = 15 min T = room temperature S/P = 40 mL/g	~28 mg/g		
	SFE with 20% co-solvent	S = CO_2 p = 100 bar t = 20 min (stat + dyn)[p] T = 35°C q = 2.5 mL/min	~22.5 mg/g		
Seeds	MAE	S = methanol t = 30 min T = 109°C S/P = 12 mL/g P = 350 W	6.09 µg/g[o]	Δ^9-Tetrahydrocannabinol, cannabidiol and cannabinol	Chang, Yen, Wu, Hsu, and Wu (2017)
	UAE	S = methanol t = 30 min S/P = 20 mL/g F = 47 kHz	3.73 µg/g		
	SOX	S = methanol t = 8 h T = boiling temperature S/P = 20 mL/g	5.81 µg/g		
	HRE[q]	S = methanol t = 4 h T = boiling temperature S/P = 20 mL/g	4.15 µg/g		
	SFE	S = CO_2 p = 340 bar t = 120 min (stat + dyn) T = 120°C q = 2.5 mL/min	3.61 µg/g		

Leaves and small inflorescence	UAE	S = methanol:chloroform 9:1 t = 15 min S/P = 10 mL/g	54.6 g/g[o]	Cannabidiolic acid	Olejar et al. (2021)
	DMAC	S = ethanol t = 2 h S/P = 10 mL/g	59.8 mg/g		
	PLE	Decarboxylation: S = water t = 2×3 min T = 140°C Extraction: S = ethanol t = 2×3 min T = 120°C	65.5 mg/g	Cannabidiol	

S, solvent; t, time; T, temperature; S/P, solvent plant ratio; p, pressure; P, power; F, frequency; q, CO_2 flow rate.

[a] Optimal process parameters.
[b] Dry weight.
[c] Supercritical fluid extraction.
[d] Maceration.
[e] Soxhlet extraction.
[f] Percolation.
[g] Ultrasound-assisted extraction.
[h] Steam distillation.
[i] Hydrodistillation.
[j] Pressurized liquid extraction.
[k] Rapid solid-liquid dynamic extraction.
[l] Dynamic maceration.
[m] Microwave-assisted extraction.
[n] Gravimetric percentage.
[o] Only cannabinoids.
[p] Static and dynamic extraction step.
[q] Heat reflux extraction.

and finally collapse followed by the liberation of the energy. It is worth mentioning that this process is usually followed by other effects, such as fragmentation, erosion, sonocapillary effect, sonoporation, local sheer stress, detexturation, and combined effects (Chemat et al., 2017; Chemat, Tomao, & Virot, 2008).

Collapse of the cavitation bubbles causes inter-particle collision and shockwaves, which is considered to be the main reason for the fragmentation of the treated material (Chemat et al., 2017). Erosion is another effect which may occur during sonication, which enhances the diffusion of the solvent improving solubilization and extraction yield (Petigny, Périno-Issartier, Wajsman, & Chemat, 2013). Sonocapillary effects (ultrasonic capillary effects) are increasing in depth and velocity of liquid's penetration into the canals and pores under the certain conditions of sonication (Mason, 2015). This process is not yet fully understood; however, there are some evidences that it is connected with the cavitation (Dezhkunov & Leighton, 2004). Mechanism of the action is proposed, where sonocapillary effect influences swelling and rehydration of the plant tissue, thus affecting desorption and diffusion positively. Therefore sonocapillary effect directly influences mass transfer process (Vinatoru, 2001). The effect of sonoporation is connected with the permeability of the cell membrane of the treated material. It can be reversible and irreversible. Sonoporation may be used for both reversible and irreversible pores, where releasing of the cell content in the extraction medium takes place (Chemat et al., 2017). Local shear stress is also an effect which accompanies sonication process. It is generated within the liquid and at the vicinity of the solids. This force with turbulences induces evolution of cavitation bubble, i.e., oscillation and collapse. Induced streaming and acoustic micro-streaming are in focus when mixing or emulsification is applied (Vilkhu, Manasseh, Mawson, & Ashokkumar, 2011). Besides mentioned effects, detexturation and physical modifications of plant structures have also been reported. Gradual degradation of cell wall has been reported after applying ultrasound, which further enhanced penetration of solvent into the cell (Chemat et al., 2017). Although such effect is quite rare, destructive effect of ultrasound on living cell, microorganisms, and/or enzyme has been reported (Mason, Paniwnyk, & Lorimer, 1996).

Two types of the equipment are generally used for the UAE. Those are ultrasonic baths and ultrasonic probes. Ultrasonic baths have been used in laboratories before probes, usually for cleaning and homogenization. Most ultrasonic baths operate at a frequency and intensity range of 25–50 kHz and 1–5 W/cm^2, respectively (Meullemiestre et al., 2015). Ultrasonic baths are cheap, allow the simultaneous treatment of different samples, but are characterized by low reproducibility and low power of ultrasound delivered to the sample (Chemat et al., 2017). Ultrasonic probes are preferred option for extraction purposes since immersed into liquid medium they allow direct delivery of ultrasound in the extraction solvent with minimal energy loss. However, their application is limited to small volumes of medium and quick increase in temperature of the medium which possibly can cause the degradation of the target compound. Thus if probe is used for extraction, reactor needs to be cooled down. In the case of industrial application, one of the main parameters is quantity of product for treatment. Since ultrasonic probe is limited to small volume, one of the solutions may be continuous flow through reactor. Another solution is application of larger radiating surface like in the case of bath. REUS company (www.etsreus.com, France) has developed different equipment for ultrasound extraction from laboratory-scaled (0.5–3 L), pilot (30–50 L) to industrial scale (500–1000 L).

9.3.1.1 UAE of hemp bioactives

It was demonstrated that UAE could be utilized for recovery of polyphenols, cannabinoids (CBDs), terpenes, fatty acids, and other phytochemicals from *Cannabis sativa*. UAE with methanol (MeOH) as an extraction medium was used in a recent study to recover CBDs and polyphenols from female inflorescence (Rožanc et al., 2021). The highest extraction yield (Y) was reported for Soxhlet extraction (SOX) with MeOH (20.5 g/100 g DW) while the highest Y among alternative extraction techniques was achieved with UAE-MeOH (~ 14 g/100 g DW). The authors reported that chemical composition of isolated hemp extracts would be greatly affected by the selection of extraction technique and process parameters. Samples recovered by UAE had the highest total polyphenols content and the strongest antioxidant activity. The highest CBDs and lowest cannabinoid acids (CBDAs) quantity was obtained with SFE and UAE-MeOH suggesting that decarboxylation process of CBDAs toward CBDs was rather insufficient due to lower applied temperatures compared to SOX-MeOH.

Since medical *Cannabis* preparations are also consumed orally, olive oil as solvent coupled with UAE could be used for sequel extraction of CBDs and terpenes (Ternelli et al., 2020). Palmieri et al. (2020) compared rapid solid-liquid dynamic extraction (RSLDE), SOX, maceration (MAC), and UAE efficiency to isolate polyphenols from hemp inflorescence. Lutein (1384.09 µg/g dry extract) and rosmarinic acid (514.33 µg/g dry extract) were predominately present in extracts obtained by UAE. These extracts had the lowest total polyphenol content and antioxidant activity determined by ABTS assay compared to other techniques but also had the highest DPPH radical scavenging activity, suggesting that differences in the antioxidant activity could be related to technique and parameters selection.

9.3.2 Microwave-assisted extraction

Microwave-assisted extraction (MAE), developed as a new approach to solving the issues connected with the conventional extraction techniques, reaches huge popularity among the scientific community due to its applicability in the extraction of wide range of compounds such as essential oils, pigments, antioxidants, aromas, etc. Apart from this, it is characterized by good efficiency, high selectivity and recovery of bioactive compounds; reduced extraction time and volume of organic solvent, improved extraction yield and product quality, because material can be rapidly heated at lower operating temperatures (Veggi, Martinez, & Meireles, 2012). The selectivity of the MAE is based on the polarity of the compounds of interest—compounds with the higher dipolar moment will interact strongly with the microwaves compared to the compounds with lower dipolar moment (Filly et al., 2014). On the other hand, disadvantages of MAE are linked with poor efficiency of microwaves if the solvent or target compounds are non-polar, possible thermal degradation of bioactive molecules, and often the requirement for the separation techniques to remove the solid residues from the extract, such as filtration and/or centrifugation.

Microwaves are non-ionizing electromagnetic irradiation in the wavelength range 0.001–1 m and the frequency range of 0.3–300 GHz. This energy in the form of waves can easily penetrate into the plant material and interact with the polar compounds in the matrix (Cardoso-Ugarte, Juárez-Becerra, SosaMorales, & López-Malo, 2013). This energy is further converted into heat by ionic conduction and dipole rotation. Ionic conduction is ionic migration during the application of the magnetic field. Resistance of the solution occurs as an answer to this flow and results as a friction that heats the extract (Veggi et al., 2012). Dipole rotation represents the rearrangements of the dipoles with the applied field (Sparr Eskilsson & Björklund, 2000).

The efficiency of the heating during the MAE is in correlation with the dissipation factor of the material, representing the ability of the matter to absorb the microwave energy and dissipate the heat to the surroundings (Chen, Siochi, Ward, & McGrath, 1993). The degree of the absorption of the microwaves increases with the dielectric constant of the solvent. Therefore using the solvent with the high dielectric constant increases the extraction power and strengthens the interaction between the matrix and the solvent.

The mechanism of the MAE is related to the heat and mass transfer (in the same direction) from matrix to solvent. In the case of the conventional extraction techniques, heat transfer goes out of the liquid phase to solid phase. On the other hand, mass transfer goes vice versa. The solvent penetrates into the solid matrix, dissolute active compound(s) which by internal diffusion migrate on the surface of solid matrix, and then migrate into the bulk solution (i.e., liquid phase of the extract) by external diffusion. During the MAE, microwaves pass through the medium, and the medium absorbs its energy and converts it into thermal energy (Joana Gil-Chávez et al., 2013). This causes the rupture of plant cell and easy extraction of active compound(s).

Apparatus for the MAE may be generally divided into two groups: open-vessel and closed-vessel systems. Open-vessel systems are operating under the atmospheric pressure and widely used in the laboratories, where the maximal temperature is limited by the boiling point of the solvent used at the atmospheric pressure. Closed-vessel systems offer the control of both temperature and pressure. There are three generations of these systems: MW equipment of first generation is made of Teflon with the low-pressure limits, MW equipment of the second generation is made of Teflon and polymers, such as polyetherimide resistant to the inner pressure of 20 atm, while the equipment of third generation is redesigned to resist the inner pressure up to 150 atm.

9.3.2.1 MAE of hemp bioactives

Similarly to UAE, MAE could also be utilized for isolation of a broad spectrum of bioactive compounds from industrial hemp. A recent study investigated efficiency of MAE in recovery of linoleic acid and α-linolenic acid oil from hempseed (Rezvankhah et al., 2019). Oil yield obtained under MAE optimal conditions (33.91% w/w) was lower compared to SOX (37.93% w/w). However, oils obtained with MAE exhibited higher tocopherol content and oxidative stability compared to SOX due to shorter extraction procedure and lower temperatures used in MAE. Brighenti et al. (2017) compared MAE, UAE, dynamic maceration (DMAC), and SFE in terms of non-psychoactive cannabinoids yield from fiber-type hemp. In the preliminary step, the authors reported that due its polar nature, higher CBD yields were achieved with polar solvents (e.g., MeOH and EtOH) comparing to less polar solvents (e.g., hexane). The highest CBDA yields was achieved with DMAC (~ 23.0 mg/g) while MAE (~ 21.0 mg/g) performed better compared to other alternative extraction techniques (~ 18 mg/g for both). On the other hand, concentration of its neutral counterpart CBD was highest in MAE (~ 4 mg/g) suggesting that part of CBDA was decarboxylated to CBD due to higher extraction temperature used in MAE compared to other applied techniques. The authors suggested that DMAC was the best technique for analytical purposes; however, MAE was significantly faster compared to DMAC (5 vs 15 min); thus it might be better for non-psychoactive cannabinoids recovery on larger scale. Another study compared performance of MAE, UAE, heat reflux extraction (HRE), SOX, and SFE in isolation of cannabinoids from hempseeds (Chang et al., 2017). The highest total CBD yield was obtained with MAE (6.09 μg/g) compared to all other extraction procedures (3.61–5.81 μg/g). This was supported with morphological observations on microstructure level, whereas hempseed cell wall breaking was significant after MAE which allowed easier diffusion of target compounds toward extraction medium. Furthermore, MAE was the fastest technique and required the smallest solvent to plant ratio, indicating that MAE is a rapid and cost-effective technology for CBD isolation.

9.3.3 Subcritical water extraction

Subcritical water extraction (SWE) is a special case of pressurized liquid extraction (PLE) which is also recognized as accelerated solvent extraction (ASE), pressurized fluid extraction (PFE), pressurized hot solvent extraction (PHSE), etc. SWE is based on the application of water in its subcritical state, i.e., the state of elevated temperature (above boiling point) and increased pressure enough to maintain water in the liquid state (Ko, Cheigh, & Chung, 2014). Such conditions influence dielectric constant of the water, decreasing it from about 80 to below 30 making it moderately polar solvent. In those conditions, subcritical water is able to moderately dissolve polar and nonpolar compounds. Temperature and pressure are the main parameters which affect the SWE. Temperature is more important and also more influential than pressure, because it directly influences the dielectric constant of the water. Thus, by changing the temperature of the system, the properties of subcritical water will change and accordingly the selectivity of the SWE process. For example, water reaches equivalent value for dielectric constant for methanol at approximately 214°C. At approximately 295°C, subcritical water has same dielectric constant as acetone. This tuning of dielectric constant of subcritical water by changing temperature of the system allows selectivity toward certain class of compounds during the extraction. In other words, less polar compounds will be extracted at higher temperatures, while moderately polar compounds will be extracted more efficiently at lower temperatures. At 200°C subcritical water may act as acid-base catalyst due to higher concentration of hydronium and hydroxyl ions than that at the ambient conditions. The reduction of polarity allows the extraction of moderately polar and non-polar compounds from different matrixes. Temperature increase causes the decrease of the strength of the hydrogen bonds among the water molecules, decrease of the viscosity enhancing the penetration of the matrix particles, and decrease of the surface tension of the subcritical water enhancing extraction (Özel & Göğüş, 2014; (Teo, Tan, Yong, Hew, & Ong, 2010). High temperature allows disruption of the interactions between compound of interest and matrix. These interactions are actually different forces such as hydrogen bonding, dipole-dipole interactions, and van der Waals forces (Richter et al., 1996).

Although the higher temperatures are desired, it should be noted that thermolabile compounds may decompose under such conditions (Özel & Göğüş, 2014). On the other hand, during the SWE new compounds might be formed due to the occurrence of different chemical reactions, such as caramelization and Maillard reaction (Ahmad & Langrish, 2012; He et al., 2012; Plaza, Amigo-Benavent, del Castillo, Ibáñez, & Herrero, 2010), and these compounds are often more potent antioxidants which may increase the activity of the extracts (Veličković et al., 2017). Besides these processes, other reactions such as hydrolysis, oxidation, methylation, and isomerization may take place under the SWE conditions (Carr, Mammucari, & Foster, 2011).

Although pressure has less significant impact on the SWE process (Ozel, Gogus, & Lewis, 2003), its role is to maintain water in the liquid state, but also to increase the permeability of the water into the matrix (Crescenzi, Di Corcia, Nazzari, & Samperi, 2000). However, newer study showed that pressure also significantly influences the solubility of the compounds in the subcritical water, where the yield of each compound depended more or less on the applied pressure, similarly to the effect of the temperature (Cvetanović et al., 2018).

9.3.3.1 *PLE of hemp bioactives*

A recent study compared PLE, UAE, and DMAC in terms of CBD recovery from hemp material (leaves and small inflorescence) and decarboxylation conversion rate (Olejar et al., 2021). Since the authors used substantially higher temperatures during PLE, decarboxylation conversion rate was almost 100% in PLE samples compared to DMAC and UAE samples. Due to low CBD solubility, water was used as an effective medium for decarboxylation while ethanol was used as extraction medium. The concentration of CBD was 61.3, 1.3, and 1.2 mg/g in PLE, UAE, and DMAC samples, respectively. On the other hand, content of CBDA was 0.2, 49.7, and 54.8 mg/g in PLE, UAE, and DMAC samples, respectively, indicating that not only conversion rate was superior when using PLE but also higher total CBD yield.

9.3.4 Supercritical fluid extraction

Supercritical fluid extraction (SFE) uses fluid in their supercritical state for the extraction of the compounds from their natural sources. Supercritical fluids are fluids maintained at the pressure and temperature above their critical values. Supercritical fluids are not in defined state, since their physical and chemical characteristics are between gases and liquids (Sahena et al., 2009). Fluid in its supercritical state has a density near the value for liquids, while the diffusivity and viscosity are close to the values for gases. Such combination of the properties allows high dissolution capacity, easy penetration, and dissolution (Pasquali, Bettini, & Giordano, 2008). Fluids in their supercritical state have significantly higher dissolution power, higher selectivity, and diffusion coefficients. Supercritical fluids have a solvent power comparable to those of the liquids, while mass transfer characteristics are near to the ones for gases. Variation in the main parameters, i.e., pressure and temperature, allows significant variation in the solvent power of these fluids. It is generally adopted that solvation power of the supercritical fluids increases with the density at the constant temperature (isothermal process) and that solvation power increases with the temperature at constant density. Carbon dioxide is most frequently used in SFE processes due to its moderate critical parameters (31.1°C and 73.8 bar), easy evaporation at the room temperature (this allows simple separation from the extract by depressurization), nontoxic, non-flammable, chemically stable, generally recognized as safe (GRAS), and inexpensive (Reverchon & De Marco, 2006). This extraction technique is quite effective for isolation of the nonpolar compounds with the medium molecular weight. If the polar compounds are of interest, their extraction is conducted with the addition of small amount of the polar solvents (co-solvent or modifier) to the supercritical fluid. This addition triggers certain changes in the properties of the supercritical fluids, where the selectivity consequently increases toward the polar compounds. The most common modifiers are water, ethanol, methanol, acetone, and dimethyl sulfoxide, whose addition triggers different interactions, e.g., dipole-dipole, dipole-induced dipole, hydrogen bonding, etc. (Lang & Wai, 2001). However, additional step for the separation of the modifier will be required because they are introduced in the liquid state (Reverchon, 1997). Additional disadvantage of the co-solvent is decreasing the selectivity of the SFE process.

The main parameters of the SFE process are pressure, temperature, particle size, flow rate, extraction time, and moisture content of the plant material. In order to achieve the maximal yield, it is essential to use the optimal extraction conditions. The pressure increase causes the increase of the solvating power followed by selectivity decrease in the (Reverchon & De Marco, 2006). Pressure elevation increases density of the supercritical carbon dioxide at the constant temperature and consequently increases the extraction yield. On the other hand, the influence of the temperature is more complex than that of the pressure. Temperature increase causes the decrease of the density of the supercritical carbon dioxide, thus reducing its solvation power. Moreover, the temperature increase also increases vapor pressure of the compounds in the system. If the pressure is sufficiently high, negative influence of the temperature on the density will be negligible, but temperature will still positively influence the solubility of the compounds into the supercritical carbon dioxide through vapor pressure. Therefore recommended temperature for the SFE process in the case of thermolabile compounds is between 35°C and 60°C (Reverchon & De Marco, 2006).

The flow rate of the supercritical fluid is also an important parameter which will significantly influence final yield of the process. This factor has direct influence on the mass transfer and on the kinetics of the SFE process. It affects axial dispersion, convective mass transfer, and quantitation of the extracted compounds in the process (de Melo, Silvestre, & Silva, 2014). The increase of the supercritical carbon dioxide flow increases mass transfer because of the shifting of the concentration gradient with the fresh solvent. However, in some cases, the excessive flow increase may reduce the extraction yield because of the shortening of the contact period between the supercritical fluid and plant material. Matrix particle size is another crucial parameter in the SFE process. Large particle sizes may prolong the SFE process simply because this process is governed by internal mass transfer. On the other hand, if the particles are too small, difficulties in the maintaining of the proper flow rate may occur (Lang & Wai, 2001). Particle size may also affect the porosity. If the particle size is not optimal, part of the solvent will flow through the channels formed inside the bed, decreasing the contact between the materials and supercritical fluid. Optimal particle sizes range from 0.25 to 2.0 mm (Reverchon & De Marco, 2006).

Summarizing all, solubility of a target compound in supercritical carbon dioxide is a major factor determining the efficiency and extraction yield. Basically, solubility depends (a) on the volatility of the substance, which is correlated to the temperature and (b) of the solvation power of supercritical fluid, which is dependent on the fluid density. According to the literature review, most of the extractions of the natural compounds are carried out at pressures of 100–400 bar and temperatures 40–60°C (de Melo et al., 2014). In this pressure and temperature interval the carbon dioxide density ranges from 200 to 900 kg/m^3, which is suitable for non-polar substance extraction.

Extraction time is determined by the thermodynamics of diffusion process during the supercritical fluid extraction. Extraction is mainly performed during fast (first) extraction period that is controlled by the solubility of the substance in the supercritical fluid and is usually interrupted when the second period of slow extraction starts. The slow extraction is controlled by internal (molecular) diffusion, where the diffusion of solute out of the matrix is usually the limiting step (Brunner, 2005). Supercritical fluid extraction can take from 20 min to several hours. Extraction time is an important factor that affects the cost of the extraction process in industrial-scale level.

Supercritical fluids may be used for the extraction of the solid or liquid raw materials. Extraction process is based on the following principles: dried and grinded plant material is placed into the cylindrical vessel (reactor) for obtaining fixed bed of particles. Supercritical fluid is introduced into the reactor through the compressor which compresses fluid to desired pressure. Reactor is also maintained at the constant and desired temperature with the heat exchanger. Mixture of the supercritical fluid and dissolved compounds from the plant material is further transferred into the separator, where the extract will be separated by simple depressurization. Separation process may occur in several steps which allow fractionation of the extract in different quantities. System for processing of the solids is batch-type system and may operate discontinuously or semi-continuously (Herrero, Cifuentes, & Ibanez, 2006). The SFE process may be applied in different scale systems. First is analytical scale which uses less than a gram to a few grams of the sample for extraction. The second one is the preparative scale which requires several hundred grams of the material for the extraction. The third is pilot scale in which kilograms of the sample may be processed, and the last one is the industrial scale for tones of the materials for extraction.

Supercritical fluid extraction process can be carried out in one-step operation or in multi-step operation, i.e., fractionation, by varying pressure and/or temperature in each step of extraction. Additionally, when the different classes of organic compounds are to be extracted from the same matrix, fractionation can be applied. In this process the fact that supercritical carbon dioxide's solvent power can be changed with pressure and temperature is used to obtain extracts with target compound(s). Coupling of several separators in series in the SFE plant allows fractionation of the obtained extracts. Fractionation is occurring by maintaining the separators at different pressures and temperatures.

High capital costs of supercritical fluid extraction equipment are considered to be one of the commonly mentioned drawbacks. It is also worth mentioning that operating costs of the supercritical fluid extraction plants are lower than those of conventional techniques. For such reason, large-scale SFE plants operate worldwide as economically competitive, usually for the purposes of food and pharmaceutical industries (Sovová & Stateva, 2011).

9.3.4.1 SFE of hemp bioactives

SFE could be used for recovery of both volatile aromatic compounds and non-volatile lipophilic bioactives from hemp. SFE has been established method for production of CBD-rich oil from *C. sativa* inflorescence. Marzorati et al. (2020) applied heat treatment (100°C, 6h) of Finola variety inflorescence for decarboxylation of CBDA to CBD which was followed by SFE at 380 bar and 60°C to obtain 14% of total extraction yield with ≈ 50% CBD content. Further purification steps (winterization and flash chromatography) could be performed to enhance the CBD purity of the extract by up to almost 80%. Furthermore, this SFE (340 bar and 55°C) can provide a viable extraction of cannabinoids from *C. sativa* strains rich in cannabinoids with high yield and efficiency (Rovetto & Aieta, 2017). However, further cannabinoid fractionation (three separators in a cascade configuration of decreasing temperature and pressure) is necessary in order to increase their concentration in crude extract.

Devi and Khanam (2019) performed in-depth comparative study of different conventional and novel extraction techniques isolation of hempseed oil with special emphasis on physical, chemical, and industrial-scale economic aspects, whereby SFE provided oil with the highest purity and was evaluated as the most economically feasible process compared to other methods. Optimization study was performed and maximum oil yield from *C. sativa* seeds of Felina cultivar (21.5%) was obtained at 40°C, 300 bar, and particle size of 0.71 mm (Da Porto et al., 2012). Grijó, Piva, Osorio, and Cardozo-Filho (2019) highlighted the advantages of pressurized *n*-propane over supercritical CO_2 in terms of lower extraction pressures, reduced solvent consumption, and lower operating cost for isolation of hempseed oil. Moreover, liquefied dimethyl ether could also be used as a green solvent for isolation of hempseed oil which allows processing at reduced pressures and temperatures preserving the integrity of the oil (Subratti, Lalgee, & Jalsa, 2019). Furthermore, pretreatment with ultrasounds could be performed on hempseeds (10 min, without additional solvent) in order to improve oil yield by 3.3% compared to SFE without pretreatment (Da Porto, Natolino, & Decorti, 2015).

Naz et al. (2017) compared SFE with traditional steam distillation and hydrodistillation for isolation of volatile aromatics from *C. indica* and *C. sativa* leaves. SFE was able to provide better oil recovery compared to applied distillation techniques, while SFE at low pressure (85 bar) and temperature of 45°C provided the highest oil yield (0.031% and 0.039%, respectively). The main compounds of *C. sativa* essential oil were caryophyllene (40.6%–50.0%) and humulene (9.51%–16.0%), while the main components of *C. indica* oil were caryophyllene (21.1%–25.1%), linalool (20.8%–22.1%), and eucalyptol (9.67%–12.10%). Besides leaves and inflorescences, hemp root is also evaluated as raw material for phytochemical exploitation in order to provide added value to hemp cultivation. SFE has provided advantages compared to isolation of bioactive compounds from hemp root, while friedelin and epifriedelinol were major triterpenoids identified in these extracts (Kornpointner et al., 2021).

Industrial hemp threshing residues (stalks and leaves) could be processed into cannabinoid and antioxidant fractions by SFE, PLE, and enzyme-assisted extractions. Results suggested that optimized SFE (465 bar, 70°C, 120 min) provides 8.3 g/100 g DW of lipophilic fraction recovering more than 93% of initial CBD and CBDA amount from raw material, while PLE with 80% ethanol at 100°C for 45 min yielded 18.9 g/100 g DW of flavonoid-containing polar fractions (Kitrytė et al., 2018). Furthermore, optimized SFE conditions for recovery of CBD from variety Santhica dust residue were 350 bar and 50°C (Attard et al., 2018). Polyphenols yield and bioactivity (antioxidant and enzyme inhibitory) could be improved by pretreatment of hempseed hull by extrusion (Leonard, Zhang, Ying, Xiong, & Fang, 2021).

9.3.5 Deep eutectic solvents

Deep eutectic solvents (DES) are new generation solvents composed of a broad range of hydrogen bond donors (HBD) and hydrogen bond acceptor (HBA). DES formation is induced by molecular charge relocation between HBA and HBD causing significant melting point depression of DES constituents. In the past few years, natural DES (NADES) are particularly interesting to the scientific community due to their sustainable "green" properties as most of the constituents are suitable for human consumption, nontoxic and present in nature, and due to appropriate physicochemical properties since generated solvents are chemically stable liquids with low volatility in the wide temperature range (Mišan et al., 2019).

DES could be used as an efficient extraction medium for isolation of industrial hemp phytochemicals when coupled with UAE. These novel solvents are particularly interesting since the vast number of different combinations (10^6) may be formulated allowing the selection of solvents that have higher affinity toward targeted compounds. In a recent study, menthol-based hydrophobic DES were used with UAE to recover cannabinoids from hemp inflorescence (Křížek et al., 2018). DES as combination of menthol and acetic acid was utilized to obtain extract with the highest quantity of cannabidiol (CBD $\simeq 5$ mg/g), tetrahydrocannabinol (THC $\simeq 75$ mg/g), and their acidic analogs compared to extracts recovered by other DES, ethanol (EtOH), MeOH, and mixture of MeOH and chloroform.

9.4 Encapsulation

Bioactive compounds isolated from botanical and natural sources can exhibit strong off-flavors/odors, poor water solubility, low chemical stability, or limited biological activity, which limit their direct utilization by the industry. Therefore the encapsulation of the bioactive compounds to improve their functional characteristics and preserve them from environmental stresses has been proposed (Gómez et al., 2018; McClements, 2018a, 2018b).

In recent years, the encapsulation techniques are getting more attention from food, health/pharmaceutical, and cosmeceutical industries due to the large number of benefits it offers (Pateiro et al., 2021). Several encapsulation techniques have been developed to encapsulate bioactive agents, which vary in the ingredients and processing methods used to assemble them (McClements, 2018a, 2018b). The encapsulation process is based on enclosing bioactive molecules in liquid, solid, or gaseous states within a matrix or inert material, usually a polymer. Aspherical wall preserves the coated compounds, which protect the sensitive molecules against deteriorative processes and increasing their stability. Moreover, the large surface area obtained through encapsulation results in several advantages, including good reactivity, aqueous solubility, or efficient absorption (Pateiro et al., 2021). The main advantages of encapsulation of bioactive compounds are summarized in Fig. 9.2. It is important to note that the encapsulation techniques could be sub-divided into three main groups, depending on the capsule particle size. The capsules obtained by nanoencapsulation techniques presented a particle size of <0.2 μm, while particle sizes of microcapsules ranged between 0.2 and 5000 μm, and particle sizes of macrocapsules are >5000 μm (Pateiro et al., 2021).

In the last years, several techniques, including nanoemulsions, nanosuspensions, nano-liposomes, biopolymer nanoparticles, micelles, nanoprecipitation, emulsification-solvent evaporation, coacervation, ionotropic gelation, supercritical fluid extraction of emulsions, electrospinning, and electrospraying, were proposed as suitable techniques for nanoencapsulation (Pateiro et al., 2021).

Additionally, techniques which have been used for microencapsulation are spray-drying, freeze-drying, liposomes, coacervation, molecular inclusion, polymeric micelles, extrusion processes, supercritical fluids, or solvent evaporation among others (Gómez et al., 2018). As can be seen, some of the technologies are shared by both nano- and microencapsulation, simply differing in particle size, but using the same technique.

FIG. 9.2   Main advantages of encapsulation of bioactive compounds.

Another important aspect is the wall materials and carriers used in these techniques, which could be divided into organic polymers (polymeric and lipid-based particles, e.g., nanoemulsions, liposomes), inorganic compounds (metallic nanostructures), or a combination of both (Pateiro et al., 2021; Yu et al., 2018). In this regard, the use of natural/organic polymers such as carbohydrates, proteins, or lipids is preferred due to their applicability in different industries, although a combination of these is generally preferable for encapsulation purposes. Moreover, proteins are perceived as suitable wall materials and emulsifiers due to their amphiphilic nature (Kurek & Pratap-Singh, 2020).

9.4.1 Encapsulation of hemp bioactives

The utilization of industrial hemp is not only limited to obtaining nutrients and bioactive compounds to be encapsulated, but also as a potential source of macromolecules to be used as carriers that allow the encapsulation of bioactives.

9.4.1.1 Encapsulation of hemp oils

In the particular case of industrial hemp (*C. sativa* L.), the highly unsaturated fatty acid composition of the oils (extracted from hempseeds) makes this oil extremely susceptible to oxidative degradation. High susceptibility of hemp oil to oxidation is associated with the high content of fatty acids with double bonds, such as linoleic and linolenic fatty acids (Domínguez et al., 2019; Esparza, Ngo, & Boluk, 2020). Thus the use of encapsulation techniques to improve the oxidative stability of high unsaturated oils has been proposed (Domínguez, Pateiro, Agregán, & Lorenzo, 2017; Vargas-Ramella et al., 2020), being of utmost importance for cosmetics, food, and pharmaceuticals industries. Moreover, the encapsulation techniques to stabilize other hydro-soluble bioactive compounds extracted from industrial hemp (e.g., obtained using hydro-alcoholic solvents) are also frequently used. With this in mind, multiple studies that have used and characterized different encapsulation methods to improve the properties of different bioactive hemp compounds are discussed later.

Esparza et al. (2020) proposed the stabilization of hempseed oil-in-water emulsions in cellulose nanocrystal particles and their encapsulation using the spray-drying technique. In this study, different parameters were evaluated, and the ratio of hempseed oil to cellulose nanocrystal was varied to find the optimal conditions. The use of 0.7 g oil/g cellulose nanocrystal presented the best encapsulations efficiency, and this efficiency dramatically decreased (from about 90%–95% to < 40%) when this ratio was 1 or higher. Additionally, and based on the results obtained by these authors, it was predicted that coverage of 1–2 μm oil droplet with 10–14 layers of cellulose nanocrystal will result in effectively microencapsulated oil (Esparza et al., 2020).

Similarly, using the spray-drying technique, Kurek and Pratap-Singh (2020) studied different combinations of maltodextrin and hemp, pea and rice protein isolates as carrier materials to encapsulate hempseed oil. The oil content in the microcapsules varied from 10% to 20%. The microcapsules size was significantly lower (11.7–34.5 μm) in the samples containing 10% of hempseed oil than those containing 20% of oil (33.9–56.23 μm). In addition, the use of rice protein in the encapsulation process also produced smaller particles (11.7–19.9 μm) in comparison with hemp protein (21.1–33.9 μm) and pea protein (34.5–56.23 μm). Regarding the morphology of microcapsules, the use of hemp and rice protein in combination with 10% of oil resulted in a smoother surface, which may indicate better retention of the core inside the capsules (Kurek & Pratap-Singh, 2020). The highest encapsulation efficiency was observed in samples with rice protein, while the lowest was with hemp protein. Additionally, the use of 10% of oil presented the highest encapsulation efficiency in all cases. Finally, these authors reported that the encapsulation of hempseed oil in a plant-based protein-maltodextrin complex successfully inhibited its lipid oxidation (Kurek & Pratap-Singh, 2020). These authors also observed that the microencapsulation of hempseed oil with pea protein-maltodextrin combination at 10% oil loading depicted the lowest oil release rates and best oxidative stability (Kurek & Pratap-Singh, 2020).

Lan, Ohm, Chen, and Rao (2021) proposed the encapsulation of hempseed oil using pea protein isolate and sugar beet pectin as wall materials in the spray-drying process. In this study, the impact of pH and wall/core ratio during the coacervation process on physicochemical properties of microcapsules was evaluated. Both parameters had an influence on the microcapsules' parameters. In this sense, the wall/core ratio affected the viscoelastic properties, while pH highly influenced the encapsulation efficiency, spray-drying efficiency, oil distribution in microparticles, and also oxidative stability, whereby pH = 3.5 appeared to be the optimal pH for coacervation process. Although higher encapsulation efficiency and co-acervates with softer structure were obtained at pH 2.5, in comparison with those prepared at pH 3.5, the microcapsules formed at 2.5 presented lower oxidative stability (due to the weaker electrostatic interaction strength between pea isolate protein and sugar beet pectin) (Lan et al., 2021). This is due to the formation of partially broken coacervates during the spray-drying process, resulting in hollow and incomplete particle shape, which promote the hempseed oil degradation.

Another promising technique to encapsulate lipophilic hemp compounds is the use of liposomes as an emerging encapsulating vector as demonstrated by Shi et al. (2021) who proposed this method to produce encapsulated hempseed oil. By using egg yolk lecithin and cholesterol to encapsulate hemp oil into liposomes, a uniform particle size distribution (122 nm) was obtained, with the encapsulation efficiency of 81.3%. Moreover, they observed that the encapsulation of hempseed oil had a clear protective effect on both, total phenolic compounds and also scavenging DPPH* activity. On the basis of the reported results, the authors concluded that the use of liposomes increased oxidative stability and protected the bioactive compounds of hempseed oil from degradative processes, which retained the nutrients and active ingredients for industrial applications (Shi et al., 2021).

The use of nanoemulsions based on hemp oil and surfactants for oral delivery of baicalein was proposed (Yin, Xiang, Wang, Yin, & Hou, 2017). In this study, the researchers used high-pressure homogenization to reduce surfactant [poly(ethylene glycol) monooleate] amount added to the nanoemulsions. These nanoemulsions presented ~ 90 nm particle sizes, and the entrapment efficiency using this procedure was extremely high (99.3%).

Similarly, other authors proposed the encapsulation of essential oils obtained from hemp inflorescences into nanoemulsions (Rossi et al., 2020). After multiple experiments, including different essential oil proportions (3%, 4%, and 6%) in water using 4% of Tween 80 as surfactant and with/without ethyl oleate, the best option for the formation of stable hemp essential oils nanoemulsions was found—the amount of oil phase was kept constant at 6% (3% essential oil and 3% ethyl oleate) with using 4% of Tween 80 and water. The emulsions were then subjected to the homogenization process with four cycles at high pressure (130 MPa). The obtained system was characterized by mean diameter of the nanoparticles of about 200 nm and stability more than 6 months (Rossi et al., 2020). Thus the development of stable nanoemulsions from essential oils obtained from hemp inflorescences presented better water dispersity and stability over time and reduced high volatility of the essential oils, which are important aspects for pharmaceuticals and cosmeceuticals commercialization (Rossi et al., 2020).

9.4.1.2 Encapsulation of lipophilic hemp-based bioactive compounds

In addition to the spray-drying technique, other procedures have been proposed for the encapsulation of bioactive compounds obtained from hemp plant materials. Due to the lipophilic character of cannabidiol, it is of very poor aqueous solubility, which limits its applications in the food, cosmetic, and pharmaceutical industry. In order to overcome this limitation, different encapsulation solutions have been proposed. In very recent research, the authors proposed the nanoencapsulation of lipophilic terpenes-cannabidiols mixture and crystalline cannabidiol by using insoluble yolk low-density lipoprotein recovered by complexing with carboxymethylcellulose as an encapsulation agent (Fei, Wan, & Wang, 2021). In this study, the effects of defatting process and type (Tween 80 or sodium lauryl sulfate) and concentration of emulsifier on dispersibility of 2% dispersions of insoluble yolk low-density lipoprotein/carboxymethylcellulose and nanoencapsulation properties were evaluated. The authors identified the use of 2% insoluble yolk low-density lipoprotein (with 42% of fat) dispersed with 1000 ppm of Tween 80 as an excellent encapsulation system to produce heat and freeze stable nanoemulsion of cannabidiol (Fei et al., 2021).

Li et al. (2021) proposed the complexation of cannabidiol using cyclodextrins (β-cyclodextrin and 2,6-di-O-methyl-β-cyclodextrin). The encapsulation was carried out using the freeze-drying method with the cannabidiol/cyclodextrin ratio of 1. With this process, the water solubility of cannabidiol dramatically increased, as well as the antioxidant activity (measured as ABTS free radical scavenging ability). Additionally, the loading efficiency was between 17.7% and 20.4%, while the complexation efficiency was 90.8%–92.4%, which implied that this procedure was suitable for the encapsulation of cannabidiol (Li et al., 2021) and its stability and solubility improvement.

The microencapsulation of the cannabidiol in liposomes was demonstrated by Valh et al. (2021), who obtained liposomes with 1910 nm particle size and reported the encapsulation efficiency of about 90%. Thus the use of liposomes (with cannabidiol) demonstrated a potential coating formulation for the development of new hygienic, sanitary, and medical products (Valh et al., 2021).

9.4.1.3 Hemp-based carrier material for encapsulation

On the other hand, in addition to bioactive compounds, hemp has other molecules of great importance for encapsulation purposes. As aforementioned, several hemp-based biopolymers, including proteins, can serve as a viable alternative to synthetic polymeric materials in the encapsulation process. Moreover, proteins obtained from hemp and used as carrier materials can provide a physical matrix and barrier to encapsulated substances and contribute to increasing the nutritive value of the systems they constitute (Belščak-Cvitanović et al., 2018). However, very limited number of studies have used plant-derived proteins (e.g., protein isolates) recovered from hemp materials, especially with regard to the formulation of encapsulated delivery systems. In very recent research, Plati, Ritzoulis, Pavlidou, and Paraskevopoulou (2021) proposed the extraction of protein from organic hemp flour for the preparation of protein-gum arabic complex coacervates for the encapsulation of bioactive compounds. In this case, the structural transition as a function of pH and protein-to-gum arabic ratio was evaluated, implying that the maximum coacervation occurred at pH 3.5 with 2:1 protein/gum arabic ratio.

Another study proposed the use of industrial hemp material to obtain both, polyphenol-rich extracts and protein isolates for developing particulate systems of those compounds in combination with alginates (Belščak-Cvitanović et al., 2018). The authors observed that the protein isolate reinforced particles without affecting the mechanical gel properties and structure, which may be beneficial to future encapsulation applications. In fact, the use of protein isolates obtained from industrial hemp as encapsulation material (alginate-protein isolate-CaCl$_2$) increased the encapsulation efficiency of polyphenols from 73.6% (control samples) to 87.3%. Not only the encapsulation efficiency was improved, but also the total content of hydroxycinnamic acid derivatives and flavonoid derivatives in the hemp extract encapsulated with alginate-protein isolate-CaCl$_2$ was also significantly higher, which demonstrated the high potential of hemp from both, the extraction of potentially bioactive compounds and also encapsulation materials (proteins). Moreover, these authors stated that interactions between proteins and polyphenols could provide additional protection for the potential bioactive ligand bound to the protein. Therefore the encapsulation study revealed potential protein-polysaccharide interactions which did not markedly affect the physico-chemical properties but improved the retention of polyphenols in the formulated particulate delivery matrix (Belščak-Cvitanović et al., 2018). Hemp meal, a by-product remaining after the cold mechanical pressing of hemp cannabinoid-free hempseeds variety Helena, could be successfully utilized for extraction and encapsulation of polyphenol-rich extracts using aqueous solutions of β-cyclodextrin (Mourtzinos, Menexis, Iakovidis, Makris, & Goula, 2018).

9.5 Challenges and opportunities

Although the interest of the scientific community for investigation of the hemp is constantly growing, there is still room for further improvements. Thus further investigation and challenges may be divided into several directions. First one is the improvement of the extraction techniques for obtaining the product enriched with the bioactive compounds in optimal ratio which ensures maximal biological activity of the product. This step may include further investigation of the extraction processes, application of different solvents, optimization of the extraction, and improvement of the downstream processes (isolation and purification of/from the crude extract). Second direction is further investigation of the encapsulation processes due to the importance of finding a way to protect bioactive compounds from any degradation process which may occur, especially polyunsaturated fatty acids with proven human health benefits. Besides the protection from the degradation, encapsulation ensures high quality of the final product regarding the odor, solubility of bioactive compounds, and improved activity (better bioavailability). It is very important that hemp itself provides necessary macromolecules for encapsulation carrier material. This may simplify the encapsulation process and reduce its costs. Therefore technologies based on encapsulation are unique and their application in the food and pharmaceutical sectors generates new marketing opportunities. However, this technology requires further development especially for the scaling up of techniques to industrial levels, so it is expected to increase in the coming years. Third direction concerns the mass production of the products and making them available to the customers. This step includes scaling-up the production processes which have to be in accordance to the good manufacturing practices (GMP) and HACCP, as well as good laboratory practices (GLP). This includes intensive funding and further investigations and market analysis which will provide necessary data for accomplishing mentioned goals. Finally, hemp-based extracts exhibited significant biological activity and health benefits against the different types of diseases and disorders, especially against the various types of cancer. Therefore the required large-scaled studies and investments will be justified by development of the different products which will help people to fight these diseases.

References

Ahmad, J., & Langrish, T. A. G. (2012). Optimisation of total phenolic acids extraction from mandarin peels using microwave energy: The importance of the Maillard reaction. *Journal of Food Engineering, 109*(1), 162–174. https://doi.org/10.1016/j.jfoodeng.2011.09.017.

Andre, C. M., Hausman, J.-F., & Guerriero, G. (2016). Cannabis sativa: The plant of the thousand and one molecules. *Frontiers in Plant Science*, (FEB2016), 19. https://doi.org/10.3389/FPLS.2016.00019.

Attard, T. M., Bainier, C., Reinaud, M., Lanot, A., McQueen-Mason, S. J., & Hunt, A. J. (2018). Utilisation of super-critical fluids for the effective extraction of waxes and Cannabidiol (CBD) from hemp wastes. *Industrial Crops and Products, 112*, 38–46. https://doi.org/10.1016/J.INDCROP.2017.10.045.

Azmir, J., Zaidul, I. S. M., Rahman, M. M., Sharif, K. M., Mohamed, A., Sahena, F., … Omar, A. K. M. (2013). Techniques for extraction of bioactive compounds from plant materials: A review. *Journal of Food Engineering, 117*(4), 426–436. https://doi.org/10.1016/j.jfoodeng.2013.01.014.

Belščak-Cvitanović, A., Vojvodić, A., Bušić, A., Keppler, J., Steffen-Heins, A., & Komes, D. (2018). Encapsulation templated approach to valorization of cocoa husk, poppy and hemp macrostructural and bioactive constituents. *Industrial Crops and Products, 112*, 402–411. https://doi.org/10.1016/j.indcrop.2017.12.020.

Brighenti, V., Pellati, F., Steinbach, M., Maran, D., & Benvenuti, S. (2017). Development of a new extraction technique and HPLC method for the analysis of non-psychoactive cannabinoids in fibre-type *Cannabis sativa* L. (hemp). *Journal of Pharmaceutical and Biomedical Analysis, 143*, 228–236. https://doi.org/10.1016/j.jpba.2017.05.049.

Brunner, G. (2005). Supercritical fluids: Technology and application to food processing. *Journal of Food Engineering, 67*(1–2), 21–33. https://doi.org/10.1016/j.jfoodeng.2004.05.060.

Cardoso-Ugarte, G. A., Juárez-Becerra, G. P., SosaMorales, M. E., & López-Malo, A. (2013). Microwave-assisted ex-traction of essential oils from herbs. *Journal of Microwave Power and Electromagnetic Energy, 47*(1), 63–72. https://doi.org/10.1080/08327823.2013.11689846.

Carr, A. G., Mammucari, R., & Foster, N. R. (2011). A review of subcritical water as a solvent and its utilisation for the processing of hydrophobic organic compounds. *Chemical Engineering Journal, 172*(1), 1–17. https://doi.org/10.1016/j.cej.2011.06.007.

Chang, C. W., Yen, C. C., Wu, M. T., Hsu, M. C., & Wu, T. Y. (2017). Microwave-assisted extraction of cannabinoids in hemp nut using response surface methodology: Optimization and comparative study. *Molecules, 22*(1894), 1–15. https://doi.org/10.3390/molecules22111894.

Chemat, F., Rombaut, N., Sicaire, A. G., Meullemiestre, A., Fabiano-Tixier, A. S., & Abert-Vian, M. (2017). Ultrasound assisted extraction of food and natural products. Mechanisms, techniques, combinations, protocols and applica-tions. A review. *Ultrasonics Sonochemistry, 34*, 540–560. https://doi.org/10.1016/j.ultsonch.2016.06.035.

Chemat, F., Tomao, V., & Virot, M. (2008). Ultrasound-assisted extraction in food analysis. In S. Ötles (Ed.), *Handbook of food analysis instruments* (pp. 85–94). Boca Raton: CRC Press.

Chemat, F., Vian, M. A., & Cravotto, G. (2012). Green extraction of natural products: Concept and principles. *International Journal of Molecular Sciences, 13*(7), 8615–8627. https://doi.org/10.3390/ijms13078615.

Chen, M., Siochi, E. J., Ward, T. C., & McGrath, J. E. (1993). Basic ideas of microwave processing of polymers. *Polymer Engineering and Science, 33*(17), 1092–1109. https://doi.org/10.1002/pen.760331703.

Crescenzi, C., Di Corcia, A., Nazzari, M., & Samperi, R. (2000). Hot phosphate-buffered water extraction coupled on-line with liquid chromatography/mass spectrometry for analyzing contaminants in soil. *Analytical Chemistry, 72*(14), 3050–3055. https://doi.org/10.1021/ac000090q.

Cvetanović, A., Švarc-Gajić, J., Zeković, Z., Gašić, U., Tešić, Ž., Zengin, G., … Đurović, S. (2018). Subcritical water extraction as a cutting edge technology for the extraction of bioactive compounds from chamomile: Influence of pressure on chemical composition and bioactivity of extracts. *Food Chemistry, 266*, 389–396. https://doi.org/10.1016/J.FOODCHEM.2018.06.037.

Da Porto, C., Natolino, A., & Decorti, D. (2015). Effect of ultrasound pre-treatment of hemp (Cannabis sativa L.) seed on supercritical CO2 extraction of oil. *Journal of Food Science and Technology, 52*(3), 1748–1753.

Da Porto, C., Voinovich, D., Decorti, D., & Natolino, A. (2012). Response surface optimization of hemp seed (*Cannabis sativa* L.) oil yield and oxidation stability by supercritical carbon dioxide extraction. *Journal of Supercritical Fluids, 68*, 45–51. https://doi.org/10.1016/J.SUPFLU.2012.04.008.

de Melo, M. M. R., Silvestre, A. J. D., & Silva, C. M. (2014). Supercritical fluid extraction of vegetable matrices: Applications, trends and future perspectives of a convincing green technology. *Journal of Supercritical Fluids, 92*, 115–176. https://doi.org/10.1016/j.supflu.2014.04.007.

Devi, V., & Khanam, S. (2019). Comparative study of different extraction processes for hemp (*Cannabis sativa*) seed oil considering physical, chemical and industrial-scale economic aspects. *Journal of Cleaner Production, 207*, 645–657. https://doi.org/10.1016/J.JCLEPRO.2018.10.036.

Dezhkunov, N. V., & Leighton, T. G. (2004). Study into correlation between the ultrasonic capillary effect and so-noluminescence. *Journal of Engineering Physics and Thermophysics, 77*(1), 53–61. https://doi.org/10.1023/B:-JOEP.0000020719.33924.aa.

Domínguez, R., Pateiro, M., Agregán, R., & Lorenzo, J. M. (2017). Effect of the partial replacement of pork backfat by microencapsulated fish oil or mixed fish and olive oil on the quality of frankfurter type sausage. *Journal of Food Science and Technology, 54*(1), 26–37. https://doi.org/10.1007/s13197-016-2405-7.

Domínguez, R., Pateiro, M., Gagaoua, M., Barba, F. J., Zhang, W., & Lorenzo, J. M. (2019). A comprehensive review on lipid oxidation in meat and meat products. *Antioxidants, 8*(10), 429. https://doi.org/10.3390/ANTIOX8100429.

Esparza, Y., Ngo, T. D., & Boluk, Y. (2020). Preparation of powdered oil particles by spray drying of cellulose nanocrystals stabilized pickering hempseed oil emulsions. *Colloids and Surfaces A: Physicochemical and Engineering Aspects, 598.* https://doi.org/10.1016/j.colsurfa.2020.124823, 124823.

Fei, T., Wan, Z., & Wang, T. (2021). Dispersing insoluble yolk low-density lipoprotein (LDL) recovered by complexing with carboxymethylcellulose (CMC) for the nanoencapsulation of hemp cannabidiol (CBD) through emulsification at neutral pH. *Food Hydrocolloids, 116.* https://doi.org/10.1016/j.foodhyd.2021.106656, 106656.

Filly, A., Fernandez, X., Minuti, M., Visinoni, F., Cravotto, G., & Chemat, F. (2014). Solvent-free microwave extraction of essential oil from aromatic herbs: From laboratory to pilot and industrial scale. *Food Chemistry, 150,* 193–198. https://doi.org/10.1016/j.foodchem.2013.10.139.

Gómez, B., Barba, F. J., Domínguez, R., Putnik, P., Bursać Kovačević, D., Pateiro, M., … Lorenzo, J. M. (2018). Microencapsulation of antioxidant compounds through innovative technologies and its specific application in meat processing. *Trends in Food Science and Technology, 82,* 135–147. https://doi.org/10.1016/j.tifs.2018.10.006.

Grijó, D. R., Piva, G. K., Osorio, I. V., & Cardozo-Filho, L. (2019). Hemp (*Cannabis sativa* L.) seed oil extraction with pressurized n-propane and supercritical carbon dioxide. *Journal of Supercritical Fluids, 143,* 268–274.

He, L., Zhang, X., Xu, H., Xu, C., Yuan, F., Knez, Ž., … Gao, Y. (2012). Subcritical water extraction of phenolic compounds from pomegranate (*Punica granatum* L.) seed residues and investigation into their antioxidant activities with HPLC–ABTS+ assay. *Food and Bioproducts Processing, 90*(2), 215–223. https://doi.org/10.1016/j.fbp.2011.03.003.

Herrero, M., Cifuentes, A., & Ibanez, E. (2006). Sub- and supercritical fluid extraction of functional ingredients from different natural sources: Plants, food-by-products, algae and microalgae: A review. *Food Chemistry, 98*(1), 136–148. https://doi.org/10.1016/j.foodchem.2005.05.058.

Joana Gil-Chávez, G., Villa, J. A., Fernando Ayala-Zavala, J., Basilio Heredia, J., Sepulveda, D., Yahia, E. M., & González-Aguilar, G. A. (2013). Technologies for extraction and production of bioactive compounds to be used as nutraceuticals and food ingredients: An overview. *Comprehensive Reviews in Food Science and Food Safety, 12*(1), 5–23. https://doi.org/10.1111/1541-4337.12005.

Kitrytė, V., Bagdonaitė, D., & Rimantas Venskutonis, P. (2018). Biorefining of industrial hemp (*Cannabis sativa* L.) threshing residues into cannabinoid and antioxidant fractions by supercritical carbon dioxide, pressurized liquid and enzyme-assisted extractions. *Food Chemistry, 267,* 420–429.

Ko, M.-J., Cheigh, C.-I., & Chung, M.-S. (2014). Relationship analysis between flavonoids structure and subcritical water extraction (SWE). *Food Chemistry, 143,* 147–155. https://doi.org/10.1016/j.foodchem.2013.07.104.

Kornpointner, C., Sainz Martinez, A., Marinovic, S., Haselmair-Gosch, C., Jamnik, P., Schröder, K., … Halbwirth, H. (2021). Chemical composition and antioxidant potential of *Cannabis sativa* L. roots. *Industrial Crops and Products, 165.* https://doi.org/10.1016/J.INDCROP.2021.113422, 113422.

Křížek, T., Bursová, M., Horsley, R., Kuchař, M., Tůma, P., Čabala, R., & Hložek, T. (2018). Menthol-based hydrophobic deep eutectic solvents: Towards greener and efficient extraction of phytocannabinoids. *Journal of Cleaner Production, 193,* 391–396. https://doi.org/10.1016/j.jclepro.2018.05.080.

Kurek, M. A., & Pratap-Singh, A. (2020). Plant-based (hemp, pea and rice) protein–maltodextrin combinations as wall material for spray-drying microencapsulation of hempseed (*Cannabis sativa*) oil. *Foods, 9*(11), 1707. https://doi.org/10.3390/foods9111707.

Lan, Y., Ohm, J. B., Chen, B., & Rao, J. (2021). Microencapsulation of hemp seed oil by pea protein isolate–sugar beet pectin complex coacervation: Influence of coacervation pH and wall/core ratio. *Food Hydrocolloids, 113.* https://doi.org/10.1016/j.foodhyd.2020.106423, 106423.

Lang, Q., & Wai, C. M. (2001). Supercritical fluid extraction in herbal and natural product studies—A practical review. *Talanta, 53*(4), 771–782. https://doi.org/10.1016/S0039-9140(00)00557-9.

Leonard, W., Zhang, P., Ying, D., Xiong, Y., & Fang, Z. (2021). Extrusion improves the phenolic profile and biological activities of hempseed (*Cannabis sativa* L.) hull. *Food Chemistry, 346.* https://doi.org/10.1016/J.FOODCHEM.2020.128606, 128606.

Li, H., Chang, S. L., Chang, T. R., You, Y., Wang, X. D., Wang, L. W., … Zhao, B. (2021). Inclusion complexes of cannabidiol with β-cyclodextrin and its derivative: Physicochemical properties, water solubility, and antioxidant activity. *Journal of Molecular Liquids, 334.* https://doi.org/10.1016/j.molliq.2021.116070, 116070.

Marzorati, S., Friscione, D., Picchi, E., & Verotta, L. (2020). Cannabidiol from inflorescences of *Cannabis sativa* L.: Green extraction and purification processes. *Industrial Crops and Products, 155,* 112816.

Mason, T. J. (2015). Some neglected or rejected paths in sonochemistry – A very personal view. *Ultrasonics Sonochemistry, 25,* 89–93. https://doi.org/10.1016/j.ultsonch.2014.11.014.

Mason, T. J., Paniwnyk, L., & Lorimer, J. P. (1996). The uses of ultrasound in food technology. *Ultrasonics Sonochemistry, 3*(3), S253–S260. https://doi.org/10.1016/S1350-4177(96)00034-X.

McClements, D. J. (2018a). Delivery by design (DbD): A standardized approach to the development of efficacious nanoparticle- and microparticle-based delivery systems. *Comprehensive Reviews in Food Science and Food Safety, 17*(1), 200–219. https://doi.org/10.1111/1541-4337.12313.

McClements, D. J. (2018b). Encapsulation, protection, and delivery of bioactive proteins and peptides using nanoparticle and microparticle systems: A review. *Advances in Colloid and Interface Science, 253,* 1–22. https://doi.org/10.1016/j.cis.2018.02.002.

Meullemiestre, A., Breil, C., Abert-Vian, M., & Chemat, F. (2015). *Modern techniques and solvents for the extraction of microbial oils.* Cham: Springer International Publishing. https://doi.org/10.1007/978-3-319-22717-7.

Mišan, A., Naðpal, J., Stupar, A., Pojić, M., Mandić, A., Verpoorte, R., & Choi, Y. H. (2019). The perspectives of natural deep eutectic solvents in agri-food sector. *Critical Reviews in Food Science and Nutrition, 60*(15), 2564–2592. https://doi.org/10.1080/10408398.2019.1650717.

Mourtzinos, I., Menexis, N., Iakovidis, D., Makris, D. P., & Goula, A. (2018). A green extraction process to recover polyphenols from byproducts of hemp oil processing. *Recycling, 3*(2), 15. https://doi.org/10.3390/RECYCLING3020015.

Naz, S., Hanif, M. A., Bhatti, H. N., & Ansari, T. M. (2017). Impact of supercritical fluid extraction and traditional distillation on the isolation of aromatic compounds from *Cannabis indica* and *Cannabis sativa*. *Journal of Essential Oil-Bearing Plants, 20*(1), 175–184. https://doi.org/10.1080/0972060X.2017.1281766.

Olejar, K. J., Hatfield, J., Arellano, C. J., Gurau, A. T., Seifried, D., Vanden, B., & Kinney, C. A. (2021). Thermo-chemical conversion of cannabis biomass and extraction by pressurized liquid extraction for the isolation of cannabidiol. *Industrial Crops and Products, 170*(June). https://doi.org/10.1016/j.indcrop.2021.113771, 113771.

Özel, M. Z., & Göğüş, F. (2014). Subcritical water as a green solvent for plant extraction. In F. Chemat, & M. A. Vian (Eds.), *Alternative solvents for natural products extraction* (pp. 73–89). Berlin: Springer-Verlag. https://doi.org/10.1007/978-3-662-43628-8_4.

Ozel, M. Z., Gogus, F., & Lewis, A. C. (2003). Subcritical water extraction of essential oils from Thymbra spicata. *Food Chemistry, 82*(3), 381–386. https://doi.org/10.1016/S0308-8146(02)00558-7.

Palmieri, S., Pellegrini, M., Ricci, A., Compagnone, D., & Lo Sterzo, C. (2020). Chemical composition and antioxidant activity of thyme, hemp and coriander extracts: A comparison study of maceration, soxhlet, UAE and RSLDE techniques. *Foods, 9*(1221), 1–18. https://doi.org/10.3390/foods9091221.

Paniwnyk, L., Beaufoy, E., Lorimer, J., & Mason, T. (2001). The extraction of rutin from flower buds of *Sophora japonica*. *Ultrasonics Sonochemistry, 8*(3), 299–301. https://doi.org/10.1016/S1350-4177(00)00075-4.

Pasquali, I., Bettini, R., & Giordano, F. (2008). Supercritical fluid technologies: An innovative approach for manipulating the solid-state of pharmaceuticals. *Advanced Drug Delivery Reviews, 60*(3), 399–410. https://doi.org/10.1016/j.addr.2007.08.030.

Pateiro, M., Gómez, B., Munekata, P. E. S., Barba, F. J., Putnik, P., Kovačević, D. B., & Lorenzo, J. M. (2021). Nanoencapsulation of promising bioactive compounds to improve their absorption, stability, functionality and the appearance of the final food products. *Molecules, 26*(6), 1547. https://doi.org/10.3390/molecules26061547.

Petigny, L., Périno-Issartier, S., Wajsman, J., & Chemat, F. (2013). Batch and continuous ultrasound assisted extraction of boldo leaves (*Peumus boldus* Mol.). *International Journal of Molecular Sciences, 14*(3), 5750–5764. https://doi.org/10.3390/ijms14035750.

Plati, F., Ritzoulis, C., Pavlidou, E., & Paraskevopoulou, A. (2021). Complex coacervate formation between hemp protein isolate and gum Arabic: Formulation and characterization. *International Journal of Biological Macromolecules, 182,* 144–153. https://doi.org/10.1016/j.ijbiomac.2021.04.003.

Plaza, M., Amigo-Benavent, M., del Castillo, M. D., Ibáñez, E., & Herrero, M. (2010). Facts about the formation of new antioxidants in natural samples after subcritical water extraction. *Food Research International, 43*(10), 2341–2348. https://doi.org/10.1016/j.foodres.2010.07.036.

Reverchon, E. (1997). Supercritical fluid extraction and fractionation of essential oils and related products. *Journal of Supercritical Fluids, 10*(1), 1–37. https://doi.org/10.1016/S0896-8446(97)00014-4.

Reverchon, E., & De Marco, I. (2006). Supercritical fluid extraction and fractionation of natural matter. *Journal of Supercritical Fluids, 38*(2), 146–166. https://doi.org/10.1016/j.supflu.2006.03.020.

Rezvankhah, A., Emam-Djomeh, Z., Safari, M., Askari, G., & Salami, M. (2019). Microwave-assisted extraction of hempseed oil: Studying and comparing of fatty acid composition, antioxidant activity, physiochemical and thermal properties with Soxhlet extraction. *Journal of Food Science and Technology, 56*(9), 4198–4210. https://doi.org/10.1007/s13197-019-03890-8.

Richter, B. E., Jones, B. A., Ezzell, J. L., Porter, N. L., Avdalovic, N., & Pohl, C. (1996). Accelerated solvent extraction: A technique for sample preparation. *Analytical Chemistry, 68*(6), 1033–1039. https://doi.org/10.1021/ac9508199.

Rossi, P., Cappelli, A., Marinelli, O., Valzano, M., Pavoni, L., Bonacucina, G., … Nabissi, M. (2020). Mosquitocidal and anti-inflammatory properties of the essential oils obtained from monoecious, male, and female inflorescences of hemp (*Cannabis sativa* L.) and their encapsulation in nanoemulsions. *Molecules, 25*(15), 3451. https://doi.org/10.3390/molecules25153451.

Rovetto, L. J., & Aieta, N. V. (2017). Supercritical carbon dioxide extraction of cannabinoids from *Cannabis sativa* L. *Journal of Supercritical Fluids, 129*, 16–27.

Rožanc, J., Kotnik, P., Milojević, M., Gradišnik, L., Hrnčič, M. K., Knez, Ž., & Maver, U. (2021). Different Cannabis sativa extraction methods result in different biological activities against a colon cancer cell line and healthy colon cells. *Plants, 10*(3), 1–16. https://doi.org/10.3390/plants10030566.

Sahena, F., Zaidul, I. S. M., Jinap, S., Karim, A. A., Abbas, K. A., Norulaini, N. A. N., & Omar, A. K. M. (2009). Application of supercritical CO_2 in lipid extraction—A review. *Journal of Food Engineering, 95*(2), 240–253. https://doi.org/10.1016/j.jfoodeng.2009.06.026.

Shi, Y., Wang, W., Zhu, X., Wang, B., Hao, Y., Wang, L., … Elfalleh, W. (2021). Preparation and physicochemical stability of hemp seed oil liposomes. *Industrial Crops and Products, 162*. https://doi.org/10.1016/j.indcrop.2021.113283, 113283.

Sovová, H., & Stateva, R. P. (2011). Supercritical fluid extraction from vegetable materials. *Reviews in Chemical Engineering, 27*(3–4). https://doi.org/10.1515/REVCE.2011.002.

Sparr Eskilsson, C., & Björklund, E. (2000). Analytical-scale microwave-assisted extraction. *Journal of Chromatography A, 902*(1), 227–250. https://doi.org/10.1016/S0021-9673(00)00921-3.

Subratti, A., Lalgee, L. J., & Jalsa, N. K. (2019). Liquified dimethyl ether (DME): A green solvent for the extraction of hemp (*Cannabis sativa* L.) seed oil. *Sustainable Chemistry and Pharmacy, 12*. https://doi.org/10.1016/J.SCP.2019.100144, 100144.

Teo, C. C., Tan, S. N., Yong, J. W. H., Hew, C. S., & Ong, E. S. (2010). Pressurized hot water extraction (PHWE). *Journal of Chromatography A, 1217*(16), 2484–2494. https://doi.org/10.1016/j.chroma.2009.12.050.

Ternelli, M., Brighenti, V., Anceschi, L., Poto, M., Bertelli, D., Licata, M., & Pellati, F. (2020). Innovative methods for the preparation of medical Cannabis oils with a high content of both cannabinoids and terpenes. *Journal of Pharmaceutical and Biomedical Analysis, 186*. https://doi.org/10.1016/j.jpba.2020.113296, 113296.

Tiwari, B. K. (2015). Ultrasound: A clean, green extraction technology. *TrAC Trends in Analytical Chemistry, 71*, 100–109. https://doi.org/10.1016/j.trac.2015.04.013.

Valh, J. V., Peršin, Z., Vončina, B., Vrezner, K., Tušek, L., & Zemljič, L. F. (2021). Microencapsulation of cannabidiol in liposomes as coating for cellulose for potential advanced sanitary material. *Coatings, 11*(1), 1–18. https://doi.org/10.3390/coatings11010003.

Vargas-Ramella, M., Pateiro, M., Barba, F. J., Franco, D., Campagnol, P. C. B., Munekata, P. E. S., … Lorenzo, J. M. (2020). Microencapsulation of healthier oils to enhance the physicochemical and nutritional properties of deer pâté. *LWT, 125*. https://doi.org/10.1016/j.lwt.2020.109223, 109223.

Veggi, P. C., Martinez, J., & Meireles, M. A. A. (2012). Fundamentals of microwave extraction. In F. Chemat, & G. Cravotto (Eds.), *Microwave-assisted extraction for bioactive compounds: Theory and practice* (pp. 15–52). New York: Springer Science+ Business Media. https://doi.org/10.1007/978-1-4614-4830-3_2.

Veličković, V., Đurović, S., Radojković, M., Cvetanović, A., Švarc-Gajić, J., Vujić, J., … Mašković, P. Z. (2017). Application of conventional and non-conventional extraction approaches for extraction of *Erica carnea* L.: Chemical profile and biological activity of obtained extracts. *Journal of Supercritical Fluids, 128*, 331–337. https://doi.org/10.1016/j.supflu.2017.03.023.

Vilkhu, K., Manasseh, R., Mawson, R., & Ashokkumar, M. (2011). Ultrasonic recovery and modification of food ingredients. In H. Feng, G. Barbosa-Canovas, & J. Weiss (Eds.), *Ultrasound technologies for food and bioprocessing* (pp. 345–368). New York: Springer. https://doi.org/10.1007/978-1-4419-7472-3_13.

Vinatoru, M. (2001). An overview of the ultrasonically assisted extraction of bioactive principles from herbs. *Ultrasonics Sonochemistry, 8*(3), 303–313. https://doi.org/10.1016/S1350-4177(01)00071-2.

Wang, L., & Weller, C. L. (2006). Recent advances in extraction of nutraceuticals from plants. *Trends in Food Science and Technology, 17*(6), 300–312. https://doi.org/10.1016/j.tifs.2005.12.004.

Yin, J., Xiang, C., Wang, P., Yin, Y., & Hou, Y. (2017). Biocompatible nanoemulsions based on hemp oil and less surfactants for oral delivery of baicalein with enhanced bioavailability. *International Journal of Nanomedicine, 12*, 2923–2931. https://doi.org/10.2147/IJN.S131167.

Yu, H., Park, J. Y., Kwon, C. W., Hong, S. C., Park, K. M., & Chang, P. S. (2018). An overview of nanotechnology in food science: Preparative methods, practical applications, and safety. *Journal of Chemistry, 2018*. https://doi.org/10.1155/2018/5427978, 5427978.

Zheljazkov, V. D., Sikora, V., Semerdjieva, I. B., Kačániová, M., Astatkie, T., & Dincheva, I. (2020). Grinding and fractionation during distillation alter hemp essential oil profile and its antimicrobial activity. *Molecules, 25*(17), 3943. https://doi.org/10.3390/MOLECULES25173943.

Industrial hemp foods and beverages and product properties

Lorenzo Nissen[a], Flavia Casciano[b], Elena Babini[c], and Andrea Gianotti[b]

[a]CIRI-Interdepartmental Centre of Agri-Food Industrial Research, Alma Mater Studiorum-University of Bologna, Cesena, Italy [b]DiSTAL-Department of Agricultural and Food Sciences, Alma Mater Studiorum-University of Bologna, Bologna, Italy [c]DiSTAL-Department of Agricultural and Food Sciences, Alma Mater Studiorum-University of Bologna, Cesena, Italy

10.1 Introduction

Hemp (*Cannabis sativa* L.) is member of the *Cannabaceae* family and a C3 crop native to Central-Northeast Asia where there is evidence of its cultivation dating back more than 5000 years (Li, 1973). Hemp, namely industrial hemp or fiber hemp, has an incredible number of possible applications and, especially, it has attracted interest for use in medical therapy and for textile uses (El-Sohly, 2002; Russo, 2001). The worldwide commercial growing of industrial hemp has been strongly limited because it can be easily confounded with high-THC hemp types, namely marijuana, which are a different breed of industrial hemp. Regardless of its origin, the currently domesticated form of *C. sativa* L. is widespread and cultivated not only across Asia but even in North America, Europe, and Africa. At present, there are no more traces of wild-type hemp, and only domesticated (i.e., species chosen and selected by humans for characteristics making them useful to people) and ruderal (i.e., forms growing outside of cultivation) hemp plants exist.

Agronomically, hemp is a versatile, sustainable, undemanding crop, either by the point of nutrition, pesticide use, space, and techniques. Moreover, it is soil refilling, has low environmental impact, and is useful for several application fields, from agricultural and phytoremediation to food and feed, cosmetic, building, and pharmaceutical industries. Today, the EU is the world's largest hemp-producing market second only to Canada, with France, the Netherlands, Lithuania, and Romania as major production centers (Johnson, 2018). Nonetheless, there is still a lack of awareness and much confusion between "industrial hemp" and "drug hemp", especially among public opinion (Farinon, Molinari, Costantini, & Merendino, 2020).

The different uses of *C. sativa* L. are based on the presence and quantity of two main phytochemicals characteristic of this crop, namely the psychoactive compound of the plant, delta-9-tetrahydrocannabinol (THC), and the nonpsychoactive cannabidiol (CBD). These compounds are cannabinoids typical of all strains of *C. sativa* L. Theoretically, all genotypes of hemp, including those of industrial hemp, contain the psychotropic agent (THC). Nonetheless, unlike marijuana, industrial hemp genotypes that have been licensed for cultivation contain THC levels of less than 0.3% or 0.2% (w/v) of the female plant's reproductive part during flowering, which is about 50 times lower than that of marijuana. However, the classification of the different phenotypes of *C. sativa* L. has distinguished three principal chemotypes based on THC and CBD content and ratio (Broséus, Anglada, & Esseiva, 2010; De Meijer, 2014; Pacifico et al., 2006; Staginnus, Zörntlein, & de Meijer, 2014). Chemotype I includes the cultivars of *C. sativa* subsp. *indica*, known as the "drug type", characterized by a low CBD/THC ratio (0.00–0.05) due to high THC content (> 0.3% of female inflorescences' dry weight). Chemotype II, known as the "intermediate type" contains CBD and THC in a ratio ranging from 0.5 to 3.0, usually with a slight prevalence of CBD. These two chemotypes are applied for recreational and medicinal purposes. Chemotype III includes cultivars from the *C. sativa* subsp. *sativa* group and is considered the "fiber type", characterized by high CBD/THC ratio (15–25) due to high CBD and low THC content (< 0.3% of dry weight). The cultivars of this chemotype are cultivated for industrial purposes, either for textile, biobuilding, feeds, and foods, and are those that we will deal with in this chapter.

Moreover, the use of hemp seed for nutrition is beyond law infringements (European Monitoring Centre for Drugs and Drug Addiction; Onofri, de Meijer, & Mandolino, 2015; Staginnus et al., 2014), because cannabinoids, including THC, are synthesized and stored in pedunculated glandular trichomes, which are absent on roots and seeds. Anyhow, in Europe, the cultivation of hemp is strictly ruled by the Commission Regulation No 206/2004. Traditionally, *C. sativa* L. plants were cultivated primarily as a fiber crop to produce textiles and ropes, especially in the Western world. So far, the seeds of this plant, regardless of their high nutritional value, were generally treated as a by-product from industrial processing of the fiber and were mainly intended for animals in the form of feeding cakes.

Today, with the surge on studies related to the recognition of hemp seeds' nutritional features and health benefits, the production of hemp seeds has increased and hemp seeds are becoming a product with an important and growing market. The global hemp seed market trends to reach about $1.7 billion USD in 2027 where 45% is addressed to the food sector. The whole hemp segment is growing, and its forecast is florid because of the stable nature of the derived product in terms of denaturation and oxidative stability. Moreover, the European market is forecasting to be that with the fastest compound annual growth rate (CAGR) of 12.1% (https://www.fortunebusinessinsights.com/ ID Report: FBI103478).

In fact, hemp seeds and hemp-based products are attractive for those consumers searching for sustainable and eco-friendly food products with high nutritional value, those affected by celiac disease and lactose intolerance, as well as for vegans and athletes. Hemp seeds have commonly been claimed as a nutritionally complete food source that can be consumed entire as whole seed, hulled seed, dehulled, or processed in different food products, including oil, flour, and protein concentrates. The seed contains 25% to 35% lipids, depending on the cultivar, latitude, and growing season with a unique balanced fatty acid (FA) composition, characterized by over 80% amount of polyunsaturated FAs (PUFAs) with the essential FAs

(EFAs) n-6 linoleic acid (LA) and n-3 alpha-linolenic acid (ALA) in the favorable ratio for human nutrition (Da Porto, Decorti, & Natolino, 2015; Irakli et al., 2019; Vonapartis, Aubin, Seguin, Mustafa, & Charron, 2015).

Among phytosterols, tocopherol is the main component of the unsaponifiable oil fraction that is crucial to protect the PUFAs from oxidation due to its renowned antioxidant activity. In hemp seeds, 65% of the total protein consists of a single storage protein, edestin, but 20% to 25% are related to proteins of high biological value, easily digestible, and rich in essential amino acids (AAs) (Farinon et al., 2020).

Hemp seeds also contain 20% to 30% carbohydrates, a great part of which are constituted mainly in insoluble dietary fiber, beneficial as substrates for intestinal microbiota in the colon. In fact, sometimes hulled hemp seed are used as an adjuvant for colon cleanse and toxin clearance, perhaps due to its ability to foster beneficial intestinal microbes that counteract the activity and the metabolism of opportunistic and pathogenic species (Irakli et al., 2019).

In addition to its nutritional value, hemp seed is also rich in natural antioxidants and other bioactive components such as peptides, phenolics, tocopherols, carotenoids, phytosterols, terpenes, and cannabinoids on seed surface, whose contents and yields depend on environmental, agronomic and processing conditions (Irakli et al., 2019), as well as on genetic inheritance (e.g., cannabivarin in South Africans strains).

Considering the antinutritional factors, for example, in respect to soybeans, hemp seeds present more phytic acid, which could negatively affect the proteins' and minerals' digestion and bioavailability, while containing very low amounts of trypsin inhibitors (Malomo, Onuh, Girgih, & Aluko, 2015) that improve digestibility (Table 10.1).

10.2 Hemp seeds for beverages

10.2.1 Hemp seed milk

Milk is one of the most consumed foods. However, the estimated prevalence of cow's milk allergy and lactose intolerance in early childhood is estimated at 2% to 3%, while a deficiency of lactase that results in gastrointestinal disorders affects 65% of the world's population (Lomer, Parkes, & Sanderson, 2008). In addition, there is growing health awareness among consumers, who are increasingly directed toward alternative dairy plant-based products (Sethi, Tyagi, & Anurag, 2016). The global market for non-dairy milk and milk alternatives is growing quickly due to interest by intolerant milk consumers and those searching for foods with added values, such as functional products (Tangyu, Muller, Bolten, & Wittmann, 2019). In addition to health reasons, the demand for plant-based alternatives to milk is also supported by environmental and ethical reasons (McClements, Newman, & McClements, 2019).

To meet consumers' demands, drinks alternative to cow's milk have been developed and produced from matrices such as nuts (e.g., almond, hazelnut, coconut), cereals (e.g., rice, oat), and legumes (e.g., soy).

Soy and rice drinks are the most successful milk substitutes. Soy boasts a high protein and vitamin content, yet is low in saturated FAs and completely cholesterol-free (Paul, Kumar, Kumar, & Sharma, 2020). The rice-based drink is also cholesterol-free but, unlike soy, it is also free of unsaturated fats and allergens (Abou-Dobara, Ismail, & Refat, 2016; Sethi et al., 2016).

TABLE 10.1   Examples of commercial hemp-based products.

Product	Producers	Country of origin
Hemp seed drink	Ecomil SA	Murcia, Spain
Hemp seed snacks	Iswari Superfood	Setubal, Portugal
Hemp seed pasta	ARES SRLS	Termoli, Italy
Flavored hemp seed beer	Tenuta Modesti	Piacenza, Italy
Hemp seed protein	Nutiva	Richmond, CA, USA
Hemp seed flour	Hanf Farm Gmbh	Melz, Germany
Hemp seed oil	Guenard	Noyers-sur-Cher, France
Hemp seed chocolate bars	Manitoba Harvest	Winnipeg, MB, Canada
Hemp seed bread	Biona organic	Kingston upon Thames, UK
Fermented hemp seed foods	Hemp OZ	Sidney, NSW, Australia
Hemp seed sweets	CBWeed SRL	Forlì, Italy
Hemp seed fibers	Chana International SRL	Salonta, Romania
Hemp seed juice	Hemp Juice Company BV	Oude Pekela, The Netherlands
Hemp seed nuts	Sky Green	Oppen Ginza, Tokyo, Japan
Hemp seed spread	Finola Oy	Kuopio, Finland
Hemp seed milk	Hudson River Foods	New York, NY, USA
Hemp seed yogurt	Evia Yogurth	St, Marrickville, NSW, Australia
Hemp seed herbal tea	Annique	Pretoria, South Africa
Hemp seed butter	Sun & Seed Ltd	London, UK
Shelled hemp seed	Probios, SPA	Calenzano, Italy

Moreover, soy-based drinks contain few saturated FAs and are completely free of cholesterol or lactose (Sethi et al., 2016). Another advantageous characteristic of rice-based drink is its high content of selenium and magnesium (Singhal, Baker, & Baker, 2017). Most alternative beverages lack nutritional balance in comparison to cow's milk but contain active functional compounds with health-promoting properties (Sethi et al., 2016). As done by several companies producing plant-based drinks, the mixtures of different type of drinks could be the solution to meet the right nutritional balance, taste, and texture. For example, according to the nutritional values, a mix of soy, rice, and hemp drink could be comparable to cow's milk due to similar proteins, sugars, and lipid content, respectively (Nissen, di Carlo, & Gianotti, 2020; Singhal et al., 2017).

Still, these types of drinks do not have the right to be labeled as "milk" based on the distinct nutritional composition. In fact, the European Commission explicitly prohibits the use of such wording agreeing to Reg. CEE/UE 22.10.2007 n. 1234. The correct term is "plant-based drink based on …" (Sethi et al., 2016), although different terms are used worldwide such as drinks, beverages, dairy alternatives, or similar wording (McClements et al., 2019). Unfortunately, the taste of most of plant-based drinks is far from that of cow's milk, and more than the addition of food technical compounds, a solution to solve this problem could be found in functional transformation of the original product. For example, various functional beverage products that contain probiotics have been launched, including nuts, cereals, soybeans, and vegetable-based beverages, as recently reviewed (Frassinetti et al., 2018).

The main strategies in designing novel functional beverages may include novel functional components, enhancing the functionality and health-promoting effects (also by new technologies), development of all-natural products, and using natural food ingredients such as emulsifiers, preservatives (e.g., bacteriocins, oligosaccharides, chitosan, essential oils, and plant extracts), and sweeteners (from fruit and honey) (Nazir et al., 2019). Hemp seeds represent a potentially suitable vegetable source for functional milk substitutes. In fact, hemp seeds are a nutritional powerhouse and contain bioactive and fiber compounds with striking prebiotic capability, capable of fostering the growth of beneficial and probiotic bacteria and limiting or inhibiting opportunistic or pathogenic species. Moreover, hemp seeds are characterized by a low content of saturated fats and a favorable percentage of PUFAs ω-3 and ω-6, are naturally free of cholesterol, and have a low glycemic index (Hartsel, Eades, Hickory, & Makriyannis, 2016). Considering the protein fractions (65% edestin and 35% albumin), hemp seeds have a well-balanced AA composition (Wang, Jiang, & Xiong, 2018) and are particularly abundant in glutamic acid, arginine, and aspartic acid (Bartkiene et al., 2016). The Protein Digestibility Corrected Amino Acid Score (PDCAAS) is calculated as the product of protein digestibility and the AA score, and hemp seeds have a score similar to that of lentils, around half of that of casein (used as the positive control), and approximately twice that of almonds (randomly used as a low digestibility sample) (House, Neufeld, & Leson, 2010). The total proportion of essential AAs in hemp protein isolate (HPI) is also significantly higher than that of soy protein isolate (SPI) (Wang, Tang, Yang, & Gao, 2008). Additionally, hemp seeds are plentiful in terpenes and terpenoid compounds, such as 1-(R)-α-pinene, Δ-3-carene, β-myrcene, and β-caryophillene (Booth & Bohlmann, 2019; Nissen et al., 2010). These molecules exhibit antioxidant and pathogen-antimicrobial effects (Booth & Bohlmann, 2019; Nissen et al., 2010), even after heat treatment during hemp seed drink production, which reduces the concentration of these compounds (King, 2019). Otherwise, in a recent study, it was reported that such compounds remain intact even in industrially processed UHT hemp seed drink (Nissen, di Carlo, & Gianotti, 2020). Terpenoids of hemp share a precursor with phytocannabinoids and are flavor and fragrance components common in human diets that have been designated as generally regarded as safe (GRAS) by the US Food and Drug Administration and other regulatory agencies (Da Cheng, Xiao-Jie, & Pei, 2015). Phytocannabinoid-terpenoid interactions could produce synergy with respect to treatment of pain, inflammation, depression, anxiety, addiction, epilepsy, and fungal and bacterial infections (Russo, 2001). It has been demonstrated that the abundance and synergistic effect of these compounds in hemp seed-based fermented drinks did not influence the ability of probiotic bacteria to ferment hemp seed-derived fibers (Nissen, di Carlo, & Gianotti, 2020). Other health-related and evidence-based issues on the properties of hemp seed drinks included studies that have shown that the consumption of hemp milk could lead to the reduction of serum triacylglycerols, cholesterol, and thyroid hormones (Chichłowska, Kliber, Kozłowska, Biskupski, & Grygorowicz, 2009). Thus, considering all these aspects, including its low allergenicity and lactose-free nature, hemp seed drink is considered as an attractive alternative to dairy, soy, rice, and nut milks. Nevertheless, hemp seed drinks are still not widespread and have a small market ($185 million USD) (Zion Market Research, 2018) due to its taste and look, which limit its acceptance to the consumer (Jeske, Bez, Arendt, & Zannini, 2019). In fact, if biogenic amines, such as cadaverine and putrescine naturally present in the matrix, are transferred to derived products, the taste could be affected (Bartkiene et al., 2016), as well as some organic acids, in particular some medium chains, e.g., hexanoic acid, that could confer a rancid taste (Nissen, di Carlo, & Gianotti, 2020). Additionally, hemp milk does not have an attractive look due to the instability of the beverage solution, structurally representing an oil-in-water (O/W) emulsion, which easily flocculates, coalesces, and creams. This also determines a loss of quality that shortens the shelf-life and limits its acceptance by the consumer (Wang et al., 2018). The principal technological reason is because the hemp protein fraction has low emulsifying capacity (Yin, Tang, Wen, & Yang, 2009) due to its compact structure and poor water solubility. Therefore, attempts to modify the native structure of hemp protein for improving the emulsifying capacity by enzymatic hydrolysis (Yin et al., 2008) and acylation (Yin et al., 2009) have been made. To obtain a balanced product, commercial emulsifiers or stabilizers are commonly added to improve the kinetic stability of emulsions. Nevertheless, the use of additives does not only increase the production cost but also leads to health concerns. For example, some studies have shown that long-term consumptions of synthetic emulsifying agents could induce chronic inflammatory diseases, obesity-associated diseases, and metabolic disorders (Cani & Everard, 2015; Chassaing et al., 2015). Alternatively, different types of plant-based emulsifiers (e.g., proteins, polysaccharides, phospholipids, and biosurfactants) can be used to improve the formation and stability of O/W emulsions (McClements et al., 2019). A typical hemp milk is processed by soaking the seeds in water (1: 5 w/v) to soften the tissues, and then the seeds are mechanically ground. The subsequent phases include centrifugation or filtration to remove coarse particles, heat treatment to microbiologically stabilize the product, possible fortification with other ingredients (e.g., fats, thickeners, stabilizers, colors, flavors, and nutrients), and finally packaging (McClements et al., 2019). However, thermal treatments may damage heat-sensitive nutrients, especially ω-3 FAs and vitamins, and likewise also reduce product freshness. Otherwise, in the future, other treatments that are less invasive and destructive can be applied, such as those applied to nonthermal production of many emerging vegetable-based beverages, such as cold-pressed-without-heat-pasteurized juices (cold-pasteurization technique to kill potentially harmful microorganisms). With these cases, it could be possible to prepare hemp milk preserving its nutrients and original clean flavor.

10.2.2 Fermented hemp seed drinks

Another solution to improve the stability of hemp seed-based drink, from both nutritional and functional points of view, could be fermentation with probiotic cultures. Recently, in a technological study, a mixture of hemp seed-, soy-, and rice-based drinks has been developed to solve and overcome nutritional and functional deficiencies of the individual drinks.

Thus, a combination of these drinks subjected to fermentation by a mixture of *lactobacilli* and *bifidobacteria* yielded a product whose pH was kept stable after a few hours, allowing prolonged fermentation and high growth of beneficial microbes (Nissen, Demircan, Taneyo-Saa, & Gianotti, 2019).

It has been demonstrated that the abundance and the synergistic effect of hemp seed terpenes and terpenoids, with antimicrobial features, in hemp seed-based drinks did not influence the ability of probiotic bacteria to ferment hemp seed-derived fibers (Nissen, di Carlo, & Gianotti, 2020). In fact, the authors have found that the high amount of beta-myrcene and alpha-bergamotene found in hemp seed drinks were positively correlated with the concentration of saturated or unsaturated fats and with probiotic activity related to both *lactobacilli* and *bifidobacteria*. This output showed that the high quantity of these two compounds did not impair but rather improved probiotic fermentation and hemp seed functionalization (Nissen, di Carlo, & Gianotti, 2020).

Fermentation of plant-based drinks is mainly conducted with lactic acid bacteria, in particular *Streptococcus* spp. and *Lactobacillus* spp. In fact, these bacteria are able to resist and grow in difficult environments prevailing in plant-based drinks due to the extreme acidity (yogurt from cow's milk is fermented to pH 4.6, while plant-based drinks to less than pH 4) (Filannino et al., 2014; Nielsen, Martinussen, Flambard, Sørensen, & Otte, 2009; Peres, Peres, Hernández-Mendoza, & Malcata, 2012) and the presence of inhibitory factors (hydroxycinnamic acids, tannins) (Filannino et al., 2014; Rodríguez et al., 2009). Therefore, plant-based drink is a challenging matrix for unspecialized microbial fermentation.

Lactobacillus spp., *Lb. plantarum*, and *Lb. fermentum* contain different alpha-glucosidases that make them ideal to ferment plant-based drinks such as soy (LeBlanc et al., 2008). *Lactobacillus plantarum* PH04 is a beneficial microbe that could be commercially claimed as a probiotic (Nguyen, Kang, & Lee, 2007). Apart from being extremely resistant to low pH and bile acids, *Lb. fermentum* ME-3 is able to lower human blood's cholesterol level and produce strong antioxidant and antipathogenic compounds (Mikelsaar & Zilmer, 2009).

Another lactic acid bacteria (LAB), different than *Lactobacillus* spp., is *Bifidobacterium bifidum*, and its strains could be applied in this kind of fermentation with beneficial effects on the host, such as those provided by MIMBb75 (Guglielmetti, Mora, Gschwender, & Popp, 2011). Once the plant-based mixtures were fermented, the hemp seed- and soy-based drinks were described by the presence of 2-heptanol, 2-methyl, and 1-octen-3-ol, while hemp seed- and rice-based drinks were described by the presence of heptanol. Generally, the alcohol compounds—heptanol, 2-heptanol, 2-methyl, 1-octen-3-ol, and cyclohexen-2-ol appeared to be typical for the hemp seed matrices. The first two were more abundant in hemp blends, while the latter was quantified in plain hemp drink (Nissen et al., 2019). These compounds were characteristic for rice and soybean raw materials or products, conferring a typical olfactory issue described as musty, pungent, leafy green, and mushroom taste, respectively (Chung, 1999). 1-octen-3-ol is reported to be a natural product derived from linoleic acid during oxidative breakdown, which can be found in different plant and mushroom species, with high antimicrobial activity against spoilage and opportunistic microbes (Xiong et al., 2017). Cyclohexen-2-ol is found in tea leaves and exhibits sweet, floral, caramelized, honey-like notes associated with high-quality tea (Kfoury et al., 2018). Probiotic fermentation of mixed plant-based drinks including hemp seed produced higher concentration of aldehydes such as nonanal and 2-heptenal (Z), mainly correlated to the matrix, while 2,4-decadienal production was driven by microbes. 2-heptenal (Z), nonanal, and decanal were the aldehydes mainly characterizing the hemp seed matrices. The presence of these compounds confers a typical scent. 2-heptenal (Z) has an almond flavor, and its abundances are linked to fermentation process, for example, during sourdough fermentation (Petel, Prost, & Onno, 2017). Nonanal and decanal has green and soapy aroma characteristics (Kim, Jang, & Lee, 2018); the former is reported to be associated with buckwheat and rice products and has a rose-orange odor while the latter is found to be associated with orange juice sensorial volatile (Baxter, Easton, Schneebeli, & Whitfield, 2005). From our results, 2-heptenal (Z) seemed to well characterize hemp matrix compared to soy and rice, while the addition of hemp to soy and rice blends improves the abundance of nonanal and decanal in the final products, respectively. Among other chemical compounds, considering the class of ketones, higher quantity of some species, such as 2-butanone, 3-hydroxy, and 2,3-butanedione, were found to be associated to probiotic fermentations of hemp seed drinks. These two compounds are renown to be microbially related to food fermentations; the former is related to that of dough fermentation (Lu et al., 2016), while the latter to fermented balsamic vinegars (Corsini, Castro, Barroso, & Durán-Guerrero, 2019) and beer (Da Silva et al., 2015). The impact of 2-butanone, 3-hydroxy, and 2,3-butanedione on the final products are associated with better texture and aroma. In fact, 2,3-butanedione is used to improve the aroma of butter or to make beer lighter to the palate, while 2-butanone, 3-hydroxy is associated with a pleasant almond aroma and dairy food texture improvement. Another interesting feature on the volatilome of fermented hemp seed-based drinks is the higher concentration of alkanes, such as octadecane, that has a fuel-like smell found in alcoholic hop-based beverages (Roberts, Dufour, & Lewis, 2004) with low odor threshold (Wang et al., 2015). Lastly, the fermentation of hemp seed drinks was characterized by higher proportion of medium-chain FAs, octanoic acid (caprylic acid), and nonanoic acid (pelargonic acid) originating mainly from hemp seed matrix, and short-chain saturated FA, propanoic acid, as well as lactic acid originating from the action of to the microbial starter. In particular, caprylic acid ranges up to 2.5% of total FAs in the lipid matter of hemp seed (Ross, El Sohly, El Kashoury, & El Sohly, 1996). Health benefits of caprylic and pelargonic acids are associated with excessive calorie burning, inducing weight loss (Rego Costa, Rosado, & Soares-Mota, 2012), and binding to -OH of serine residues of ghrelin to activate the hormone and regulate hunger (St-Onge & Jones, 2002). Caprylic acid was recently used in combination with antioxidant compounds, such as trans-resveratrol, to produce esters lipophenols that have stronger and more stable host antioxidant activity (Oh & Shahidi, 2018). Pelargonic acid is found in whey-fermented alcoholic beverages (Dragone, Mussatto, Oliveira, & Teixeira, 2009). However, the aroma of these compounds is not very pleasant. While pelargonic has a waxy, dirty, and cheesy odor and taste (Mosciano, 1989), caprylic acid is known to have a slightly unpleasant rancid-like smell and taste (Martena, Pfeuffer, & Schrezenmeir, 2006).

10.2.3 Hemp flavored beer

(Ascrizzi, Ceccarini, Tavarini, Flamini, & Angelini, 2019) have recently published research data on utilization of a blend of *C. sativa* L. inflorescences for aromatizing artisanal beer. In particular, the proposed concept offered the utilization of inflorescences that are normally regarded as seed threshing residues as further exploitable by-products. The conducted study was based on the fact that hemp could be an ideal flavoring agent in beer brewing due to its peculiar aroma bouquet of hemp essential oil and due to botanical proximity *C. sativa sativa* has to *Humulus lupulus* L. Thus, hemp could reflect *H. lupulus* uses, and in that study was applied to flavor artisanal beer, produced by malting grains instead of using concentrated malt extract (Ascrizzi, Ceccarini, et al., 2019). The addition of the *C. sativa* inflorescences was performed in three stages: (i) at the beginning of the rinsing of the threshers, directly on them, as in the "mash-hemping" technique; (ii) at the end of the boiling phase; and (iii) during the last 10 min of the whirlpooling phase. The development of the beer aroma is a complex process, not only attributable to the presence of flavor compounds and their precursors and their metabolization and inactivation by the yeasts during the brewing process (Briggs, Brookes, Stevens, & Boulton, 2004), but also to a synergistic effect of several compounds and the contribution of minor compounds (Nickerson & Van Engel, 1992). The authors observed that the volatile emission profiles of hemp and control beer were characterized by molecules other than terpenes, which were contained in a quantity of more than 83%. The addition of hemp inflorescences to the mixture generated a richer profile of volatile compounds due to the increment of the monoterpene's relative content. Other compounds, such as ethyl octanoate and hexanoate, were reduced while isopentyl acetate increased with the addition of hemp. The latter compound exhibited a more fragrant and sweeter odor contribution. Myrcene, which was the monoterpene hydrocarbon with the highest relative concentration, increased from 5.6% to 9.8% in the hemp flavored beer. The obtained profile clearly demonstrated the impact of the hemp essential oil, abundantly present in hemp inflorescences. The authors observed that this compound was responsible for an aromatic contribution to the final product, described as sweet and balsamic (Ascrizzi, Iannone, et al., 2019).

10.2.4 Hemp alcoholic beverages

Recently, a group of researchers (Romano et al., 2021) produced mead by adding different parts of industrial hemp plant (inforescences, leaves, and steams) to mead fermentation. Traditional mead is prepared by diluting honey in water and adding yeasts, while alternatives with aromatic herbs, spices, fresh fruit, and juices added during fermentation are also available. Mead contains between 8% and 18% (v/v) ethanol (Iglesias et al., 2014; Pereira, Mendes-Ferreira, Oliveira, Estevinho, & Mendes-Faia, 2013; Vidrih & Hribar, 2016). The fermentation of mead is similar to that of other alcoholic beverages, but the process is slower due to the high fructose concentration (Iglesias et al., 2014). This beverage contains beneficial nutritional compounds; it has favorable effects on digestion and metabolism and has been used in the treatment of anemia and chronic diseases of the gastrointestinal tract (Gupta & Sharma, 2009). The composition and physicochemical parameters of mead fermented by *Saccharomyces cerevisiae* biotype M3/5 and produced with the addition of various parts (inflorescences, leaves, and stems) of *C. sativa* L. at different concentrations were reported (Romano et al., 2021). The polyphenol and CBD level determined at 227 to 256 mg GAE/l and 0.26 and 0.49 mg/kg, respectively, indicated the anxiolytic and neuroprotective properties of the final product. Moreover, the increase in volatile organic compounds (VOCs) in comparison to the control was reported, mainly alcohols, esters, and terpenes. In particular, the addition of hemp accelerated the fermentation with a higher content of ethyl alcohol in the final product (approximately 13.5%). Thus, the addition of hemp-based ingredients yielded a product with an augmented content of bioactive compounds and with a particularly fresh and fruity flavor.

Ascrizzi, Iannone, et al. (2019) demonstrated the utilization of hemp inflorescences to obtain an alcoholic beverage (i.e., liqueur) and its influence on the resulting volatile features. The liqueur was obtained by maceration of hemp inflorescences in pure ethyl alcohol to allow the extraction of aromatic compounds by filtration of the alcohol extract and its dilute to a volume of alcohol equal to 28% with glucose syrup. The volatile compounds of obtained liqueur were dominated by monoterpene hydrocarbons, whose relative content accounted for up to 90.4%. Among these compounds, α-pinene (38.8%) and myrcene (28.0%) made up more than 50% of the total composition, conferring a balsamic nuance, reported as pine-like for the former and sweet for the latter. Also, a woody and herbal scent contribution to this liqueur bouquet was due to β-pinene, which appeared as the third most abundant (12.4%) VOC detected in this headspace. Moreover, the other two more abundant terpenes were β-caryophyllene and limonene, associated with a spicy, clove-like odor note, and a pleasant lemon-like, citrusy aroma, respectively. All previously mentioned terpenes were the result of the hemp flower's contribution to this beverage aroma. On the contrary, the volatile emission profiles of the average commercial liqueurs are mainly composed of non-terpene, especially esters such as ethyl octanoate and decanoate, which are by-products of the fermentation of carbohydrates, with low relative contents of terpenes (Christoph & Bauer-Christoph, 2007; Vázquez-Araújo, Rodríguez-Solana, Cortés-Diéguez, & Domínguez, 2013).

10.3 Hemp protein

10.3.1 Hemp protein flour

Commercial hemp flour (HPF), also known as hemp protein powder, is the refined fraction of hemp meal (HPM). HPM, or press cake, is produced in the form of small cylindrical bars after mechanical pressing of the seed to extract the oil. These bars, after being ground, undergo a series of sieving steps to separate the HPF from the larger particles, which constitute the hemp bran (HPB). While the flour and oil are food products with commercial value, the HPB is normally discarded or used for animal feed. Depending on the kind of seed (whole or dehulled) and the sieving mesh size (usually 350 mm or 250 mm), it is possible to refine flours to different degrees. Hemp seeds are rich in proteins, whose amount and AA composition vary depending on the hemp variety, environmental and agricultural conditions, and postharvest processing (such as shelling), which alters the ratio of seed components (Galasso et al., 2016; House et al., 2010; Malomo, He, & Aluko, 2014; Pojić et al., 2014; Russo & Reggiani, 2015). After the seeds are pressed to extract the oil, almost all the proteins remain in the meal. As evaluated by House et al. (2010), the meal's protein content ranges from 30% to 50% dry weight for different hemp varieties. Depending on the sieving mesh size, proteins can reach higher percentages in the flour. Pojić et al. (2014) separated the HPM into four fractions (> 350, > 250, > 180, and < 180 μm): the nutritional values of meal and fractions are reported in Table 10.2. The two cotyledon-containing fractions (> 180 and < 180 μm) were found to be significantly richer in proteins (41.2% and 44.4% w/w, respectively) than the hull-containing fractions (> 350 and > 250 μm: 10.6% and 20.3% w/w, respectively) and the whole meal (27.9% w/w). Other nutritional components, which are present in lower percentages than proteins, include lipids, carbohydrates, and fiber (Table 10.2). The FA distribution of

TABLE 10.2 Proximate analysis of hemp meal fractions (Pojić et al., 2014).

(%)	Whole meal[a]	Hemp meal fractions[a]			
		> 350 m	> 250 m	> 180 m	> 180 m
Moisture content	6.98 ± 0.01b	6.63 ± 0.04a	7.39 ± 0.04c	7.34 ± 0.02c	7.88 ± 0.06d
Protein content	10.6 ± 0.10a	20.3 ± 0.25b	41.2 ± 0.04d	44.4 ± 0.02e	27.9 ± 0.12c
Lipid content	8.26 ± 0.02a	10.0 ± 0.05b	15.1 ± 0.02d	18.6 ± 0.04e	11.8 ± 0.01c
Total sugar content	< 0.05 a	0.56 ± 0.08a	4.96 ± 0.11d	3.46 ± 0.08c	1.49 ± 0.08b
Ash content	3.46 ± 0.02a	5.51 ± 0.06b	9.60 ± 0.01d	9.83 ± 0.01e	6.74 ± 0.02c
Crude fiber content	29.5 ± 0.04e	21.3 ± 0.03d	7.13 ± 0.04b	4.96 ± 0.01a	17.3 ± 0.03c

[a] *Values are means ± SD. Means in the same row, followed by different letters are significantly different (P < .05).*

the lipid component was similar across the different fractions, with a relatively high content of the two EFAs—linoleic acid (18:2 ω-6, 54.1%–55.4%) and α-linolenic acid (18:3 ω-3, 17.3%–18.4%). Other FAs were the oleic (18:1 ω-9, 12.7%–13.9%), palmitic (16:0, 6.48%–7.05%), stearic (18:0, 3.18%–3.86%), and γ-linolenic (18:3 ω-6, 2.57%–2.75%). According to WHO/FAO/UNU (2007) guidelines, the ratio of polyunsaturated to saturated FAs (PUFA/SFA) in a balanced diet should be above 0.40; the mean ratio of all meal fractions ranged from 6.02 to 7.14, indicating a highly favorable PUFA/SFA ratio (Pojić et al., 2014). Some antinutritional compounds (primarily phytic acid—but also trypsin inhibitor, glucosinosilates, and condensed tannins) have been isolated in HPM, mostly in the cotyledon fractions (Pojić et al., 2014). These compounds may interfere with protein digestibility, organoleptic properties, and bioavailability of macro- and micro-elements. The adverse effect of phytic acid has been attributed to its strong ability to chelate and precipitate multivalent metal ions, decreasing the availability of these minerals. However, certain health-promoting and disease-preventing properties of phytic acid have recently been recognized (Omoruyi et al., 2013), so its overall effect (and that of other antinutritional compounds) should be further evaluated. In any case, selecting dioecious hemp varieties—those that contain lower amounts of antinutritional compounds than monoecious varieties (Russo & Reggiani, 2015)—and fractionating them by sieving or by other dry fractionation techniques (Pojić et al., 2018) could reduce the deleterious effects of these compounds.

10.3.2 Nutritional quality of hemp proteins

The two main hemp seed proteins are edestin (a legumin class of protein), accounting for approximately 60% to 80% of the total protein content, and albumin, for the remaining 20% to 40%. Edestin is a globular protein with a molecular weight (MW) of approximately 300 kDa (Wang et al., 2008). It is composed of six identical subunits, each consisting of an acidic (AS) and basic (BS) subunit linked by one disulfide bond. The AS is approximately 34 kDa and relatively homogeneous, while the BS consists mainly of two subunits of about 20 and 18 kDa (Wang & Xiong, 2019). Albumin is also a globular protein, with fewer disulfide bonds than edestin. Hence it has a more open (flexible) structure (Malomo & Aluko, 2015a). Additionally, a methionine- and cysteine-rich seed protein (10 kDa protein, 2S albumin) has also been isolated from hemp seed (Odani & Odani, 1998). The protein consists of two polypeptide chains, a small one with 27 AA residues and a large one with 61; the two are held together by two disulfide bonds. In human nutrition, the quality of a protein is defined by (1) the relative contribution of AAs to an individual's AA requirement and (2) the protein's digestibility (House et al., 2010). The AA composition of hemp proteins (%, g/100 g of protein) in whole and dehulled seeds and in the meal from different hemp varieties is reported in Table 10.3 (no data are

TABLE 10.3 Amino acid composition (%, w/w) of hemp seed (whole and dehulled) and hemp meal.

Amino acids	Hemp seed			Hemp meal			FAO/WHO/UNU[b]
	Whole[c]	Dehulled[c]	HPM1[c]	HPM2[d]	HPM[e]		
Ala	4.5	4.4	4.7	4.2	5.2		
Arg	10.7	13.3	11.4	13.2	12		
Asx	11.2	10.7	10.7	10.8	10.9		
Cys	1.9	1.9	2.1	1.5	1.7		
Glx	17.6	18.2	17.6	18.5	17.2		
Gly	5	4.7	4.9	4.7	4.9		
His[a]	2.6	2.8	2.7	3.3	2.9	1.8	
Ile[a]	3.6	3.8	4.2	3.6	4.1	3.1	
Leu[a]	7	6.2	6.9	6.9	6.9	6.3	
Lys[a]	4	3.7	3.9	3.9	4.2	5.2	
Met[a]	2.6	2.7	2.6	2.3	2.4		
Phe[a]	4.8	4.2	4.7	4.7	4.7		
Pro	4.2	4.7	4.6	4.4	4.7		
Ser	5.6	5	5.1	5.7	5.3		
Thr[a]	4.8	3.7	3.9	3.8	3.3	2.7	
Trp[a]	1.1	1.1	1.1	1.1	0.8	0.7	
Tyr	3.2	3.7	3.4	3	3.5		
Val[a]	5.4	5.2	5.6	4.7	5.3	4.2	
AAA	8	7.9	8.1	7.7	8.2	4.6	
SAA	4.6	4.6	4.6	3.8	4.1	2.6	

[a] Essential amino acids.
[b] Essential amino acid recommendation by WHO/FAO/UNU for pre-school child (2007).
[c] House et al. (2010) (values are the mean from hemp varieties: USO 31, USO 14, Crag, Finola).
[d] Malomo and Aluko (2015b).
[e] Russo and Reggiani (2015) (values are the mean from hemp varieties: Carmagnola, CS, Fibranova, Futura 75, Felina 32, Ferimon).
Asx, Glx, sum of Asn and Asp, and Gln and Glu, respectively; AAA, aromatic amino acids, Phe and Tyr; SAA, sulfur amino acids, Met and Cys.

available for HPM fractions). Hemp proteins contain all essential AAs required by humans: threonine, methionine, valine, phenylalanine, isoleucine, leucine, tryptophan, histidine, and lysine. A comparison of protein AA profile from egg white, hemp seed, and soybean shows that hemp seed protein is comparable to these other high-quality proteins (Callaway, 2004). Plant storage proteins are often low in lysine (cereals) and sulphurated AAs (legumes). As evaluated by Russo and Reggiani (2015), the mean content of lysine in HPM is 4.2%, while the content of methionine and cysteine is 2.4% and 1.7%, respectively. The Fibranova variety showed a methionine content of 2.8%, significantly above the mean.

The mean amount of lysin (4.2%) is below the WHO/FAO/UNU (2007) reference for pre-school children (5.2%). House et al. (2010) calculated the respective AA scores of hemp protein (a value that reflects the extent to which a dietary protein meets the needs of an individual for a particular AA: scores of 1.0 or greater for an AA indicate that this AA is not limiting relative to requirements). Lysine (score 0.5 to 0.62) was the most limiting AA in hemp protein, followed by leucine (0.91) and tryptophan (0.91). It should be noted that the low lysine content of hemp protein positions it below the reference protein casein (1.51) but in the same range as the main cereal grains (whole wheat = 0.44; corn = 0.54). The mean value of total sulfur-containing AA (4.1%), as estimated by Russo and Reggiani (2015), is higher than the reference parameter established by WHO/FAO/UNU (2007) (2.6%). Similarly, the methionine + cysteine score (1.46 to 1.72) is higher than 1 and in the range of the reference casein (1.50) (House et al., 2010). Of the other nonlimiting essential AAs, threonine was higher in the monoecious varieties (Futura 75, Felina 32, Ferimon) than the dioecious varieties (Carmagnola, CS, Fibranova), as evaluated by Russo and Reggiani (2015). Among the nonessential AAs, glutamate and arginine are predominant. Arginine accounts for approximately 12% of hemp seed proteins compared to less than 7% for most of other food proteins, including the proteins from potato, wheat, maize, rice, soy, rapeseed, egg white, and whey (Callaway, 2004). Arginine is a dietary precursor for the formation of nitric oxide (NO), a potent mediator of vascular tone and therefore very important for the health of the cardiovascular system. The Arg/Lys ratio, a determinant of the cholesterolemic and antherogenic effects of a protein, ranges from 3.0–5.5 in hemp protein, remarkably higher than that of SPI (1.41) or casein (0.46). Thus, hemp protein is particularly valuable as a nutritional and bioactive ingredient for the formulation of foods that promote cardiovascular health (Wang & Xiong, 2019).

The degree of digestion of dietary proteins depends on enzyme accessibility, which is affected by the molecular structure as well as other components associated with proteins (Wang & Xiong, 2019). Using a rat bioassay, House et al. (2010) evaluated the true protein digestibility percentage (TDP) of HPM proteins along with their protein digestibility-corrected AA score (PDCAAS), which is the TDP corrected for the limiting AA (lysine, in hemp's case). The PDCAAS parameter has been adopted by WHO/FAO/UNU (2007) as the preferred method for measuring a protein's value in terms of human nutrition. Casein (TDP = 97.6%; PDCAAS = 100%) was used as the reference protein. The mean values for three hemp varieties (USO-31, Finola, and Crag) were 86.7% (TDP) and 48% (PDCAAS). These values are similar to those of major pulse protein sources (such as lentils and pinto beans) and above those of cereal grain products (such as whole wheat). The differences in the mean TDP and PDCCAAS values of whole hemp seeds (85.2% and 51%, respectively) and dehulled hemp seeds (94.9% and 61%, respectively) indicate that the hull's presence significantly decreases protein digestibility. The fractionation of hemp seed material is a good strategy to separate the fractions that reduce the protein digestibility (e.g., hull), which consequently greatly increase the protein digestibility of the flour. Hemp proteins have an interesting AA profile showing good digestibility with all essential AAs (although slightly poor in lysine) and elevated levels of arginine. Therefore, breeding efforts to enhance lysine content of hemp proteins, but also to decrease the level of antinutritional factors, would increase the value of hemp proteins for human consumption (Russo & Reggiani, 2015).

10.3.3 New products from HPM processing: Protein concentrate, isolate, and hydrolysates

The protein content of HPF varies depending on different parameters, as reported in the previous sections. Several biotransformation processes, starting from the meal, have been developed to obtain products with higher protein content. One of them consists of the removal of the fiber by treating the flour with carbohydrases and then passing the digested material through a 10 kDa ultrafiltration membrane (Malomo & Aluko, 2015b). The retentate, hemp protein concentrate (HPC), reached a protein content of 74% (about double that of the original meal) and showed a protein digestibility significantly higher than that of HPM. The addition of phytase to the carbohydrases removed phytic acid, one of the antinutritional compounds in the meal. The most purified and enriched form of hemp protein product, with a protein content > 90%, is HPI, which is generally produced by alkaline extraction followed by isoelectric precipitation. The pH of the alkaline extraction is generally 9–10, higher than that of legume protein extraction (pH = 8), because native hemp seed proteins are tightly compacted and may be closely integrated with other components such as phenolic compounds (Wang & Xiong, 2019). The protein extraction techniques yield proteins with different denaturation rates, so the resulting proteins differ in their technofunctional properties (solubility and surface/interfacial activity). These differences are reflected in their gelling, emulsifying, and stabilizing properties, and thus in their applications (Dapčević-Hadnađev et al., 2019; Dapčević-Hadnađev, Hadnađev, Lazaridou, Moschakis, & Biliaderis, 2018). Hadnađev et al. (2018) emphasized that hemp proteins isolated by micellization are well suited for use in high-protein food products (e.g., protein supplements for physically active people) due to their light color and bland taste. On the other hand, proteins isolated by alkaline extraction/isoelectric precipitation appear to be more suitable as thickening and texturizing agents (in food and non-food systems) due to their ability to retain large amounts of water. Furthermore, the slightly acidic pH conditions (5.5–6.5) did not alter the subunit composition of hemp protein (at high pH, the disulfide bonds between some subunits may be destroyed). The recovery yield was, however, lower than that obtained with the common alkaline extraction/isoelectric precipitation method. In addition to their nutritional role, proteins are responsible for various physiological activities relevant for health promotion. Most of these activities are performed by peptide sequences encrypted in the parent protein that become active when cleaved intact. The release of these peptides can be obtained through the action of digestive enzymes (gastrointestinal digestion), during food processing (ripening, fermentation, cooking) or storage, or by in vitro hydrolysis—by proteolytic enzymes or microbial fermentation (Rizzello et al., 2016). Most of the studies on bioactive peptides from hemp proteins have focused on protein extracts from hemp seed or meal and their hydrolysis by proteolytic enzymes. The effects detected for hemp protein enzymatic hydrolysates, by both *in vitro* and *in vivo* studies, include antioxidant, antihypertensive, antiproliferative, hypocholesterolemic, anti-inflammatory, and neuroprotective (Farinon et al., 2020; Hadnađev et al., 2018, and references therein). Recently, three biorefinery processes (enzymatic hydrolysis, liquid fermentation by *Lactobacillus* spp., and solid-state fermentation by *Pleurotus ostreatus*) transformed HPB, a by-product of the hemp seed food-processing chain, into products with antioxidant and antihypertensive properties (Setti et al., 2020).

10.3.4 HPF food application

In the past, food prepared from hemp used to be part of the diet of the low and middle class. Since hemp seed proteins and dietary fibers promote a satiation effect, hemp food was used as suitable food for manual workers (Švec & Hrušková, 2015). Today, hemp-based food products have created a new market niche in favor of consumers who require natural, sustainable, healthy products with high nutritional value. The increased attention given to these products has pushed food manufacturers to develop a wide range of HPF-containing food for retail, particularly bakery products such as bread, cookies, and crackers. HPF and other hemp seed derivatives have also been used in other categories of food, such as energy bars, beverages, processed meat products, dairy products, and infant formula. The addition of HPF to wheat flour significantly ameliorated the nutritional value of bread, particularly by increasing the total protein content (Pojić, Dapčević Hadnađev, Hadnađev, Rakita, & Brlek, 2015; Švec & Hrušková, 2015). Enrichment with HPF also leads to a higher content of macroelements (especially Mg and Ca) and micro-elements (such as Mn, Cu, Fe, and Zn). At the same time, the amounts of Na, complex carbohydrates, and total metabolizable energy of the bread significantly decreased (Pojić et al., 2015). This nutritional improvement was more marked for the 20% hemp seed flour (HSF)-enriched bread than bread enriched with lower percentages. Other advantages of HPF addition include increased dietary fiber (which, however, was higher when using HPM) (Švec & Hrušková, 2015) and total phenolic content, particularly epicatechin, ferulic acid, and protocatechuic acid (Mikulec et al., 2019). However, the addition of HPF or HPM can negatively affect the rheological properties of dough and sensory properties of bread (the latter due to the bitter taste of the hemp source). These effects will not be explored here, but the literature generally agrees that using 10% to 15% HSF should provide an optimal balance between rheological qualities of dough and sensory and nutritional properties of bread (Farinon et al., 2020).

10.4 Hemp seeds for food products

10.4.1 Gluten-free bakery products

Celiac disease (CeD) is a permanent inflammatory disease of the small intestine caused by an autoimmune reaction to gluten in genetically susceptible individuals. This disorder represents the most frequent food intolerance and affects about 1% of the population in Europe (Allen & Orfila, 2018). People suffering from this disease are forced to avoid products containing gluten, replacing them with gluten-free (GF) products. However, GF products are characterized by poor consumer acceptability and qualitative and nutritional deficiencies

(Naqash, Gani, Gani, & Masoodi, 2017), as they are low in proteins and fibers and rich in fats, sugars, and salt (Melini & Melini, 2019). To increase the nutritional value of GF baked goods, naturally GF cereals, such as rice, or pseudo-cereals, such as quinoa, rich in essential minerals and AAs, can be used (Houben, Höchstötter, & Becker, 2012). Another strategy is to use various seeds, such as chia or hemp seeds. Chia seeds (*Salvia hispanica* L.) boast a high nutritional value thanks to the presence of bioactive compounds, such as vitamin E, carotenoids, omega-3 fats, and high fiber content. Although the European Commission (2013) has approved the use of chia seeds as a novel food ingredient, there is a limitation on their use in bread products, set at no more than 10%. HSF represents a potential ingredient for GF bakery products. In addition to its high protein content (33% w/w), HSF is characterized by a low saturated fat content (4.5% w/w) and a good percentage of PUFAs (38% w/w), including ω-3 (8% w/w), and ω-6 (27% w/w). HSF is naturally cholesterol-free, has a low glycemic index, 4% (w/w) of dietary fiber, and is rich in magnesium (0.7% w/w) (Ross et al., 1996). Furthermore, HSF contains around 300 mg/kg of polyphenols and a wide range of bioactive molecules such as flavonoids, terpenes, and potentially cannabinoids, which have surprising antioxidant activity *in vitro* and *ex vivo* (Frassinetti et al., 2018). Sourdough fermentation provides optimum pH for enzymatic degradation of antinutrients—phytic acid, condensed tannins, and total saponins resulting in better bioavailability of iron, zinc, and calcium and better digestibility (Gobbetti, Cagno, & De Angelis, 2010; Nionelli et al., 2018). Depending on the flour used, during fermentation, the LAB in the sourdough can produce a remarkable mixture of organic acids that, together with the other metabolites, can inhibit the development of pathogenic bacteria and show fungicidal or fungistatic activity against some fungi or yeasts responsible for the deterioration of bread (Cizeikiene, Juodeikiene, Paskevicius, & Bartkiene, 2013).

In addition to organic acids, during fermentation, microorganisms synthesize numerous other VOCs that interact with the food matrix. VOCs include molecules belonging to different chemical classes, such as esters, acids, alcohols, aldehydes, etc., which possess, in addition to the known flavoring properties, beneficial properties for human health, including anticancer and antiobesity properties (Ayseli & Ipek Ayseli, 2016; Mota-Gutierrez, Barbosa-Pereira, Ferrocino, & Cocolin, 2019). Nionelli et al. (2018) showed that it is possible to increase *in vitro* digestibility of HSF proteins by fermenting the industrial HPF with LAB. Furthermore, subjects suffering from CeD also show problems with malabsorption of B vitamins, iron, copper, zinc, and other minerals (Kupper, 2005). It is therefore of fundamental importance to increase the bioavailability of these minerals in GF products. To confirm this, a study by Di Cagno et al. (2008) showed that sourdough bread, characterized by a greater phytase activity, was distinguished by a greater availability of micronutrients. In fact, sourdough represents an excellent strategy to improve the quality of GF baked goods thanks to its ability to increase volume and improve flavor, thus meeting the acceptability of consumers.

HPF is suitable for celiac patients, because of the lack of the gluten fractions of protein. In fact, 10% to 20% of the GF corn starch in GF bread can be replaced with HPF or HPC (Korus, Witczak, Ziobro, & Juszczak, 2017); in GF cookies, even 20% to 60% can be replaced with HPF (Korus, Gumul, Krystyjan, Juszczak, & Korus, 2017), leading to significant improvements in the nutritional value of the GF food. In particular, this change greatly increased the amount of total proteins and total dietary fiber (both soluble and insoluble, but especially the insoluble fraction) as well as fats and minerals. Furthermore, HSF-substituted cookies showed significantly higher values of total phenolic and flavonoid compounds compared to the control cookies—reflected in up to three times higher total antioxidant capacity (Korus, Witczak, et al., 2017). Similar results regarding nutritional and antioxidant properties were obtained for GF crackers in which brown rice flour was partially substituted with 10% to 40% of HSF (Radočaj, Dimić, & Tsao, 2014); the addition of HPF increased the protein, fiber, mineral, and EFA content more than that obtained by pulse flour enrichment. In another work where different fermentation processes were compared in a GF bakery (Nissen, Bordoni, & Gianotti, 2020), the addition of HSF and the use of different fermentation types gave rise to specific VOCs profile predicting the organoleptic characteristics of bread. In sourdough bread, pleasant almond-like and lemon-like notes could be related to HSF addition or *S. cerevisiae* fermentation, respectively. In addition, increased traits of herbal, fruity, and floral notes (minor alcohols) could be provided by HSF addition both in yeast- and LAB-fermented bread, and sour and buttery nuances due to propanoic acid could be predicted in sourdough bread (Nissen, Bordoni, & Gianotti, 2020). Therefore, HSF or proteins may be considered valuable ingredients for the formulation of improved GF products.

10.4.2 Conventional bakery products

The market is increasingly directed toward the placing of innovative versions of bakery products, based on the integration of traditional flours with nontraditional ones rich in bioactive compounds (Dini, Garcia, & Viña, 2012). Food industry by-products lend themselves well to the fortification of baked goods (Kohajdová, Karovičová, & Jurasová, 2012; Kumar & Kumar, 2011; O'Shea, Doran, Auty, Arendt, & Gallagher, 2013) as sources of protein, antioxidants, and dietary fiber. For example, residues from cold-pressed oil are a valuable source of protein and FAs (Parry, Cheng, Moore, & Yu, 2008) and, among these, one of the most promising cold-pressed seed cakes is what remains from the pressing of hemp seeds. Pojić et al. (2015) examined the possibility of integrating this by-product in bread making with the aim of enriching white bread with proteins and other health-beneficial compounds. Although the addition of HPF reduced the amount of gluten in the final flour, the stability and strength of the dough were not affected by the addition of hemp up to 10%, while a portion of 20% resulted in the processing. Furthermore, the addition of HPF raised the nutritional value of the bread, bringing in particular proteins and iron (Pojić et al., 2015).

10.4.3 Hemp-based snack products

Energy bars, which mainly provide proteins, fats, and carbohydrates, are widely consumed among athletes to meet their calorie needs. The bars are commonly produced with corn, wheat, soy, and rice in an extrusion process. Norajit, Gu, and Ryu (2011) added extruded and defatted hemp in different percentages to the rice flour and evaluated its physicochemical properties, the effect on nutritional compounds, and the antioxidant activity of rice flour. The different levels of hemp powder resulted in variations in the physical and mechanical properties and chemical composition of the extruded rice/hemp products. In particular, the total content of phenols and flavonoids increased following the addition of hemp, reaching the best result of antioxidant activity when 40% of whole hemp was added. While the flavor scores of all hemp powder-blended extruded rice energy bars were higher than those of the hemp-free control bar, higher hemp levels (20%–40%) led to lower taste scores, although no significant differences were observed in overall acceptance. Overall, the energy bar was more appreciated when formulated with whole hemp rather than defatted hemp seed powder. Expanded snack products are also very popular among children and, as these are high-calorie but low-nutritional products (Paraman, Sharif, Supriyadi, & Rizvi, 2015), there are widespread attempts to improve their nutritional value by adding value-added ingredients, like polyphenols, vitamins, fibers, fats, etc. (Jozinović, Šubarić, Ačkar, Babić, & Miličević, 2016). Hemp is well suited for this purpose thanks to its excellent nutritional composition. Jozinović et al. (2017) evaluated the effect of adding defatted hemp cake in the production of corn snacks by varying hemp cake level added to corn grits, the moisture content of the mixtures and the temperature in the ejection zone of extruder. The results showed that the increase in the moisture content and the hemp cake level reduced the fracturability of the extrusions and increase the total color change of the extrudates, while the temperature did not affect this parameter. In conclusion, the defatted hemp cake can be included with excellent results in corn grits to produce snacks, while further studies are needed to evaluate the impact on nutritional value.

10.4.4 Hemp-based meat analogs

The Western diet is based on high consumption of products of animal origin, questioned for their negative impact on consumer health and on the environment (Battaglia Richi et al., 2015). In fact, the high consumption of meat increases the risks of incidence of cardiovascular diseases, type 2 diabetes, and various types of cancer (Battaglia Richi et al., 2015). Furthermore, animal production is responsible for 14.5% of global greenhouse gas emissions (Gerber et al., 2013). A plausible solution to these challenges is given by the vegan/vegetarian diet, with high potential for reducing the environmental impact and promoting a healthier lifestyle (EAT-Lancet Commission, 2019; Gerber et al., 2013). The increased consumers' awareness of the advantages of a plant-based diet resulted in increased turning to vegan or vegetarian diets, thus promoting the market growth in the plant-based food sector. Vegan/vegetarian alternatives are now available for different animal-based foods varying from milk to meat analogues. However, while the market for milk analogues is already widespread, the market for meat substitutes is still quite small, but with projected growth in the future. Thus, it is expected that meat substitutes market will reach 2.4 billion euros in 2025, led by Europe with 40% of the total market, compared to 1.5 billion euros in 2018 (Deloitte, 2019).

The most popular meat substitutes on the market are produced from soy (i.e., tofu, tempeh) and wheat protein (i.e., seitan), but there are also products based on mycoproteins and textured vegetable proteins. A new alternative product to meat, although still not widespread on the market, is high humidity meat analogue (HMMA), a plant-based product designed to mimic whole muscle meat. The HMMA exhibits similar characteristics to meat for moisture, protein, and fat contents; has a rubbery consistency similar to that of meat; and has a densely layered, somewhat fibrous structure. Most meat replacement products are composed of soy, which is imported in large quantities to meet the large demand, while only a small part is produced with other vegetable protein sources, such as wheat, peas, broad beans, lupins, or mushrooms. This means that the advantages deriving from the sustainability of the production of plant matrices can be lost, and therefore it is necessary to find other crops that can be cultivated in most of the world to reduce the food import dependency (Zahari et al., 2020).

Since the most important ingredient for obtaining HMMA is protein, industrial hemp is of high potential to replace soy in HMMA. In particular, HPC could be substituted for SPI as demonstrated by Zahari et al. (2020), who replaced SPI with HPC, in different percentages, specifically 20%, 40%, and 60% by weight. The results showed that HPC was able to absorb less water and required higher processing temperatures than SPI to achieve denaturation. HMMA is produced by high-moisture extrusion processing, during which protein denaturation, unfolding, realignment, and cross-linking occur due to the shear, heat, pressure, and cooling applied. Zhang et al. (2019) emphasized the complexity of conformational changes and molecular interactions occurring between proteins, carbohydrates, lipids, and other nutrients being responsible for the final quality of texturized plant proteins, essential for consumer acceptability (Zahari et al., 2020; Zhang et al., 2019). Hemp appeared to be a suitable raw material for this type of application as hemp proteins are characterized by high quality and functionality (Raikos, Neacsu, Russell, & Duthie, 2014; Tang, Ten, Wang, & Yang, 2006; Wang & Xiong, 2019).

10.4.5 Hemp-based biorefinery products

The agri-food industry produces large quantities of waste with negative implications for the environment, economy, and society (Torres-León et al., 2018). However, agri-food waste is rich in nutrients , namely proteins, polysaccharides, lipids, dietary fibers, and other bioactives, which can be reutilized through biorefinery processes. The recovery of compounds from food by-products is an excellent strategy to overcome the problems related to the depletion of natural resources, caused by overpopulation and the lack of arable land, climate change, and environmental pollution. The higher food demands caused by a steady increase in global population, reaching 10 billion by 2050, as well as the necessity to assure sustainability in both agricultural and industrial production, caused the search for alternative and sustainable protein sources (Pojić, Mišan, & Tiwari, 2018). For the recovery of proteins from by-products, different techniques are available—such as chemical extraction, based on the solubility of proteins at alkaline pH, liquid fermentation by microorganisms, and solid-state fermentation (SSF) by fungi (Baiano, 2014). After extraction, the protein extracts can be further enzymatically hydrolyzed to bioactive peptides, which increases their functionality, in terms of anticancer, antimicrobial, hypocholesterolemic, and immunostimulant effects (Martínez-Alvarez, Chamorro, & Brenes, 2015; Rizzello et al., 2016). The release of bioactive peptides can be carried out by microbial fermentation, for example, with *Lactobacillus* spp. These microorganisms have an efficient proteolytic system and a high adaptability to different environments and food matrices (Hafeez et al., 2014; Raveschot et al., 2018; Rizzello et al., 2016). It was shown that the lactic acid fermentation was able to increase the functionality of cocoa beans by improving the procyanidin level and their antiradical activity (Di Mattia et al., 2013), as well as to increase the prebiotic potential of fibers from different plant sources (Garcia-Amezquita, Tejada-Ortigoza, Serna-Saldivar, & Welti-Chanes, 2018; Nissen, Bordoni, & Gianotti, 2020; Sánchez-Zapata et al., 2013).

A final example of biotransformation is solid state fermentation (SSF) by fungi. For example, some studies have shown that SSF with fungi can increase the nutritional value and biological capacity of cereals (Subramaniam, Sabaratnam, Kuppusamy, & Tan, 2014; Zhai, Wang, & Han, 2015). Furthermore, this biotechnology allows the production of enzymes from food industry by-products, such as wheat bran, rice straw, banana waste, and tea waste (Oberoi, Chavan, & Bansal, 2010). In particular, *Pleorotus ostreatus*, through the SSF fermentation of solid food waste, can obtain specific enzymes (Akpinar & Urek, 2012; Fernández-Fueyo et al., 2014; Zilly et al., 2012). Setti et al. (2020) compared the effectiveness of the three biorefinery bioprocesses (chemical extraction, liquid fermentation by microorganisms, solid fermentation by fungi) for the recovery of bioactive compounds from an industrial by-product of hemp (*C. sativa* L.). In fact, at an industrial level, hemp seeds are pressed to produce edible oil and flour, which is then sieved several times to separate the particles of different sizes. While the small particles are collected to be marketed as HPM, the large particles or HPB constitute a by-product, which is discarded or used for animal feed despite being particularly rich in proteins and other nutrients (Pojić et al., 2014). In the study by Setti et al. (2020), liquid fermentation was carried out with *Lactobacillus* spp., while SSF was carried out with *P. ostreatus*. The results showed different levels of antioxidant activity (including ABTS radical scavenging, Fe^{2+} chelating capacity, and ferric reducing power) and antihypertensive activity (evaluated by means of in vitro assays). In particular, the best results were those of the antihypertensive activity, as the values obtained were comparable to those of the protein hydrolysates of the whole seeds. This indicates that industrial processing does not impact bioactivity, and these compounds are also retained in the final products. Although the best result of ACE inhibition was obtained thanks to the treatment with pepsin, a good result was also observed with the samples fermented by *Lactobacillus* spp. and with SSF by *P. ostreatus* (even after 9 days of incubation). These results highlight the potential of HPB for obtaining products with high added value, which would drive further growth in the hemp seed market in the food, cosmetic, and pharmaceutical sectors.

10.5 Challenges and opportunities

The hemp seed food market sector is growing fast, but processing on a large scale should not be in contrast with the environmentally friendly reputation that characterizes this feedstock. In this view, LAB transformation represents an excellent sustainable strategy to ameliorate hemp seed-based foods (Nissen et al., 2019; Nissen, Bordoni, & Gianotti, 2020; Nissen, di Carlo, & Gianotti, 2020). Safety procedures, such as microbial decontamination, should be conducted with treatments less invasive and destructive than UHT treatments, such as those offered by emerging thermal and nonthermal processing technologies already applied for vegetable-based beverages, such as high-pressure processing, pulsed electric fields, ultrasound, pulsed-light, ohmic heating, etc. Thus, it could be possible to prepare hemp seed-based milk with preserved nutrients and original clean flavor. Moreover, the novel processing technologies could also be applicable as alternative technologies for the modification of the structure of proteins and modulating their technofunctional properties. The potential of high-pressure homogenization (HPH) has already been demonstrated to aid the preparation of stable O/W food emulsions (Fernández-Ávila, Escriu, & Trujillo, 2015). HPH allows mechanical dispersion of oil into fine droplets resulting in an increased total surface area and a uniform size distribution with an improved stability (Briviba, Graf, Walz, Guamis, & Butz, 2016; Lee, Lefèvre, Subirade, & Paquin, 2009). The protein solubility is one of the most important properties in developing hemp-based food products since it affects the applicability of hemp-based materials in food products, as well as other functional properties (Yin et al., 2008).

Since hemp proteins are characterized by low solubility in comparison to proteins from other sources of plant origin, technological solutions that will increase the protein isolation technique, the enzymatic processing—hydrolyses with trypsin (Yin et al., 2008), cellulase, hemicellulose, xylanase, and phytase (Malomo & Aluko, 2015a)—have been demonstrated to successfully increase hemp protein solubility. Thus, the higher implementation of novel processing technologies in the design of hemp-based food will satisfy the demands for whole or minimally processed foods and *free-from* foods—trends that have dominated food markets in recent years (Barry-Ryan, Vassallo, & Pojić, 2020). The demand for cost-effective alternative technologies and additive-free food products is on a constant rise.

10.6 Conclusion

Hemp seeds represent a versatile source of food nutrients, so the derived products are nutritionally impressive, too, characterized not only by a proper composition of nutrients but also a favorable balance. The products obtained are nutritionally impressive, bearing not solely a proper composition of nutrients but even a fine balance of these. Bioactive compounds are found to be plentiful, and the most of them are renowned to contribute to well-being. The fiber composition is able to foster the growth of beneficial microbes in the human gut as well as contain the growth of microbial pathogens. Hemp-based products are niche market products, especially suitable for consumers with special needs—lactose intolerant consumers due to similarity of the hemp-based drink with cow's milk, celiac patients due to the absence of gluten, and athletes due to perfectly tuned AA composition. Notwithstanding, even if niche consumers are keen to buy such products, the everyday consumers are not. The reasons are to be found in the general misinterpretation of the nature of hemp seeds, as well as on the uneasy distinction between hemp chemotypes. The fact that cannabinoids are not present in hemp seeds and that cultivation and commercialization of chemotype 3 (the "fiber type") is not illegal is still not understood by most, and another generation may pass prior to global acceptance.

The second challenge is unacceptable sensory properties, because the taste and product appearance is far from that of normal standards. Moreover, with an aim to improve the product acceptability, the sustainable and eco-friendly nature of hemp seed should not be corrupted during food transformation by chemicals or invasive or energy-demanding processing.

References

Abou-Dobara, M., Ismail, M., & Refat, N. (2016). Chemical composition, sensory evaluation and starter activity in cow, soy, peanut and rice milk. *Journal of Nutritional Health & Food Engineering, 5*, 1–8.

Akpinar, M., & Urek, R. O. (2012). Production of ligninolytic enzymes by solid-state fermentation using *Pleurotus eryngii*. *Preparative Biochemistry & Biotechnology, 42*(6), 582–597.

Allen, B., & Orfila, C. (2018). The availability and nutritional adequacy of gluten-free bread and pasta. *Nutrients, 10*(10), 1370. https://doi.org/10.3390/nu10101370.

Ascrizzi, R., Ceccarini, L., Tavarini, S., Flamini, G., & Angelini, L. G. (2019). Valorisation of hemp inflorescence after seed harvest: Cultivation site and harvest time influence agronomic characteristics and essential oil yield and composition. *Industrial Crops and Products, 139*, 111541. https://doi.org/10.1016/j.indcrop.2019.111541.

Ascrizzi, R., Iannone, M., Cinque, G., Marianelli, A., Pistelli, L., & Flamini, G. (2019). "Hemping" the drinks: Aromatizing alcoholic beverages with a blend of *Cannabis sativa* L. flowers. *Food Chemistry, 325*. https://doi.org/10.1016/j.foodchem.2020.126909, 126909.

Ayseli, M. T., & Ipek Ayseli, Y. (2016). Flavors of the future: Health benefits of flavor precursors and volatile compounds in plant foods. *Trends in Food Science & Technology, 48*, 69–77.

Baiano, A. (2014). Recovery of biomolecules from food wastes—A review. *Molecules, 19*(9), 14821–14842. https://doi.org/10.3390/molecules190914821.

Barry-Ryan, C., Vassallo, M., & Pojić, M. (2020). The consumption of healthy grains. In M. Pojić, & U. Tiwari (Eds.), *Innovative processing technologies for healthy grains* Wiley-Blackwell. https://doi.org/10.1002/9781119470182.ch10.

Bartkiene, E., Schleining, G., Krungleviciute, V., Zadeike, D., Zavistanaviciute, P., & Dimaite, I. (2016). Development and quality evaluation of lacto-fermented product based on hulled and not hulled hempseed (*Cannabis sativa* L.). *LWT-Food Science and Technology, 72*, 544–551. https://doi.org/10.1016/j.lwt.2016.05.027.

Battaglia Richi, E., Baumer, B., Conrad, B., Darioli, R., Schmid, A., & Keller, U. (2015). Health risks associated with meat consumption: A review of epidemiological studies. *International Journal for Vitamin and Nutrition Research, 85*(1–2), 70–78. https://doi.org/10.1024/0300-9831/a000224.

Baxter, I. A., Easton, K., Schneebeli, K., & Whitfield, F. B. (2005). High pressure processing of Australian navel orange juices: Sensory analysis and volatile flavor profiling. *Innovative Food Science & Emerging Technologies, 4*, 372–387. https://doi.org/10.1016/j.ifset.2005.05.005.

Booth, J. K., & Bohlmann, J. (2019). Terpenes in *Cannabis sativa* – From plant genome to humans. *Plant Science, 284*, 67–72. https://doi.org/10.1016/j.plantsci.2019.03.022.

Briggs, D., Brookes, P., Stevens, R., & Boulton, C. (2004). An outline of brewing. In *Brewing* (pp. 1–10). https://doi.org/10.1533/9781855739062.1.

Briviba, K., Graf, V., Walz, E., Guamis, B., & Butz, P. (2016). Ultra high-pressure homogenization of almond milk: Physico-chemical and physiological effects. *Food Chemistry, 192*, 82–89.

Broséus, J., Anglada, F., & Esseiva, P. (2010). The differentiation of fibre and drug type Cannabis seedlings by gas chromatography/mass spectrometry and chemometric tools. *Forensic Science International, 200*, 87–92.

Callaway, J. (2004). Hempseed as a nutritional resource: An overview. *Euphytica, 140*, 65–72.

Cani, P. D., & Everard, A. (2015). Keeping gut lining at bay: Impact of emulsifiers. *Trends in Endocrinology and Metabolism, 26*, 273–274.

Chassaing, B., Koren, O., Goodrich, J. K., Poole, A. C., Srinivasan, S., Ley, R. E., et al. (2015). Dietary emulsifiers impact the mouse GM promoting colitis and metabolic syndrome. *Nature, 519*, 92–96. https://doi.org/10.1038/nature14232.

Chichłowska, J., Kliber, A., Kozłowska, J., Biskupski, M., & Grygorowicz, Z. (2009). *Insulin, thyroid hormone levels and metabolic changes after treated rats with hemp milk.* http://www.hempreport.com/pdf/HempMilkStudy%5B1%5D.pdf.

Christoph, N., & Bauer-Christoph, C. (2007). Flavour of Spirit Drinks: Raw Materials, Fermentation, Distillation, and Ageing. In R. G. Berger (Ed.), *Flavours and Fragrances.* Berlin, Heidelberg: Springer. https://doi.org/10.1007/978-3-540-49339-6_10.

Chung, H. Y. (1999). Volatile components in fermented soybean (*Glycine max*) curds. *Journal of Agricultural and Food Chemistry, 47*(7), 2690–2696. https://doi.org/10.1021/jf981166a.

Cizeikiene, D., Juodeikiene, G., Paskevicius, A., & Bartkiene, E. (2013). Antimicrobial activity of lactic acid bacteria against pathogenic and spoilage microorganism isolated from food and their control in wheat bread. *Food Control, 31*(2), 539–545. https://doi.org/10.1016/j.foodcont.2012.12.004.

Corsini, L., Castro, R., Barroso, C. G., & Durán-Guerrero, E. (2019). Characterization by gas chromatography-olfactometry of the most odour-active compounds in Italian balsamic vinegars with geographical indication. *Food Chemistry, 272*, 702–708. https://doi.org/10.1016/j.foodchem.2018.08.100.

Da Cheng, H., Xiao-Jie, G., & Pei, G. X. (2015). Phytochemical and biological research of Cannabis pharmaceutical resources. *Medicinal Plants, 431*–464. https://doi.org/10.1016/B978-0-08-100085-4.00011-6.

Da Porto, C., Decorti, D., & Natolino, A. (2015). Potential oil yield, fatty acid composition, and oxidation stability of the hempseed oil from four *Cannabis sativa* L. cultivars. *Journal of Dietary Supplements, 12*(1), 1–10.

Da Silva, G. C., da Silva, A. A. S., da Silva, L. S. N., Godoy, R. L., Nogueira, L. C., & Quitério, S. L. (2015). Method development by GC–ECD and HS-SPME–GC–MS for beer volatile analysis. *Food Chemistry, 167*, 71–77.

Dapčević-Hadnađev, T., Dizdar, M., Pojić, M., Krstonošić, V., Zychowski, L. M., & Hadnađev, M. (2019). Emulsifying properties of hemp proteins: Effect of isolation technique. *Food Hydrocolloids, 89*, 912–920.

Dapčević-Hadnađev, T., Hadnađev, M., Lazaridou, A., Moschakis, T., & Biliaderis, C. G. (2018). Hempseed meal protein isolates prepared by different isolation techniques. Part II. Gelation properties at different ionic strengths. *Food Hydrocolloids, 81*, 481–489.

De Meijer, E. (2014). The chemical phenotypes (chemotypes) of Cannabis. In *Handbooks in psychopharmacology* (1st ed.). Oxford, UK; New York, NY, USA: Oxford University Press, ISBN:978-0-19-966268-5.

Deloitte. (2019). *Plant-based alternatives-driving industry M&A contents*. London, UK: Deloitte LLP.

Di Cagno, R., Rizzello, C. G., De Angelis, M., Cassone, A., Giuliani, G., & Benedusi, A. (2008). Use of selected sourdough strains of Lactobacillus for removing gluten and enhancing the nutritional properties of gluten-free bread. *Journal of Food Protection, 71*(7), 1491–1495. https://doi.org/10.4315/0362-028x-71.7.1491.

Di Mattia, C., Martuscelli, M., Sacchetti, G., Scheirlinck, I., Beheydt, B., & Mastrocola, D. (2013). Effect of fermentation and drying on procyanidins, antiradical activity and reducing properties of cocoa beans. *Food and Bioprocess Technology, 6*, 3420–3432.

Dini, C., Garcia, M. A., & Viña, S. Z. (2012). Non-traditional flours: Frontiers between ancestral heritage and innovation. *Food & Function, 3*(6), 606–620.

Dragone, G., Mussatto, S. I., Oliveira, J. M., & Teixeira, J. A. (2009). Characterization of volatile compounds in an alcoholic beverage produced by whey fermentation. *Food Chemistry, 112*, 929–935.

EAT-Lancet Commission. (2019). *Healthy diets from sustainable food systems: Food planet health*. Stockholm, Sweden: EAT-Lancet Commission.

El-Sohly, M. A. (2002). Chemical constituents of Cannabis. In F. Grotenhermen, & E. Russo (Eds.), *Cannabis and cannabinoids. Pharmacology, toxicology, and therapeutic potential* (pp. 27–36). New York: The Haworth Press.

European Commission. (2013). Commission implementing decision of 22 January 2013 authorizing an extension of use of chia (*Salvia hispanica*) seed as a novel food ingredient under regulation (EC) no 258/97 of the European Parliament and of the council (2013/50/EU). *Official Journal of the European Union*. Retrieved from https://eur-lex.europa.eu/eli/dec_impl/2013/50/oj.

Farinon, B., Molinari, R., Costantini, L., & Merendino, N. (2020). The seed of industrial hemp (*Cannabis sativa* L.): Nutritional quality and potential functionality for human health and nutrition. *Nutrients, 12*(7), 1935. https://doi.org/10.3390/nu12071935.

Fernández-Ávila, C., Escriu, R., & Trujillo, A. J. (2015). Ultra-high-pressure homogenation enhances physicochemical properties of soy protein isolate-stabilized emulsions. *Food Research International, 75*, 357–366.

Fernández-Fueyo, E., Castanera, R., Ruiz-Dueñas, F. J., López-Lucendo, M. F., Ramírez, L., & Pisabarro, A. G. (2014). Ligninolytic peroxidase gene expression by *Pleurotus ostreatus*: Differential regulation in lignocellulose medium and effect of temperature and pH. *Fungal Genetics and Biology, 72*, 150–161.

Filannino, P., Cardinali, G., Rizzello, C. G., Buchin, S., De Angelis, M., & Gobbetti, M. (2014). Metabolic responses of *Lactobacillus plantarum* strains during fermentation and storage of vegetable and fruit juices. *Applied and Environmental Microbiology, 80*(7), 2206–2215. https://doi.org/10.1128/AEM.03885-13.

Frassinetti, S., Moccia, E., Caltavuturo, L., Gabriele, M., Longo, V., & Bellani, L. (2018). Nutraceutical potential of hemp (*Cannabis sativa* L.) seeds and sprouts. *Food Chemistry, 262*, 56–66.

Galasso, I., Russo, R., Mapelli, S., Ponzoni, E., Brambilla, I. M., & Battelli, G. (2016). Variability in seed traits in a collection of *Cannabis sativa* L. genotypes. *Frontiers in Plant Science, 7*, 688.

Garcia-Amezquita, L. E., Tejada-Ortigoza, V., Serna-Saldivar, S. O., & Welti-Chanes, J. (2018). Dietary fiber concentrates from fruit and vegetable by-products: Processing, modification, and application as functional ingredients. *Food and Bioprocess Technology, 11*, 1439–1463.

Gerber, P., Steinfeld, H., Henderson, B., Mottet, A., Opio, C., & Dijkman, J. (2013). *Tackling climate change through livestock—A global assessment of emissions and mitigation opportunities*. Rome, Italy: Food and Agriculture Organization of the United Nations (FAO).

Gobbetti, M., Cagno, R. D., & De Angelis, M. (2010). Functional microorganisms for functional food quality. *Critical Reviews in Food Science and Nutrition, 50*(8), 716–727. https://doi.org/10.1080/10408398.2010.499770.

Guglielmetti, S., Mora, D., Gschwender, M., & Popp, K. (2011). Randomised clinical trial: *Bifidobacterium bifidum* MIMBb75 significantly alleviates irritable bowel syndrome and improves quality of life—A double-blind, placebo-controlled study. *Alimentary Pharmacology & Therapeutics, 33*, 1123–1132. https://doi.org/10.1111/j.1365-2036.2011.04633.x.

Gupta, J. K., & Sharma, R. (2009). Production technology and quality characteristics of mead and fruit-honey wines: A review. *Natural Products Radiance, 8*(4), 345–355.

Hadnađev, M., Dapčević-Hadnađev, T., Lazaridou, A., Moschakis, T., Michaelidou, A. M., & Popović, S. (2018). Hempseed meal protein isolates prepared by different isolation techniques. Part I. Physicochemical properties. *Food Hydrocolloids, 79*, 526–533.

Hadnađev, M., Dizdar, M., Dapčević-Hadnađev, T., Jovanov, P., Mišan, A., & Sakač, M. (2018). Hydrolyzed hemp seed proteins as bioactive peptides. *Journal on Processing and Energy in Agriculture, 22*, 90–94.

Hafeez, Z., Cakir-Kiefer, C., Roux, E., Perrin, C., Miclo, L., & Dary-Mourot, A. (2014). Strategies of producing bioactive peptides from milk proteins to functionalize fermented milk products. *Food Research International, 63*, 71–80.

Hartsel, J. A., Eades, J., Hickory, B., & Makriyannis, A. (2016). *Cannabis sativa* and hemp. *Nutraceuticals, 735–754.* https://doi.org/10.1016/b978-0-12-802147-7.00053-x.

Houben, A., Höchstötter, A., & Becker, T. (2012). Possibilities to increase the quality in gluten-free bread production: An overview. *European Food Research and Technology, 235*, 195–208. https://doi.org/10.1007/s00217-012-1720-0.

House, J. D., Neufeld, J., & Leson, G. (2010). Evaluating the quality of protein from hemp seed (*Cannabis sativa* L.) products through the use of the protein digestibility-corrected amino acid score method. *Journal of Agricultural and Food Chemistry, 58*, 11801–11807. https://doi.org/10.3389/fpls.2016.00688.

Iglesias, A., Pascoal, A., Choupina, A. B., Carvalho, C. A., Feás, X., & Estevinho, L. M. (2014). Developments in the fermentation process and quality improvement strategies for mead production. *Molecules, 19*(8), 12577–12590. https://doi.org/10.3390/molecules190812577.

Irakli, M., Tsaliki, E., Kalivas, A., Kleisiaris, F., Sarrou, E., & Cook, C. M. (2019). Effect of genotype and growing year on the nutritional, phytochemical, and antioxidant properties of industrial hemp (*Cannabis sativa* L.) seeds. *Antioxidants, 8*(10), 491.

Jeske, S., Bez, J., Arendt, E. K., & Zannini, E. (2019). Formation, stability, and sensory characteristics of a lentil-based milk substitute as affected by homogenisation and pasteurisation. *European Food Research and Technology, 245*, 1519–1531. https://doi.org/10.1007/s00217-019-03286-0.

Johnson, R. (2018). *Hemp as an agricultural commodity; CRS report RL32725.* Washington, DC, USA: Congressional Research Service. Available from: https://fas.org/sgp/crs/misc/RL32725.pdf. (Accessed 12 March 2020).

Jozinović, A., Ačkar, D., Jokić, S., Babić, J., Panak Balentić, J., & Banožić, M. (2017). Optimisation of extrusion variables for the production of corn snack products enriched with defatted hemp cake. *Czech Journal of Food Science, 35*, 507–516.

Jozinović, A., Šubarić, D., Ačkar, D., Babić, J., & Miličević, B. (2016). Influence of spelt flour addition on properties of extruded products based on corn grits. *Journal of Food Engineering, 172*, 1–37.

Kfoury, N., Morimoto, J., Kern, A., Scott, E. R., Orians, C. M., & Ahmed, S. (2018). Striking changes in tea metabolites due to elevational effects. *Food Chemistry, 264*, 334–341.

Kim, M. K., Jang, H. W., & Lee, K. G. (2018). Sensory and instrumental volatile flavor analysis of commercial orange juices prepared by different processing methods. *Food Chemistry, 267*, 217–222.

King, J. W. (2019). The relationship between cannabis/hemp use in foods and processing methodology. *Current Opinion in Food Science, 28*, 32–40. https://doi.org/10.1016/j.cofs.2019.04.007.

Kohajdová, Z., Karovičová, J., & Jurasová, M. (2012). Influence of carrot pomace powder on the rheological characteristics of wheat flour dough and on wheat rolls quality. *Acta Scientiarum Polonorum. Technologia Alimentaria, 11*(4), 381–387.

Korus, A., Gumul, D., Krystyjan, M., Juszczak, L., & Korus, J. (2017). Evaluation of the quality, nutritional value and antioxidant activity of gluten-free biscuits made from corn-acorn flour or corn-hemp flour composites. *European Food Research and Technology, 243*, 1429–1438.

Korus, J., Witczak, M., Ziobro, R., & Juszczak, L. (2017). Hemp (*Cannabis sativa* subsp. sativa) flour and protein preparation as natural nutrients and structure forming agents in starch-based gluten-free bread. *LWT-Food Science and Technology, 84*, 143–150.

Kumar, K., & Kumar, N. (2011). Development of vitamin and dietary fibre enriched carrot pomace and wheat flour based buns. *Journal of Pure and Applied Science & Technology, 2*, 108–116.

Kupper, C. (2005). Dietary guidelines and implementation for celiac disease. *Gastroenterology, 128*(4 suppl. 1). https://doi.org/10.1053/j.gastro.2005.02.024.

LeBlanc, J. G., Ledue-Clier, F., Bensaada, M., de Giori, G. S., Guerekobaya, T., & Sesma, F. (2008). Ability of *Lactobacillus fermentum* to overcome host alpha-galactosidase deficiency, as evidenced by reduction of hydrogen excretion in rats consuming soya alpha-galacto-oligosaccharides. *BMC Microbiology, 8*, 22. https://doi.org/10.1186/1471-2180-8-22.

Lee, S. H., Lefèvre, T., Subirade, M., & Paquin, P. (2009). Effects of ultra-high pressure homogenization on the properties and structure of interfacial protein layer in whey protein-stabilized emulsion. *Food Chemistry, 113*, 191–195.

Li, H. L. (1973). The origin and use of Cannabis in eastern Asia linguistic-cultural implications. *Economic Botany, 28*(3), 293–301.

Lomer, M. C., Parkes, G. C., & Sanderson, J. D. (2008). Review article: Lactose intolerance in clinical practice—Myths and realities. *Alimentary Pharmacology & Therapeutics, 27*(2), 93–103. https://doi.org/10.1111/j.1365-2036.2007.03557.x.

Lu, Z. M., Liu, N., Wang, L. J., Wu, L. H., Gong, J. S., & Yu, Y. J. (2016). Elucidating and regulating the acetoin production role of microbial functional groups in multispecies acetic acid fermentation. *Applied and Environmental Microbiology*, 82(19), 5860–5868. https://doi.org/10.1128/AEM.01331-16.

Malomo, S., Onuh, J., Girgih, A., & Aluko, R. (2015). Structural and antihypertensive properties of enzymatic hemp seed protein hydrolysates. *Nutrients*, 7, 7616–7632.

Malomo, S. A., & Aluko, R. E. (2015a). A comparative study of the structural and functional properties of isolated hemp seed (*Cannabis sativa* L.) albumin and globulin fractions. *Food Hydrocolloids*, 43, 743–752.

Malomo, S. A., & Aluko, R. E. (2015b). Conversion of a low protein hemp seed meal into a functional protein concentrate through enzymatic digestion of fibre coupled with membrane ultrafiltration. *Innovative Food Science & Emerging Technologies*, 31, 151–159.

Malomo, S. A., He, R., & Aluko, R. E. (2014). Structural and functional properties of hemp seed protein products. *Journal of Food Science*, 79, C1512–C1521.

Martena, B., Pfeuffer, M., & Schrezenmeir, J. (2006). Medium-chain triglycerides. *International Dairy Journal*, 16, 1374–1382.

Martínez-Alvarez, O., Chamorro, S., & Brenes, A. (2015). Protein hydrolysates from animal processing by-products as a source of bioactive molecules with interest in animal feeding: A review. *Food Research International*, 73, 204–212.

McClements, D. J., Newman, E., & McClements, I. F. (2019). Plant-based milks: A review of the science underpinning their design, fabrication, and performance. *Comprehensive Reviews in Food Science and Food Safety*, 18, 2047–2067. https://doi.org/10.1111/1541-4337.12505.

Melini, V., & Melini, F. (2019). Gluten-free diet: Gaps and needs for a healthier diet. *Nutrients*, 11(1), 170. https://doi.org/10.3390/nu11010170.

Mikelsaar, M., & Zilmer, M. (2009). *Lactobacillus fermentum* ME-3. An antimicrobial and antioxidative probiotic. *Microbial Ecology in Health and Disease*, 21, 1–27. https://doi.org/10.1080/08910600902815561.

Mikulec, A., Kowalski, S., Sabat, R., Skoczylas, Ł., Tabaszewska, M., & Wywrocka-Gurgul, A. (2019). Hemp flour as a valuable component for enriching physicochemical and antioxidant properties of wheat bread. *LWT-Food Science and Technology*, 102, 164–172.

Mosciano, G. (1989). Nonanoic acid. *Porfumes Flavorist*, 14, 47.

Mota-Gutierrez, J., Barbosa-Pereira, L., Ferrocino, I., & Cocolin, L. (2019). Traceability of functional volatile compounds generated on inoculated cocoa fermentation and its potential health benefits. *Nutrients*, 11(4), 884. https://doi.org/10.3390/nu11040884.

Naqash, F., Gani, A., Gani, A., & Masoodi, F. A. (2017). Gluten-free baking: Combating the challenges – A review. *Trends in Food Science & Technology*, 66, 98–107. https://doi.org/10.1016/j.tifs.2017.06.004.

Nazir, M., Arif, S., Rao Khan, R. S., Nazir, W., Khalid, N., & Maqsood, S. (2019). Opportunities and challenges for functional and medicinal beverages: Current and future trends. *Trends in Food Science and Technology*, 88, 513–526.

Nguyen, T. D. T., Kang, J. H., & Lee, M. S. (2007). Characterization of *Lactobacillus plantarum* PH04, a potential probiotic bacterium with cholesterol-lowering effects. *International Journal of Food Microbiology*, 113(3), 358–361. https://doi.org/10.1016/j.ijfoodmicro.2006.08.015.

Nickerson, G. B., & Van Engel, L. (1992). Hop aroma component profile and the aroma unit. *Journal of the American Society of Brewing Chemists*, 50(3), 77–81. https://doi.org/10.1094/ASBCJ-50-0077.

Nielsen, M. S., Martinussen, T., Flambard, B., Sørensen, K. I., & Otte, J. (2009). Peptide profiles and angiotensin-I-converting enzyme inhibitory activity of fermented milk products: Effect of bacterial strain, fermentation pH, and storage time. *International Dairy Journal*, 19(3), 155–165. https://doi.org/10.1016/j.idairyj.2008.10.003.

Nionelli, L., Montemurro, M., Pontonio, E., Verni, M., Gobbetti, M., & Rizzello, C. G. (2018). Pro-technological and functional characterization of lactic acid bacteria to be used as starters for hemp (*Cannabis sativa* L.) sourdough fermentation and wheat bread fortification. *International Journal of Food Microbiology*, 279, 14–25. https://doi.org/10.1016/j.ijfoodmicro.2018.04.036.

Nissen, L., Bordoni, A., & Gianotti, A. (2020). Shift of volatile organic compounds (VOCs) in gluten-free hemp-enriched sourdough bread: A metabolomic approach. *Nutrients*, 12(4), 1050.

Nissen, L., Demircan, B., Taneyo-Saa, D. L., & Gianotti, A. (2019). Shift of aromatic profile in probiotic hemp drink formulations: A metabolomic approach. *Microorganisms*, 7(11), 509. https://doi.org/10.3390/microorganisms7110509.

Nissen, L., di Carlo, E., & Gianotti, A. (2020). Prebiotic potential of hemp blended drinks fermented by probiotics. *Food Research International*, 131. https://doi.org/10.1016/j.foodres.2020.109029, 109029.

Nissen, L., Zatta, A., Stefanini, I., Grandi, S., Sgorbati, B., & Biavati, B. (2010). Characterization and antimicrobial activity of essential oils of industrial hemp varieties (*Cannabis sativa* L.). *Fitoterapia*, 81(5), 413–419. https://doi.org/10.1016/j.fitote.2009.11.010.

Norajit, K., Gu, B. J., & Ryu, G. H. (2011). Effects of the addition of hemp powder on the physicochemical properties and energy bar qualities of extruded rice. *Food Chemistry, 129*(4), 1919–1925. https://doi.org/10.1016/j.foodchem.2011.06.002.

O'Shea, N., Doran, L., Auty, M., Arendt, E., & Gallagher, E. (2013). The rheology, microstructureand sensory characteristics of a gluten-free bread formulation enhanced with orange pomace. *Food & Function, 4*, 1856–1863.

Oberoi, H. S., Chavan, Y., & Bansal, S. (2010). Production of cellulases through solid state fermentation using kinnow pulp as a major substrate. *Food and Bioprocess Technology, 3*, 528–536.

Odani, S., & Odani, S. (1998). Isolation and primary structure of a methionine-and cystine-rich seed protein of *Cannabis sativa*. *Bioscience, Biotechnology, and Biochemistry, 62*, 650–654.

Oh, W. Y., & Shahidi, F. (2018). Antioxidant activity of resveratrol ester derivatives in food and biological model systems. *Food Chemistry, 261*, 267–273. https://doi.org/10.1016/j.foodchem.2018.03.085.

Omoruyi, F. O., Budiaman, A., Eng, Y., Olumese, F. E., Hoesel, J. L., & Ejilemele, A. (2013). The potential benefits and adverse effects of phytic acid supplement in streptozotocin-induced diabetic rats. *Advances in Pharmacological and Pharmaceutical Sciences, 2013*. https://doi.org/10.1155/2013/172494, 172494.

Onofri, C., de Meijer, E., & Mandolino, G. (2015). Sequence heterogeneity of cannabidiolic- and tetrahydrocannabinolic acid-synthase in *Cannabis sativa* L. and its relationship with chemical phenotype. *Phytochemistry, 116*, 57–68.

Pacifico, D., Miselli, F., Micheler, M., Carboni, A., Ranalli, P., & Mandolino, G. (2006). Genetics and marker-assisted selection of the chemotype in *Cannabis sativa* L. *Molecular Breeding, 17*, 257–268.

Paraman, I., Sharif, M. K., Supriyadi, S., & Rizvi, S. S. H. (2015). Agro-food industry byproducts into value-added extruded foods. *Food and Bioproducts Processing, 96*, 78–85.

Parry, J. W., Cheng, Z., Moore, J., & Yu, L. L. (2008). Fatty acid composition, antioxidant properties, and Antiproliferative capacity of selected cold-pressed seed flours. *Journal of the American Oil Chemists' Society, 85*, 457–464. https://doi.org/10.1007/s11746-008-1207-0.

Paul, A. A., Kumar, S., Kumar, V., & Sharma, R. (2020). Milk analog: Plant based alternatives to conventional milk, production, potential and health concerns. *Critical Reviews in Food Science and Nutrition, 60*(18), 3005–3023.

Pereira, A. P., Mendes-Ferreira, A., Oliveira, J. M., Estevinho, L. M., & Mendes-Faia, A. (2013). High-cell-density fermentation of *Saccharomyces cerevisiae* for the optimisation of mead production. *Food Microbiology, 33*(1), 114–123. https://doi.org/10.1016/j.fm.2012.09.006.

Peres, C. M., Peres, C., Hernández-Mendoza, A., & Malcata, F. X. (2012). Review on fermented plant materials as carriers and sources of potentially probiotic lactic acid bacteria – With an emphasis on table olives. *Trends in Food Science & Technology, 26*(1), 31–42. https://doi.org/10.1016/j.tifs.2012.01.006.

Petel, C., Prost, C., & Onno, B. (2017). Sourdough volatile compounds and their contribution to bread: A review. *Trends in Food Science and Technology, 59*, 105–123.

Pojić, M., Dapčević Hadnađev, T., Hadnađev, M., Rakita, S., & Brlek, T. (2015). Bread supplementation with hemp seed cake: A by-product of hemp oil processing. *Journal of Food Quality, 38*, 431–440.

Pojić, M., Mišan, A., Sakač, M., Dapčević Hadnađev, T., Šarić, B., & Milovanović, I. (2014). Characterization of byproducts originating from hemp oil processing. *Journal of Agricultural and Food Chemistry, 62*(51), 12436–12442.

Pojić, M., Mišan, A., & Tiwari, B. (2018). Eco-innovative technologies for extraction of proteins for human consumption from renewable protein sources of plant origin. *Trends in Food Science & Technology, 75*, 93–104.

Radočaj, O., Dimić, E., & Tsao, R. (2014). Effects of hemp (*Cannabis sativa* L.) seed oil press-cake and decaffeinated green tea leaves (*Camellia sinensis*) on functional characteristics of gluten-free crackers. *Journal of Food Science, 79*, C318–C325.

Raikos, V., Neacsu, M., Russell, W., & Duthie, G. (2014). Comparative study of the functional properties of lupin, green pea, fava bean, hemp, and buckwheat flours as affected by pH. *Food Science & Nutrition, 2*(6), 802–810. https://doi.org/10.1002/fsn3.143.

Raveschot, C., Cudennec, B., Coutte, F., Flahaut, C., Fremont, M., & Drider, D. (2018). Production of bioactive peptides by Lactobacillus species: From gene to application. *Frontiers in Microbiology, 9*, 2354. https://doi.org/10.3389/fmicb.2018.02354.

Rego Costa, A. C., Rosado, E. L., & Soares-Mota, M. (2012). Influence of the dietary intake of medium chain triglycerides on body composition, energy expenditure and satiety: A systematic review. *Nutricion Hospitalaria, 27*(1), 103–108. https://doi.org/10.1590/S0212-16112012000100011.

Rizzello, C. G., Tagliazucchi, D., Babini, E., Rutella, G. S., Saa, D. L. T., & Gianotti, A. (2016). Bioactive peptides from vegetable food matrices: Research trends and novel biotechnologies for synthesis and recovery. *Journal of Functional Foods, 27*, 549–569.

Roberts, M. T., Dufour, J. P., & Lewis, A. C. (2004). Application of comprehensive multidimensional gas chromatography combined with time-of-flight mass spectrometry (GC x GC-TOFMS) for high resolution analysis of hop essential oil. *Journal of Separation Science, 27*(5–6), 473–478. https://doi.org/10.1002/jssc.200301669.

Rodríguez, H., Curiel, J. A., Landete, J. M., de Las Rivas, B., López de Felipe, F., & Gómez-Cordovés, C. (2009). Food phenolics and lactic acid bacteria. *International Journal of Food Microbiology, 132*(2–3), 79–90.

Romano, R., Aiello, A., De Luca, L., Sica, R., Caprio, E., & Pizzolongo, F. (2021). Characterization of a new type of mead fermented with *Cannabis sativa* L. (hemp). *Journal of Food Science.* https://doi.org/10.1111/1750-3841.15614.

Ross, S. A., EI Sohly, H. N., EI Kashoury, E. A., & El Sohly, M. A. (1996). Fatty acids of Cannabis seeds. *Phytochemical Analysis, 7,* 279–283.

Russo, E. B. (2001). Hemp for headache: an in-depth historical and scientific review of Cannabis in migraine treatment. *Journal of Cannabis Therapeutics, 1*(2), 21–92.

Russo, R., & Reggiani, R. (2015). Evaluation of protein concentration, amino acid profile and antinutritional compounds in hempseed meal from dioecious and monoecious varieties. *American Journal of Plant Sciences, 6,* 14–22.

Sánchez-Zapata, E., Fernández-López, J., Pérez-Alvarez, J. A., Soares, J., Sousa, S., & Gomes, A. M. P. (2013). In vitro evaluation of "horchata" co-products as carbon source for probiotic bacteria growth. *Food and Bioproducts Processing, 91*(3), 279–286.

Sethi, S., Tyagi, S. K., & Anurag, R. (2016). Plant-based milk alternatives an emerging segment of functional beverages: A review. *Journal of Food Science and Technology, 53*(9), 3408–3423. https://doi.org/10.1007/s13197-016-2328-3.

Setti, L., Samaei, S. P., Maggiore, I., Nissen, L., Gianotti, A., & Babini, E. (2020). Comparing the effectiveness of three different biorefinery processes at recovering bioactive products from hemp (*Cannabis sativa* L.) byproduct. *Food and Bioprocess Technology, 13,* 2156–2171. https://doi.org/10.1007/s11947-020-02550-6.

Singhal, S., Baker, R. D., & Baker, S. S. (2017). A comparison of the nutritional value of cow's milk and non-dairy beverages. *Journal of Pediatric Gastroenterology and Nutrition, 64*(5), 799–805. https://doi.org/10.1097/MPG.0000000000001380.

Staginnus, C., Zörntlein, S., & de Meijer, E. (2014). A PCR marker linked to a THCA synthase polymorphism is a reliable tool to discriminate potentially THC-rich plants of *Cannabis sativa* L. *Journal of Forensic Sciences, 59,* 919–926.

St-Onge, M. P., & Jones, P. J. (2002). Physiological effects of medium-chain triglycerides: Potential agents in the prevention of obesity. *Journal of Nutrition, 132,* 329–332.

Subramaniam, S., Sabaratnam, V., Kuppusamy, U. R., & Tan, Y. S. (2014). Solid-substrate fermentation of wheat grains by mycelia of indigenous species of the genus Ganoderma (higher basidiomycetes) to enhance the antioxidant activities. *International Journal of Medicinal Mushrooms, 16,* 259–267.

Švec, I., & Hrušková, M. (2015). Properties and nutritional value of wheat bread enriched by hemp products. *Potravinarstvo Scientific Journal of Food Industry, 9,* 304–308.

Tang, C.-H., Ten, Z., Wang, X.-S., & Yang, X.-Q. (2006). Physicochemical and functional properties of hemp (*Cannabis sativa* L.) protein isolate. *Journal of Agricultural and Food Chemistry, 54,* 8945–8950.

Tangyu, M., Muller, J., Bolten, C. J., & Wittmann, C. (2019). Fermentation of plant-based milk alternatives for improved flavour and nutritional value. *Applied Microbiology and Biotechnology, 103*(23–24), 9263–9275. https://doi.org/10.1007/s00253-019-10175-9.

Torres-León, C., Ramírez-Guzman, N., Londoño-Hernandez, L., Martinez-Medina, G. A., Díaz-Herrera, R., & Navarro-Macias, V. (2018). Food waste and byproducts: an opportunity to minimize malnutrition and hunger in developing countries. *Frontiers in Sustainable Food Systems, 2,* 52. https://doi.org/10.3389/fsufs.2018.0005.

Vázquez-Araújo, L., Rodríguez-Solana, R., Cortés-Diéguez, S. M., & Domínguez, J. M. (2013). Study of the suitability of two hop cultivars for making herb liqueurs: Volatile composition, sensory analysis, and consumer study. *European Food Research and Technology, 237*(5), 775–786. https://doi.org/10.1007/s00217-013-2050-6.

Vidrih, R., & Hribar, J. (2016). Mead: The oldest alcoholic beverage. In *Traditional foods* (pp. 325–338). Boston, MA: Springer. https://doi.org/10.1007/978-1-4899-7648-2_26.

Vonapartis, E., Aubin, M.-P., Seguin, P., Mustafa, A. F., & Charron, J.-B. (2015). Seed composition of ten industrial hemp cultivars approved for production in Canada. *Journal of Food Composition and Analysis, 39,* 8–12.

Wang, Q., Jiang, J., & Xiong, Y. L. (2018). High pressure homogenization combined with pH shift treatment: A process to produce physically and oxidatively stable hemp milk. *Food Research International, 106,* 487–494. https://doi.org/10.1016/j.foodres.2018.01.021.

Wang, Q., & Xiong, Y. L. (2019). Processing, nutrition, and functionality of hempseed protein: A review. *Comprehensive Reviews in Food Science and Food Safety.* https://doi.org/10.1111/1541-4337.12450.

Wang, X., Xie, K., Zhuang, H., Ye, R., Fang, Z., & Feng, T. (2015). Volatile flavor compounds, total polyphenolic contents and antioxidant activities of a China gingko wine. *Food Chemistry, 182*, 41–46.

Wang, X.-S., Tang, C.-H., Yang, X.-Q., & Gao, W.-R. (2008). Characterization, amino acid composition and in vitro digestibility of hemp (Cannabis sativa L.) proteins. *Food Chemistry, 107*, 11–18.

WHO/FAO/UNU. (2007). *Protein and amino acid requirements in human nutrition. Report of a joint WHO/FAO/UNU expert consultation*. WHO technical report series no. 935 Geneva: WHO.

Xiong, C., Li, Q., Li, S., Chen, C., Chen, Z., & Huang, W. (2017). In vitro antimicrobial activities and mechanism of 1-octen-3-ol against food-related bacteria and pathogenic fungi. *Journal of Oleo Science, 66*(9), 1041–1049. https://doi.org/10.5650/jos.ess16196.

Yin, S. W., Tang, C. H., Cao, J. S., Hu, E. K., Wen, Q. B., & Yang, X. Q. (2008). Effects of limited enzymatic hydrolysis with trypsin on the functional properties of hemp (*Cannabis sativa* L.) protein isolate. *Food Chemistry, 106*, 1004–1013.

Yin, S. W., Tang, C. H., Wen, Q. B., & Yang, X. Q. (2009). Functional and structural properties and in vitro digestibility of acylated hemp (*Cannabis sativa* L.) protein isolates. *International Journal of Food Science and Technology, 44*, 2653–2661.

Zahari, I., Ferawati, F., Helstad, A., Ahlström, C., Östbring, K., & Rayner, M. (2020). Development of high-moisture meat analogues with hemp and soy protein using extrusion cooking. *Food, 9*, 772. https://doi.org/10.3390/foods9060772.

Zhai, F. H., Wang, Q., & Han, J. R. (2015). Nutritional components and antioxidant properties of seven kinds of cereals fermented by the basidiomycete Agaricus blazei. *Journal of Cereal Science, 65*, 202–208.

Zhang, J., Liu, L., Liu, H., Yoon, A., Rizvi, S., & Wang, Q. (2019). Changes in conformation and quality of vegetable protein during texturization process by extrusion. *Critical Reviews in Food Science and Nutrition, 59*(20), 3267–3280. https://doi.org/10.1080/10408398.2018.1487383.

Zilly, A., dos Santos Bazanella, G. C., Helm, C. V., Vaz Araùjo, C. A., Marques de Souza, C. G., & Bracht, A. (2012). Solid-state bioconversion of passion fruit waste by white-rot fungi for production of oxidative and hydrolytic enzymes. *Food and Bioprocess Technology, 5*, 1573–1580.

Zion Market Research. *Global hemp milk market: Industry size, share, trends, growth, applications, analysis, and forecast, 2018-2026*. https://www.zionmarketresearch.com/report/hemp-milk-market.

Industrial hemp-based dietary supplements and cosmetic products

Anna Bakowska-Barczak[a], Yussef Esparza[b], Harmandeep Kaur[c], and Tomasz Popek[a]

[a]Radient Technologies Inc, Edmonton, AB, Canada [b]BioNeutra, Edmonton, AB, Canada
[c]Advanced BioInnovations Inc, AB, Edmonton, Canada

11.1 Introduction

Industrial hemp (*Cannabis sativa* L., Cannabaceae) is a versatile herbaceous crop that has been used for fiber, food, and medicinal purposes (Andre, Hausman, & Guerriero, 2016; Crini, Lichtfouse, Chanet, & Morin-Crini, 2020; Fike, 2016; Vonapartis, Aubin, Seguin, Mustafa, & Charron, 2015). The cultivation of hemp began in China around 2800 BCE (Dewey, 1914; Dunford, 2015; Pengilly, 2003) making its way to Europe around 1500 BCE (Callaway, 2004; Simmonds, 1976). Historically, a multitude of products have been derived from the seeds, fiber, and wooden core of the hemp plant (Ranalli & Venturi, 2004). As a traditional fiber crop, hemp became an important material for production of rope and canvas (the word being derived from *Cannabis*) (Fortenberry & Bennett, 2001; Roulac, 1997). In 1938 the *Popular Mechanics* magazine described hemp as "the new billion-dollar crop," with about 30,000 different products derived from the hemp fiber and the stalk alone. The hempseed has been well documented as a source of dietary fiber, protein, and fat, with high nutritional value (Callaway, 2004; Crescente et al., 2018). Furthermore, properties of hemp have been used to assist in treating and preventing illnesses for thousands of years in traditional oriental medicine (Callaway, 2004; Jones, 1996). In recent years, the interest in investigating the potential use of industrial hemp in food and nutraceuticals reemerged largely due to the European and Canadian decisions to allow hemp production. By 2018 over 50,000 ha were dedicated to industrial hemp in Europe, a 614% increase compared to 1993 (European Industrial Hemp Association (EIHA), 2020). On the other side of the Atlantic over 20,000 ha were licensed for hemp cultivation in Canada in 2020, an 833% increase from 1998 (Health Canada, 2021; Laate, 2012). Shortly after Europe and Canada, several states in the United States began authorizing feasibility studies to determine its potential value as a crop. Since the implementation of the

2018 Farm Bill, the total acreage used for hemp cultivation in the United States has increased to over 465,000 (about 188,500 ha) in 2020 (Drotleff, 2020).

New opportunities for hemp opened very recently around the world with the increased interest in cannabidiol (CBD), a nonpsychoactive cannabinoid present in its flowers and leaves, and CBD-related products. CBD was first extracted from Cannabis plant in 1940 by chemist Roger Adams while its chemical structure was described twenty-three years later by Dr. Raphael Mechoulam (Pertwee, 2006). The knowledge of CBD's chemical basis opened the door for the pharmacological research and small clinical trials, which gathered momentum in the 1980s when Mechoulam and other scientists began conducting game-changing research on CBD for epilepsy (Weedmaps, 2020). More and more promising preclinical research on CBD started emerging in the early 2000s, making CBD-rich strains more popular and stimulating growth of hemp-related products. In a report released by Vote Hemp and Hemp Business Journal, the retail value of hemp-related products reached $820 million in 2017, a 30% increase in 2 years (Vote Hemp, 2017). The hemp-derived CBD products constituted 23% ($190 million) of total sales and had a 66% increase in sales value in 2 years (Hemp Industries Association, 2016). The global cannabidiol market size was valued at USD 2.8 billion in 2020 and is expected to expand at a compound annual growth rate of 21.2% from 2021 to 2028 (Grand View Research, 2021a).

Use of cannabidiol has become the newest consumer trend due to the increasing number of publications promoting the benefits related to its use. CBD is widely used in personal and skincare industries as well in food and beverage industry. According to Single Care survey (SingleCare, 2020), nearly 80% of CBD users preferred cosmetic and personal care products such as lotions/balms, soaps, oil sprays, skincare products, and others (Fig. 11.1). The adoption of cannabidiol-based cosmetic products, food, and supplements is expected to significantly increase in the future.

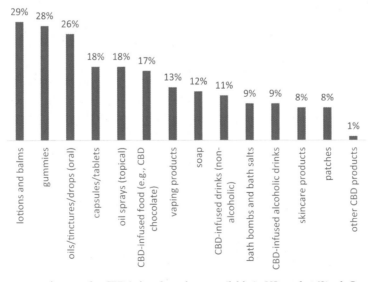

FIG. 11.1   Consumer preferences for CBD-infused products available in US market (SingleCare, 2020).

Moreover, the increasing consumer preference for plant-based ingredients, especially in dietary supplements and cosmetics, will further boost the demand for the hemp and hemp-derived CBD products market. Growing awareness about nutritional value of hempseeds and hemp-derived CBD and increasing number of preclinical and clinical trials involving hemp-based products are the key factors driving market growth. The consumers are looking for more natural and effective supplements and cosmetics that can fulfill all their nutritional requirements and body care, which is further supporting the adoption of these products.

11.1.1 Hemp-based dietary supplements and marketing

Consumers have become increasingly interested in the way their diet can address health deficits and well-being. Currently, most of the grocery shoppers base their purchases on the pursuit of preventing, managing, or treating a specific health condition. With the increased trend for plant-based food and supplements, some unconventional plant-derived oils, such as hempseed oil, have earned a reputation of providing not only culinary and alimentary services but also providing medicinal and nutraceutical potential (Uluata & Özdemir, 2012). Currently, hempseed oil is primarily advertised as a natural health product for body care purposes, as oil for salad dressings, or to be taken directly as a dietary supplement. Very recently the cannabinoids (mainly CBD) present in hemp flowers have captured consumers' attention due to their ability to support bodily comfort and increase the subjective measures of quality of life.

The use of different parts of the hemp plant as food/medicine dates to at least 6000 years ago (Li, 2012). Although hemp leaves, sprouts, and flower can be consumed as a raw food by preparing juices or salads, hempseeds are perhaps the most widely used part of the plant as dietary supplement and foods. Among supplements, softgel capsules and oil are predominant, with powders rich in plant-based protein and/or fiber are gaining consumer's interest. Hempseed composition varies among different cultivars, but they are typically about 25%–30% oil, 25%–30% protein, 30%–40% fiber, and about 6%–7% moisture (Leonard, Zhang, Ying, & Fang, 2020).

Because hemp prohibition was only lifted about 20 years ago, it is only recently that hempseed and hemp flowers extract have been investigated for its applications in the food and nutraceutical industry for its benefits beyond basic nutrition and recreational use.

11.2 Hemp active ingredients for dietary supplements

11.2.1 Hempseed oil

Hempseed oil is rich in essential fatty acids, linoleic acid, alpha-linolenic acid, gamma-linolenic acid (GLA), and oleic acid (Leizer, Ribnicky, Poulev, Dushenkov, & Raskin, 2000). Linolenic and alpha-linolenic acids are precursors of arachidonic and eicosapentaenoic acids, respectively. Extensive research has demonstrated the beneficial effect of these fatty acids against cardiovascular diseases, obesity, and diabetes (Sokoła-Wysoczańska et al., 2018). In addition, GLA, being a precursor of eicosanoids, has a role in the regulation of inflammatory response (Kapoor & Huang, 2006). Hempseed oil has gained presence in the stores as dietary supplement in the form of refined and unrefined oil and softgel capsules. Most companies center their marketing highlighting the beneficial properties of a balanced omega-6 to omega-3 ratio of hempseed oil. It is well known that diet, especially in Western countries, is richer in omega-6 fatty acids. The incorporation of omega-3, from hempseed oil for example, has been proposed to influence numerous conditions, including cardiovascular, cancer, arthritis, diabetes, and neurological disorders (Simopoulos, 2002).

In a human-based study trial, serum fatty acid levels were compared after supplementation with hempseed and flaxseed oil at 30 mL/day for 4 weeks. Results showed that hempseed oil resulted in higher concentrations of linoleic acid, GLA, arachidonic acid, and lower total-to-HDL cholesterol compared to flaxseed oil supplementation (Table 11.1) (Schwab et al., 2006). A double-blind, placebo-controlled clinical trial evaluated the effect of daily consumption of two 1 g capsules, containing placebo, fish oil, flaxseed oil, or hemp oil on plasma levels of 86 volunteers for 12 weeks. The authors found an increase in DHA and EPA plasma levels after fish oil supplementation but did not find differences in placebo and hempseed oil regarding plasma fatty acid and lipid profile after 12 weeks (Kaul et al., 2008). A study conducted on hyperlipidemic children and adolescents showed that daily supplementation of hempseed oil (3 g) for 8 weeks reduced the red blood cell content of total saturated and monounsaturated fatty acids and increased the levels of total omega-3 and omega-6. However, hempseed oil supplementation did not affect the serum lipid profile (total cholesterol, HDL cholesterol, LDL cholesterol) (Table 11.1) (Del Bo et al., 2019). Besides fatty acids, hempseed oil contains around 2% of phytosterols such as β-sitosterol and campesterol, and tocopherols—compounds with antioxidant activity (Montserrat-De La Paz, Marín-Aguilar, García-Giménez, & Fernández-Arche, 2014,). Other antioxidants identified in hempseeds include polyphenols (Teh & Birch, 2013) and lignanamides (Yan et al., 2015). Besides antioxidant activity, lignanamides showed significant antineuroinflammatory activities in in vivo studies (Yan et al., 2015) suggesting their contribution in overall hempseed health benefits.

11.2.2 Hempseed proteins

Hempseed contains about 25%–30% protein and is currently gaining a big interest as a source of plant-based protein isolates. Hempseed contains mainly the storage proteins, albumin (25%–37%) and the legumin called edestin (67%–75%), and low quantities of protease inhibitors. The essential amino acid content of hempseed proteins is superior to that of soybean and is sufficient for humans who are 10 years of age or older. It was demonstrated (House, Neufeld, & Leson, 2010) that in vitro protein digestibility values were up to 92% for dehulled hempseed. Albumin and edestin contained in hempseeds are both proteins, rich in methionine and arginine. High arginine content makes hempseed protein especially valuable as a nutritional ingredient to formulate foods or dietary supplements that enhance cardiovascular health. This is because arginine is a precursor of nitric oxide, the vasodilating agent that enhances blood flow and contributes to maintenance of normal blood pressure (Wu & Meininger, 2002). Moreover, a recent study suggests that hempseed protein may improve chronic inflammatory states and promote regenerative processes (Rodriguez-Martin et al., 2020). In addition, hempseed protein hydrolysates have shown antioxidant activity and blood pressure-lowering effect. Hempseed protein hydrolysates have been increasingly studied to develop bioactive peptides with ACE inhibitory activity, renin inhibition, antihypertensive

TABLE 11.1 Human-based trials with hemp-based dietary supplements.

Supplement type	Purpose	Experimental design	Experimental group	Dosing	Treatment duration	Inclusion criteria	Exclusion criteria	Results	Reference
Hempseed oil capsule	Effects on cardiovascular parameters (lipid profile, LDL oxidation, inflammatory markers, and platelet aggregation).	Double-blinded, placebo-controlled clinical trial	Male ($n = 34$) and female ($n = 54$) healthy volunteers were randomly divided into four groups (placebo, fish oil, flaxseed oil, and hempseed oil)	Oral daily supplementation with 2 capsules each containing 1 g of the relative treatment	12 weeks	Fasting TC levels	N/A	No significant effect on the ALA and LA plasma levels; TC, TG, LDL, and HDL amounts, or inflammatory (CRP and TNF-α) and platelet aggregation analyzed markers (collagen and thrombin-induced platelet aggregation)	Kaul et al. (2008)
Hempseed oil soft capsule	Modulation of hyperlipidemia and CVD risk in children and adolescents.	Randomized, controlled, two-arm parallel-group study	Children and adolescents ($n = 36$) with diagnosis of primary hyperlipidemia randomly divided into two groups: test and control group	Oral daily supplementation with one hempseed oil soft capsule containing 3 g of oil	8 weeks	Primary hyperlipidemia; 6–16 years old	Secondary hyperlipidemia, and/or renal, endocrine, lipid, neurologic, onco-hematologic disorders requiring drug treatment; BMI > 85th percentile; smokers.	No significant effect on the levels of TC, LDL-C, HDL-C, TGs, body weight, BMI, and blood pressure. Significantly modified the RBC FAs composition increasing the total n-3 PUFAs, n-3 LCPUFAs, and n-6 LCPUFAs, and decreasing the n-6/n-3 PUFAs ratio as well as the SFAs and MUFAs.	Del Bo et al. (2019)

Continued

TABLE 11.1 Human-based trials with hemp-based dietary supplements—cont'd

Supplement type	Purpose	Experimental design	Experimental group	Dosing	Inclusion criteria	Exclusion criteria	Results	Reference
Hempseed oil and evening primrose oil (9:1) cosupplementation	Therapeutic and protective effect of the hot-natured diet cosupplemented with hempseed/evening primrose oils or of the hempseed/evening primrose oils dietary cosupplementation on MS patients	Double-blind, randomized trial	Female ($n = 42$) and male ($n = 23$) patients with diagnosis of RMSS were randomly divided into three groups: A, B, and C	Group A: Hot-natured diet cosupplemented with 18–21 g of hempseed/evening primrose oils Group B: 18–21 g olive oil supplementation Group C: 18–21 g of hempseed/evening primrose oils cosupplementation	Diagnosis of RRMS (EDSS < 6); 14–55 years old	Secondary or primary progressive MS; pregnancy; corticosteroid treatment; and diagnosis of other chronic neurological and inflammatory diseases such as cancer, rheumatic diseases, and heart diseases	Significant improvement of the MS clinical condition and inflammatory status of the patients, without any adverse effects. Significant improvement of the degree of Th2/Th1 ratio and the value of EDSS and of relapse rate; Significant increase of the level of antiinflammatory IL-4 cytokine; Hot-natured diet cosupplementation with hempseed/evening primrose oils significantly decreased the level of pro-inflammatory IFN-γ and IL-17 cytokines; Cosupplementation of hempseed/evening primrose oils decreased (but not statistically significant) trend in the level of IFN-γ and IL-17 cytokines	Rezapour-Firouzi et al. (2013)

Material	Objective	Design	Participants	Dose	Duration	Inclusion criteria		Results	Reference
Hempseed oil (HO) and flaxseed oil (FO)	Effect on serum lipids and fasting concentrations of serum total and lipoprotein lipids, plasma glucose and insulin, and hemostatic factors	A randomized, double-blind crossover design	Fourteen healthy volunteers	30 mL/day	4 weeks for HO + 4 weeks for washout period + 4 weeks for FO	N/A	N/A	The HO period resulted in higher proportions of LA and gamma-linolenic acid in serum cholesteryl esters (CE) and triglycerides (TG) as compared with the FO period ($P < 0.001$), whereas the FO period resulted in a higher proportion of ALA in both serum CE and TG as compared with the HO period ($P < 0.001$). The proportion of arachidonic acid in CE was lower after the FO period than after the HO period ($P < .05$). The HO period resulted in a lower total-to-HDL cholesterol ratio compared with the FO period ($P = .065$).	Schwab et al. (2006)
Hempseed protein (HSP), or HSP plus HSP hydrolysate (HSP+)	Hypotensive potential of consuming HSP and HSP+ in comparison to casein protein	A randomized, double-blind, crossover clinical trial	Thirty-five hypertensive participants aged 18–75 years	Consumption of 25 g of casein, HSP, or HSP+ twice a day	6 weeks for casein + 2 weeks for washout + 6 weeks for HSP + 2 weeks for washout + 6 weeks for HSP +	BMI between 18.5 and 40 kg/m², systolic blood pressure (SBP) between 130 and 160 mmHg, and diastolic blood pressure (DBP) ≤110 mmHg	N/A	Ongoing study	Samsamikor, Mackay, Mollard, and Aluko (2020)

Continued

TABLE 11.1 Human-based trials with hemp-based dietary supplements—cont'd

Supplement type	Purpose	Experimental design	Experimental group	Dosing	Treatment duration	Inclusion criteria	Exclusion criteria	Results	Reference
Hempseed protein	Benefits of supplementation during resistance training	A randomized, double-blind crossover study	Forty healthy volunteers (12 females) divided into two groups	Group 1: 60 g/d of hemp powder (containing 40 g protein and 9 g oil) Group 2: 60 g/d of soy powder (matched for macronutrients and calories)	8 weeks of resistance training (1 h/d, 6 d/week)	N/A	N/A	Strength and muscle thickness increased in females, but not males; Drop in twitch torque post fatigue was attenuated in males; Attenuation of inflammation may contribute to the beneficial effects of hemp powder on muscle mass and strength, and fatigue.	Kaviani, Chilibeck, Toles, and Candow (2016)
Hemp CBD oil extract	Effects on stress resilience, perceived recovery from normal and physical stress. Secondary purpose: effects on perceived appetite, mood, feelings of well-being, sleep quality, body composition and safety	A randomized, placebo-controlled, double-blind study	65 overweight, but otherwise healthy men and women divided into two groups	Group 1: Hemp Oil Extract [Hemp, 60 mg/d Plus CBDTM Extra Strength Hemp Extract Oil (15 mg hemp-derived CBD)] Group 2: Placebo (PLA)	6 weeks	Age of 18–55 years, BMI 25–34.99 kg/m² , systolic pressure 100–139 mmHg, diastolic pressure 65–89 mmHg with a normal resting heart rate (90 beats/min)	Athletes, participants using weight loss medications, participant regularly taking a thyroid medication or have with a metabolic disorders, hepatorenal syndrome, cancer, chronic inflammatory conditions, autoimmune, dyssomnia, or other sleep disorders, neurologic, gastrointestinal, or musculoskeletal disease, participants with allergies, smokers, and nursing, lactating, and postpartum women.	HDL cholesterol significantly improved; psychometric measures of perceived sleep, stress response, and perceived life pleasure were supported and with no clinically relevant safety concerns.	Lopez et al. (2020)

effect, acetylcholinesterase inhibition, antioxidant activity, metal binding properties, cellular growth, hypocholesterolemic effect, and serum glucose regulation (Cattaneo et al., 2021; Wang & Xiong, 2019). Other compounds of interest from hempseed protein processing are polyphenols, tannins, flavonoids, and soluble proteins. A study shows that the soluble fraction after isoelectric precipitation of hempseed protein showed higher antioxidant properties, antiproliferative properties against two human colon cancer cell lines, and antiinflammatory activities in lipopolysaccharide-induced RAW 264.7 macrophages (Dia, Wang, Lin, & Pangloli, 2019). This is interesting from the processing point of view, where the supernatant from isoelectric precipitation of proteins usually receives less attention.

Plant-based proteins are gaining popularity as a dietary supplement in training diets. Hempseed protein supplementation during resistance training increased strength and muscle thickness in females and reduced inflammation in muscles after training in general (Table 11.1) (Kaviani et al., 2016). According to the authors, attenuation of inflammation may contribute to the beneficial effects of hemp protein powder on muscle mass and strength, and fatigue. Furthermore, Guang and Wenwei (2008) developed a process for using hemp protein powder in treating anemia by improving the levels of human hemoglobin and erythrocyte, which increased oxygen storage and improved oxygen resistance. Currently, the hempseed protein powders enriched with omega-3 and omega-6 fatty acids and fiber are available on the market as an easy to use and dose dietary supplement (Table 11.2).

The high digestibility, composition of essential amino acids, and low allergenicity of hempseed proteins have drawn great interest of both academia and industry. The use of hempseed protein products as an alternative to the commonly used casein, whey, wheat, and soy protein is on a rise and is projected to grow at a CAGR of 15.2% during the forecasted period 2019–2024 (Research and Markets, 2020). The market demand for hempseed protein is also driven by the lactose intolerant population as it is free from lactose and also by the consumers who are allergic to peanuts, soybeans, and other legumes. The demand for hempseed protein is especially higher in the dietary supplement industry due to its high fiber and omega-3 and omega-6 content.

11.2.3 Cannabidiol and other cannabinoids

As a plant-derived, easily extractable, biologically active compound with large therapeutic index and overall excellent safety profile, CBD is currently marketed both as a mainly hemp-derived dietary supplement, subjected to evolving legislation and regulatory actions, and as a drug, such as Epidiolex (CBD only) and Sativex (THC and CBD in 1:1 ratio), with specific approval indications worldwide (Cerino et al., 2021). As per the report by Businesswire in April 2021, the global CBD nutraceuticals market size is expected to reach USD 19.25 billion by 2028. The demand for CBD nutraceuticals is expected to increase during the forecast period owing to increasing obesity and related disorders, growing health consciousness, and greater awareness regarding benefits associated with cannabidiol nutraceuticals (Businesswire, 2021).

CBD nutritional supplements account for approximately one-third of the global 1.34 billion dollar market for CBD. The claimed wide application against variety of diseases and immense health benefits have attracted the nutraceutical industries to integrate CBD in their current product portfolio. There has been a dramatic rise in edibles or super foods containing CBD. For instance, edibles like cliché special brownie, beverages, cookies, candy, and even

TABLE 11.2 Examples of hemp-based or hemp-infused dietary supplements available on the market.

Supplement type	Product name	Active ingredient	Concentration of active	Manufacturer claim	Manufacturer	Reference
Softgel/Capsule	Hemp Oil +	Hempseed oil, hemp stalk oil	990 mg of hemp oil/capsule	A synergistic, clean blend of phytocannabinoids that supports and regulates one of the body's most important, yet little-known, systems—the Endocannabinoid System. By nourishing this system, Hemp Oil + can benefit users by relieving stress, easing discomfort, and improving health and regularity of the digestive system.	Thorne	https://www.thorne.com/
	Hemp Seed Oil Softgels	Cold-pressed hempseed oil	Omega 3–185 mg, Omega 6–535 mg, Omega 9–105 mg/capsule	N/A	Manitoba Harvest	https://manitobaharvest.ca/
	Hemp Advantage Plus	Broad-spectrum hemp extract (aerial parts), palmitoylethanolamide (PEA), black pepper fruit and clove bud oils	300 mg PEA, 70 mg hempseed oil/capsule	Supports bodily comfort and increase the subjective measures of quality of life.	Metagenics	https://www.metagenics.com/
	Swisse Hemp Seed Oil	Hempseed oil, vitamin E	Hempseed oil (66%), vitamin E	Providing a balanced source of Omega 3 and 6 essential fatty acids necessary for a healthy diet, Swisse Hemp Seed Oil supports nutritional well-being.	Swisse	https://swisse.com.au/
	Swisse Hemp Seed Oil + Turmeric	Hempseed oil, turmeric, vitamin E	Hempseed oil (49%), Turmeric Powder (9%)	Swisse Hemp Seed Oil + Turmeric provides a balanced source of Omega 3 and 6 essential fatty acids and antioxidants to support nutritional well-being.	Swisse	https://swisse.com.au/

Product	Ingredients	Dosage	Brand	Benefits	Website
Vthrive The Vitamin Shoppe CBD broad spectrum hemp extract + BioPerine black pepper extract	Broad spectrum hemp extract (aerial parts), BioPerine black pepper extract (fruit), terpene blend (black pepper essential oil (*Piper nigrum*) (fruit); orange essential oil (*Citrus sinensis*) (peel))	Broad spectrum hemp extract (CBD 55, 35 or 15mg), piperine 5mg, terpenes 4mg per capsule	The Vitamin Shoppe	Includes BioPerine Black Pepper Extract for superior bioavailability and enhanced absorption	https://www.vitaminshoppe.com/
Balance CBD Capsules	CBD hemp extract, *Bacopa monnieri*, *Rhodiola rosea*, sulbutiamine, vitamin B12, 5-HTP, CDP-choline, valerian root + chamomile	20mg CBD/capsule	Dr Love CBD	*Bacopa monnieri*: Helps increase resistance to stress. *Rhodiola rosea*: Helps improve mood and concentration. Sulbutiamine: Protects against symptoms of anxiety. B12: Fights against the development of depression. 5-HTP: Amino acid building block for serotonin. CDP-choline: Neuroprotectant against chronic stress. Valerian root + chamomile: Promotes a relaxed mind.	https://drlovecbd.com/
Entourage High-CBD Softgels	CBD hemp extract	50mg CBD/capsule	Dr Love CBD	Aids in the management of mental health, pain, and inflammation.	https://drlovecbd.com/

Continued

TABLE 11.2 Examples of hemp-based or hemp-infused dietary supplements available on the market—cont'd

Supplement type	Product name	Active ingredient	Concentration of active	Manufacturer claim	Manufacturer	Reference
	Focus CBD Capsules	5-HTP, CDP-choline, acetyl-ʟ-carnitine, alpha GPC + forskolin, artichoke	20 mg CBD/capsule	5-HTP: Building block for other neurotransmitters. CDP-choline: Neuroprotectant and antiinflammatory. Acetyl-ʟ-carnitine: Supports overall brain function. Alpha GPC + forskolin: Aids in memory support. Artichoke: Healthy brain function via gut-brain axis.	Dr Love CBD	https://drlovecbd.com/
	Relief G CBD + CBG Capsules	Cannabidiol hemp extract, cannabigerol hemp extract, acai berry extract, Curcumin C3 Complex, Bioperine, glucosamine, chondroitin.	25 mg CBD, 25 mg CBG per capsule	Glucosamine: Helps repair soft tissue like ligaments and cartilage Chondroitin: Helps repair and grow cartilage C3 Complex: Antioxidant, cardio support, immune booster + more Acai berry extract: Antioxidant, protectant, brain booster	Dr Love CBD	https://drlovecbd.com/
	Rest N CBD + CBN Softgels	Hemp derived CBN, CBD hemp extract, melatonin	10 mg CBD, 10 mg CBN, 5 mg melatonin per capsule	CBD: Antiinflammatory, pain reliever, reported sleep aid/anxiety reducer CBN: Sedative Melatonin: Naturally occurring sleep aid	Dr Love CBD	https://drlovecbd.com/

Tincture/ Liquid oil	Liquid hemp oil	Hempseed oil	925 mg/1 mL	Liquid herbal formula is easy to take and absorbs quickly	Nature's Truth	https://www.naturestruth.com/
	Water soluble CBD, full spectrum hemp oil	Concentrated cannabinoid and terpene-rich hemp oil	10 mg CBD per 1 mL	This fast acting, water-soluble delivery system is up to 10 times more bioavailable than oil. Water-soluble CBD is CBD oil that has been broken down using nanotechnology to make the particles smaller. These tiny particles are easier for the body to absorb, helping the benefits of CBD take effect faster and allowing it to easily mix with water or food.	CBD American shaman	https://cbdamericanshaman.com/water-soluble-full-spectrum-hemp-oil-30ml
	CBD Oil Tincture Drops	Hemp extract (cannabidiol, cannabigerol, cannabinol)	10, 16.5, 25, 50, 100, 125, 166.5 and 250 mg CBD per serving	Support a sense of calm, manage signs of daily stress, enhance exercise recovery, complement a wellness routine	CBD MD	https://www.cbdmd.com/cbd-oil-tincture-dropper
	CALM hemp extract oil blend (Chocolate Mint, Mandarin Orange, Cinnamon Spice)	Whole hemp extract oil	10 mg CBD/mL	N/A	Restorative Botanicals	https://www.restorativebotanicals.com/
	Original formula CBD Oil	Charlotte's Web premium hemp extract	50 mg CBD/mL	Support a sense of calm for focus, manage everyday stresses, recovery from exercise-induced inflammation, maintain healthy sleep cycles	Charlotte's Web Stainley Brothers	https://www.charlottesweb.com/
	Broad Spectrum CBD Oil	Hemp-derived broad spectrum cannabinoid extract, hempseed oil	50 mg CBD/mL	Strong source of support for daily wellness routine.	Green Roads	https://greenroads.com/

Continued

TABLE 11.2 Examples of hemp-based or hemp-infused dietary supplements available on the market—cont'd

Supplement type	Product name	Active ingredient	Concentration of active	Manufacturer claim	Manufacturer	Reference
	Extra Strength CBN + CBD Sleep Tincture 1:3	Full spectrum hemp extract (aerial parts), cannabinol (CBN), natural terpenes	300 mg CBN, 900 mg CBD in 30 mL bottle	Formulated for restful sleep, relief, and relaxation.	Balanced Health Botanicals (BHB), CBDistillery	https://www.balancedhealthbotanicals.com/#brands
	CBD + CBG Oil Tincture 1:1	Full spectrum hemp extract (aerial parts), cannabigerol (CBG), natural terpenes	500 mg CBG, 500 mg CBD in 30 mL bottle	Helps to achieve a healthy equilibrium and homeostasis.	Balanced Health Botanicals (BHB), CBDistillery	https://www.balancedhealthbotanicals.com/#brands
	Ultra Broad Spectrum Cannabinoid tinctures	Ultra Broad Spectrum extract, naturally occurring cannabinoids: CBD, CBG, CBN, CBC, CBDV, natural terpenes	1000 mg = 40 mg of Ultra Broad Spectrum extract and 33 mg of naturally occurring cannabinoids (CBD, CBG, CBN, CBC, CBDV) per 1 mL dropper	Medterra's Ultra Broad Spectrum Cannabinoid tinctures utilize our potent full-plant hemp extract without THC. Unlike our isolate tinctures which contains only CBD, our Ultra Broad Spectrum tinctures contain CBD with additional beneficial compounds such as CBG, CBN, CBC, CBDV, and natural terpenes.	Medterra	https://medterracbd.com/product-cbd-broad-spectrum
	Plus CBD Oil Gold Drops	Extra Strength Formula full spectrum extract, extra virgin olive oil, monk fruit, *Quillaja saponaria*	10 mg CBD per serving	CBD and the other ingredients in full spectrum hemp extract are helpful in enhancing health in healthy people and to restore balance and homeostasis to the human system.	CV Sciences + Plus CBD Oil	https://www.pluscbdoil.com/cbd-products/drops/extra-strength-formula
Powder	Hemp Yeah! Protein Powder	Hempseed protein	15 g plant protein, 8 g fiber, 2 g Omegas 3 and 6 per 30 g serving	N/A	Manitoba Harvest	https://manitobaharvest.ca/
	Organic Hemp Yeah! Max Protein Unsweetened	Organic hemp protein concentrate	20 g plant protein, 3 g fiber, 4.5 g Omegas 3 and 6 per 32 g serving	N/A	Manitoba Harvest	https://manitobaharvest.ca/

Category	Product Name	Ingredients	Amount	Description	Company	Website
	Organic Hemp Yeah! Max Fiber (Unsweetened, Vanilla, Chocolate)	Organic hemp protein concentrate	13 g plant protein, 13 g fiber per 33 g serving	N/A	Manitoba Harvest	https://manitobaharvest.ca/
	Broad Spectrum CBD Formulation Powder	Cannabidiol (CBD) from hemp extract (aerial parts)	0.93 g CBD per 1 g	Easy to add to food and beverages and specifically formulated for those looking to cook with CBD.	Balanced Health Botanicals (BHB), CBDistillery	https://www.thecbdistillery.com/
	CBG Isolate High Purity Powder	Hemp-based extract (aerial parts)	100% CBG powder (1-g size)	Easy to add to lotions, creams, foods, and drinks.	Balanced Health Botanicals (BHB), CBDistillery	https://www.thecbdistillery.com/
	Ground Hemp Root	Hemp root	N/A	N/A	Raw Acres	https://rawacres.com/
Gummies	Bliss CBD Gummies	Phytocannabinoid hemp extract	20 mg CBD per gummy	Supports the mechanisms that contribute to overall wellness, such as aiding in the management of mental health, pain, and inflammation	Dr Love CBD	https://drlovecbd.com/
	Rest N CBN Gummies	Hemp-derived CBN, melatonin	10 mg CBN, 5 mg Melatonin per gummy	CBN: Sedative Melatonin: Naturally occurring sleep aid	Dr Love CBD	https://drlovecbd.com/
	Extra Strength Relax Bear	Hemp-derived cannabinoid extract	25 mg CBD per gummy	Helps to manage daily stress	Green Roads	https://greenroads.com/
	Sleepy Zs gummies	Hemp extract, melatonin	25 mg CBD, 5 mg melatonin per gummy	Supports natural sleep cycle.	Green Roads	https://greenroads.com/
	Rise N' Shines Immune Support gummies	Hemp-derived full spectrum cannabinoid extract, elderberry, vitamin B12	25 mg CBD per gummy	Vitamin B12 to support healthy bones and blood. Vitamin C to support immune system, CBD to manage daily stressors. Elderberry extract—known to be rich in antioxidants	Green Roads	https://greenroads.com/

Continued

TABLE 11.2 Examples of hemp-based or hemp-infused dietary supplements available on the market—cont'd

Supplement type	Product name	Active ingredient	Concentration of active	Manufacturer claim	Manufacturer	Reference
	CBD Gummies with Turmeric and Spirulina 1500 mg	Broad spectrum CBD extract, organic spirulina, turmeric	50 mg of CBD, 50 mg of turmeric, 20 mg of spirulina per serving (2 gummies)	The health-boosting one-two punch of "superfoods" Turmeric and Spirulina in this agave-sweetened CBD gummy with beneficial antioxidant and antiinflammatory properties.	CBD Fx	https://cbdfx.com/products/cbd-gummies-turmeric-spirulina
	CBD Gummies	Hemp extract (cannabidiol cannabigerol, cannabinol)	10 mg, 25 mg or 50 mg per gummy	Support a sense of calm, manage signs of daily stress, enhance exercise recovery, complement a wellness routine	cbd MD	https://www.cbdmd.com/cbd-gummie
	CBD Sleep Tight Gummies	*Melissa officinalis* (Lemon balm), *Matricaria chamomilla* (Chamomile), L-theanine, *Passiflora* (Passion flower), Cannabidiol (hemp extract), 5-HTP (5-hydroxytryptophan), melatonin.	25 mg CBD per gummy	Medterra's CBD Sleep Tight Gummies will help you relax before bed, promote a restful night's sleep, and to have you ready to take on tomorrow.	MedTerra	https://medterracbd.com/product-cbd-sleep-gummies
	Plus CBD Extra Strength Gummies (2 flavors)	Full strength hemp extract	10 mg CBD per gummy	Infused with our potent Extra Strength Formula extract, these gummies contain much more than CBD. They have an array of minor cannabinoids, terpenes, and perhaps most importantly, fatty acids. The plant uses those fatty acids as building blocks to produce its cannabinoids—and it turns out that we use those same building blocks to produce our own endocannabinoids that help us with health and homeostasis.	CV Sciences + Plus CBD Oil	https://www.pluscbdoil.com/cbd-products/cbdoilgummies

Category	Product	Ingredients	Dosage	Benefits	Company	Website
Tea	Hemp + Herb Joint Health	Organic meadowsweet, hemp full spectrum extract, organic licorice root, organic orange peel oil.	20 mg hemp extract per bag	Supports a healthy response to inflammation associated with an active lifestyle	Traditional Medicinals	https://www.traditionalmedicinals.com/
	Hemp + Herb Mental Focus	Organic eleuthero root aqueous extract, hemp soft extract, guayusa leaf, organic maté leaf, organic roasted, organic spearmint leaf, organic peppermint leaf	20 mg hemp extract per bag	Helps support mental alertness	Traditional Medicinals	https://www.traditionalmedicinals.com/
	Hemp + Herb Stress Relief	Organic chamomile flower, organic skullcap herb, hemp soft extract, organic hawthorn berries, organic licorice root	20 mg hemp extract per bag	Supports relaxation and soothes occasionally jangled nerves.	Traditional Medicinals	https://www.traditionalmedicinals.com/
	Soul tonic hemp tea/valerian root tea for sleep	Decarbed hemp, lemongrass, peppermint, chamomile, rosehip, spearmint, valerian root, hibiscus, cornflower petals	60 mg CBD per serving	Promoting calmness and relaxation this herbal blend is designed for a better night's sleep	MagicLeaf Tea LLC	https://magicleaftea.com/
	Hemp Tea (Available in Lemon Verbena, Apple Mint, Hops, Thyme)	Hemp leaves, herbs, dry fruits—depending on the flavor selection	12.5%–25% hemp leaves, natural herbs	N/A	My Herbs	https://www.fromaustria.com/en/my-herbs
	Organic Hemp Tea (available in Ginger, Peppermint, Original)	Organic Cannabis sativa L. plant and seeds	40 mg per 1.5 g/cup	Supports everyday holistic health.	Body and Mind Botanicals	https://bodyandmindbotanicals.com/

dried meat containing CBD as one of the ingredients (Grand View Research, 2021a). Growing competition in the CBD dietary supplements and nutraceuticals space is pressurizing companies to launch new products and increase their geographical reach. For instance, in April 2019, Aphria Inc. entered the German market by launching the company's first cannabidiol-based nutraceuticals product range—CannRelief. In April 2019, Australian biopharmaceutical firm MGC Pharmaceuticals entered into a partnership with the Chinese e-commerce platform YuShop Global for gaining entry in the Chinese CBD nutraceuticals space. The increasing demand for hemp CBD oil to provide relief from general pain, nerve pain, chronic pain, anxiety, and sleep disorders will drive the hemp oil in dietary supplements market in the near future. However, the scientific data currently available to support this demand is very limited. The results of recently finished clinical trial examining the effect of a commercially available CBD-containing hemp oil extract on stress resilience, perceived recovery, mood, feelings of well-being, sleep quality, body composition, and clinical safety markers in 65 healthy overweight human subjects were published (Lopez et al., 2020). The supplementation with hemp CBD oil improved HDL cholesterol, supported psychometric measures of perceived sleep, stress response, and life pleasure and was well tolerated with no clinically relevant safety concerns (Table 11.1). A new clinical trial was initiated by the group of scientists from the University of Texas and it will be the first trial to evaluate whether cannabidiol (CBD) can interfere with the reconsolidation of naturally acquired pathological interoceptive fear memory in humans (Clinical Trial No NCT04726475, n.d.). Researchers from Center for Applied Health Sciences started the clinical trial exploring the effects of 8 weeks of CBD supplementation (50 mg CBD/day) on mental and physical health, sleep measures, and NK cell number and cytotoxic function. The results of both trials are expected in early 2022.

On the basis of product, the market is segmented into CBD capsules, CBD tinctures, and others. The tinctures segment accounted for a major share in 2019. Health benefits associated with the tinctures, increasing health consciousness among users, and shifting trend toward diet and nutritious food products boost the growth of this segment (Businesswire, 2021). CBD tinctures available on the market are primarily based on the medium chain triglyceride (MCT) oil as a carrier and differ in the CBD potency (from 10 mg to 50 mg of CBD per ml) as presented in Table 11.2. The tinctures are usually offered in the bottle with glass or plastic pipette for dosing which might be challenging for some costumers.

In 2020 the CBD supplement market was dominated by CBD gummies (Fig. 11.1) with 28% share in the market. The convenience of using gummies, the full control of CBD dosing (standardized amount per gummy), and broad selection of flavors and sizes available on the market are probably the main factors stimulating the growth of that supplement category. The global CBD gummies market size is expected to reach USD 13.9 billion by 2028, according to a new report by Grand View Research (2021b). The marketing strategy emphasizing "the exceptional medicinal properties without any harmful side effects" is expected to be a major factor contributing to the growth of the gummies market. The consumption of CBD gummies to relieve anxiety, depression, pain, inflammation, and to induce sleep is expected to boost the growth of the market and be the key marketing catchphrase. Very recently, product formulators added a vitamins or herbal extract to the CBD gummies to emphasize or expand the possible health-related benefits of their products. For example, there are "Sleepy CBD gummies" that along with CBD contain 5 mg of melatonin, a known hormone commonly used in supplements promoting sleep, and there are "Immune Support gummies" with CBD, elderberry extract, vitamins B12 and C, and various other antioxidants to boost immune system (Table 11.2).

However, in view of the lack of stringent controls of marketed products by the competent authorities, mislabeling of CBD supplements is still frequent, which may represent a fraud for consumers, as well as pose a risk for their health (Cerino et al., 2021).

In a study involving 88 CBD liquid products, including oil and vaporization liquid, marketed by 31 different companies, CBD median levels were 9.45 mg/mL (range 0.10–655.27 mg/mL), which was lower compared to the median labeled concentration of 15.00 mg/mL (range 1.33–800.00). CBD-containing oil to be consumed orally was inaccurately labeled in 55% of cases (Bonn-Miller, Loflin, & Thomas, 2017). The FDA has sent warning letters to several companies requesting them to stop claiming that their CBD products may treat or even cure serious diseases, including cancer (Corroon & Kight, 2018).

The widespread use of CBD products has been poorly investigated. One study recruited 2409 individuals who participated in an online survey designed to assess the reasons, risks, and modalities behind CBD consumption (Corroon & Phillips, 2018). Approximately a third of participants reported some nonserious adverse event, with the most frequently occurring adverse events being dry mouth (11.12%), euphoria (6.43%), hunger (6.35%), red eyes (2.74%), and sedation/fatigue (1.78%). Among the 1483 users who reported using CBD to treat a medical condition, which most frequently included pain, anxiety, depression, and sleep disorders, approximately one-third of participants stated that CBD alone could manage their medical condition by itself. Importantly, ~ 40% of participants were motivated to consume CBD as part of a healthy lifestyle. Despite its numerous limitations, including the lack of data regarding the dose consumed and timing of consumption, as well as the absence of a control group, this study has the merit to capture consumers' motivations behind CBD use, which is essential to direct further research and investigations, as well as regulatory interventions.

Besides CBD, hemp contains other minor phytocannabinoids such as cannabigerol (CBG), cannabinol (CBN), and cannabichromene (CBC), many of which also have beneficial effects on human health. Little is known about these nonpsychoactive cannabinoids. Research on the beneficial health effects of CBG, CBN, and CBC has mostly been carried out on animals (rats and mice) and not yet in humans (Zagožen, Čerenak, & Kreft, 2021). Thus studies on the effects of minor cannabinoids in humans are urgently needed since dietary supplements containing these interesting compounds are available already on the market (Table 11.2). Existing online information has described the potential health benefits of CBG as a pain relief, anticancer and antiinflammatory agent. Moreover, CBG might raise the mood, increase clarity of mind, regulate blood pressure, or improve digestive health . CBG containing tinctures, capsules, or high purity CBG isolates, that are available on the market, are promoted as products that may have a positive effect on neurocognition, can help reduce inflammation, and can assist with relieving anxiety and sadness. Other minor cannabinoid offered more frequently as the dietary supplement is CBN. Companies that sell CBN products often market it as a sleep aid (Table 11.2), but currently there is only anecdotal evidence that CBN could be sedative. Growing popularity and demand for supplements containing minor cannabinoids has grown exponentially in the last few years, and more and more cultivators are using new genetics to express these cannabinoids in larger amounts supporting the development of new hemp-based supplements.

Despite the large consumption of cannabinoids as a nutritional supplement, evidence from preclinical and clinical studies exploring the effects of cannabinoid-based dietary supplements is limited to nonexisting. The need for further research is becoming more compelling as many of the multinational companies are planning to supplement some of their most popular products with CBD.

11.3 Hemp-based cosmetic products

The skin is not only the largest organ of human body, but it is also its most visible part. Its function is to protect and insulate the body against external agents and to activate the first line of the immune defense system. Any structural or physiological alteration of the skin may have consequences for our whole health, and conversely, our skin state may reveal a disorder located elsewhere in the body (Kanitakis, 2002). Realizing the importance of the skin since early stages, humans have used their ethnobotanical knowledge to care for it, keeping it moist and nourished. They also tried to improve their appearance with the use of cosmetics.

The skincare market is estimated to be worth between $135 and $155 billion by 2021 (Jhawar, Schoenberg, Wang, & Saedi, 2019). That is a great market for innovation and introduction of new products. Hence, topical and cosmetic products containing hempseed oil, hemp root, and CBD have gained great attention from industrial and consumer's perspective. Aside from the antiinflammatory potential and treatment of dermatitis, hempseed oil has shown to improve skin moisture while being well tolerated by consumers in terms of textural properties and absorption upon application on skin (Kowalska, Wozniak, & Pazdzior, 2017; Ligeza, Wygladacz, Tobiasz, Jaworecka, & Reich, 2016). Thus the use of hempseed oil and extracts containing CBD has extended to numerous topical products such as hand creams, facial products, arthritis targeted products, transdermal patches, lip balms, and massage gels and oils.

Most manufacturers claim benefits of using hempseed oil emphasizing its moisturizing, hydrating, and skin rejuvenating properties. On the other hand, CBD claims are more focused on arthritis and muscle relief, and antiacne properties. The therapeutic effect of CBD on skin is explained by cannabinoid receptors on many different cell types of the skin, including, but not limited to epidermal keratinocytes, melanocytes, mast cells, fibroblasts, sebocytes, sweat gland cells, as well as certain cell populations of hair follicles (Tóth, Ádám, Bíró, & Oláh, 2019). Scientific evidence supporting claims of commercial topical formulations is limited or at least is not specified by manufacturer's studies. Claims are more likely attributed to beneficial properties of individual ingredients of hempseed oil, hemp root, and CBD.

11.3.1 Hempseed oil

Perhaps one of the most attractive features of hempseed oil for dermal products is its fatty acid profile rich in essential omega-3 and omega-6 polyunsaturated fatty acids (PUFAs) such as alpha-linolenic acid (ALA) and linoleic acid (LA) (Kolodziejczyk, Kozlowska, & Ozimek, 2012). As keratinocytes in the epidermis have limited ability to synthesize them, PUFAs from the diet and topical products are important nutrients for skin's normal function and appearance, its photoprotective properties, and immune functions (Hansen & Jensen, 1985; Huang, Wang, Yang, Chou, & Fang, 2018; Pilkington, Watson, Nicolaou, & Rhodes, 2011).

The epidermis's stratum corneum—the outermost layer of the skin—has a critical barrier function. The hydration of the stratum corneum is required to maintain skin plasticity, desquamation, and optimal barrier functions. Optimal hydration is often accompanied by aesthetic properties such as skin softness and lack of flaking. Intercellular lipids in the stratum corneum are fundamental in the regulation of water transport, barrier properties, and stress and relaxation of skin. While keratinocytes can synthesize most of the epidermal fatty acids, certain fatty acids like ALA and LA cannot be synthesized (Chapkin, Ziboh, Marcelo, & Voorhees, 1986).

Linoleic acid is the most abundant PUFA in the epidermis, and it is directly correlated with barrier and permeability of the skin. Acylceramides containing linoleic acid are essential for their barrier function. Deficiencies in linoleic acid result in the utilization of nonessential oleic acid as a substitute in ceramides synthesis; however, this leads to barrier defects and epidermal changes. Typical changes in the epidermis as a consequence of essential fatty acid deficiencies (EFAD) include atopic dermatitis, scaly dermatosis, psoriasis, skin permeability to water, and hair loss (Khnykin, Miner, & Jahnsen, 2011). Dietary supplementation and topical application of PUFA topical products has demonstrated to ameliorate clinical symptoms in skin conditions (Table 11.3).

Due to its composition, the hempseed oil is classified as a noncomedogenic oil, with the noncomedogenic rating of zero, meaning that it will neither clog skin pores nor cause the skin to break out in acne. Moreover, it is "dry oil" which do not leave oily layer on the skin after application. It is recommended as an ingredient in cosmetic product for all types of skin (dry, oil, combination, normal), but especially for the skins prone to eczema, psoriasis, or atopic dermatitis.

Atopic dermatitis (AD), with a worldwide prevalence of about 20% in children and 3% adults (Maghfour et al., 2021), is a chronic skin disease characterized by itchy, inflamed, and dry skin caused by abnormal epidermal barrier. People suffering from AD cannot tolerate even basic soothing or moisturizing formulas (Metwally et al., 2021). It has been shown that dietary hempseed oil, in contrast to olive oil, increases the plasma levels of LA, ALA, and GLA (gamma-linolenic acid) and also alleviates clinical symptoms of atopic dermatitis (Table 11.3) (Callaway et al., 2005). In a different study, 96% of patients showed improvement of AD symptoms after 5 months of daily doses of GLA (Senapati, Banerjee, Nath Gangopadhyay, Ghosh Road, & Naihati, 2008). Gamma-linolenic acid acts as a powerful antiinflammatory agent while simultaneously promoting skin growth and new cell generation (Bojarowicz & Woźniak, 2008).

The effect of hempseed oil administration on skin morphology and antiaging properties of mice was studied. Results showed that hempseed oil improved skin moisture content, dermal thickness, and increased the number of hair follicles (Li et al., 2012). Most studies have investigated the effect of dietary supplementation with essential fatty acids. Effects of topical administration have been studied less. In one such study, an emollient formulation containing GLA resulted in improved cutaneous hydration in children with atopic dermatitis (Ferreira, Fiadeiro, Silva, & Soares, 1998). Some studies have suggested that hempseed oil can help relieve the symptoms of dry skin and itchiness in atopic dermatitis or be helpful in treating acne (Jin & Lee, 2019; Kurek-Górecka, Balwierz, Mizera, Nowak, & Żurawska-Płaksej, 2018). However, other studies have failed to demonstrate a significant effect of essential fatty acids from hempseed oil administration on atopic dermatitis. Guidelines for atopic dermatitis developed by an interdisciplinary European committee concluded that oral supplementation of unsaturated fatty acids for treatment of atopic dermatitis lacks consistent evidence, while topical application in emollient may be tried in some cases (Wollenberg et al., 2018). Huang, Pei, Gu, and Wang (2020) pointed out that degradation products of highly unsaturated fatty

TABLE 11.3 Human-based trials with hemp-based cosmetic products.

Product type	Purpose	Experimental design	Experimental group	Dosing	Treatment duration	Inclusion criteria	Exclusion criteria	Results	Reference
Cold-pressed hempseed oil	Effects of dietary hempseed oil and olive oil on plasma lipid profiles, skin quality, and dermal medication usage	Controlled, randomized single-blind crossover study	Patients ($n = 20$) with a diagnosis of atopic dermatitis divided into two groups: olive oil group and hemp oil group	Oral consumption of 30 mL/day of olive or hempseed oils	Two intervention periods of 8 weeks separated by a 4-week washout period	Patients with atopic dermatitis BMI <30 kg/m², Age 25–60 years	Lipid-lowering, antihypertensive or asthma medications, nutrient supplements, steroid-containing skin creams, oral cyclosporine, and use of solarium	Significant improvement of skin quality and the atopic symptomology of the patients without any negative side effects or adverse reactions. Significant increase of LA, ALA, and GLA in plasma. Significant decrease of patients' skin dryness, itchiness, and dermal medication usage	Callaway et al. (2005)
C. sativa seed extract cream (3%)	Effects of topical applications on the reduction of skin sebum and erythema content	Controlled, single blinded and comparative study	11 patients applied the cannabis extract cream on the right cheek and control cream on a left cheek twice per day	Twice a day	12 weeks	Healthy males with an age between 20 and 35 years with no identified skin diseases or allergy to substances in formulations.	N/A	Significant decrease in sebum level in comparison to a control cream ($P < .05$) Significant decrease in erythema	Ali and Akhtar (2015)
Topical CBD gel (1%)	Effects of topical CBD for atopic dermatitis (AD)	A pre-post observational study	14 patients	N/A	2 weeks, application frequency was tracked using the eczema care log.	Age of 18 or older Diagnosis of AD/eczema Previous consideration of the use of topical CBD Not pregnant or breastfeeding women	N/A	Average decrease in intensity of pruritus. Average reduction in the impact of pruritus on the quality of life	Maghfour et al. (2021)

Product	Aim	Study design	Subjects	Dosage	Duration		Results	Reference
Hydrogel containing hemp flowers extract (0.5%, 1%)	Restoring the hydrolipid balance and rebuilding the hydrolipid barrier of the skin, damaged in a cleaning process by sodium lauryl sulfate (SLS)	Six fields (2×2cm) were marked on the forearm skin. 0.2mL of 1% SLS solution was applied to 5 fields. One field (control field) was not treated with any sample. After rinsing and drying, after 10min, 0.2g of hydrogels were applied on 4 fields of skin treated with SLS. Dry-out hydrogels were removed from the skin with a paper towel 30min after application.	15 healthy volunteers in the age of 28–36	N/A	N/A	N/A	Significant improvement of the skin condition An influence to restoring of the hydrolipid balance and rebuilding of the hydrolipid barrier of the skin was demonstrated	Zagórska-Dziok, Bujak, Ziemlewska, and Nizioł-Łukaszewska (2021)
CBD-enriched ointment	Effect on severe skin chronic diseases and/or their outcome scars	A spontaneous, anecdotal, retrospective study	20 patients with skin disorders: psoriasis ($n=5$), atopic dermatitis ($n=5$), and resulting outcome scars ($n=10$)	Twice a day	3 months	N/A	Improvement of the skin parameters, the symptoms, and the PASI index score. No irritant or allergic reactions were documented	Palmieri, Laurino, and Vadalà (2019)

acids from hempseed oil could have negative effects in consumers. Formulation of stable microemulsions could help improving the stability of unsaturated hemp oil in topical products (Esparza, Ngo, & Boluk, 2020; Pei, Luo, Gu, & Wang, 2020).

Besides essential fatty acids, hempseed oil contains carotenoids that are delaying the skin aging process due to their antioxidative and UV absorption properties. Moreover, carotenoids improve skin hydration and regeneration and stimulate fibroblasts for collagen and elastin production (Igielska-Kalwat, Gościańska, & Nowak, 2015). Phytosterols present in the hempseed oil affect the hydration of the dry skin and improve the skin barrier function improving skin elasticity and skin roughness (Sikorska, Nowicki, & Wilkowska, 2015).

Hempseed oil has been shown to have mild to weak antimicrobial properties especially against Gram-positive bacteria. It has been suggested that unrefined hempseed oil may contain higher concentrations of tocopherols, ALA, and cannabinoids that can act as more potent antimicrobials than unsaturated fatty acids (Mikulcová et al., 2017). Other authors have suggested a pronounced antimicrobial activity of hempseed oil against *Bacillus subtilis* and *Staphylococcus aureus* (Ali, Almagboul, Khogali, & Gergeir, 2012).

There are numerous hempseed oil-based cosmetic products, such as lotions, oils, creams, shampoos, toothpastes, mouthwashes, including claimed therapeutic products for various diseases/symptoms (Table 11.4). For example, hempseed oil is marketed as an effective cosmetic treatment for hair, with claims that direct application of the oil to hair has moisturizing benefits, can aid hair growth, may protect the hair, and aid in damage repair, and the oil may add shine to the hair. These claims are currently unproved but there are a substantial number of online retailers selling various hemp oil-based products intended for direct application on hair. Although many hemp-based cosmetic products offered for sale on low-cost e-commerce internet websites appear amateurish and hastily formulated to attract consumers, with hemp often being just a minor component, there are also manufacturers with reliable formulations producing quality soaps, shampoos, creams, personal care products, with claimed beneficial action to the skin, scalp, or hair, which are advertised in the mainstream e-commerce sites or their own websites.

11.3.2 Hemp roots

The resinous flower tops and seeds are not the only part of the cannabis plant filled with therapeutic promise. Cannabis roots have also provided relief for various ailments in traditional cultures.

The first mention of the curative qualities of cannabis roots dates back to 77 CE, as described in the Natural Histories by Latin naturalist Pliny the Elder (Pliny the Elder, 1856). Since then, cannabis roots have been utilized by herbalists and physicians not only in Europe, but in many regions stretching from China to Argentina. Hemp roots were valued for treating a wide range of conditions, including fever, infections, gout, arthritis, joint pain, and even postpartum hemorrhage (Ryz, Remillard, & Russo, 2017).

The therapeutic properties of hemp roots are most likely related to the presence of different active compounds, including triterpenoids (friedelin and epifriedelanol), monoterpenes (carvone, dihydrocarvone), alkaloids (cannabisativine, anhydrocannabisativine), or phytosterols (sitosterol, campesterol, stigmasterol) (Ryz et al., 2017). Within cannabis roots, there are also residues of atropine and nerine, as well as choline, which is not an alkaloid

TABLE 11.4 Examples of hemp-based or hemp-infused cosmetics available on the market.

Cosmetic product type	Product name	Ingredient used	Concentration of active	Manufacturer claim	Manufacturer	Reference
Cream	CBD Cream	Argan Kernel Oil, Fructooligosaccharides, Beet Root Extract, Olive Fruit Oil, Tocopheryl Acetate, Sodium Hyaluronate, Hemp Extract, Caryophyllene.	300 mg CBD	The best-selling retail brand for CBD-infused topical cream	Reliva CBD Wellness	https://relivacbd.com/product/cbd-cream-300mg
	CBD Relief Stick	Beeswax, Hempseed oil, Shea Butter, Hemp Extract, Tocopherol, Eucalyptus Leaf, Cornmint Herb Oil, Tea Tree Oil, *Arnica montana* Flower Extract, Beta Caryophyllene.	250 mg CBD	Conveniently pocket sized, the Reliva CBD Stick soothes and moisturizes your skin when you're on the go with an easy-to-use roll on application. The CBD Stick is infused with peppermint essential oil to help uplift and inspire.	Reliva CBD Wellness	https://relivacbd.com/product/cbd-stick-250mg
	Reverse Anti-Aging CBD Cream	*Aloe barbadensis* (Organic Aloe) Juice, Rosa damascena (Rose) Distillate, *Persea Americana* (Avocado) Oil, *Simmondsia chinensis* (Jojoba) Oil, Cannabidiol Hemp Extract, Matrixyl 3000TM, PhytoCellTecTM, Argireline NP, Tocopherol (Vitamin E), Retinol Palmitate (Vitamin A), Rosa mosqueta (Rosehip) Seed Oil, Ascorbyl Palmitate (Vitamin C Ester), dl-Panthenol (Pro-Vitamin B5), Allantoin (Comfrey), *Triticum vulgare* (Wheat Germ) Oil, *Daucus carota* (Carrot Seed) Essential Oil, *Salix nigra* (Black Willow Bark) Extract, *Rosmarinus officinalis* (Rosemary), *Azadirachta indica* (Neem) Oil, Sea Kelp Extract, Tocopherol (Vitamin E), Citral, Limonene	1 oz. cream contains 100 mg of CBD	Promotes vibrant skin and regrowth of new cells. Smooths away fine lines and helps support a healthy, youthful glow.	Dr Love CBD	https://drlovecbd.com/

Continued

TABLE 11.4 Examples of hemp-based or hemp-infused cosmetics available on the market—cont'd

Cosmetic product type	Product name	Ingredient used	Concentration of active	Manufacturer claim	Manufacturer	Reference
	Skin Relief CBD cream	*Carthamus tinctorius* (Safflower) Oleosomes, *Persea gratissma* (Avocado) Oil, *Lavandula angustifolia* (Lavender) Oil, Phenoxyethanol, Hemp- Derived Cannabinoid Extract, Tocopherol (Vitamin E).	1 oz. cream contains 250 mg of CBD	CBD soothes the skin; Avocado oil, lavender oil, and vitamin E leave the skin glow.	Green Roads	https://greenroads.com/
	Face cream	*Cannabis sativa* Seed Oil, *Butyrospermum parkii* Butter, Cera Alba, Tocopheryl Acetate, Limonene,	N/A	Hydration	The Body Shop	https://www.thebodyshop.com/
	Hemp seed oil Face cream	Enriched hemp oil extract, turmeric, Vitamin A, Vitamin C, Vitamin D, Vitamin E, *Arnica montana*, menthol	N/A	Antiaging and wrinkle	Melao	https://hemp-genics.com/
	Hemp Code Face cream	*Prunus amygdalus dulcis* (sweet almond) oil, *Cannabis sativa* seed oil, *Lycium barbarum* fruit extract, tocopherol, limonene, linalool, hexyl cinnamal, coumarin.	N/A	Skin nourishment	Hemp code	https://beautysense.ca/
	Acne Treatment Medicated Cream	Bentonite, Hemp Extract, Colloidal Oatmeal, *Eucalyptus globulus* Leaf Oil, *Melaleuca Alternifolia* (Tea Tree) Leaf Oil, *Mentha Piperita* (Peppermint) Oil, Silver Oxide, *Simmondsia chinensis* (Jojoba) Seed Oil, Zinc Oxide.	Salicylic Acid 1%	Helps clear up blemishes, pimples, blackheads, and whiteheads, allows the skin to heal through everyday use.	Abacus Health Products Inc.	https://www.charlottesweb.com/
Serum	Face Serum	*Cannabis sativa* Seed Oil, Enantia Chlorantha Bark Extract	N/A	Antiblemish	Sephora	https://www.sephora.com/
	Renew Anti-Aging CBD Serum	Hyaluronic acid, Vitamin C, Matrixyl 3000, CBD hemp extract,	1 fl oz. contains 250 mg CBD, 60% Hyaluronic acid, 4.5% Vitamin C, 1.3% Matrixyl 3000,	Helps lock away skin moisture, promotes increased collagen production and a firm, more even-looking complexion.	Dr Love CBD	https://drlovecbd.com/

Category	Product	Ingredients	CBD Content	Description	Brand	URL
Muscle Cream	Medterra CBD Pain Relief Cream	*Arnica montana* 1X HPUS 7%, Menthol 4% Aloe Barbadensis (*Aloe vera*) Leaf Juice, Argania Spinosa Kernel Oil, Butyrospermum Parkii (Shea) Butter Cannabidiol (CBD), *Cocos nucifera* (Coconut) OilHelianthus Annuus (Sunflower) Extract, Mentha Piperita (Peppermint) Oil, *Oryza sativa* (Rice) Bran Extract, *Rosmarinus officinalis* (Rosemary) Leaf Extract Simmondsia Chinensis (Jojoba) Seed Oil, Tocopherol,	500 mg and 1000 mg of CBD	Perfect for sore backs and stiff necks, this Pain Cream will provide fast relief for arthritis and joint pain. Registered as an OTC product, Medterra's Pain Cream breaks the stereotypes of traditional pain creams that only rely on menthol.	Medterra	https://medterracbd.com/product-pain-cream
	CBD American Shaman Topical Cream	Hemp Extract (Aerial Parts), Cannabidiol, Cannabigerol, α-Bisabolol, Terpineol, α-Pinene, Limonene, β-Caryophyllene, Linalool, Humulene.	500 mg CBD per jar (250 mg CBD per oz)	It helps in dealing with muscle pains, joint pains, and smoother skins.	CBD American shaman	https://cbdamericanshaman.com/topical-cream
	CBD Muscle & Joint Relief Cream	*Carthamus tinctorius* (Safflower) Oleosomes, *Cucumis sativus* (Cucumber) Fruit Extract, Ethylhexylglycerin, Hemp-Derived (*Cannabis sativa*) Cannabinoid Extract, *Lavandula angustifolia* (Lavender) Oil, *Persea gratissima* (Avocado) Oil, *Rosmarinus officinalis* (Rosemary) Oil, Tocopherol (Vitamin E)	1 oz. bottle contains 750 mg CBD, Methyl Salicylate 10%, Menthol 4%	Pain relief for backaches, arthritis, muscle aches and pain, strains or sprains.	Green Roads	https://greenroads.com/
	Hemp-EaZe Hemp Root And Honey Deep Healing Body Butter	N/A	N/A	Treatment for bruising, sore muscles, skin irritations, burns, scrapes, and cuts.	Tierra Sol Farm	https://www.tierrasolfarm.com/
	Extra Strength Hemp Root—Muscle Cream	Beeswax, Shea Butter, Arnica oil, Hempseed oil, Hemp Root extract, Essential Oils; Copaiba Balsam, Eucalyptus, Lavender, Peppermint, Wintergreen, Rosemary.	400 mg root extract in 60 mL	Soothes aching joints and muscles, works on nerve, joint and muscular pain.	Raw Acres	https://rawacres.com/

Continued

TABLE 11.4 Examples of hemp-based or hemp-infused cosmetics available on the market—cont'd

Cosmetic product type	Product name	Ingredient used	Concentration of active	Manufacturer claim	Manufacturer	Reference
Body cream/butter	Hemp-infused cream with CBD	*Aloe vera*, coconut oil, oat extract, sea buckthorn oil, and full-spectrum hemp extract	2.5 oz. tube contains 750 mg CBD	Revitalize, restore, and rejuvenate your skin with our body cream	Charlotte's Web Stanley Brothers	https://www.charlottesweb.com/
	UNSCENTED: Massage Cream	Jojoba seed oil, shea butter, cottonseed oil, beeswax, and CBD hemp oil	900 mg Hemp Oil Extract per 200 g Jar	This product utilizes the natural health benefits of CBD hemp extracts, including pain and anxiety relief. It locks in the body's natural moisture and nourish the skin with essential vitamins and minerals.	CBD Clinic	https://cbdclinic.co/shop/massage-cream-unscented-
Balm	Hemp-infused balm with CBD	Menthol, peppermint oil, eucalyptus oil and full-spectrum hemp extract	1.5 oz. tin contains 450 mg of CBD		Charlotte's Web Stanley Brothers	https://www.charlottesweb.com/
Roll-On	HEMP-INFUSED ROLL-ON WITH CBD (PEPPERMINT, Lavender)	Peppermint oil, menthol, eucalyptus oil and full-spectrum hemp extract	0.34 oz. bottle contains 100 mg CBD		Charlotte's Web Stanley Brothers	https://www.charlottesweb.com/
	Plus CBD Roll on (extra strength Hemp Extract)	Aucklandia Lappa Root Extract, *Achyranthes bidentata* Root Extract, *Cinnamomum cassia* Bark Extract, Poria Cocos Extract, *Glycyrrhiza glabra* (Licorice) Root Extract, Menthol, Hemp Oil (Aerial Plant Parts), Hydrolyzed Jojoba Esters,	200 mg CBD	PlusCBD Oil Hemp CBD Roll-On provides high intensity support without any extra dyes or fragrances—just hemp CBD oil and natural ingredients. Camphor, a warming agent, has been shown to increase circulation and readily absorbs into the skin. When applied topically, menthol has a cooling effect and increases blood flow to the skin.	CV Sciences + Plus CBD Oil	https://www.pluscbdoil.com/cbd-products/rollons/cbdoilrollon

Product	Ingredients	Amount/CBD	Benefits	Brand	Website
Freeze CBD Roller	Natural Menthol 3.5%, *Ilex paraguariensis* Leaf Extract, Uncaria Tomentosa Extract, *Boswellia serrata* Extract, Camphor Terpene	3.04 fl oz. contains 500 mg CBD	Menthol: Produces a cool, soothing sensation. Eucalyptus Oil: Helps reduce joint pain and inflammation.	Dr Love CBD	https://drlovecbd.com/
Cool Relief CBD Roll-On	*Carthamus tinctorius* (Safflower) Oleosomes, Bentonite, *Persea gratissima* (Avocado) Oil, Mentha piperita (Peppermint) Oil, Hemp Derived Broad Spectrum Cannabinoid Extract, Menthol, *Rosmarinus officinalis* (Rosemary) Oil, Tocopherol (Vitamin E), *Eucalyptus globulus* (Leaf Oil)	3 oz. bottle contains 750 mg CBD	With 750 mg of supportive CBD Peppermint, a cool friend to sore joints and limbs everywhere Rosemary oil: a centuries-old aroma used to awaken the senses and chase off those 'blah' feelings Bentonite clay: loaded up with wonderful nutrients to support your skin Avocado oil: keeps your skin glowing and moisturized!	Green Roads	https://greenroads.com/
Heat Relief CBD Roll-On	*Carthamus tinctorius* (Safflower) Oleosomes, Bentonite Clay, Xanthan Gum, *Persea gratissima* (Avocado) Oil, Hemp- Derived Broad Spectrum Cannabinoid Extract, Menthol, *Rosmarinus officinalis* (Rosemary) Oil, Tocopherol (Vitamin E), Capsaicin and *Capsicum annuum* (Cayenne) Extract.	4 oz. bottle contains 750 mg CBD	Pepper compounds for warm, cozy feeling. Rosemary oil: a centuries-old aroma used to awaken the senses and chase off those 'blah' feelings. Bentonite clay: loaded up with wonderful nutrients to support your skin. Avocado oil: keeps your skin glowing and moisturized!	Green Roads	https://greenroads.com/
Gel — Hemp multi-restore Gel cream	*Butyrospermum parkii* butter, shea butter, *Aloe barbadensis* leaf juice powder, *Helianthus annuus* seed oil, sunflower seed oil, *Cannabis sativa* seed oil, linalool, limonene, citral,	na	Skin moisturizing/ nourishing	Garnier organic	https://www.garnier.co.uk/

Continued

TABLE 11.4 Examples of hemp-based or hemp-infused cosmetics available on the market—cont'd

Cosmetic product type	Product name	Ingredient used	Concentration of active	Manufacturer claim	Manufacturer	Reference
	Full Spectrum Hemp Extract Cooling Gel	Aloe barbadensis (Aloe vera) Leaf Juice, Menthol, Hemp Extract Oil, Butyrospermum parkii (Shea) Butter, Oryza sativa (Rice) Bran Extract, Helianthus annuus (Sunflower) Extract, Rosmarinus officinalis (Rosemary) Leaf Extract, Tocopherol, Mangifera indica (Mango) Seed Butter, Hippophae rhamnoides (Seabuckthorn) Fruit Oil, Calophyllum inophyllum Oil, Calendula officinalis Flower Extract, Arnica montana Flower Extract, Camellia sinensis (Green Tea) Leaf Extract, Boswellia serrata Extract, Anthemis nobilis Flower Extract,	1.7 oz. pump contains 510 mg CBD	Our Cooling Gel is formulated for a more mindful wellness routine and it's hemp-infused with menthol and arnica so you can hit the gym, trail and personal goals.	Charlotte's Web Stanley Brothers	https://www.charlottesweb.com/hemp-infused-cooling-gel
	Hemp-Infused Cooling Gel with CBD	Menthol, arnica, and full-spectrum hemp extract	1.7 oz. pump contains 510 mg CBD	na	Charlotte's Web Stanley Brothers	https://www.charlottesweb.com/
Ointment	Arthritis Aches and Pain Relief Ointment	Menthol, camphor, beeswax, clove oil, cottonseed oil, eucalyptus oil, frankincense oil, hemp extract (THC-free), jojoba seed oil, lavender oil,	Camphor 10%, Menthol 10%, 200 mg CBD Hemp Extract per 40 g Tube	Arthritis Aches and Pain Relief Ointment was specifically formulated for effective absorption, offering penetrating short-term arthritis pain relief from arthritic symptoms including joint pain and stiffness. Rub this topical pain relief arthritis ointment deeply into your skin regularly for more comfort and flexibility every day	Charlotte's Web Stanley Brothers	https://www.charlottesweb.com/
	Back and Neck Pain Relief Ointment	Menthol, camphor, argan oil, cottonseed oil, frankincense oil, hemp extract (THC-free), honeysuckle oil, jojoba seed oil, and myrrh oil.	Camphor 10%, Menthol 15%, 200 mg CBD Hemp Extract per 40 g Tube	Helps soothe minor upper and lower back discomfort, including general muscle soreness, strains, sprains, and bruises. Use on neck, back, shoulders, or waistline.	Abacus Health Products Inc.	https://www.charlottesweb.com/

	Active Sport Pain Relief Ointment	Menthol, camphor, cottonseed oil, beeswax, hemp oil (THC-free), frankincense oil, jojoba seed oil, myrrh oil, honeysuckle oil,	Camphor 3.1%, Menthol 4%, 200 mg CBD Hemp Extract per 40 g Tube	Provides advanced, temporary pain relief for tired and aching muscles and joints.	Abacus Health Products Inc.	https://www.charlottesweb.com/
	Pain Relief Ointment—Level 5	16% Menthol, 11% Camphor	400 mg Hemp Oil Extract per 44 g Jar	CBD CLINIC, LEVEL 05 Pain Relief Ointment is specially designed for intense muscle and joint pains associated with backaches, arthritis, sprains, strains, and bruises.	CBD Clinic	https://cbdclinic.co/shop/level-5-pro-sport-deep-muscle-joint-pain-single-jar
	Eczema Therapy Medicated Ointment	Colloidal oatmeal, cottonseed oil, beeswax, hemp oil (THC-free), jojoba seed oil, peppermint oil, chamomile oil, eucalyptus oil, zinc oxide, silver oxide	Colloidal oatmeal 1%, 200 mg CBD Hemp Extract per 40 g Tube	Helps relieve irritation while locking in the body's natural moisture	Abacus Health Products Inc.	https://www.charlottesweb.com/
Massage oil	Massage Therapy Pain Relief Oil	Menthol, camphor, jojoba seed oil, cottonseed oil, frankincense oil, CBD hemp oil (THC-free), honeysuckle oil, myrrh oil, argan oil,	Camphor 10%, Menthol 15%, 200 mg CBD Hemp Extract per 100 g Tube	Reduces stiffness and discomfort associated with tired, sore, or aching muscles.	Abacus Health Products Inc.	https://www.charlottesweb.com/
	Touch CBD Massage Oil	High Oleic, Expeller Pressed Sunflower Seed Oil, Cannabidiol Hemp Extract, Rice Bran Oil, Expeller Pressed Sweet Almond Oil, Grapeseed Oil, Limonene, Citral, Orange Oil, Natural Vitamin E.	1000 mg of CBD in 8 oz	Creates a soothing, relieving massage experience	Dr Love CBD	https://drlovecbd.com/
Shampoo and Conditioner	Deep Moisture Repair Hemp Shampoo & Conditioner	Hempseed oil, green tea essential, green tea extract, Aloe vera, ginseng extract, camphor essential oil, borage seed oil, yucca extract, dermochlorella extract (seaweed based), nettle extract, Coco glucoside, jojoba oil, awapuhi, soluable natural plant collagen, wheat germ glycerides (gluten free), echinacea extract, licorice root extract, cocoa butter, ivy extract, soy protein (non-GMO), ascorbyl tetraisopalmitate (Palm based), golden seal extract, behenic acid (natural based), ylang ylang extract, sea salt.	N/A	Promotes shine, reduces frizz, restores hair conditions, and helps soften coarse hair.	Organic Pure Oil Inc.	https://www.organicpureoil.com/

Continued

TABLE 11.4 Examples of hemp-based or hemp-infused cosmetics available on the market—cont'd

Cosmetic product type	Product name	Ingredient used	Concentration of active	Manufacturer claim	Manufacturer	Reference
	Fortifying + Hemp Seed Oil Shampoo & Conditioner	*Aloe barbadensis* Leaf Juice, Water (Aqua), Sodium C14–16 Olefin Sulfonate, Cocamidopropyl Betaine, Coco-Glucoside, *Cocos nucifera* (Coconut) Water, *Cannabis sativa* Seed Oil, *Theobroma cacao* (Cocoa) Seed Butter, *Passiflora incarnata* Seed Oil, Polyquaternium-10, Polyquaternium-6, Acrylates Crosspolymer-4, Glycol Distearate, Laureth-4, PEG-150 Pentaerythrityl Tetrastearate, PPG-2	N/A	Bestows hair with supple strength and delicate softness.	Maui Moisture	https://www.mauimoisture.com/
	Original Herbal Shampoo & Conditioner	Cocamidopropyl Betaine, Cocamide MEA, *Cannabis sativa* Seed Oil, *Helianthus annuus* (Sunflower) Seed Oil, *Punica granatum* Extract, *Helianthus annuus* (Sunflower) Seed Extract, Leuconostoc/Radish Root Ferment Filtrate	N/A	Helps limit heat damage and preserve moisture; Nourishes and strengthens hair, making it look fuller and less likely to break, as well as encouraging new growth	Hempz	https://hempz.com/
	HEMP ME! Hair Shampoo & Conditioner	*Cannabis sativa* Seed Extract,	N/A	Hempseed Extract nourishes the hair and scalp and promotes hair growth. The hair is revitalized, vibrant, and shiny.	REVUELE	https://revuele.eu/
	Calm the Frizz Shampoo & Conditioner	Hempseed oil	N/A	Tames unruly hair, making it noticeably luscious and smooth.	Aussie	https://aussiehair.com/
	Calm the Frizz 3 Minute Miracle deep conditioning treatment	Hempseed oil	N/A	Keeps unruly hair in its place	Aussie	https://aussiehair.com/

	Ingredients		Description	Brand	Website
MILL CREEK BOTANICALS Hemp Shampoo & Conditioner	Organic Barbadensis (*Aloe vera*) Leaf Juice, *Cannabis sativa* (Hemp) Seed Oil, Argania Spinosa (Argan) Oil, OrganicPrunus Serotina (Wild Cherry) Bark Extract, Organic*Chamomilla recutita* (Matricaria) Flower Extract, Organic Hamaelis Virginiana (Witch Hazel) Leaf Extract, Organic *Hydrastis canadensis* (Golden Seal) Extract, Organic Calendula Ocinalis Flower Extract, Organic *Humulus lupulus* (Hops) Extract, Organic *Equisetum arvense* (Field Horsetail) Extract, Panthenol (Provitamin B5), Folic Acid (VitaminB9), Niacin (Vitamin B3), Simmondsia Chinesis (Jojoba) Oil,	N/A	Hemp oil maintains hairs' natural texture and supports scalp health leaving your hair soft, manageable, and full of life.	MILL CREEK BOTANICALS	https://millcreekbotanicals.com/
Essential Moisture Shampoo & Conditioner	*Olea europaea* (Olive) Fruit Oil, *Cannabis sativa* Seed Oil, Hydrolyzed Quinoa, Silk Amino Acids, Hydrolyzed Silk	N/A	Nourishing	Marc Anthony	https://marcanthony.com/
Botanique Hemp + Hydration Shampoo & Conditioner	*Cannabis sativa* (Hemp) Seed Oil, *Hibiscus sabdariffa* Flower Extract, Coumarin, Hexyl Cinnamal, Limonene, Linalool	N/A	Helps to revive dry and dull hair, leaving it softer, shinier, and stronger	TRESemmé	https://www.tresemme.com/
MELAO herbal moisturizing hemp oil hair shampoo	Hempseed oil	N/A	Omega-6 and Omega-3 Essential fatty acids from hempseed oil strengthen the scalp and hair-shaft.	Guangzhou Yilong Cosmetics Co., Ltd	https://www.globalsources.com/
Medical Hemp Cannabis Formula Shampoo	Hempseed oil, D-panthenol, Vitamin E	N/A	Revitalize, moisturize, and restores hair strength.	Medical Hemp	https://medicalhemp.pl/
Liquid oil — Hemp multi-restore Facial sleeping oil	*Olea europaea* fruit oil (olive fruit oil), *Cannabis sativa* seed oil, *Helianthus annuus* seed oil (sunflower seed oil), *Prunus armeniaca* kernel oil (apricot kernel oil), tocopherol, linalool, limonene, citral	N/A	Skin nourishment	Garnier organic	https://www.garnier.co.uk/

Continued

TABLE 11.4 Examples of hemp-based or hemp-infused cosmetics available on the market—cont'd

Cosmetic product type	Product name	Ingredient used	Concentration of active	Manufacturer claim	Manufacturer	Reference
	Hemp Oil for Hair & Skin	*Prunus amygdalus dulcis* Oil (Sweet Almond), *Helianthus annuus* Seed Oil (Sunflower), *Cannabis sativa* Seed Oil (Hemp), Fragrance, Limonene, Linalool.	N/A	Used to strengthen the hair on a cellular level, hemp oil produces a protective layer to prevent and repair damage. Hemp oil can also be used to treat scalp conditions such as dandruff, and can stimulate the hair follicles and speed growth due to naturally occurring vitamin E. On the skin, hemp oil increases elasticity and has excellent hydration properties, bringing down the redness and irritation of acne or eczema outbreaks.	Nature Spell	https://naturespell.co.uk/
Lip Balm	Balmy CBD Lip Balm	Beeswax, Coconut Oil, Shea Butter, Emu Oil, CBD Hemp Extract	Beeswax: locks in moisture, protects from the air Coconut Oil: a super-substance in itself, moisturizing Shea Butter: antiinflammatory moisturizer Emu Oil: a rare compound that penetrates all 7 layers of skin for deep moisturizing	25mg CBD	Dr Love CBD	https://drlovecbd.com/
	Hemp Root—Lip Balm	Beeswax, Shea Butter, Hempseed oil, Hemp Root extract, Essential Oils; Peppermint, Wintergreen	Protects and nourishes dry chapped lip.	N/A	Raw Acres	https://rawacres.com/

Type	Product	Ingredients	CBD Content	Description	Brand	Website
Bath Bomb	Soak CBD Bath Bomb	Hemp CBD, Epsom Salt, Cream of Tarter, Witch Hazel, Eucalyptus Oil, Lavender Oil.	50 mg CBD	N/A	Dr Love CBD	https://drlovecbd.com/
	CBD Refresh bath bomb	Epsom Salt, Kaolin Clay, Mentha piperita (Peppermint) Oil, *Citrus limonum* (Lemon) Oil, Hemp Derived Cannabinoid Extract.	150 mg CBD	Soothing and refreshing	Green Roads	https://greenroads.com/
Transdermal Patch	CBD Infused Patch	Polymer Blend, CBD Hemp Extract	20 mg, 60 mg or 100 mg per Patch	Targeted relief of body craves	Social CBD	https://socialcbd.com/
	CBD Transdermal Patches	Organic Aloe Butter, Sunflower Lecithin, Full Spectrum Hemp Extract Apricot Kernel Oil, Red Palm Oil, Shea Olein, Safflower Oil, Organic Sage Essential Oil	60 mg full-spectrum CBD per patch	Convenient way to monitor daily CBD intake.	PureKana	https://purekana.com/
	Relief Wrap	Menthol, *Centella asiatica* Extract, *Polygonum cuspidatum* Root Extract, Scutellaria Baicalensis Root Extract, *Camellia sinensis* Leaf Extract, Glycyrrhiza Glabara (Licorice Root Extract), Sodium Hyaluronate, Full Spectrum CBD, Phytocannabinoid-Rich (PCR) Hemp Oil	50 mg Full Spectrum CBD extract per wrap	Designed to relieve discomfort in various areas of the body	Envy CBD	https://envycbd.com/
	Pure Ratios 96-h CBD Patch	Avocado Oil, Almond Oil, *Aloe vera*, Shea Olein.	40 mg CBD per patch	Targets inflammation and chronic aches, improves mood and increase activity. Supports back pain, knee pain, and nerve pain as well as relieving the symptoms of issues like Rheumatoid Arthritis and Fibromyalgia.	Handpicked CBD	https://handpickedcbd.com/

Continued

TABLE 11.4 Examples of hemp-based or hemp-infused cosmetics available on the market—cont'd

Cosmetic product type	Product name	Ingredient used	Concentration of active	Manufacturer claim	Manufacturer	Reference
Salve	CBDistillery CBDol Topical CBD Salve	*Butyrospermum parkii* (Shea) Butter, *Cannabis sativa* (Hemp) Seed Oil, Cera Alba (Beeswax), *Gaultheria procumbens* (Wintergreen) Leaf Oil, *Cannabis sativa* (Hemp) Full Spectrum Extract, Mentha Piperita (Peppermint) Oil, *Olea europaea* (Olive) Fruit Oil, *Prunus armeniaca* (Apricot) Kernel Oil, *Helianthus annuus* (Sunflower) Seed Oil, *Prunus amygdalus dulcis* (Sweet Almond) Oil, *Lavandula angustifolia* (Lavender) Flower Oil, *Eucalyptus globulus* Leaf Oil, *Cinnamomum camphora* (Camphor) Bark Oil, *Rosmarinus officinalis* (Rosemary) Leaf Oil, *Arnica montana* Flower Oil, Boswellia Carterii (Frankincense) Oil, Melaleuca Alternifolia (Tea Tree) Leaf Oil, Lactobacillus Ferment, Lactobacillus, *Cocos nucifera* (Coconut) Fruit Extract, *Citrus grandis* (Grapefruit) Peel Extract, *Camellia sinensis* (Green Tea) Leaf Extract, *Chamomilla recutita* (Matricaria) Flower Extract, *Cinnamomum zeylanicum* (Cinnamon) Leaf Oil, *Salix nigra* (Willow) Bark Extract, *Populus tremuloides* (Aspen) Bark Extract, *Lonicera caprifolium* (Honeysuckle) Extract, *Aloe barbadensis* (Aloe) Extract	500 mg CBD in 1 oz. container	Provides targeted relief exactly where body needs.	Balanced Health Botanicals (BHB)	https://www.balancedhealthbotanicals.com/#brands
	Hemp Root—Hand Salve	Beeswax, Shea Butter, Arnica oil, Hempseed oil, Hemp Root extract, Essential Oils; Tea Tree, Eucalyptus, Lavender, Peppermint, Wintergreen	N/A	Nourishes and moisturizes skin.	Raw Acres	https://rawacres.com/

floss	Hello hemp seed oil + coconut oil infused floss	N/A	N/A	N/A	hello	https://www.hello-products.com/
floss picks	Hello hempseed oil + coconut oil infused floss picks	N/A	N/A	floss picks taste fab, and they're infused with hempseed oil and coconut oil to thrill your grill	hello	https://www.hello-products.com/
toothpaste	Hello activated charcoal + hempseed oil fluoride free toothpaste	N/A	Charcoal powder, *Cannabis sativa* seed oil (hempseed oil), titanium dioxide, coconut oil, *Stevia rebaudiana* leaf extract	Formulated with activated charcoal (made from sustainable bamboo), an epic whitener and that removes surface stains, and our farm grown mint, which provides mind-blowing freshness. If this paste soothed and moisturized any more, it'd be illegal (coconut oil + hempseed oil are a soothing dream team).	hello	https://www.hello-products.com/
Salt	Hemp Root—Foot Soak	N/A	Himalayan Pink Salt, Dead Sea Salt, Hemp Roots, Sodium Bicarbonate, Essential Oils; Peppermint, Rosemary	Restore, Refresh and Rejuvenate achy feet.	Raw Acres	https://rawacres.com/
Salt	Hemp Root—Bath Salts	N/A	Himalayan Pink Salt, Dead Sea Salt, Hemp Roots, Sodium Bicarbonate, Malva Flower, Hibiscus Flower, Lavender	Helps reduce pain, swelling, and inflammation. Soothes and relaxes body before bed.	Raw Acres	https://rawacres.com/
Face mask	CBD Face Mask—5 Variations (charcoal, cucumber, *aloe vera*, rose and lavender)	CBD 50 mg	Broad Spectrum Hemp Extract, (Mentha Piperita (Peppermint) Oil, Sodium Hyaluronate, *Cucumis sativus* (cucumber) distillate, *Sambucus nigra* Flower Extract, *Prunus amygdalus dulcis* (Sweet Almond) Extract, *Aloe barbadensis* Leaf Juice, Sodium Hyaluronate, *Malva sylvestris* Extract, *Primula veris* Extract, *Veronica officinalis* Extract, *Melissa officinalis* Leaf Extract, *Achillea millefolium* Extract, *Lavandula angustifolia* (Lavender) Oil, Enantia Chlorantha Bark Extract), *Citrus reticulata* Fruit Extract, *Citrus aurantium* Amara Fruit Extract, *Citrus aurantium* Sinensis Peel Extract	The sheet masks are available in five formulations designed to treat specific skincare challenges such as dry, stressed, or congested skin. Our CBD Face Masks contain an assortment of active botanicals and essential oils that when coupled with CBD create a blissful home-spa experience.	CBD Fx	https://cbdfx.com/products/cbd-face-mask

but is very important for the formation of acetylcholine, lipid bilayers, and cellular signaling (Mole & Turner, 1973). Carvone and dihydrocarvone give spearmint its distinct aroma and modulate the TRPM8 ionotropic receptor, which is sensitive to cooling and pain (Calixto, Kassuya, Andre, & Ferreira, 2005). Carvone's antinociceptive activity has been confirmed by several in vivo tests, and that is why spearmint essential oil, which contains up to 70% of carvone, is currently under investigation as a treatment for osteoarthritis (Goncalves et al., 2008). Friedelin has shown potent antiinflammatory activity in in vivo experiments, reducing edema and swelling in any area of the body tested (Antonisamy, Duraipandiyan, Ignacimuthu, & Kim, 2015). As a treatment for inflammation, cannabis root has historically been administered as a topical preparation following a boiled water extraction or decoction. Nowadays the hemp root extract can be found in topical products (creams, butters) that are claimed to soothe aching joints and muscles, work on nerve, joint, and muscular pain, to treat bruising, skin irritations, burns, scrapes, and cuts (Table 11.4).

11.3.3 Cannabidiol (CBD) and other cannabinoids

The market for cannabidiol-infused products is ever growing, with the topical products as one of the hottest product formats. The Brightfield Group reported, in 2019, US sales of CBD topicals exceeded $703 million, second only to CBD tinctures. By 2025 topicals are projected to reach $4.5 billion. In 2020 CWI Consulting Services—a consumer packaged goods (CPG) planning and development consultancy that focuses on CBD products for the mainstream market—conducted research and gathered data on the topicals CBD market. It surveyed 360 hemp-based topical CBD products from 112 brands. They found that the top 53 brands (company that sold more than $350,000 in topical CBD products in 2019) comprised almost 60% of all sales in 2019 in the CBD sector (Marrapodi, 2021).

CBD has made its way into almost every personal care topical format—from lotions to face masks. As per report by CWI Consulting Services, lotions and creams largely outweighed all other topical formats in US in 2020, accounting for 43% of all formats. Next were balms, making up 26% of the stock keeping units (SKUs). Together, these two categories made up 69% of the topical products reviewed. Other major formats were sticks (7.5%), roll-ons (6%), oils (4%), gels (3.6%), and patches (3%), making up a total of 24.1% (Marrapodi, 2021).

The importance of CBD in topical products resides in its antagonist/inverse agonist effect on cannabinoid receptor 1 (CB1) which is of great potential downregulating an overexpressed endocannabinoid system (ECS) (Lu, Dopart, & Kendall, 2016; Scopinho, Guimarães, Corrêa, & Resstel, 2011). The same downregulating effect of CBD on CB2 receptor can explain the wide range of skin antiinflammatory and antiarthritis properties of CBD (Pertwee, 2008).

Besides ECS receptors, CBD is able to activate several other cellular receptors. Transient receptor potential channels (TRP) are expressed in sensory nerve endings, but also in many neuronal cell populations including keratinocytes and skin-resident immune cells. They participate in the formation and maintenance of skin barrier, hair follicle growth, and cutaneous immunological and inflammatory processes (Yang, Feng, Luo, Madison, & Hu, 2019).

CBD is able to desensitize the TRPV1 (transient receptor potential vanilloid-1) receptor leading to an analgesic effect suggesting its potential use as an antiinflammatory and for chronic pain (De Petrocellis et al., 2011). There are few studies done to explore this effect of CBD. In one such study, CBD activation of TRPV4 resulted in inhibition of sebocytes lipogenesis which is beneficial for acne treatment (Oláh et al., 2014). A study evaluated the application of a CBD ointment on 20 participants suffering from psoriasis and atopic dermatitis. The commercial product evaluated was "Hemptouch organic skin care ointment" (Hemptouch Ltd., Novo Mesto, Slovenia) and listed the following ingredients: CBD seed oil, *Mangifera indica*, *Calendula officinalis*, *Lavendula officinalis*, *Chamomile*, *Amyris Balsamifera*, and butyrospermum (shea butter). After 90 days of treatment, improvements were reported in hydration and skin elasticity. More notoriously, the CBD ointment reduced cutaneous blemishes and scars (Table 11.3) (Palmieri et al., 2019). In another study, Zagórska-Dziok et al. (2021) tested the efficiency of hydrogel containing 0.5% and 1% of hemp flowers extract on restoring the hydrolipid balance and rebuilding of the hydrolipid barrier of the skin damaged by excessive usage of sodium lauryl sulfate (SLS), a surfactant commonly used in cosmetics. After application of hydrogels containing hemp flower extracts on (SLS)-treated skin for 30 min, the skin condition improved significantly, and authors concluded that hemp hydrogels eliminated adverse effect of analyzed surfactant on the skin moisture (Table 11.3).

Furthermore, the ability of cannabinoids to stimulate cannabinoid (CB) receptors, which might be relevant in the treatment of psoriasis, a chronic skin disorder that causes skin cells to multiply faster than normal, was also studied (Ramot et al., 2013; Wilkinson & Williamson, 2007). Results obtained demonstrated that cannabinoids and their receptors have the potential to treat psoriasis by modulation of keratins K6 and K16 expression in human skin. These findings invite the speculation that the therapeutic down-modulation of K6 and/or K16 expression by CB1 agonists might become exploitable for the management of other dermatoses besides psoriasis, such as pachyonychia congenita and acne (American Academy of Dermatology Association, 2021), and could be used to modulate keratinocyte migration-dependent reepithelialization in wound healing, similar to related findings in periodontal wound repair (Kozono et al., 2010). Recent observational study reported the effectiveness of topical CBD oil use in patients with epidermolysis bullosa, a rare blistering skin disorder that is challenging to manage because skin fragility and repeated wound healing cause itching, pain, limited mobility, and recurrent infections (Chelliah, Zinn, Khuu, & Teng, 2018). All patient participating in the study reported faster wound healing, less blistering, and amelioration of pain confirming effectiveness of topically applied CBD. Further studies are therefore needed to exploit putative therapeutic potential of the CBD and endocannabinoid signaling in the clinical management of skin-related diseases.

A limited number of studies have investigated the effect of hemp constituents on hair growth properties. However, it is becoming more accepted that the endocannabinoid system is a key regulator of hair follicle cell growth. In this context cannabinoid receptor CB1 agonist like THC has shown to inhibit hair shaft elongation, decrease proliferation of hair matrix keratinocytes, and induce intraepithelial apoptosis and premature hair follicle regression. This can be used to manage unwanted hair loss in certain areas. In contrast, CB1 antagonist like CBD from hemp can be used to promote hair growth (Smith & Satino, 2021). Szabó et al. (2017) showed that lower doses of CBD resulted in hair shaft elongation. However, higher doses can suppress hair shaft production, which can be explained by the concentration-dependent activation of other cellular receptors (Szabó et al., 2020). Smith and Satino (2021) proposed that CBD mechanism can be beneficial to treat androgenetic alopecia in combination with existing treatment drugs (Smith & Satino, 2021).

CBD and other cannabinoids such as cannabichromene (CBC), cannabigerol (CBG), and cannabinol (CBN) have shown wide antimicrobial properties, including activity against methicillin-resistant *S. aureus* strains (Appendino et al., 2008). Terpenes including limonene and β-caryophyllene have also shown antimicrobial properties by altering the cell membrane leading to lysis (Schofs, Sparo, & Sánchez Bruni, 2021). The antimicrobial properties of CBD and other hemp components are promising in the formulation of chemical-free creams and lotions where synthetic preservatives such as phenoxyethanol need to be used. In addition, the antimicrobial properties of extracts from different parts of the hemp plant could be of clinical relevance in the development of wound dressings, transdermal patches, biomaterials, and ointments to treat recurrent skin infections (Khan, Warner, & Wang, 2014; Metwally et al., 2021).

European health and beauty product manufacturers can now safely use hemp-derived CBG in their cosmetics and skin care lines with approval from a key market regulator for trade in the European Union. The European Commission added CBG in April 2021 as a safe ingredient for skin conditioning to CosIng, its cosmetics ingredient database. The EU added hemp-derived CBD to the CosIng database in February 2021, and synthetic CBD and hemp leaves are also among the approved ingredients for cosmetics.

CBG is believed to work with the CB1 and CB2 endocannabinoid receptors located in the skin. Proponents say the hemp-derived compound has antiinflammatory, antibacterial, and antioxidant properties that help the endocannabinoid system maintain healthy skin function. Once CBG enters the epidermis, it penetrates to the basal and subcutaneous skin layers where it helps calm inflammation, reduces blemishes, purifies pores, balances oil production, and enhances cellular turnover (Hemp Today, 2021).

The increasing interest in cosmetic and personal care products produced from natural, organic, and plant-based ingredients, including hemp- and CBD-based products, drives the small business takeovers by worldwide giants. In 2020 Colgate-Palmolive acquired Hello Products—a natural, oral care specialist with a range of CBD-infused toothpaste and mouth rinses and a series of hempseed oil-based personal care products. Earlier this year, Colgate-Palmolive filed patent applications on antibacterial CBD oral care compositions prepared using hempseed oil and hemp CBD oil. The new product line of personal care products will include antiperspirant, mouthwash, toothpaste, oral gel, chewing gums, oral sprays among others (Arora et al., 2021a, 2021b, 2021c; Hernandez et al., 2021) and will complement the existing line of Hello products (Table 11.4).

In another takeover, Canadian cannabis processor "The Valens Company" bought Green Roads in a deal potentially valued at $60 million. Based in South Florida, Green Roads is the largest privately owned CBD company in the United States. UK CBD firm Dragonfly Biosciences has formed a nonexclusive distribution agreement with Health House International Australia to sell its CBD oils over the counter in Australia. The Australian Therapeutic Goods Administration recently changed the regulations to allow CBD products to be sold in Australian Pharmacies.

In conclusion, hemp-based cosmetic products are gaining momentum in the current market. Although the properties of CBD and hempseed-based ingredients are promising, the evidence in favor of hemp-derived commercial products and clinical applications is limited (Fitzcharles, Clauw, & Hauser, 2020).

Moreover, the scientific research related to hemp-based cosmetics should extend to the other ingredients often found in cannabis topicals, such as arnica, lavender, mango butter, turmeric, ginger, white willow bark, aloe, wintergreen, and others. The healing properties and interactions of these ingredients with cannabis are generally not well understood. In fact, there is a general misconception that natural ingredients in a topical will work well together. Charlene Freedom, a certified natural health practitioner in Toronto, Canada, who holds workshops on making homemade cannabis topicals, says that is not the case. For example, she has found that beeswax blocks cannabinoids from penetrating the skin. "We find it's good in a balm, not a cream" (Christine Giraud, 2019).

11.4 Challenges and opportunities

There are many challenges that popularization of industrial hemp (and derived products) faces today. They touch every facet of this new economic *gem*: plant growing, processing, production, importation, transport, marketing, sale, consumption, research, etc. There is a plethora of underlying problems, including very tight global regulations, local laws that are either unfavorable or leave too much room for individual interpretation, frequent changes of governing laws, historically negative connotations related to "weed," consumer's perception and limited acceptance, lack of confidence in industry as a whole, financial risks for growers and processors, lack of understanding and appreciation for the current quality standards required in development and production of medicinal products, somewhat limited state of scientific knowledge and evidence-based clinical data, strong conflicts of interests, sometimes even at, seemingly, the same side of the "barricade" (Sabet, 2020; Tao, 2013). Finally, as it could be expected, there is also a fair share of misinformation in the media, which makes the whole pool more murky, and where truth rarely has a chance to come to the surface (LaCapria, 2016).

Undoubtedly, many of these challenges have roots in numerous controversies and fables that were built around cannabis genus during the "Cannabis Great Prohibition" that started in 1930s of the last century. But there are also fundamental questions that, 100 years later, are still not answered and thus complicate things. It starts from a very basic problem around taxonomic classification of Cannabis plants (Chouvy, 2019).

11.4.1 Genetic classification and psychoactive and nonpsychoactive cannabinoid content

The place of Cannabis in the International Code of Nomenclature for Algae, Fungi, and Plants (ICN) is not clear as scientists still strongly disagree about how many species and subspecies the genus Cannabis have and how to distinguish them. This disagreement circulates around importance of plant's morphological features or phytochemical composition (biologically same plants produce significantly different quantities of both psychoactive and nonpsychoactive cannabinoids) on its taxonomy. These two facts have many further implications, but eventually it all comes to one thing: for regulatory and law enforcement purposes, there is no other discriminant feature that would be easy to test for and that would allow to decidedly distinguish hemp from marijuana, than a demarcation line drawn at somewhat arbitrarily chosen content of THC per dry weight of the plant. Thanks to work of two Canadian plant scientists, Ernest Small and Arthur Cronquist (Small & Cronquist, 1976), the most widely used limit in the world, for hemp, is 0.3% THC on a dry weight basis; any plant material that contains more THC than that, for legal purposes, is considered as marijuana, regardless of all other considerations. There are exceptions of course, for example, a number of European countries have either higher or lower limit, however in the November of 2020, European Parliament voted for increasing this limit to 0.3%, to bring it in line with regulations in most other countries. To add to the confusion, in some languages, such as it is the case of many German and Slavic languages, the same word "hemp" may refer, at least in the common speech, to either hemp or to marijuana. For example, this is the case for German *Hanf*, Dutch *hennep*, Swedish *hampa*, Norwegian and Danish *hamp*, Russian *Конопля*, Polish *Konopie*. And this is despite the fact, that before introduction and popularization of marijuana, all these were for centuries grown exclusively for a variety of industrial uses, such as animal feed, textiles fiber, rope fiber, building and construction materials.

11.4.2 Regulations and regulatory issues

But while there is, and always was (at least in countries where it was historically legal), wide acceptance for industrial hemp, it does not mean that there is the same acceptance for other uses of cannabis, with particularly strong feelings being often expressed against recreational use of cannabis, which co-notes with "narcotic" and "illegal." Consequently, use of such term, especially instinctively, can lead to unintentional but bitter misunderstanding.

Three international legal instruments, Single Convention on Narcotic Drugs of 1961, Convention on Psychotropic Substances of 1971, United Nations Convention Against Illicit Traffic in Narcotic Drugs and Psychotropic Substances of 1988, along with local, national, or state-level laws and regulations govern cultivation, manufacturing, and trade of drug and narcotic substances (United Nation, 2013), and it is widely accepted interpretation that they govern all cannabis and related products as well. And since the national-level laws are not harmonized, the resulting international legal situation is quite convoluted and due to recent changes, fluid. Therefore businesses that would like to engage in an international trade may look first at the most regulated and therefore stable and accessible markets, such as Canadian, Australian, and perhaps European. US market, while huge, has it quirks related to differences in legality of cannabis at the state and federal levels.

The global prohibition on Cannabis has been challenged worldwide for over a decade now, and as a result, it was decriminalized, abolished, or even legalized in a number of countries. In comparison, on international level, the progress is relatively slow. Not surprisingly, international conventions classified all cannabis, including hemp, as harmful drug. However, on 2nd December 2020, the United Nations Commission on Narcotic Drugs (UNC) partially followed recommendations made by the WHO's Expert Committee on Drug Dependence (ECDD) and reclassified cannabis and cannabis resin within the international drug conventions (UN Commission on Narcotic Drugs, 2020), thus removing it from the list of drugs that are seen as exceptionally harmful to public health. Nonmedical use of cannabis and related products has not been affected by this reclassification and remains under very strict international control. However, this step is very important in that way that it has lifted some international procedural barriers for cannabis-related medicinal research and development.

Authorities in various countries have recognized it and already modified their legislation to accept CBD as a medicinal product. However, the same authorities are not convinced that the current body of knowledge on CBD is sufficient to let it be marketed without any restrictions. For example, US FDA is far more critical than the broad pro-cannabis community. In its Note for Consumers, current of 03 May 2020 (FDA, 2020) it reminds that it is illegal to market CBD by adding it to food or label it as dietary supplement. Moreover, FDA lists some serious potential problems that were already reported, such as evidence for risk of liver injury, synergistic enhancement of action of alcohol and other sedative substances, potential risk of damage of male fertility system or male offspring of women that were exposed to CBD (Payne, Mazur, Hotaling, & Pastuszak, 2019). The last one is based on animal studies, but this is one of the reasons why animal studies are conducted: to give us early warning of a potential problem.

Therefore both industrial hemp and CBD products are universally subjected to strict local legislations and regulations, even if, like it is the case with hemp in the United States, they do not fall into controlled substance category anymore (Government of Canada, 2018; Medicines and Healthcare products Regulatory Agency (MHRA), 2016; Government of UK, 2014; Australian Government, 2020; New Zealand's Medicinal Cannabis Agency, 2020; Swissmedic, 2019; FDA, 2021b).

11.4.3 Lacking body of knowledge and its consequences

Some may see the UNC's cautious approach as a testimony to not how difficult is to get over the old preconceptions, but rather how little is known, overall, about the plant and its constituents: years of relatively strict prohibition hindered studies to the point that we still do not know much about plant's medicinal potential, safety, and overall risks associated with its use or exposure to it (Biasutti, Leffers, & Callaghan, 2020; Huber, Newman, & LaFave, 2016; Lafaye, Karila, Blecha, & Benyamina, 2017; National Academies of Sciences, Engineering, and Medicine, 2017). At the same time, in many of the signatory states, products containing cannabis or cannabinoids are still strictly regulated, if not illegal, regardless of their narcotic potential or risk they may pose for public health.

But what do we really know about cannabis and cannabinoids, how complete is our knowledge? Perhaps the most striking example is CBD, since this cannabinoid is the most, and compared to others, well researched. There is no obvious evidence of its recreational use, its safety profile is good, and, to date, there are no reports of health-related problems associated with it, if it is not abused and used alone (Chesney, Oliver, & Green, 2020; Fasinu, Phillips, ElSohly, & Walker, 2016; Iffland & Grotenhermen, 2017). Moreover, CBD is the only cannabinoid with quite well documented legitimate uses (Lopez et al., 2020; Maghfour et al., 2021; Thiele et al., 2018).

The rise of the brand new dynamic economy, fuelled by appetites for building lucrative businesses and flow of cash from various actors, gave impulse to quite unprecedented growth of research and publications on cannabis. Also, with relaxed or lifted restrictions came time to meet various regulatory requirements and standards. This calls for reliable scientific data and analytical methods.

Since the beginning, commercialization of cannabis products has been accompanied with very big claims but little to no scientific and clinical evidence to support them (Thompson, 2018).

The proliferation of false claims regarding the therapeutic potential of cannabis provoked reaction of authorities, popularized by plethora of different information sources (Federal Trade Commision, 2013). Selling, otherwise legal products, either physically labeled with unsanctioned medical claim or advice, or likewise advertised online, violates governing acts (like, in the United States, Federal Food, Drug, and Cosmetic Act) and sooner or later meets with reaction of authorities and regulatory bodies (FDA, 2019, 2021a).

There might be no agreement, neither on true medicinal value of cannabis, nor on consequences of its prolonged, especially recreational use. However, there is general agreement that all cannabis production, but particularly the one which is health or medicinal products oriented, needs to be lifted to Current Good Manufacturing Practice (cGMP) standards, with adequate quality assurance, testing by independent third-party laboratories, and in general should be subject to the same scrutiny and control as other equivalent industries are.

11.4.4 Potential for abuse

Besides the real safety concerns, another reason might be a risk, or perhaps a fear, of potential abuse. There are a number of ways, some well documented and some anecdotal, how hemp or CBD might be abused. The most straightforward one is using CBD as a relatively cheap, and readily available in large quantities, precursor in synthesis of psychoactive Δ^8-THC, Δ^{10}-THC, and other THC isomers (Cooke, 2021; Gaoni & Mechoulam, 1966; Webster, Sarna, & Mechoulam, 2004). Also, while it would be tedious work, it is technically feasible to fortify the minute content of THC, present in every hemp and nonhighly rectified hemp extract, to the level that can be considered ready for recreational use. However, because of general good availability and low price of THC originating from marijuana, such method would make very little practical sense.

While contested there is still very popular belief, particularly supported by marijuana growers, that changes in environmental conditions may lead to appreciable changes in chemical composition of the cannabis plant, including THC level (Place, 2019; Toth et al., 2020). Assuming that the belief is correct, it might be also assumed that this leaves room for abuse. However, the more important implication is a raise of legitimate concern that planted hemp may, without grower's intention, exceed mandated by law limit of THC. Naturally, this would lead to all kind of legal and financial problems that even honest grower may face. If a local law observes particular limit of THC that makes a difference between hemp and marijuana, for example 0.3% THC on a dry weight basis, and there are two identical plants or even plantations, grown from the same batch of seeds, and side by side, but one will test for 0.2% and another one for 0.4% of THC on a dry weight basis, the local law enforcers will most likely classify them as hemp and marijuana, respectively. And the crop that is tested "high" will have to be destroyed. And numerous real world evidence shows that such situation, if not addressed on time, may result in facing legal charges and certainly will result in heavy financial losses (Kotowski, 2021; Marijuana Moment, 2020; Matney, 2021; Earlenbaugh, 2020).

However, if this is not a response to stress conditions, but rather plant's genetics that determines how much THC is formed, , then growing hemp cultivars that are guaranteed to yield less THC than the limit set by regulators, might be a valuable solution. There is a research suggesting that plants showing specific genetic traits, which are homozygous for functional CBDA synthase, are not capable to yield more than 0.3% THC. The authors suggested that this finding should be considered as a method to distinguish hemp from marijuana (Weiblen et al., 2015). Indeed, if this finding is confirmed, if screening for the presence of nonfunctional

CBDA synthase is practical, and it can be performed before planting, this would save a lot of heartburn and billions of dollars to farmers.

11.4.5 Analytical requirements and challenges

To meet the demand for related analytical requirements, many existing laboratories added cannabis testing to the portfolio of their services, and even more, new and focused solely on cannabis, laboratories emerged. With the new cannabis legislation being still in its infancy, some laboratories provided service legally, operations of others were borderline illegal. But one thing most of these businesses had in common: they had no or little expertise in analyzing cannabis and related products. The cannabis plants have notoriously rich and of complex chemical composition, therefore, especially when cannabis is combined with various matrix material (in a form of candy, cookie, mixture of oils, etc.), the development and validation of quantitative analytical methods is extremely difficult even for very seasoned cannabis researchers and analysts. Because in such finished products all kind of interactions occur, they inevitably lead to the formulation of new chemical entities which may interfere with process of analysis and affect its results. In an ideal case, the analytical method would have to be validated for each individual product individually, which is absolutely not feasible. Not surprisingly, it was nothing unusual to receive two certificates of analysis (sometimes from by the same laboratory!) for two independent samples of the very same, carefully homogenized material, with results that were 50% or more, off (Hazekamp & Gieringer, 2011).

It is easy to blame lack of guidance of established agencies, such as FDA or EMA, but reality is, that body of knowledge for such complex matter needs time to develop. And before the recent changes in regulations, specific regulatory barriers would not allow to develop such knowledge, because in many parts of the world it was either impossible or at least difficult for researchers to get formal permit to work with cannabis, and to gain access to the quantities and types of cannabis necessary for experiments that would allow to address specific research questions. Therefore the available, tested, and validated methods were usually very limited both in numbers and in scope (Leghissa, Hildenbrand, & Schug, 2018).

Cannabis analytical testing industry has certainly matured over time, and now there are hundreds of laboratories, with decent knowledge and relevant experience, that provide services to growers, producers, and consumers. However, there are no standardized analytical methods available throughout the whole cannabis industry, as most of the earlier mentioned laboratories use their own, in-house developed methods. While this is perfectly good solution, practiced and well understood in research environment, unfortunately it does not meet requirements of various laboratories that analyze material for potency for regulatory purposes (such as labeling). The problem is well recognized, and there are already initiatives that aim to remedy such situation. One of them is "NIST Tools for Cannabis Laboratory Quality Assurance"—a program that aims at development of integrated services for cannabis industry, which is imitated by the National Institute of Standards and Technology (NIST, 2021).

The standardized and reliable analytical methods for quantification and identification of cannabinoids along with actual utilization of these methods, are fundamental for winning a wide acceptance of the industry, thus allowing expansion of hemp-based food, nutraceutical and cosmetic products. To ensure safety, regulatory compliance, and genuineness of hemp-derived food supplements, it is important to quantify the amount of THC in the final product and include it in the label. Since the impact of CBD and other cannabinoids is dose dependent, an acceptable limit of cannabinoids needs to be established for inclusion in the labels of dietary supplements. The use of CBD (and other hemp-based cannabinoids) infused dietary supplements should be assessed in broader perspective to cover all ingredients used in the preparation, as well as any contaminants that are already known to be common in recreational cannabis. The establishment of the pharmacokinetics of cannabinoids, when incorporated in different matrixes of dietary supplements or cosmetics, while challenging, needs to be understood to reassert safety and effectiveness of these products.

11.5 Conclusions

The potential of hemp, as an extraordinary source of supplement and or cosmetic ingredients, is tremendous: seed are high in omega-3 and omega-6 content and contain a large amount of well digestible protein, flowers are very rich source of cannabinoids, especially CBD. However, this potential is highly dependent on political and economic framework around the world. Legislative and regulatory interventions should aim at promoting adequate clinical research to prove or disprove safety of hemp-derived products, which has not been done so far independently by the industry or by the scientific community. Despite the large consumption of CBD as a nutritional supplement, evidence from preclinical and clinical studies exploring the effects of CBD as a dietary ingredient is scarce to nonexisting. Moreover, the emerging interest in minor cannabinoids such as CBG, CBN, or CBC requires immediate attention of scientific community to confirm their efficacy and safety. Innovation of novel hemp-derived dietary supplements and cosmetics requires precise identification and quantification of major bioactives and standardization of the products. The analytical methods required for bioactive components such as CBD need to be standardized. To ensure the authenticity and safety of hemp-derived products, it is important to quantify the amount of THC and other cannabinoids in the final product and include it in the label.

Hemp use in the dietary supplement and cosmetic industries is predicted to expand in the coming years as demand grows driven by rising awareness among consumers regarding health, growing number of vegan and vegetarian population, and innovation in product formulations. The projected substantial revenue has a real chance to revolutionize the food, cosmetic, and pharmaceutical industries.

References

Ali, A., & Akhtar, N. (2015). The safety and efficacy of 3% Cannabis seeds extract cream for reduction of human cheek skin sebum and erythema content. *Pakistan Journal of Pharmaceutical Sciences, 28*(4), 1389–1395.

Ali, E. M. M., Almagboul, A. Z. I., Khogali, S. M. E., & Gergeir, U. M. A. (2012). Antimicrobial activity of *Cannabis sativa* L. *Chinese Medicine, 3*, 61–64. https://doi.org/10.4236/cm.2012.31010.

American Academy of Dermatology Association. (2021). *Public, researchers demonstrate growing interest in cannabis treatment.* June https://www.aad.org/news/topical-cannabis.

Andre, C. M., Hausman, J. F., & Guerriero, G. (2016). *Cannabis sativa*: The plant of the thousand and one molecules. *Frontiers in Plant Science, 7*, 19. https://doi.org/10.3389/fpls.2016.00019.

Antonisamy, P., Duraipandiyan, V., Ignacimuthu, S., & Kim, J.-H. (2015). Anti-diarrhoeal activity of friedelin isolated from Azima tetracantha Lam. in Wistar rats. *South Indian Journal of Biological Sciences, 1*, 34–37.

Appendino, G., Gibbons, S., Giana, A., Pagani, A., Grassi, G., Stavri, M., et al. (2008). Antibacterial cannabinoids from *Cannabis sativa*: A structure-activity study. *Journal of Natural Products, 71,* 1427–1430. https://doi.org/10.1021/np8002673.

Arora, P., Potnis, S., Martinetti, M., Haskel, A., Suriano, D., Gregson, C., et al. (2021a). *WIPO International patent no WO/2021/071867. Oral care compositions comprising a cannabinoid.* Inventor Colgate Palmolive co.

Arora, P., Potnis, S., Martinetti, M., Haskel, A., Xu, Y., & Suriano, D. (2021b). *WIPO International patent no WO/2021/072422. Oral care compositions and a method of use.* Inventor Colgate Palmolive Co.

Arora, P., Potnis, S., Martinetti, M., Haskel, A., Xu, Y., & Suriano, D. (2021c). *WIPO International patent no WO/2021/072423. Oral care compositions and a method of use.* Inventor Colgate Palmolive Co.

Australian Government. (2020). *Notice of final decision to amend (or not amend) the current poisons standard – Cannabidiol.* https://www.tga.gov.au/scheduling-decision-final/notice-final-decision-amend-or-not-amend-current-poisons-standard-cannabidiol.

Biasutti, W. R., Leffers, K. S. H., & Callaghan, R. C. (2020). Systematic review of cannabis use and risk of occupational injury. *Substance Use Misuse, 55,* 1733–1745.

Bojarowicz, H., & Woźniak, B. (2008). Wielonienasycone kwasy tłuszczowe oraz ich wpływ na skórę polyunsaturated fatty acids and their infl uence on skin condition. *Problemy Higieny i Epidemiologii, 89*(4), 471–475.

Bonn-Miller, M. O., Loflin, M. J. E., & Thomas, B. F. (2017). Labeling accuracy of cannabidiol extracts sold online. *Journal of the American Medical Association, 318,* 1708–1709.

Businesswire. (2021). *CBD nutraceuticals market share, size, trends, industry analysis report, by product; by distribution channel; by region; segment forecast, 2021–2028.* May https://www.businesswire.com/news/home/20210423005232/en/World-CBD-Nutraceuticals-Market-Share-Size-Trends-Industry-Analysis-Report-2021-2028--ResearchAndMarkets.com.

Calixto, J. B., Kassuya, C. A. L., Andre, E., & Ferreira, J. (2005). Contribution of natural products to the discovery of the transient receptor potential (TRP) channels family and their functions. *Pharmacology & Therapeutics, 106,* 179–208.

Callaway, J., Schwab, U., Harvima, I., Halonen, P., Mykkänen, O., Hyvönen, P., et al. (2005). Efficacy of dietary hempseed oil in patients with atopic dermatitis. *Journal of Dermatological Treatment, 16,* 87–94. https://doi.org/10.1080/09546630510035832.

Callaway, J. C. (2004). Hempseed as a nutritional resource: An overview. *Euphytica, 140,* 65–72.

Cattaneo, C., Givonetti, A., Leoni, V., Guerrieri, N., Manfredi, M., Giorgi, A., et al. (2021). Biochemical aspects of seeds from *Cannabis sativa* L. plants grown in a mountain environment. *Scientific Reports, 11*(1), 3927. https://doi.org/10.1038/s41598-021-83290-1.

Cerino, P., Buonerba, C., Cannazza, G., D'Auria, J., Ottoni, E., Fulgione, A., et al. (2021). A review of Hemp as food and nutritional supplement. *Cannabis and Cannabinoid Research, 6*(1). https://doi.org/10.1089/can.2020.0001.

Chapkin, R. S., Ziboh, V. A., Marcelo, C. L., & Voorhees, J. J. (1986). Metabolism of essential fatty acids by human epidermal enzyme preparations: Evidence of chain elongation. *Journal of Lipid Research, 27,* 945–954.

Chelliah, M. P., Zinn, Z., Khuu, P., & Teng, J. M. C. (2018). Self-initiated use of topical cannabidiol oil for epidermolysis bullosa. *Pediatric Dermatology, 35,* 224–227. https://doi.org/10.1111/pde.13545.

Chesney, E., Oliver, D., & Green, A. (2020). Adverse effects of cannabidiol: A systematic review and meta-analysis of randomized clinical trials. *Neuropsychopharmacology, 45,* 1799–1806. https://doi.org/10.1038/s41386-020-0667-2.

Chouvy, P. A. (2019). Cannabis cultivation in the world: Heritages, trends and challenges. *EchoGéo, 48.* https://journals.openedition.org/echogeo/17591.

Clinical Trial No NCT04726475. n.d. https://clinicaltrials.gov/ct2/show/NCT04726475?cond=cannabidiol+AND+%22Anxiety+Disorders%22&draw=2&rank=9.

Cooke, J. (2021). *How to make Delta 8 THC: CBD to Δ8 THC step-by-step.* https://dailycbd.com/en/how-to-make-delta-8-thc.

Corroon, J., & Kight, R. (2018). Regulatory status of cannabidiol in the United States: A perspective. *Cannabis and Cannabinoid Research, 3,* 190–194.

Corroon, J., & Phillips, J. A. (2018). A cross-sectional study of cannabidiol users. *Cannabis and Cannabinoid Research, 3,* 152–161.

Crescente, G., Piccolella, S., Esposito, A., Scognamiglio, M., Fiorentino, A., & Pacifico, S. (2018). Chemical composition and nutraceutical properties of hempseed: An ancient food with actual functional value. *Phytochemical Reviews, 17,* 733–749.

Crini, G., Lichtfouse, E., Chanet, G., & Morin-Crini, N. (2020). Applications of hemp in textiles, paper industry, insulation and building materials, horticulture, animal nutrition, food and beverages, nutraceuticals, cosmetics and hygiene, medicine, agrochemistry, energy production and environment: A review. *Environmental Chemistry Letters*, *18*(1), 1451–1476.

De Petrocellis, L., Ligresti, A., Moriello, A. S., Allarà, M., Bisogno, T., Petrosino, S., et al. (2011). Effects of cannabinoids and cannabinoid-enriched Cannabis extracts on TRP channels and endocannabinoid metabolic enzymes. *British Journal of Pharmacology*, *163*, 1479–1494. https://doi.org/10.1111/j.1476-5381.2010.01166.x.

Del Bo, C., Deon, V., Abello, F., Massini, G., Porrini, M., Riso, P., et al. (2019). Eight-week hempseed oil intervention improves the fatty acid composition of erythrocyte phospholipids and the omega-3 index, but does not affect the lipid profile in children and adolescents with primary hyperlipidemia. *Food Research International*, *119*, 469–476.

Dewey, L. H. (1914). *Hemp. USDA 1913 Yearbook* (pp. 283–346). United State Department of Agriculture, Government Printing Office.

Dia, V. P., Wang, Z., Lin, Y., & Pangloli, P. (2019). Comparative biological activities determination of aqueous extracts of hempseed oil and hempseed protein isolate production coproducts. *Journal of the American Oil Chemists' Society*, *96*(11), 1265–1274. https://doi.org/10.1002/aocs.12282.

Drotleff, L. (2020). https://hempindustrydaily.com/2020-outlook-licensed-u-s-hemp-acreage-falls-9-from-2019-but-grower-numbers-increase-27/.

Dunford, N. T. (2015). Hemp and flaxseed oil: Properties and application for use in food. In G. Talbot (Ed.), *Specialty oils and fats in food and nutrition* (pp. 39–63). Oxford: Woodhead Publishing Ltd.

Earlenbaugh, E. (2020). *Hemp testing over 0.3% THC must be destroyed—here's why*. Leafy. https://www.leafly.com/news/industry/hemp-testing-hot-must-be-destroyed.

Esparza, Y., Ngo, T. D., & Boluk, Y. (2020). Preparation of powdered oil particles by spray drying of cellulose nanocrystals stabilized Pickering hempseed oil emulsions. *Colloids and Surfaces A: Physicochemical and Engineering Aspects*, *598*. https://doi.org/10.1016/j.colsurfa.2020.124823.

European Industrial Hemp Association (EIHA). (2020). *Hemp cultivation & production in Europe in 2018*. https://eiha.org/wp-content/uploads/2020/06/2018-Hemp-agri-report.pdf.

Fasinu, P. S., Phillips, S., ElSohly, M. A., & Walker, L. A. (2016). Current status and prospects for Cannabidiol preparations as new therapeutic agents. *Pharmacotherapy*, *36*, 781–796.

FDA. (2019). *FDA, FTC warn company marketing unapproved cannabidiol products with unsubstantiated claims to treat teething and ear pain in infants, autism, ADHD, Parkinson's and Alzheimer's disease*. https://www.fda.gov/news-events/press-announcements/fda-ftc-warn-company-marketing-unapproved-cannabidiol-products-unsubstantiated-claims-treat-teething.

FDA. (2020). *Consumer update: What you need to know (and what we're working to find out) about products containing cannabis or cannabis-derived compounds, including CBD*. https://www.fda.gov/consumers/consumer-updates/what-you-need-know-and-what-were-working-find-out-about-products-containing-cannabis-or-cannabis.

FDA. (2021a). *FTC announces latest enforcement action halting deceptive CBD product marketing*. https://www.ftc.gov/news-events/press-releases/2021/05/ftc-announces-latest-enforcement-action-halting-deceptive-cbd.

FDA. (2021b). *Regulation of Cannabis and Cannabis-derived products, including cannabidiol (CBD)*. https://www.fda.gov/news-events/public-health-focus/fda-regulation-cannabis-and-cannabis-derived-products-including-cannabidiol-cbd#relatedinfo.

Federal Trade Commision. (2013). *FTC cracks down on misleading and unsubstantiated environmental marketing claims*. https://www.ftc.gov/news-events/press-releases/2013/10/ftc-cracks-down-misleading-unsubstantiated-environmental.

Ferreira, M. J., Fiadeiro, T., Silva, M., & Soares, A. P. (1998). Topical γ-linolenic acid therapy in atopic dermatitis. A clinical and biometric evaluation. *Allergo Journal*, *7*, 213–216. https://doi.org/10.1007/BF03360571.

Fike, J. (2016). Industrial Hemp: Renewed opportunities for an ancient crop. *Critical Reviews in Plant Sciences*, *35*, 406–424.

Fitzcharles, M., Clauw, D. J., & Hauser, W. (2020). A cautious hope for cannabidiol (CBD) in rheumatology care. *Arthritis Care & Research (Hoboken)*. https://doi.org/10.1002/acr.24176.

Fortenberry, T. R., & Bennett, M. (2001). *Is industrial Hemp worth further study in the US? A survey of the literature. Staff papers 12680*. University of Wisconsin-Madison, Department of Agricultural and Applied Economics.

Gaoni, Y., & Mechoulam, R. (1966). Hashish-VII. The isomerization of cannabidiol to tetrahydrocannabinols. *Tetrahedron*, *22*, 1481–1488.

Giraud, C. (2019). *Cannabis topicals—What are they and do they really work?*. https://cannigma.com/products/cannabis-topicals-what-are-they-and-do-they-really-work/.

Goncalves, J. C., Oliveira, F. S., Benedito, R. B., Sousa, D. P., Almeida, R. N., & Araújo, D. A. M. (2008). Antinociceptive activity of (−)-carvone: Evidence of association with decreased peripheral nerve excitability. *Biological and Pharmaceutical Bulletin, 31*, 1017–1020.

Government of Canada. (2018). *Cannabis act.* Available at: https://laws-lois.justice.gc.ca/eng/acts/C-24.5/.

Government of UK. (2014). *Industrial hemp: Licensing guidance.* https://www.gov.uk/government/publications/industrial-hemp-licensing-guidance.

Grand View Research. (2021a). *Cannabidiol market size, share & trends analysis report by source type (hemp, marijuana), by distribution channel (B2B, B2C), by end-use (medical, personal use), by region, and segment forecasts, 2021–2028.* https://www.grandviewresearch.com/industry-analysis/cannabidiol-cbd-market. (Accessed 8 May 2021).

Grand View Research. (2021b). *CBD gummies market worth $13.9 billion by 2028.* https://www.grandviewresearch.com/press-release/global-cbd-gummies-market.

Guang, H., & Wenwei, C. (2008). *Application of fructus Cannabis protein in preparation of health food for relieving nutritional Anemia. China patent no. 100433990 C.* https://patents.google.com/patent/CN100433990C/en.

Hansen, H. S., & Jensen, B. (1985). Essential function of linoleic acid esterified in acylglucosylceramide and acylceramide in maintaining the epidermal water permeability barrier. Evidence from feeding studies with oleate, linoleate, arachidonate, columbinate and alpha-linolenate. *Biochimica et Biophysica Acta, 834*, 357–363.

Hazekamp, A., & Gieringer, D. (2011). *How accurate is potency testing? O'Shaughnessy's online autumn* (pp. 17–18). https://pdfs.semanticscholar.org/bb55/b0ba86710d01a8cc28c6db79445283bb4064.pdf.

Health Canada. (2021). *Industrial hemp licensing statistics for 2020.* https://www.canada.ca/en/health-canada/services/drugs-medication/cannabis/producing-selling-hemp/about-hemp-canada-hemp-industry/statistics-reports-fact-sheets-hemp.html.

Hemp Industries Association. (2016). *2015 annual retail sales for Hemp products.* https://thehia.org/HIAhemppressreleases/4010402#.

Hemp Today. (2021). *CBG gets safety approval for inclusion in EU cosmetics database.* https://hemptoday.net/cbg-gets-safety-approval-for-inclusion-in-eu-cosmetics-database/.

Hernandez, E. H., Kennedy, S., Wu, Q., Maksimovic, S., Morgan, A., Potechin, K., et al. (2021). *WIPO International patent no WO/2021/072419. Personal care compositions and methods.* Inventor Colgate Palmolive Co.

House, J. D., Neufeld, J., & Leson, G. (2010). Evaluating the quality of protein from hemp seed (*Cannabis sativa* L.) products through the use of the protein digestibility-corrected amino acid score method. *Journal of Agricultural and Food Chemistry, 58*(22), 11801–11807.

Huang, T. H., Wang, P. W., Yang, S. C., Chou, W. L., & Fang, J. Y. (2018). Cosmetic and therapeutic applications of fish Oil's fatty acids on the skin. *Marine Drugs, 16*, 256.

Huang, Y., Pei, L., Gu, X., & Wang, J. (2020). Study on the oxidation products of hemp seed oil and its application in cosmetics. *Tenside, Surfactants, Detergents, 57*, 230–236. https://doi.org/10.3139/113.110679.

Huber, A., III, Newman, R., & LaFave, D. (2016). Cannabis control and crime: Medicinal use, depenalization and the war on drugs. *The B.E. Journal of Economic Analysis & Policy, 16*(4), 20150167.

Iffland, K., & Grotenhermen, F. (2017). An update on Safetyand side effects of Cannabidiol: A review of clinical data and relevant animal studies. *Cannabis and Cannabinoid Research, 2*(1), 39–154.

Igielska-Kalwat, J., Gościańska, J., & Nowak, I. (2015). Karotenoidy jako naturalne antyoksydanty. *Postepy Higieny I Medycyny Doswiadczalnej, 69*, 418–428.

Jhawar, N., Schoenberg, E., Wang, J. V., & Saedi, N. (2019). The growing trend of cannabidiol in skincare products. *Clinics in Dermatology, 37*, 279–281. https://doi.org/10.1016/j.clindermatol.2018.11.002.

Jin, S., & Lee, M. Y. (2019). The ameliorative effect of hemp seed hexane extracts on the Propionibacterium acnes induced inflammation and lipogenesis in sebocytes. *PLoS One, 13*. https://doi.org/10.1371/journal.pone.0202933.

Jones, K. (1996). *Nutritional and medicinal guide to Hemp seed.* Gibson: Rain Forest Botanical Laboratory.

Kanitakis, J. (2002). Anatomy, histology and immunohistochemistry of normal human skin. *European Journal of Dermatology, 12*(4), 390–401.

Kapoor, R., & Huang, Y.-S. (2006). Gamma linolenic acid: An antiinflammatory omega-6 fatty acid. *Current Pharmaceutical Biotechnology, 7*(6), 531–534. https://doi.org/10.2174/138920106779116874.

Kaul, N., Kreml, R., Austria, J. A., Richard, M. N., Edel, A. L., Dibrov, E., et al. (2008). A comparison of fish oil, flaxseed oil and hempseed oil supplementation on selected parameters of cardiovascular health in healthy volunteers. *Journal of the American College of Nutrition, 27*, 51–58.

Kaviani, M., Chilibeck, P. D., Toles, K., & Candow, D. G. (2016). The benefits of hemp powder supplementation during resistance training. In *Proceedings of the Canadian society for exercise physiology annual general meeting—From health to high performance. Victoria, Canada.*

Khan, B., Warner, P., & Wang, H. (2014). Antibacterial properties of hemp and other natural fibre plants: A review. *Bioresources, 9,* 3642–3659.

Khnykin, D., Miner, J. H., & Jahnsen, F. (2011). Role of fatty acid transporters in epidermis: Implications for health and disease. *Dermatoendocrinology, 3*(2), 53–61. https://doi.org/10.4161/derm.3.2.14816.

Kolodziejczyk, P. P., Kozlowska, J., & Ozimek, L. (2012). The application of flax and hemp seeds in food, animal feed, health and cosmetics. In R. Kozlowski (Ed.), *Handbook of natural textile fibres: Vol. 2: Types, properties and factors affecting breeding and cultivation* (pp. 329–366). Cambridge, UK: Woodhead Publishing Limited.

Kotowski, J. (2021). Man behind $1 billion hemp lawsuit faces drug charges as federal judge considers motion to dismiss the suit. *KGET News.* https://www.kget.com/news/local-news/man-behind-1-billion-hemp-lawsuit-faces-drug-charges-as-federal-judge-considers-motion-to-dismiss-the-suit/.

Kowalska, M., Wozniak, M., & Pazdzior, M. (2017). Assessment of the sensory and moisturizing properties of emulsions with hemp oil. *Acta Polytechnica Hungarica, 14,* 183–195.

Kozono, S., Matsuyama, T., Biwasa, K. K., Kawahara, K., Nakajima, Y., Yoshimoto, T., et al. (2010). Involvement of the endocannabinoid system in periodontal healing. *Biochemical and Biophysical Research Communications, 394*(4), 928–933.

Kurek-Górecka, A., Balwierz, R., Mizera, P., Nowak, M., & Żurawska-Płaksej, E. (2018). Therapeutic and cosmetic importance of hemp oil. *Farmacja Polska, 74,* 704–708. https://doi.org/10.32383/farmpol/118618.

Laate, E. A. (2012). *Industrial Hemp production in Canada.* https://www1.agric.gov.ab.ca/$department/deptdocs.nsf/all/econ9631/$file/Final%20-%20Industrial%20Hemp%20Production%20in%20Canada%20-%20June%2025%20 2012.pdf?OpenElement.

LaCapria, K. (2016). *Drug law lobbying by corrections Corporation of America.* https://www.snopes.com/fact-check/drug-law-lobbying-by-corrections-corporation-of-america/.

Lafaye, G., Karila, L., Blecha, L., & Benyamina, A. (2017). Cannabis, cannabinoids, and health. *Dialogues in Clinical Neuroscience, 19*(3), 309–316. https://doi.org/10.31887/DCNS.2017.19.3/glafaye.

Leghissa, A., Hildenbrand, Z. L., & Schug, K. A. (2018). A review of methods for the chemical characterization of cannabis natural products. *Journal of Separation Science, 41,* 398–415.

Leizer, C., Ribnicky, D. M., Poulev, A., Dushenkov, D., & Raskin, I. (2000). The composition of hemp seed oil and its potential as an important source of nutrition. *Journal of Nutraceuticals, Functional & Medical Foods, 2*(4), 35–53. https://doi.org/10.1300/J133v02n04_04.

Leonard, W., Zhang, P., Ying, D., & Fang, Z. (2020). Hempseed in food industry: Nutritional value, health benefits, and industrial applications. *Comprehensive Reviews in Food Science and Food Safety, 19*(1), 282–308. https://doi.org/10.1111/1541-4337.12517.

Li, H., Ma, Y., Miao, J., Zhang, J., Ren, H., & Li, G. (2012). *The influence of the hemp seed oil on the related parameters of the skin in aging mice.* http://en.cnki.com.cn/Article_en/CJFDTOTAL-ZSFX201209063.htm.

Li, H.-L. (2012). The origin and use of Cannabis in eastern Asia: Their linguistic-cultural implications. In *Cannabis and culture* (pp. 51–62). De Gruyter Mouton. https://doi.org/10.1515/9783110812060.51.

Ligeza, M., Wygladacz, D., Tobiasz, A., Jaworecka, K., & Reich, A. (2016). Natural cold pressed oils as cosmetic products. *Family Medicine & Primary Care Review, 18,* 443–447.

Lopez, H. L., Cesareo, K. R., Raub, B., Kedia, A. W., Sandrock, J. E., Kerksick, C. M., et al. (2020). Effects of Hemp extract on markers of wellness, stress resilience, recovery and clinical biomarkers of safety in overweight, but otherwise healthy subjects. *Journal of Dietary Supplements, 17*(5), 561–586.

Lu, D., Dopart, R., & Kendall, D. A. (2016). Controlled downregulation of the cannabinoid CB1 receptor provides a promising approach for the treatment of obesity and obesity-derived type 2 diabetes. *Cell Stress & Chaperones, 21,* 1–7. https://doi.org/10.1007/s12192-015-0653-5.

Maghfour, J., Rundle, C. W., Rietcheck, H. R., Dercon, S., Lio, P., Mamo, A., et al. (2021). Assessing the effects of topical cannabidiol in patients with atopic dermatitis. *Dermatology Online Journal, 27*(2), 15–19. https://escholarship.org/uc/item/8h50x2vs.

Marijuana Moment. (2020). *Wyoming judge dismisses marijuana charges against hemp farmers.* https://www.marijuana-moment.net/wyoming-judge-dismisses-marijuana-charges-against-hemp-farmers/.

Marrapodi, A. (2021). A market overview: CBD topicals. *Cannabis Science and Technology, 4*(1), 52–54.

Matney, M. (2021). *Tax dollars hard at work': SLED arrests 2 SC Hemp farmers on felony marijuana charges.* Fit News. https://www.fitsnews.com/2021/04/26/tax-dollars-hard-at-work-sled-arrests-2-sc-hemp-farmers-on-felony-marijuana-charges.

Medicines and Healthcare products Regulatory Agency (MHRA). (2016). *Statement on products containing Cannabidiol (CBD).* https://www.gov.uk/government/news/mhra-statement-on-products-containing-cannabidiol-cbd.

Metwally, S., Ura, D. P., Krysiak, Z. J., Kaniuk, Ł., Szewczyk, P. K., & Stachewicz, U. (2021). Electrospun PCL patches with controlled fiber morphology and mechanical performance for skin moisturization via long-term release of hemp oil for atopic dermatitis. *Membranes, 11,* 1–13. https://doi.org/10.3390/membranes11010026.

Mikulcová, V., Kašpárková, V., Humpolíček, P., Buňková, L., Mikulcová, V., Kašpárková, V., et al. (2017). Formulation, characterization and properties of Hemp seed oil and its emulsions. *Molecules, 22,* 700. https://doi.org/10.3390/molecules22050700.

Mole, M. L., Jr., & Turner, C. E. (1973). Phytochemical screening of Cannabis sativa L. II. Choline and neurine in the roots of a Mexican variant. *Acta Pharmaceutica Jugoslavica, 23,* 203–205.

Montserrat-De La Paz, S., Marín-Aguilar, F., García-Giménez, M. D., & Fernández-Arche, M. A. (2014). Hemp (*Cannabis sativa* L.) seed oil: Analytical and phytochemical characterization of the unsaponifiable fraction. *Journal of Agricultural and Food Chemistry, 62*(5), 1105–1110. https://doi.org/10.1021/jf404278q.

National Academies of Sciences, Engineering, and Medicine. (2017). *The health effects of Cannabis and cannabinoids: The current state of evidence and recommendations for research.* Washington, DC: The National Academies Press. https://doi.org/10.17226/24625.

New Zealand's Medicinal Cannabis Agency. (2020). *Cannabidiol (CBD) products.* https://www.health.govt.nz/our-work/regulation-health-and-disability-system/medicinal-cannabis-agency/medicinal-cannabis-agency-information-industry/medicinal-cannabis-agency-working-medicinal-cannabis/medicinal-cannabis-agency-cannabidiol-cbd-products.

NIST. (2021). *NIST tools for Cannabis laboratory quality assurance.* https://www.nist.gov/programs-projects/nist-tools-cannabis-laboratory-quality-assurance.

Oláh, A., Tóth, B. I., Borbíró, I., Sugawara, K., Szöllősi, A. G., Czifra, G., et al. (2014). Cannabidiol exerts sebostatic and antiinflammatory effects on human sebocytes. *Journal of Clinical Investigation, 124,* 3713–3724. https://doi.org/10.1172/JCI64628.

Palmieri, B., Laurino, C., & Vadalà, M. (2019). A therapeutic effect of cbd-enriched ointment in inflammatory skin diseases and cutaneous scars. *Clinical Therapeutics, 170*(2), 93–99.

Payne, K. S., Mazur, D. J., Hotaling, J. H., & Pastuszak, A. W. (2019). Cannabis and male fertility: A systematic review. *Journal of Urology, 202,* 674–681.

Pei, L., Luo, Y., Gu, X., & Wang, J. (2020). Formation, stability and properties of hemp seed oil emulsions for application in the cosmetics industry. *Tenside, Surfactants, Detergents, 57,* 451–459. https://doi.org/10.3139/113.110712.

Pengilly, N. (2003). Traditional food and medicinal use of flaxseed. In A. D. Muir, & N. D. Westcott (Eds.), *Flax: The genus Linum* CRC Press.

Pertwee, R. G. (2006). Cannabinoid pharmacology: The first 66 years. *British Journal of Pharmacology, 147*(Suppl 1), S163–S171. https://doi.org/10.1038/sj.bjp.0706406.

Pertwee, R. G. (2008). The diverse CB 1 and CB 2 receptor pharmacology of three plant cannabinoids: Δ 9-tetrahydrocannabinol, cannabidiol and Δ 9-tetrahydrocannabivarin. *British Journal of Pharmacology, 153*(2), 199–215. https://doi.org/10.1038/sj.bjp.0707442.

Pilkington, S. M., Watson, R. E. B., Nicolaou, A., & Rhodes, L. E. (2011). Omega-3 polyunsaturated fatty acids: Photoprotective macronutrients. *Experimental Dermatology, 20,* 537–543. https://doi.org/10.1111/j.1600-0625.2011.01294.x.

Place, G. (2019). *Hemp production – Keeping THC levels low.* https://catawba.ces.ncsu.edu/2018/11/hemp-production-keeping-thc-levels-low/.

Pliny the Elder. (1856). *The natural history. Volume 4, books 12–16. Translated by H. Rackham. Loeb classical library 371.* Cambridge, MA: Harvard University Press.

Ramot, Y., Sugawara, K., Zákány, N., Tóth, B. I., Bíró, T., & Paus, R. (2013). A novel control of human keratin expression: Cannabinoid receptor 1-mediated signaling down-regulates the expression of keratins K6 and K16 in human keratinocytes in vitro and in situ. *PeerJ, 1,* e40. https://www.ncbi.nlm.nih.gov/pmc/articles/PMC3628749/pdf/peerj-01-40.pdf.

Ranalli, P., & Venturi, G. (2004). Hemp as a raw material for industrial applications. *Euphytica, 140,* 1–6.

Research and Markets. (2020). Global Hemp Protein Market - Growth, Trends and Forecasts (2020–2025). https://www.researchandmarkets.com/reports/4402834/global-hemp-protein-market-growth-trends-and?utm_source=GNOM&utm_medium=PressRelease&utm_code=3mswl9&utm_campaign=1435164+-+World+Market+for+Hemp+Protein+Market+to+2025+-+Global+Market+Forecast+to+Grow+at+a+CAGR+-of+15.2%25+During+2019-2024&utm_exec=cari18prd. 2020. Accessed 21 June 2021.

Rezapour-Firouzi, S., Arefhosseini, S. R., Mehdi, F., Mehrangiz, E.-M., Baradaran, B., Sadeghihokmabad, E., et al. (2013). Immunomodulatory and therapeutic effects of hot-nature diet and co-supplemented hemp seed, evening primrose oils intervention in multiple sclerosis patients. *Complementary Therapies in Medicine, 21*, 473–480.

Rodriguez-Martin, N. M., Montserrat-de la Paz, S., Toscano, R., Grao-Cruces, E., Villanueva, A., Pedroche, J., et al. (2020). Hemp (*Cannabis sativa* L.) protein hydrolysates promote anti-inflammatory response in primary human monocytes. *Biomolecules, 10*, 803–815.

Roulac, J. (1997). *Hemp horizons: The comeback of the World's Most promising plant*. White River Junction, VT, USA: Chelsea Green Pub.

Ryz, N. R., Remillard, D. J., & Russo, E. B. (2017). Cannabis roots: A traditional therapy with future potential for treating inflammation and pain. *Cannabis and Cannabinoid Research, 2*(1), 210–216.

Sabet, K. (2020). *The marijuana lobby spends big on bills Americans Don't want opinion*. Newsweek. https://www.newsweek.com/marijuana-lobby-spends-big-bills-americans-dont-want-opinion-1538842.

Samsamikor, M., Mackay, D., Mollard, R. C., & Aluko, R. E. (2020). A double-blind, randomized, crossover trial protocol of whole hemp seed protein and hemp seed protein hydrolysate consumption for hypertension. *Trials, 21*(354), 1–13.

Schofs, L., Sparo, M. D., & Sánchez Bruni, S. F. (2021). The antimicrobial effect behind *Cannabis sativa*. *Pharmacology Research & Perspectives, 9*(2). https://doi.org/10.1002/prp2.761, e00761.

Schwab, U. S., Callaway, J. C., Erkkilä, A. T., Gynther, J., Uusitupa, M. I. J., & Järvinen, T. (2006). Effects of hempseed and flaxseed oils on the profile of serum lipids, serum total and lipoprotein lipid concentrations and haemostatic factors. *European Journal of Nutrition, 45*(8), 470–477.

Scopinho, A. A., Guimarães, F. S., Corrêa, F. M. A., & Resstel, L. B. M. (2011). Cannabidiol inhibits the hyperphagia induced by cannabinoid-1 or serotonin-1A receptor agonists. *Pharmacology, Biochemistry, and Behavior, 98*, 268–272. https://doi.org/10.1016/j.pbb.2011.01.007.

Senapati, S., Banerjee, S., Nath Gangopadhyay, D., Ghosh Road, H., & Naihati, P. (2008). Evening primrose oil is effective in atopic dermatitis: A randomized placebo-controlled trial. *Indian Journal of Dermatology, Venereology and Leprology, 74*(5), 447–452.

Sikorska, M., Nowicki, R., & Wilkowska, A. (2015). Pielęgnacja skóry suchej i wrażliwej. *Alergologia Polska - Polish Journal of Allergology, 2*, 158–161.

Simmonds, N. (1976). Hemp: *Cannabis sativa* (Moraceae). In *1976. Evolution of crop plants* (pp. 203–204). Wiley.

Simopoulos, A. P. (2002). The importance of the ratio of omega-6/omega-3 essential fatty acids. *Biomedicine and Pharmacotherapy, 56*(8), 365–379. https://doi.org/10.1016/S0753-3322(02)00253-6.

SingleCare. (2020). *The 2020 CBD survey*. https://www.singlecare.com/blog/cbd-survey/.

Small, E., & Cronquist, A. (1976). A practical and natural taxonomy for cannabis. *Taxon, 25*, 405–435. https://doi.org/10.2307/1220524.

Smith, G., & Satino, J. (2021). Hair regrowth with cannabidiol (CBD)-rich hemp extract. *Cannabis, 4*, 53–59. https://doi.org/10.26828/cannabis/2021.01.003.

Sokoła-Wysoczańska, E., Wysoczański, T., Wagner, J., Czyż, K., Bodkowski, R., Lochyński, S., et al. (2018). Polyunsaturated fatty acids and their potential therapeutic role in cardiovascular system disorders—A review. *Nutrients, 10*(10). https://doi.org/10.3390/nu10101561.

Swissmedic. (2019). *Products containing cannabidiol (CBD) – overview*. https://www.swissmedic.ch/swissmedic/en/home/news/mitteilungen/products-containing-cannabidiol- -cbd- - -overview.html.

Szabó, I. L., Herczeg-Lisztes, E., Szollosi, A. G., Szegedi, A., Bíró, T., & Oláh, A. (2017). 263 (-)-cannabidiol differentially influences hair growth. *Journal of Investigative Dermatology, 137*, 238.

Szabó, I. L., Lisztes, E., Béke, G., Tóth, K. F., Paus, R., Oláh, A., et al. (2020). The phytocannabinoid (−)-Cannabidiol operates as a complex, differential modulator of human hair growth: Anti-inflammatory submicromolar versus hair growth inhibitory micromolar effects. *Journal of Investigative Dermatology, 140*, 484–488.e5. https://doi.org/10.1016/j.jid.2019.07.690.

Tao, B. (2013). *Big marijuana fights legalization*. https://www.politico.com/story/2013/07/big-marijuana-lobby-fights-legalization-efforts-094816.

Teh, S. S., & Birch, E. (2013). Physicochemical and quality characteristics of cold-pressed hemp, flax and canola seed oils. *Journal of Food Composition and Analysis, 30*, 26–31. https://doi.org/10.1016/j.jfca.2013.01.004.

Thiele, E. A., Marsh, E. D., French, J. A., Mazurkiewicz-Beldzinska, M., Benbadis, S. R., Joshi, C., et al. (2018). Cannabidiol in patients with seizures associated with Lennox-Gastaut syndrome (GWPCARE4): A randomised, double-blind, placebo-controlled phase 3 trial. *Lancet, 391*, 1085–1096.

Thompson, D. (2018). *CBD oil: All the rage, but is it safe & effective?*. Webmd. https://www.webmd.com/pain-management/news/20180507/cbd-oil-all-the-rage-but-is-it-safe-effective.

Toth, J. A., Stack, G. M., Cala, A. R., Carlson, C. H., Wilk, R. L., Crawford, J. L., et al. (2020). Development and validation of genetic markers for sex and cannabinoid chemotype in *Cannabis sativa* L. *GCB Bioenergy, 12*, 213–222. https://doi.org/10.1111/gcbb.12667.

Tóth, K. F., Ádám, D., Bíró, T., & Oláh, A. (2019). Cannabinoid signaling in the skin: Therapeutic potential of the "c(ut)annabinoid" system. *Molecules, 24*(5), 918–964. https://doi.org/10.3390/molecules24050918.

Uluata, S., & Özdemir, N. (2012). Antioxidant activities and oxidative stabilities of some unconventional oilseeds. *Journal of the American Oil Chemists' Society, 89*, 551–559.

UN Commission on Narcotic Drugs. (2020). *Press statement – 2 December 2020*. https://www.unodc.org/documents/commissions/CND/CND_Sessions/CND_63Reconvened/Press_statement_CND_2_December.pdf.

United Nation. (2013). *The international drug control conventions*. Publishing and Library Section, United Nations Office at Vienna. https://www.unodc.org/documents/commissions/CND/Int_Drug_Control_Conventions/Ebook/The_International_Drug_Control_Conventions_E.pdf.

Vonapartis, E., Aubin, M. P., Seguin, P., Mustafa, A. F., & Charron, J. B. (2015). Seed composition of ten industrial hemp cultivars approved for production in Canada. *Journal of Food Composition and Analysis, 39*, 8–12.

Vote Hemp. (2017). *US Hemp crop report*. https://www.votehemp.com/wp-content/uploads/2018/09/Vote-Hemp-2017-US-Hemp-Crop-Report.pdf.

Wang, Q., & Xiong, Y. L. (2019). Processing, nutrition, and functionality of hempseed protein: A review. *Comprehensive Reviews in Food Science and Food Safety, 18*(4), 936–952. https://doi.org/10.1111/1541-4337.12450.

Webster, G. R., Sarna, L., & Mechoulam, R. (2004). *Conversion of CBD to delta8-THC and delta9-THC. Patent US20040143126A1*. https://patentimages.storage.googleapis.com/dc/a0/4e/c9714dbad03d4f/US20080221339A1.pdf.

Weedmaps. (2020). https://weedmaps.com/learn/cbd/who-discovered-cbd.

Weiblen, G. D., Wenger, J. P., Craft, K. J., ElSohly, M. A., Mehmedic, Z., Treiber, E. L., et al. (2015). Gene duplication and divergence affecting drug content in *Cannabis sativa*. *New Phytologist, 208*, 1241–1250.

Wilkinson, J. D., & Williamson, E. M. (2007). Cannabinoids inhibit human keratinocyte proliferation through a non-CB1/CB2 mechanism and have a potential therapeutic value in the treatment of psoriasis. *Journal of Dermatological Science, 45*, 87–92.

Wollenberg, A., Barbarot, S., Bieber, T., Christen-Zaech, S., Deleuran, M., Fink-Wagner, A., et al. (2018). Consensus-based European guidelines for treatment of atopic eczema (atopic dermatitis) in adults and children: Part II. *Journal of the European Academy of Dermatology and Venereology, 32*, 850–878. https://doi.org/10.1111/jdv.14888.

Wu, G., & Meininger, C. J. (2002). Regulation of nitric oxide synthesis by dietary factors. *Annual Review of Nutrition, 22*, 61–68.

Yan, X., Tang, J., dos Santos Passos, C., Nurisso, A., Simões-Pires, C. A., Ji, M., et al. (2015). Characterization of lignan-amides form hemp (Cannabis sativa) seed and their antioxydant and acetylcholine esterase inhibtory activities. *Journal of Agricultural and Food Chemistry, 63*(49), 10611–10619.

Yang, P., Feng, J., Luo, J., Madison, M., & Hu, H. (2019). A critical role for TRP channels in the skin. In *Neurobiology of TRP channels* (pp. 95–111). CRC Press. https://doi.org/10.4324/9781315152837-6.

Zagórska-Dziok, M., Bujak, T., Ziemlewska, A., & Nizioł-Łukaszewska, Z. (2021). Positive effect of *Cannabis sativa* L. herb extracts on skin cells and assessment of cannabinoid-based hydrogels properties. *Molecules, 26*, 802–824.

Zagožen, M., Čerenak, A., & Kreft, S. (2021). Cannabigerol and cannabichromene in Cannabis sativa L. *Acta Pharmaceutica, 71*, 355–364.

12

Industrial hemp by-product valorization

Vita Maria Cristiana Moliterni[a], Milica Pojić[b], and Brijesh Tiwari[c]

[a]Council for Agricultural Research and Economics (CREA), Research Centre for Genomics and Bioinformatics, Fiorenzuola d'Arda, Piacenza, Italy [b]Institute of Food Technology (FINS), University of Novi Sad, Novi Sad, Serbia [c]Food Chemistry & Technology Department, Teagasc Food Research Centre, Ashtown, Dublin, Ireland

12.1 Introduction

The need to increase the circularity of agricultural and food production and in an effort to address limited natural resources, reduction of arable land, rise in world population, environment protection, and climate change have prompted the development of the agri-food waste biorefinery concept (Caldeira et al., 2020; Setti et al., 2020). Although the rising amount of agri-food waste is a global problem that requires organized waste valorization planning, it is especially accentuated in developing countries due to the large waste quantities of agri-food produced and the lack of well-defined recycling policies (Searle & Malins, 2013; Tongwane & Moeletsi, 2018). Biorefinery implies a combination of different processes and technologies for the conversion of biomass to value-added products such as biochemicals, food products, animal feed, biobased products and materials, fuels, electricity, and heat (Peck, Bennett, Bissett-Amess, Lenhart, & Mozaffarian, 2009). These processes most commonly include combustion, pyrolysis, gasification, fermentation, and different pretreatments (Ferreira, 2017) that utilize either traditional or innovative technologies.

Although the utilization of an integrated biorefinery approach is recognized as essential for the development of a sustainable industrial society and effective greenhouse gas (GHG) emission management, different aspects of biorefinery must be considered such as the technical feasibility of the processes on an industrial scale, the technoeconomic potential, determination of available quantities of agri-food waste, and lifecycle-based environmental assessments (Caldeira et al., 2020).

Industrial hemp (*Cannabis sativa* L.) cultivation in the last decade has significantly increased around the world. This trend is associated with its environmental friendliness and sustainability characterized by relative low input cultivation and high multipurpose biomass yield (Finnan & Burke, 2013) and also with the versatile utilization paths (Adesina, Bhowmik, Sharma, & Shahbazi, 2020; Schluttenhofer & Yuan, 2017). Used in crop rotation, industrial hemp has the potential to achieve a positive environmental impact in terms of soil restructuration and fertilization (Adesina et al., 2020) and also in terms of climate change mitigation due to its high photosynthesis rate (CO_2 fixation) at low nitrogen input (Tang, Struik, Amaducci, Stomph, & Yin, 2017). It has been estimated that one hectare of hemp can absorb 22 tons of CO_2, more than other industrial crops and forests (Adesina et al., 2020).

Simultaneously with the spreading of hemp cultivation, specific agricultural processing mechanizations are developing that are contributing to the growth of modern hemp agro-industrial chains capable of exploiting the multipurpose nature of this crop and increase the profitability of both cultivation and processing. Combined harvesters have been the most relevant innovations developed in the EU in recent years able to cut stems, separate panicles from straws, and valorize industrial hemp to produce fibers and seeds (Pari, Baraniecki, Kaniewski, & Scarfone, 2015). Collected panicles, depending on harvesting period (blooming or seed set), can also be used for essential oils extraction (Pari et al., 2015).

Depending on the use of the crop—dual purpose (seeds/inflorescences and fiber) or single purpose (cannabidiol (CBD) or hemp seed)—different agro-industrial and processing chains have been developed. Each segment of the hemp agro-industrial chain (e.g., harvest, primary processing, and secondary processing) possesses its own peculiarities in terms of applied machinery, procedures, and technologies for the conversion of raw materials into final products. The agricultural or processing wastes and by-products originating from those segments are diverse (Fig. 12.1) and can be a source of different classes of bioactive compounds, such as bioactive polysaccharides, bioactive peptides, unsaturated fatty acids, carotenoids, steroids, vitamins, pigments, essential oils, alkaloids, and phenolics as summarized in Table 12.1.

By valorization of hemp's agricultural and processing by-products and wastes, the profitability and sustainability of hemp production and processing is achieved, which is in line with the requirements of a circular economy toward the achievement of the UN's sustainable development goals and European Green Deal. Moreover, the valorization of hemp-based agricultural wastes and by-products supported by eco-innovations contributes to the development of new bio-based value chains for sustainable economic development (Polova & Bruskova, 2018).

12.2 Hemp harvesting wastes

The nature and quantity of hemp agricultural waste is connected to the intended uses of the crop, which affects the choice of variety (monoecious or dioecious), sowing density, harvest time, and harvesting techniques. In general, a double-purpose monoecious hemp crop at maturity produces 1%–10% seeds, 60%–70% stalks, and 20%–30% leaves and inflorescences (Matassa, Esposito, Pirozzi, & Papirio, 2020). Stalks are used for fiber/hurd extraction and seeds for oil extraction for use in food or cosmetics; produced as waste is a small amount of leaves, falling down during the life span of the crop, and roots, which both contribute to soil organic fertilization. On the other hand, during hemp seed or inflorescence cultivation,

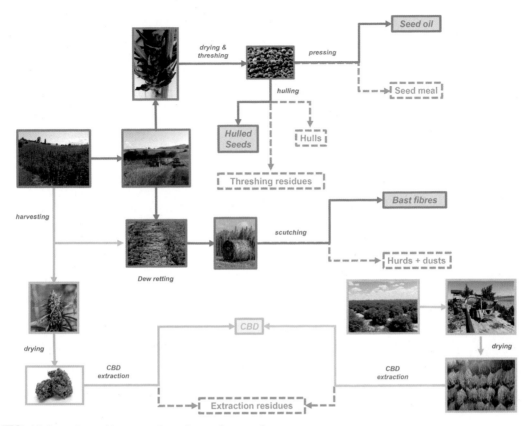

FIG. 12.1 Industrial hemp products, by-products, and wastes.

agricultural waste represents nearly 90% of the crop's biomass (stalks, leaves, and roots). Stalks and leaves are composed of lignocellulosic biomass that could be utilized for energy needs, either as a direct source of energy or as a processed source of energy, converted to biofuels like bioethanol, biodiesel, and biogas. If harvested during the flowering phase to obtain optimal fiber quality, inflorescences can represent a valorizable agricultural waste from fiber hemp cultivation (Adesina et al., 2020). The growing demand for bioactive molecules such as CBD, terpenoids, and flavonoids, which are also discretely accumulated in the inflorescences of fiber hemp cultivars, makes inflorescences a by-product of fiber hemp cultivation, actually releasing them from the definition of agricultural waste. In most industrial hemp cultivation purposes, the roots are left in the soil after harvesting and are considered agricultural waste, although they should be considered a positive externality of the crop. In fact, the hemp root system constitutes almost a half of the total crop's dry biomass, regardless of the growing conditions (Amaducci, Zatta, Raffanini, & Venturi, 2008) thus representing a good source of soil organic matter when decomposed. Furthermore, their ability to grow deeply in loosely packed soils play an important role in restructuring and stabilizing the soil (Amaducci et al., 2008). In hemp cultivation for CBD production, mechanical harvesting of the whole plant makes part of the root system available for further valorization.

TABLE 12.1 An overview of different hemp-based by-products and the routes of their valorization.

Utilized by-product	Valorized product	Pretreatment	Solvent	Technology used
Hemp roots	Bioactive extract	Air-drying, grinding	Ethyl acetate	US, maceration
Hemp roots	Bioactive extract	Washing, chopping, freezing, milling	Ethanol, *n*-hexane	Conventional extraction; supercritical CO_2 extraction
Hemp roots	Bioactive extract	N/A	N/A	N/A
Inflorescence	Volatile hydrodistillate	Frozen fresh	Water	Conventional and microwave-assisted hydrodistillation
Inflorescence	Essential oil	Drying		Hydrodistillation
Inflorescence	Essential oil for pest management	Fresh material used	N/A	Steam distillation
Threshing residues	Cannabinoids	Cutting (2 mm)	Ethanol	Pressurized liquid extraction
Threshing residues	Lipophilic extract; flavonoid-containing polar extract; mono- and disaccharides	Drying, grinding (0.2 mm)	Acetone; mixture of ethanol/water; ViscozymeL	Supercritical CO_2 extraction + pressurized liquid extraction + enzyme-assisted extraction
Threshing residues	Added-value compounds	Drying, milling	Methanol	Ultrasonic-assisted extraction

Recovered compounds	Remark	Scalability	References
Sterols, triterpenoids	Rediscovered therapeutic potential of hemp root	Laboratory scale	Jin, Dai, Xie, and Chen (2020)
Friedelin, epifriedelinol, ethyl palmitate, ethyl linoleate; ethyl elaidate; ethyl stearate; oleamide; stigmasta-3,5,22-triene; stigmasta-3,5-diene; campesterol; stigmasterol; β-sitosterol; stigmastanol; fucosterol; β-amyrone; 4-campestene-3-one; β-amyrin; stigmasta-4,22-dien-3-one; glutinol; stigmast-4-ene-3-one	Identified undescribed triterpenoids and phytosterols in *Cannabis sativa* roots so far	Laboratory scale	Kornpointner et al. (2021)
Triterpenoids: friedelin, epifriedelanol Alkaloids: cannabisativine, anhydrocannabisativine; carvone and dihydrocarvone; *N*-(*p*-hydroxy-*b*-phenylethyl)-*p*-hydroxy-*trans*-cinnamamide Sterols: sitosterol, campesterol, stigmasterol Choline	Review article	N/A	Ryz, Remillard, and Russo (2017)
Polyphenols; luteolin-7-*O*-glucoside; apigenin-7-*O*-glucoside; monoterpenes, sesquiterpenes, phytocannabinoids	Hydrodistillate yield of MAHD procedure was three times higher than conventional hydrodistillation yield	Pilot scale	Gunjević et al. (2021)
Sesquiterpenes (β-caryophyllene and its oxidized derivatives, α-humulene), monoterpene hydrocarbons (α- and β-pinene, myrcene)	Although later harvest yielded higher total dry, inflorescence and stem mass, earlier harvest is associated with higher EO extraction yield	Laboratory scale	Ascrizzi, Ceccarini, Tavarini, Flamini, and Angelini (2019)
Monoterpenes, sesquiterpene hydrocarbons. (*E*)-Caryophyllene, myrcene and α-pinene were the most abundant	Hemp EO is environmental-friendly botanical insecticide, particularly efficient for the management of aphids and houseflies	Laboratory scale	Benelli et al. (2018)
Cannabidiol	Extraction efficiency of 99.3% was achieved	Laboratory scale	Serna-Loaiza et al. (2020)
Cannabidiol, cannabidiolic acid; flavonoids; glucose; maltose	Biorefining potential demonstrated	Laboratory scale	Kitrytė, Bagdonaitė, and Rimantas Venskutonis (2018)
CBD, CBG, cannflavin A	Cannflavin A was affected by genotype and air temperature. CBD was affected by climatic conditions and harvest time	Laboratory scale	Calzolari et al. (2017)

Continued

TABLE 12.1 An overview of different hemp-based by-products and the routes of their valorization—cont'd

Utilized by-product	Valorized product	Pretreatment	Solvent	Technology used
Threshing residues	Cannabinoids enriched extracts	Grinding, pelleting	CO_2	Supercritical carbon dioxide
Threshing residues	Cannabinoids enriched extracts	Grinding (2 mm)	Ethanol	Pressurized liquid extraction
Threshing residues	Cannabidiol and cannabidiolic acid enriched extracts	Grinding, sieving	Ethanol (co-solvent)	Supercritical CO_2 Subcritical CO_2
Hemp seed hulls	Bioactive extract	Grinding	Ethanol (60%)	Ultrasonic-assisted extraction
Hempseed hull	Polyphenol enriched extract	Co-rotating, twin-screw extrusion	Methanol (80%); ethyl acetate	Conventional extraction (free phenolics) + aqueous acid extraction
Hemp bran	Hemp bran protein isolate	Milling (0.35 μm), defatting	NaOH, HCl	Aqueous alkaline extraction + isoelectric precipitation
Hemp bran protein isolate	Hemp bran protein hydrolysate	Dissolution in water	Pepsin, trypsin, pancreatin, chymotrypsin, Flavourzyme, Alcalase, Protamex, Neutrase	Enzymatic hydrolysis
		N/A	*Lactobacillus* spp.	Bacterial fermentation (liquid)
		N/A	*Pleurotus ostreatus*	Solid-state fermentation
Oil hemp seed cake	Hydrolysis products	Milling (0.5 mm), defatting (SC-CO_2)	Exopeptidase, protease; methanol	Enzymatic hydrolysis, solid-liquid extraction
Defatted hemp seed cake	Polyphenol enriched extracts	Milling (450 μm) + microwaves (deionized water)	Mixed solvents: methanol, acetone and water	Ultrasonication
		Milling (450 μm) + pulsed electric field (ethanol)		

Recovered compounds	Remark	Scalability	References
Cannabidiol, cannabidiolic acid, cannabinol, cannabigerol, cannabichromene	The amount of cannabinoids depended on the quality of the threshing residue; the amount of acid forms and degradation products depended on storage times	Pilot scale	Vági et al. (2019)
Cannabinoids (mainly cannabidiol)	Extraction efficiency of 99.3% (low pressures, 100°C, 60 min)	Laboratory scale	Serna-Loaiza et al. (2020)
Cannabinoids	Threshing residues are suitable for producing value-added extracts for cosmetic/food industry	Pilot scale	Vági, Balázs, Komoczi, Mihalovits, and Székely (2020)
N-trans-caffeoyltyramine; cannabisin B	Hempseed hull extract can be used as dietary supplement to prevent oxidative stress	Laboratory scale	Chen et al. (2012)
26 Phenylpropionamides including hydroxycinnamic acid amides and lignanamides	Total phenylpropionamide content of hempseed hull increased after extrusion. Lowering the screw speed from 300 to 150 rpm increased TPC, antioxidant activities and in vitro enzyme activity	Pilot scale (extrusion), laboratory scale (extraction)	Leonard, Zhang, Ying, Xiong, and Fang (2021)
Proteins Bioactive peptides	The bioactivity of the resulting products was dependent on the proteases used for enzymatic digestion, the bacterial strain, and the duration of the fermentation processes	Laboratory scale	Setti et al. (2020)
Lignanamides (cannabisin- and grossamide-type), cannabidiolic acid	Sediment and liquid fractions were obtained of different nutritional value, antinutrients content and sensory properties	Pilot scale (defatting), laboratory scale (enzyme hydrolysis)	Pap et al. (2020)
Polyphenolics, flavonoids	Optimum pretreatment conditions were defined using RSM for maximum recovery of polyphenols and antioxidant capacity	Laboratory scale	Teh, Niven, Bekhit, Carne, and Birch (2014)

Continued

TABLE 12.1   An overview of different hemp-based by-products and the routes of their valorization—cont'd

Utilized by-product	Valorized product	Pretreatment	Solvent	Technology used
Defatted hemp seed cake	Polyphenol enriched extract	Grinding to powder	Mixed solvents: methanol, acetone and water	Ultrasonic treatment + heating
Hemp seed meal	Protein and fiber enriched fractions	Grinding (350, 250, 180 μm)	N/A	Sieving
Hemp seeds	Bioactive extract	Drying, milling	Water in supercritical form	Pressurized hot water extraction
Fresh hemp biomass	Essential oil	Grinding/ non-grinding		Hydrodistillation
Hemp leaves and buds	Cannabinoid enriched extracts	Grinding (63, 125, 250, 500, 2000, 4000 μm)	Ethanol (co-solvent)	Supercritical carbon dioxide extraction
Hemp stems and branches	Sugar flatform processing	Drying, grinding (0.595 mm), acid pretreatment (HCl)	Endo-glucanase, β-glucosidase	Enzymatic hydrolysis
Hemp hurds/dust	Hydroxycinnamic acids	Twin-screw extrusion + milling (0.5 mm)	NaOH, water + ethanol	Thermo-mechano-chemical extraction
Hemp hurds	Fermentable sugars/ethanol	Retting, drying, milling + steam pretreatment	N/A	Enzymatic hydrolysis + saccharification and fermentation
Hemp hurds	Poly-3-hydroxybutyrate bioplastics	Washing, drying, grinding (1 mm), final drying, autoclaving in water, H_2SO_4 solution, and NaOH solution	Cellulase, β-glucosidase, cellobiase, xylanase	Enzymatic hydrolysis + ultrasonication + fermentation (*Ralstonia eutropha*)

Recovered compounds	Remark	Scalability	References
Phenolics and flavonoids	Ultrasonication increased polyphenol extraction yields and enhanced antioxidant capacity of the seed cake extracts. Heating additionally assisted the extraction	Laboratory scale	Teh and Birch (2014)
Protein, fibers, cannabisin B and N-*trans*-caffeoyltyramine, catechin, *p*-hydroxybenzoic acid, ω-6 fatty acids, antinutrients (trypsin inhibitors, phytic acid, glucosinolates, condensed tannins)	Fractionation is useful to concentrate valuable target compounds and facilitate their recovery	Laboratory scale	Pojić et al. (2014)
Cannabinol, cannabidiol, cannabichromene, cannabigerol	PHWE is suitable technique for selective extraction of cannabinoids	Laboratory scale	Nuapia, Tutu, Chimuka, and Cukrowska (2020)
β-Pinene, myrcene, δ-3-carene, limonene, eucalyptol; β-caryophyllene, α-(*E*)-bergamotene, (*Z*)-β-farnesene, α-humulene, caryophyllenyl alcohol, germacrene D-4-ol, spathulenol, caryophyllene oxide, humulene epoxide 2, β-bisabolol, α-bisabolol, sesquiterpenes, and cannabidiol	Monoterpenes in the hemp EO can be increased by grinding material prior to distillation and collecting the EO in the first 10 min. Grinding affected a slight but significant decrease in the CBD concentration of the EO	Laboratory scale	Zheljazkov et al. (2020)
Cannabinoids (THC, THCA)	Process extraction efficiency was 92%	Pilot scale	Rovetto and Aieta (2017)
Glucose	Supplementation with bovine serum albumin before cellulose hydrolysis boosted the glucose yield	Laboratory scale	Kim et al. (2021)
Ferulic, *p*-coumaric acid	Extraction by extrusion reduced extraction time, energy, solvent, and reagent consumptions	Pilot scale	Candy, Bassil, Rigal, Simon, and Raynaud (2017)
Glucan, xylan, glucose xylose, ethanol	Hemp hurds are valuable feedstock for ethanol production	Pilot scale	Barta et al. (2010)
N/A	Ultrasonic-assisted SDS digestion recovered 94% of P(3HB) directly from broth cell concentrate	Laboratory scale	Khattab and Dahman (2019)

12.3 Hemp processing by-products and waste

As shown in Fig. 12.1, the main harvested products of industrial hemp crop are stalks, seed panicles, flower panicles (inflorescences), and the whole plant in monopurpose CBD cultivation. According to their uses, these products undergo different processes, generating products, by-products, and wastes of different kind, quantity, and chemical composition.

12.3.1 Stalk processing by-products and waste

Depending on the final purpose of hemp fiber, stalks undergo different pretreatments and processes (Lee, Khalina, Lee, & Liu, 2020). In most cases, for textile and technical applications, stalks are subjected to a retting pretreatment that releases the fiber bundles by removing the surrounding pectin layer. Retting can be carried out by various methods, the most common is dew retting, which is carried out on the ground by the pectinolytic activity of the microbial communities that naturally live on hemp stalks (Lee et al., 2020). The pretreated stalks are subjected to mechanical decortication, which allows the separation of the outer fiber layer (bast fiber) from the inner woody core (hemp hurds) with the production of a discrete amount of dust (Lee et al., 2020). Hurds can be considered waste regarding bast fiber production but could be considered a processing by-product. In fact, hemp hurd can be primarily used as a livestock bedding, but due to its quantity and quality (70%–80% of the stalk, 20%–30% of lignin), it represents a valuable lignocellulosic biomass material for further valorization (Adesina et al., 2020).

12.3.2 Seed processing by-products and wastes

Seed panicles mechanically harvested with the apical portion of the plant are processed to produce seeds and food derivatives. Seed panicles, dried at temperatures below 30°C to preserve the organoleptic properties of their products (Callaway & Pate, 2009), are then mechanically threshed, leaving threshing residues, which represents a seed processing waste, and calibrated for further processing. In most cases hemp seeds are cold-pressed to produce hemp seed oil for food uses or cosmetics, leaving seed cake as a processing by-product. Oil seed cakes in general and hemp seed cake in particular, are rich in proteins, dietary fibers, and a variety of bioactive compounds (Álvarez, Mullen, Pojić, Hadnađev, & Papageorgiou, 2021; Pojić, 2021), therefore it has been identified as an interesting by-product suitable for valorization as either human food or animal feed. Threshed seeds could also be commercialized as animal feed or hulled and commercialized as food. Hulling implies the separation of the seed from its hard shell (hull), producing hulls as waste. The hemp hull accounts for about 24% of the hemp seed, and the traditional utilization of hemp hulls includes bedding for animals and livestock feeding. However, hemp hull has a valuable nutritional and nutraceutical composition, which makes it a significant and economic source of nutraceutical compounds such as natural dietary fibers and polyphenols. Different pretreatments are often needed to optimize the value addition to hemp hull; likewise, the direct utilization of hemp hull is also possible.

12.3.3 Inflorescences processing by-products and waste

Flower panicles collected at the blooming phase are freshly trimmed and subjected to distillation or chemical extraction for essential oil production, leaving vegetal debris at the end of extraction as a processing waste. Similarly, in CBD production, hemp panicles are collected at the blooming phase and dried in controlled conditions, trimmed, and subjected to chemical CBD extraction, producing extraction residues as waste. A multistep extraction process in which residues belonging to extraction phases are further processed for the recovery of valuable molecules—cannabinoids after essential oil extraction (Gunjević et al., 2021) or fermentable sugars after cannabinoids extraction (Kitrytė et al., 2018)—could be a valuable system for further valorization of extracted residues.

12.4 Valorization of hemp harvesting waste

12.4.1 Valorization of hemp lignocellulosic biomass

In one-purpose hemp cultivation, after seeds or inflorescences are harvested, the remaining crop biomass can be sold as hemp straws, if not used for fiber extraction. This lignocellulosic biomass, regardless of the variety and growing area, has been estimated to be composed of cellulose (36.5%–75.5%), hemicellulose (10.1%–32.8%), and lignin (8.0%–22.9%) (Zhao, Xu, Wang, Griffin, Roozeboom, & Wang, 2020). When compared to corn fiber, corn stover, and sorghum bagasse, hemp biomass contains a relatively higher cellulose content, has a larger range of variation in hemicellulose content depending of hemp variety and area of cultivation, and a relatively lower range of variation in lignin content compared to corn stover and sorghum bagasse, but higher in comparison to corn fiber (Das et al., 2017). Cellulose and hemicellulose could be converted to bioethanol, biobutanol, and other fermentation products through various biological pathways (Zhao et al., 2020a). The complexity of the structure and chemical composition of plant cell walls makes lignocellulosic biomass recalcitrant to conversion. In particular, lignin content and its composition is a crucial recalcitrance factor, acting as a physical barrier, and as a toxic inhibitor in the conversion process (Ji, Jia, Kumar, & Yoo, 2021; Yoo, Meng, Pu, & Ragauskas, 2020). Therefore, several pretreatment methods have been developed to reduce recalcitrance and enhance biomass conversion. An effective pretreatment process should decrystallize the cellulose without causing its hydrolysis, depolymerize hemicellulose, and avoid or limit the formation of inhibitors of carbohydrate conversion, requiring low energy input, and allowing the recovery of valuable compounds. Finally, it should be cost-effective (Yoo et al., 2020). Furthermore, to increase the surface area available for the pretreatment and conversion process, a step of biomass mechanical grinding and sieving could be added upstream of the whole process.

12.4.1.1 Hemp biomass pretreatment

Chemical pretreatment

Lignin solubilization and partial removal of hemicellulose from hemp biomass could be carried out by chemical or physical methods. Diluted acid pretreatment, the most widely used industrial pretreatment method for hemp biomass, is generally carried out with diluted sulfuric acid (0.5%–3.0%) at a temperature ranging from 150°C to 180°C for 10–20 min (Gunnarsson, Kuglarz, Karakashev, & Angelidaki, 2015; Ji et al., 2021; Kuglarz, Alvarado-Morales, Karakashev, & Angelidaki, 2016). It effectively reduces the recalcitrance of hemp biomass, thus increasing cellulose accessibility to enzymes, but it also allows carbohydrate decomposition with the production of furfurals and hydroxymethylfurfurals (HMF), which have an inhibitory effect on microorganisms, affecting their fermentation activity and the overall conversion process yield (Jönsson, Alriksson, & Nilvebrant, 2013). Diluted acid pretreatment (H_2SO_4 1%–1.5% w/v, 180°C) of hemp biomass allowed the recovery of > 95% glucans, 41%–51% xylans, and the removal of 35%–41% lignin (Kuglarz et al., 2016). On the other hand, pretreatment of hemp biomass with H_2SO_4 1%–2% resulted in glucan (87%–92%) and xylan (11%–22%) recovery but also caused the production of HMF (1.15–0.35 g/L) and furfural (10–0.25 g/L) (Gunnarsson et al., 2015).

Alkaline pretreatment, mainly carried out by diluted NaOH (1%), allows the cleavage of ester and ether bound between lignin and cellulose or hemicellulose, increasing cellulose accessibility to enzymes (Ji et al., 2021). Furthermore, hydroxide anions (OH^-), probably due to the neutralization of the reaction surround, substantially reduce sugar degradation and the formation of the toxic furan compounds (Ji et al., 2021). Alkaline pretreatment carried out with 1% and 3% NaOH at 121°C for 1 h led to the recovery of > 95% glucans and > 55% xylans without the formation of any inhibitor compounds (Jönsson et al., 2013). The main advantages of alkali pretreatment are the higher glucans recovery (66.9%–75.0%) compared to the acidic one (54.3%–63.8%) and better efficiency of lignin removal from biomass. In fact, alkali pretreated hemp biomass exhibited a significantly lower lignin content (8.4%–12.3%) than the acid pretreated one (28.9%–37%), probably due to its higher efficiency in breaking down lignin in small subunits (Ji et al., 2021; Zhao, Xu, Wang, Griffin, & Wang, 2020b).

Liquid hot water (LHW) pretreatment is an alternative to the diluted acid process, which uses water at high temperatures (130–240°C) and high pressure. In these conditions, water shows acid properties acting as a weak acid catalyst (Ji et al., 2021). LHW pretreatment has been found to partially hydrolyze hemicellulose and disrupt the lignin and cellulose structures. Furthermore, it promotes the release of acylic groups from hemicellulose with the production of acetic acid as a by-product (Ji et al., 2021). Another advantage of LHW pretreatment is the low cost of the reaction medium (i.e., water) and the low cost of washing steps generally used for the neutralization of diluted acid/basic pretreated biomass (Gunnarsson et al., 2015). During LHW pretreatment, sugar degradation and the formation of toxic compounds can be limited by maintaining the surrounding pH between 4 and 7 (Gunnarsson et al., 2015); therefore LHW pretreatment efficiency can be improved by the addition of acids or bases to stabilize the pH of the reaction mixture. On the other hand, the disadvantages of LHW pretreatment include relatively high energy consumption, low concentration of hemicellulosic sugar in the pretreatment hydrolysate, and the production of a large amount of wastewater by downstream processing (Gunnarsson et al., 2015). LHW pretreatment of hemp biomass originating from four different varieties allowed the recovery of 42.6%–54% glucans, starting from a raw biomass content of 40.1%–42.7% (Zhang, Zhou, Liu, & Zhao, 2020). Furthermore, the lignin content of the LHW pretreated biomass was lower than that pretreated with diluted acid but higher in comparison to the diluted alkali pretreated one (Zhang et al., 2020).

There are two further chemical pretreatment methods for hemp lignocellulosic biomass: an oxidative, carried out with hydrogen peroxide (H_2O_2 1% or 3% w/v) in an autoclave at 121°C (Jönsson et al., 2013), and the organosolv pretreatment (Gandolfi, Ottolina, Consonni, Riva, & Patel, 2014). Organosolv pretreatment involves the cooking of lignocellulosic biomass in a mixture of water and an organic solvent that leads to the deconstruction of both lignin and hemicellulose and their dissolution in the cooking liquor (Nitsos, Rova, & Christakopoulos, 2018). Lignin is typically retrieved as a precipitate by dilution of the liquor with water; therefore this process produces three distinct streams: a cellulose-rich pulp (1), a lignin-rich solid precipitate (2), and a hemicellulose-rich liquid (Nitsos et al., 2018).

Physical pretreatment

Steam explosion Among physical pretreatment methods, steam explosion is a process in which biomass is treated with hot steam (180–240°C) under pressure (1–3.5 MPa) followed by an explosive decompression of the biomass to atmospheric pressure that results in the rupture of the rigid structure of biomass fibers, changing the starting material into a fibrous dispersed solid (Bandyopadhyay-Ghosh, Ghosh, & Sain, 2015). The sudden release of pressure generates shear force that hydrolyzes the glycosidic bond and hydrogen bonds between the glucose chains. In this process, pressure, temperature, and the residence time of the biomass in an autoclave are the effective parameters (Islam & Rahman, 2019). Steam explosion of hemp hurds at a temperature of 200–230°C allowed the recovery of > 82% of the original content of glucan and 18%–66% of the original content of xylan, whereas lignin removal was low. Furthermore, the steam explosion caused the solubilization of most hemicellulose and partial cellulose (Barta et al., 2010). Hemicellulose degradation could be associated with an increase of glucan content in the biomass and the formation of toxic inhibitors as furans and insoluble products that could interact with the residual lignin components to form pseudolignin complex (Kumar et al., 2013).

Electron beam irradiation Electron beam irradiation (EBI) has been investigated to overcome hemp lignocellulosic biomass recalcitrance and increase cellulose accessibility. Although relatively energy-intensive, EBI pretreatment is characterized by short treatment time, no need for specific sample preparation and/or chemicals, and no residual chemicals in the sample (Henniges, Hasani, Potthast, Westman, & Rosenau, 2013). Ionizing irradiation can either originate from a radioactive source or from highly accelerated electrons, both able to replace a chemical treatment for cellulose modification. Industrial electron accelerators are classified according to their energy range as low (80–300 keV), medium (300 keV–5 MeV), and high energy (above 5 MeV) (Henniges et al., 2013). The high energy irradiation has a strong effect on the physical and chemical properties of the biomass, depending on the applied dose (Sung & Shin, 2011). It has been demonstrated that high energy irradiation reduces the degree of polymerization and crystallization of cellulose, thus increasing the surface area available for further conversion (Sung & Shin, 2011). EBI of hemp biomass with an irradiation dose of 450 kGy followed by enzymatic extraction resulted in a higher recovery of xylan and cellulose compared to hot water or 1% sodium hydroxide pretreatment (Shin & Sung, 2008). Carbohydrate degradation in hemp biomass induced by increasing irradiation doses (150, 300, 450 kGy) has been quantified, revealing higher irradiation to produce a more severe degradation of structural components; among them, xylan was more sensitive to EBI than cellulose (Sung & Shin, 2011).

12.4.1.2 Conversion of hemp biomass to biofuels

At the beginning of the 21st century, a great deal of research has been focused on renewable energies in general and valorization of crop biomass for energy in particular. Because hemp generally produces large quantities of biomass (i.e., stalks, retted and unretted hurds, bast fiber, inflorescences, and a mix of leaves and flowers) with low inputs, several research groups have studied the potential hemp has for biofuel production, namely *bioethanol* and *biomethane*. Therefore, certain research has been directed toward the identification of optimal hemp growing conditions and harvesting time for the maximization of dry biomass yield (Bandyopadhyay-Ghosh et al., 2015; Nitsos et al., 2018), as well as toward the identification of optimal pretreatment conditions and/or additive concentrations to maximize hemp-based biofuel potential (Bandyopadhyay-Ghosh et al., 2015; Gandolfi et al., 2014; Nitsos et al., 2018). These studies provided preliminary data needed for the identification of the best production pathways from an energy balance and an economic point of view.

Conversion of hemp biomass to bioethanol

Second-generation bioethanol, as well as other second-generation biofuels, are produced by lignocellulosic biomasses such as wood or agricultural and organic wastes. The biodegradability and reduced toxicity of bioethanol, for which biomass is used as a primary substrate, are its main advantages over fossil fuels (Vasić, Knez, & Leitgeb, 2021). The conventional process for bioethanol production by lignocellulosic biomasses includes three steps: *pretreatment, hydrolysis,* and *fermentation.* As mentioned previously, the purpose of pretreatment is removing the lignin and facilitating the penetration of the hydrolyzing agents (chemical or enzymatic) converting hemicellulose and cellulose polysaccharides in fermentable sugars used as substrates for fermentation and production of biofuels or other bioproducts (Ji et al., 2021).

Pretreatment: The methods for the pretreatment of hemp lignocellulosic biomass have been described in a previous section; each method has a different effectiveness, so to obtain optimal results, a combination of chemical and physical methods is recommended (Tran, Le, Mai, & Nguyen, 2019).

Hydrolysis: Lignocellulose can be hydrolyzed to single sugars (hexoses and pentoses) by enzymatic or chemical hydrolysis (saccharification). Enzymatic hydrolysis is generally preferred to a chemical process (acid, alkaline, or salt) due to its greater feasibility and compatibility with the subsequent fermentation step. In fact, chemical hydrolysis often alters the reaction environment rendering it unsuitable for the fermentative microflora (Tran et al., 2019; Vasić et al., 2021). A commercial mix of three kinds of cellulolytic enzymes (endoglucanase, exoglucanase, and hemicellulases) can be used for enzymatic hydrolysis, although a custom-made mix of the desired enzymes is also possible. Endoglucanases initiate the enzymatic hydrolysis of cellulose, reducing its degree of polymerization; exoglucanases catalyze the production of cellobiose at reducing and non-reducing ends of cellulose; and hemicellulases are specific for the degradation of polysaccharides (Vasić et al., 2021). A significant variation in the hydrolysis yield was reported in the comparison of 11 different hemp varieties processed in the same conditions, which could be ascribed to the genetic differences in the degree of biomass recalcitrance and/or cellulose accessibility to hydrolytic enzymes between varieties (Das et al., 2017).

Microbial fermentation: Fermentation microorganisms employed for sugar conversion in bioethanol production mainly belong to the *Saccharomyces cerevisiae* strain or *Escherichia coli*, *Zymomonas mobilis*, and *Klebsiella oxytoca* strains. Each one has peculiarities in its ability to ferment various types of sugars produced by the hydrolysis of the pretreated biomass. An exhaustive comparison of their specificities in transformation of glucose substrates into ethanol was provided by Tran et al. (2019). Hemp biomass conversion at an experimental or industrial level was carried out by *Saccharomyces cerevisiae* strains, which is the most commonly used microorganism for bioethanol production (Zhao et al., 2020a). Tailor-made bacterial strains obtained by genetic engineering are also commercially available to optimize bioethanol or other biocompound production yields by lignocellulosic biomass fermentation, contributing to the economic sustainability of production (Vasić, Knez, & Leitgeb, 2021; Zhao et al., 2020a). Moreover, the potential of thermophilic anaerobic bacteria for bioethanol production has been investigated, such as *Thermoanaerobacter ethanolicus*, *Thermoanaerobacter mathranii*, *Thermoanaerobacter brockii*, and others. Although these bacteria are able to withstand extreme temperatures and can survive thermochemical pretreatment, they possess low ethanol tolerance (Vasić et al., 2021). Among thermophilic anaerobic bacteria, the *Clostridium thermocellum* strain has been tested for the production of biohydrogen and bioethanol by hemp straws and hurds biomass due to its ability to grow well on amorphous and crystalline cellulose (Agbor et al., 2014).

Microbial fermentation can be carried out using two main approaches: by separate hydrolysis and fermentation (SHF) and/or by simultaneous saccharification and fermentation (SSF). In the case of SHF, the fermentation microorganisms are added when hydrolysis is completed. In the case of SSF, hydrolysis and fermentation are carried out simultaneously in the same reaction environment and with the same equipment. Apart from being time and labor consuming, the drawbacks of the SHF method are the possibility of contamination and the formation of inhibitors. Because the utilization of the SSF method reduces the risk of contamination and is faster and less labor intense, it is generally the method of choice for industrial bioethanol production (Tran et al., 2019). Apart from these processes, there is also the direct microbial conversion process (DMC) in which the biomass pretreatment, hydrolysis, and fermentation are carried out in a single step. DMC simplifies the whole production process and reduces the number of reactors, leading to a cost reduction (Vasić et al., 2021), but it is more time consuming and of lower conversion yields compared to SSF, probably due to the simultaneous presence of mixed substrates and products in the same reaction mixture (Tran et al., 2019). The bioethanol yield of the SSF process is affected by cellulose content and its availability, hydrolytic enzyme efficiency, and bacterial strain ability to convert available sugars to ethanol. Solid loading (e.g., the ratio between biomass weight and enzyme concentration) is a further factor affecting yield in SSF bioethanol production. High solid loading would be advantageous in the conversion of lignocellulosic biomass as it would guarantee high ethanol concentration while reducing water consumption. On the other hand, SSF at higher solid loading can also decrease the conversion efficiencies from pretreated biomass to sugar and ethanol (Zhao, Xu, Wang, Griffin, & Wang, 2020b). The relationship between mechanical-chemical pretreatment, enzyme and bacterial strain loading, and further additive concentration (Tween 80) for optimal bioethanol yield by SSF fermentation of hemp biomass was experimentally assessed, providing a mathematical model for the calculation of solid loading (Zhao et al., 2020c).

Some examples of hemp biomass conversion to bioethanol by SSF and SHF are available in the literature. SHF of untreated hemp biomass at low solid loading with *Saccharomyces cerevisiae* led to the production of 7.2 g/100 g of initial biomass (Kuglarz et al., 2016) and 2.89 g/L bioethanol (Kuglarz et al., 2014). In the same conditions, SHF of acid-assisted, steam-pretreated biomass produced 4.62–10.00 g/L bioethanol (Kuglarz et al., 2014) and 14.9–15.5 g bioethanol per 100 g of initial hemp biomass (Kuglarz et al., 2016). SHF of alkaline-oxidative-pretreated biomass produced 16.6–17.5 g bioethanol per 100 g of initial hemp (Kuglarz et al., 2014). Considering that these experiments have been carried out on different hemp varieties and different environments, it was pointed out that more efficient SHF conversion of pretreated hemp biomass to bioethanol with *Saccharomyces cerevisiae* should be carried out at low solid loading. They also confirmed alkaline oxidative pretreatment to be more efficient than acid steam pretreatment in hemp biomass conversion to bioethanol by SHF (Kuglarz et al., 2014). On the other hand, SSF of hemp biomass originating from different hemp varieties using *Saccharomyces cerevisiae* at 5% solid loading confirmed alkali pretreatment (diluted NaOH) to be more effective in bioethanol production when compared to LHW pretreatment and diluted acid pretreatment (Zhao et al., 2020b). Ensiled and dry hemp biomass of Futura 75 variety, pretreated under optimal conditions (2% SO$_2$ followed by steam pretreatment at 210°C) and subjected to SSF with *Saccharomyces cerevisiae* at 7.5% solid loading, resulted in similar bioethanol yield, revealing no significant effect of the ensiling process on the conversion efficiency (Sipos et al., 2010). The highest bioethanol concentration (77 g/L) has been obtained from alkali pretreatment followed by particle size reduction (< 0.2 mm) of hemp biomass by SSF carried out with 30 FPU-cellulase and 140 FXU-hemicellulase/g-solid, at 31% solid loading (Zhao et al., 2020b).

SSF used for bioethanol conversion of dry hemp hurds revealed the steam pretreatment to be one of the most efficient pretreatments for this lignocellulosic substrate to obtain high polysaccharide conversion by *Saccharomyces cerevisiae* (4%). Steam pretreatment at 210°C followed by enzymatic hydrolysis and fermentation at 10% solid loading led to an ethanol yield of 141 g/kg dry hurds (Barta et al., 2010).

Conversion of hemp biomass to biomethane

Among the multitude of technologies available for energy content of organic feedstock valorization, anaerobic digestion (AD) still represents the most widely applied process due to its high efficiency, operational flexibility, and overall environmental benefits (Gopal, Sivaram, & Barik, 2019). AD is a sequence of processes that involve microbial breakdown of organic matter in the absence of oxygen. Methane production by AD is generally carried out by a mixed microbial culture that is able to anaerobically degrade most components of the lignocellulosic biomass except lignin. Lignin content, cellulose crystallinity, and particle size of the substrate could substantially affect the conversion rate (Kreuger, Sipos, Zacchi, Svensson, & Björnsson, 2011). To obtain hemp-based biomethane, whole crop biomass (stems and leaves) is coarsely chopped and/or finely ground (Kreuger, Prade, et al., 2011; Kreuger, Sipos, et al., 2011) and subjected to AD without any pretreatment; digestate originating from a biogas plant has been used as inoculum. With a yield of 16 tons dry matter per hectare, it was possible to achieve an average gross methane yield of 136 ± 24 GJ per hectare (Kreuger, Prade, et al., 2011). Relatively low conversion rate (47%), despite relatively high carbohydrate content and low lignin content, highlighted the potential for increasing methane yield per hectare thus improving the conversion technique.

Biomass pretreatment and co-production (ethanol and biomethane) experimentally showed substantial improvement of biofuel production from hemp biomass (Kreuger, Sipos, et al., 2011). Co-production of ethanol and methane from heat-pretreated hemp biomass (stems) gave a total biofuel yield of 11.1–11.7/MJ/kg stem dry matter. It has been estimated that co-production of the whole hemp plant would give 2600–3000 L ethanol and 2800–2900 m^3 methane per 10 ha, based on a biomass yield of 16 tons per hectare (Kreuger, Sipos, et al., 2011).

Different possible hemp biomass residues (HBRs) from agricultural and industrial production (e.g., stalks, retted and unretted hurds, bast fiber, inflorescences, and a mix of leaves and flowers) and their combinations have been assessed for their biomethane potential (Matassa et al., 2020). Furthermore, chemical-mechanical pretreatment was evaluated for their effect on HBR biomethane potential valorization by small scale batch test. The highest biochemical methane potential (BMP) was obtained with the raw fibers (422 ± 20 mL $CH_4/g_{volatile\ solids}$), while unretted hemp hurds, accounting for the 52% of the whole plant dry biomass, showed a BMP value of 239 ± 10 mL $CH_4/g_{volatile\ solids}$. The mix of leaves and inflorescences and inflorescences alone showed the lowest BMP values (i.e., 118 ± 8 and 26 ± 5 mL $CH_4/g_{volatile\ solids}$, respectively) and a prolonged inhibition of methanogenesis. Alkali pretreatment of unretted hurds and mechanical grinding of retted hurds enhanced the BMP of both substrates by 15.9% (Matassa et al., 2020).

Only one pilot to large-scale experiment has been carried out to date to demonstrate the potential of hemp straw residue for biogas production through AD by reproducing the real operating conditions of an industrial plant (Asquer, Melis, Scano, & Carboni, 2019) where the effects of enzymatic treatment, organic loading rate, and specific bioenhancers on biogas production by hemp straws were investigated. High loading rate (5% wt/wt) of hemp straw increased the median values of the gas production rate of biogas by 92.1%; likewise the addition of bioenhancers increased this value to 116.6%. At higher loading rate (5% wt/wt), the increase of the specific gas production (SGP) rate due to enzymatic treatment was + 129.8%. The best management of the biodigester was therefore found in the combination of higher values of hemp straw loading coupled with enzymatic treatment, reaching 0.248 $Nm^3/kg_{volatile\ solids}$ of specific biogas production (Asquer et al., 2019). Although the results of gas production rate and SGP (biogas/methane) were promising, the SGP of biogas/methane was lower compared to corn silage commonly used in industrial plants of AD. However, the low cost of hemp straw residues and their properties in the AD characterized this by-product as a good process moderator when using other types of biomass (Asquer et al., 2019). Further studies on sustainability of hemp straw residue conversion to biogas will allow defining achievable costs and economic benefits.

Another opportunity of HBR valorization for biomethane production is anaerobic co-digestion (coAD), commonly employed to rebalance too high carbon-to-nitrogen (C/N) ratio of the lignocellulosic matrix and, in turn, overcome its complex lignin- and carbohydrate-based structure, which significantly hinders the substrate's bioaccessibility (Papirio, Matassa, Pirozzi, & Esposito, 2020). CoAD is performed by adding animal manure or anaerobic sludge from industrial processes to the digesting mixture. Papirio et al. (2020) demonstrated the utilization of abundant industrial by-products of cheese whey and hemp hurds, which mixed in 70:30 ratio allowed enhancement of biomethane production by 10.7%, as compared to the cheese whey and hemp hurds alone that yielded 446 ± 66 and 242 ± 13 mL $CH_4/g_{volatile\ solids}$, respectively. The valorization of hemp hurds as biomethane brings significant economic profitability, which for dual purpose hemp (fiber-seed) potentially amounts cca 3929 €/ha, and in the case of utilization of coAD with cheese whey could rise to 6124 €/ha (Papirio et al., 2020).

12.4.1.3 Conversion of hemp biomass to succinic acid

Succinic acid is a naturally occurring four-carbon dicarboxylic acid and is one of the most promising chemicals with various potential applications (e.g., food and beverages, pharmaceuticals, polymers, resins, paints, cosmetics, inks). It can be produced via catalytic hydrogenation of maleic acid or maleic anhydride, a process that is both expensive and harmful to the environment (Ladakis, Papapostolou, Vlysidis, & Koutinas, 2020), Succinic acid also accounts for up to the 90% of the nonvolatile acids produced during alcoholic fermentation; therefore bio-based production of succinic acid through microbial fermentation is a more attractive perspective from an economic and ecological point of view. Several bacterial strains (*Anaerobiospirillum succiniciproducens*, *Actinobacillus succinogenes*, *Mannheimia succiniciproducens*, *Basfia succiniciproducens*, and also genetically engineered strains of *E. coli*) are able to produce succinic acid through the fermentation of different carbon sources as agricultural waste (e.g., lignocellulosic biomass) or industrial by-products (Ladakis et al., 2020). *Actinobacillus succinogenes* is one of the most promising strains for industrial applications (Kuglarz et al., 2016). The potential of the strain *Actinobacillus succinogenes* 130Z was investigated for the production of succinic acid by fermentation of hemp biomass hydrolysates (Gunnarsson et al., 2015). Enzymatic hydrolysis and fermentation of hemp biomass pretreated with 3% H_2O_2 at 121°C resulted in the highest overall sugar yield (73.5%), maximum succinic acid titer (21.9 g/L), as well as highest succinic acid yield (83%) (Gunnarsson et al., 2015). These experiments confirmed hemp biomass to be a promising feedstock for succinic acid production, laying the foundations for its industrial application.

Succinic acid can be produced by hemp biomass fermentation, along with bioethanol, in an integrated biofuel and biochemical production chain with a biorefinery approach (Kuglarz et al., 2016). During fermentation with *Saccharomyces cerevisiae*, only C6 sugars from lignocellulosic biomass can be converted to ethanol due to the low conversion rate of C5 sugars by this yeast strain. To optimize substrate utilization and further valorize hemp lignocellulosic wastes, *Actinobacillus succinogenes* could be used, leading to the production of succinic acid along with bioethanol. *Actinobacillus succinogenes* is, in fact, able to convert a wide variety of C5 sugars originating from hemicellulose hydrolysis (Kuglarz et al., 2016). Pretreated biomass has been separated in a solid and liquid fraction by filtration. Solid fractions are subjected to enzymatic hydrolysis, and the slurry produced has been supplemented with minerals and with 5% (v/v) of *Saccharomyces cerevisiae* inoculum. After fermentation in anaerobic condition, bioethanol has been distilled to a purity of 96%–98%, leaving stillage as a by-product. Stillage and liquid fraction produced by pretreated biomass filtration have been mixed and subjected to *Actinobacillus succinogenes* fermentation with the production of succinic acid. Alkaline pretreatment appeared to be superior to the acid-based method with respect to the rate of enzymatic hydrolysis and ethanol productivity. On the other hand, the highest succinic acid productivity has been obtained after fermentation of the liquid fraction produced by steam-assisted 1.5% acid pretreatment. Calculated yield of the combined process revealed that up to 149 kg of bioethanol and 115 kg of succinic acid can be produced from 1 ton of dry hemp biomass.

12.4.1.4 *Production of poly-3-hydroxybutyrate [P(3HB)] from hemp biomass*

Poly(3-hydroxybutyrate) [P(3HB)] is a highly crystalline, linear polyester of 3-hydroxybutyric acid produced by a wide variety of bacteria as a carbon reserve (Sastri, 2010). [P(3HB)] is a renewable, biodegradable, and biocompatible substitute for conventional non-degradable plastics. It has many properties that are attractive for biomedical applications; it is a polyester-like polyglycolide that is biodegradable within the body but can also undergo degradation in soil, making it attractive as a degradable packaging material (Polotti, 2020). Among bacterial strains able to produce [P(3HB)], *Cupriavidus necator* (known also as *Ralstonia eutropha*) has been widely studied due to its potential for producing significant amounts of P(3HB) from simple carbon substrates such as glucose, lactic acid, and acetic acid (Akaraonye, Keshavarz, & Roy, 2010). Recombinant strains of *Cupriavidus necator* and/or *E. coli* are currently available, which are able to use C5 and C6 carbon sources for [P(3HB)] biosynthesis and to grow very rapidly on common culture media (Kim et al., 2016).

Hemp hurds hydrolysate obtained from hot water, 2% H_2O_2, and 2% NaOH pretreatment, was tested for the production of [P(3HB)] with *R. eutropha*. Maximum hydrolysis yield (72.4%) was achieved with hydrolysates from alkali-pretreated hurds biomass, showing a total sugar concentration (glucose and xylose) of 53.0 g/L (Khattab & Dahman, 2019). Although *R. eutropha* showed preference for glucose metabolism over xylose, in optimal conditions it accumulated [P(3HB)] polymer in a quantity up to 56.3% of the dry cell weight, which corresponds to a total production of 13.4 g/L. The recovery of [P(3HB)] from concentrated growing media was carried out by SDS-assisted ultrasonication (Khattab & Dahman, 2019).

12.4.2 Valorization of hemp roots

The therapeutic properties of hemp roots have been exploited and handed down for millennia, but over time the use of hemp roots to relieve different types of pain has been completely abandoned (Ryz et al., 2017). By identifying and quantifying different types of bioactive compounds in hemp roots, its potential for further valorization has been confirmed, although hemp roots have been used in traditional medicine for centuries (Jin et al., 2020; Kornpointner et al., 2021; Ryz et al., 2017). The most dominant compounds in hemp roots are triterpenoids including friedelin, epifriedelanol, and β-amyrin, but hemp root also contain alkaloids: cannabisativine and anhydrocannabisativine, carvone and dihydrocarvone; *N*-(*p*-hydroxy-*b*-phenylethyl)-*p*-hydroxy-*trans*-cinnamamide; different phytosterols: sitosterol, campesterol, and stigmasterol; and minor compounds such as choline. Recently, Kornpointner et al. (2021) identified 20 different bioactive compounds in *Cannabis sativa* root extracts, among which two triterpenoids (glutinol, β-amyrone), four phytosterols (stigmastanol, fucosterol, stigmasta-3, 5-diene, stigmasta-3,5,22-triene), and one aliphatic (oleamide) compound were identified for the first time. It must be noted that hemp roots do not represent a significant source of phytocannabinoids including THC and CBD. The aboveground portion of hemp root has also been investigated for papermaking; in fact, its high cellulose content (46.6%) and low lignin content (17.6%) suggests that hemp root bast to be a high-quality material for papermaking. Further pulping and bleaching confirmed that hemp root bast can be valorized for papermaking due to the high yield of this row material and the high-quality physical properties of its pulp (Miao, Hui, Liu, & Tang, 2013).

12.5 Valorization of hemp stalk processing by-products and wastes

12.5.1 Hurd valorization

Hurd, the woody core of the hemp stalk, is considered a by-product of stalk decortication even though they represent the largest fraction produced (bast fiber 20%–30%, hurds and dusts 70%–80%) (Crini, Lichtfouse, Chanet, & Morin-Crini, 2020). Chemical composition of hurds has been assessed by different authors for different hemp varieties. With reference to the Carmagnola cultivar, it has been established that its hurds are composed of 69% polysaccharides (cellulose 44% and hemicellulose 25%), 23% lignin, and 1.2% ash, while total oil, proteins, amino acids, and pectins make up the 4% of dry weight (Gandolfi, Ottolina, Riva, Fantoni, & Patel, 2013). These data are in line with those of other hemp varieties available in the literature (Gandolfi et al., 2013). Hemp hurds are considered a cheap cellulose source and are generally used in a range of applications such as construction materials, paper, animal bedding, absorbents, and mulch (Crini et al., 2020), although they could be valorized for the production of energy, biofuels, and other valuable products.

12.5.1.1 Thermal valorization of hurds: Pyrolysis

Hemp hurds can undergo thermal valorization by pyrolysis. Pyrolysis is the thermal decomposition of organic matter that occurs prior to combustion, in the absence of oxidizing agents (O_2), and within a temperature range of 280–1000°C. Pyrolysis produces solid or carbonized products (biochar), liquid products (bio-oils, tars, and water), and a gas mixture (syn-gas) composed mainly of CO_2, CO, H_2, and CH_4 (Hu & Gholizadeh, 2019). The factors that influence the distribution of the pyrolysis products are: heating rate, final temperature, composition of the raw material, and pressure (Uddin et al., 2018). Depending on the environmental conditions, pyrolysis processes can be classified as slow pyrolysis (characterized by slow heating rates, temperatures less than 500°C, residence time > 20 s), moderate pyrolysis (temperatures of 500°C and residence time of 10–20 s), and fast pyrolysis (fast heating rates, temperatures greater than 500°C, and residence time less than 2 s) (Uddin et al., 2018). The bio-oil is a renewable liquid fuel in which its chemical composition depends on the nature of pyrogassified biomass. Bio-oil from lignocellulosic biomass is mainly composed by acids, phenols, ketones, aldehydes, anhydrosugars, alcohols, furans, and esters (Lyu, Wu, & Zhang, 2015). The syn-gas consists mostly of CO and H_2 with a small amount of CH_4, but depending on the raw material used as feedstock and the form in which the oxygen is delivered, it can also incorporate variable quantities of CO_2 and N_2 (Uddin et al., 2018). The biochar (or charcoal, 90% carbon content) is a powerful soil conditioner. Its high porosity increases soil water and nutrient retention, rendering them available for plant nutrition for a long time. It also improves the soil structure and mechanical properties (Valentim, Guedes, Rodrigues, & Flores, 2011). Many studies have already demonstrated the positive impact of biochar application to soil on agricultural yields, due to the reduction of water and nutrient input required for production (Chan, Van Zwieten, Meszaros, Downie, & Joseph, 2007; Genesio, Miglietta, Baronti, & Vaccari, 2015; Major, Rondon, Molina, Riha, & Lehmann, 2010; Rondon, Lehmann, Ramírez, & Hurtado, 2007; Vaccari et al., 2011). The compact structure of the biochar renders it unbiodegradable by soil microorganisms, therefore biochar is considered a carbon storage system. Carbon stored in soil as biochar cannot be returned to the atmosphere in the form of

CO_2 (Majumder, Neogi, Dutta, Powel, & Banik, 2019) as in the case of hemp waste (hurds, sheaves, or straws) pellets or briquettes combustion (Kraszkiewicz et al., 2019; Pedrazzi, Morselli, Puglia, & Tartarini, 2020). Carbon sequestration through biochar by pyrolysis or gasification in low-oxygen conditions of plant waste is considered a carbon negative process (Marris, 2006). The use of biochar in agricultural practices has been demonstrated to also reduce soil N_2O emissions (Cayuela et al., 2013), a greenhouse gas with a Global Warming Potential 296 times greater than CO_2 (Myhre et al., 2013).

In slow pyrolysis, the biomass is pyrolyzed with low heating rates to maximize the coal production, yielding minimal liquid and gaseous products (Goyal, Seal, & Saxena, 2008). Slow pyrolysis of wood used to be a common technology in industries until the early 1900s, where coal, acetic acid, methanol, and ethanol were obtained from wood (Huber, Iborra, & Corma, 2006). Slow pyrolysis can be used to convert hemp hurds into biochar, liquids (distillates), and gases (Salami et al., 2020). Liquid distillates are considered a by-product and are often burned or dumped. However, these liquid distillates contain bioactive compounds and could be collected to generate new valuable products. Raw distillates collected at each stage of slow pyrolysis of hemp hurds (drying, torrefaction, and pyrolysis) carried out in a pilot plant in a temperature range from room temperature to 350°C led to the production of some potentially valuable molecules (Salami et al., 2020). Acetic acid was the main component of all distillates, while other interesting compounds included guaiacol and syringol derivatives such as 2,6-dimethoxyphenol, vanillin, and eugenol. Such compounds could be separated and purified for nutritional, pharmaceutical, and agricultural purposes. By varying the processing conditions—temperature, heating rates, and residence times—the whole process can be optimized to increase the production of the most valuable products. It has been estimated that slow pyrolysis of a ton of hemp hurds can produce cca. 300 kg of biochar and cca. 40 kg of acetic acid (Salami et al., 2020).

12.5.1.2 Thermomechanical valorization of hemp hurds

Extrusion is a widely used process in the industry to produce foods with specific technological properties (color, texture, aroma, etc.) (Vandenbossche, Candy, Evon, Rouilly, & Pontalier, 2019). An extrusion process that combines mechanical, thermal, and chemical actions in a single step and a continuous mode can be used for the selective extraction of hemicellulose and lignins from lignocellulosic residues (Candy et al., 2017). Hemp hurds represent lignocellulosic residues containing lignin-carbohydrates complex (LCC). Moreover, phenolic compounds found in hemp hurds are mainly composed of hydroxycinnamic acids (HCA) (i.e., coumaric, ferulic, sinapic, caffeic, chlorogenic, and rosmarinic acid), which are mainly ester-bound to hemicellulose and ester- and ether-bound to lignin, while a small part of HCA can be found in free form in the LCC structure (Acosta-Estrada, Gutiérrez-Uribe, & Serna-Saldívar, 2014). Due to their antioxidative, antimicrobical, and photoprotective properties, they are currently used as food preservatives (Abramovič, 2015) or in cosmetics (Taofiq, González-Paramás, Barreiro, & Ferreira, 2017). Furthermore, their beneficial effects in preventing various human diseases has been demonstrated (Stalmach, 2014). Ferulic acid obtained from agricultural by-products is known as a potential precursor for the production of vanillin or vanillic acid (Mathew & Abraham, 2004). Continuous twin-screw extrusion of hemp hurds and dusts, carried out in a pilot plant with mild temperature (50°C) and elution conditions (alkaline aqueous or hydroalcoholic solvent) and at low liquid/solid ratio, allows the recovery of 50% and 33% of free and bound para coumaric acid and ferulic acid, respectively (Candy et al., 2017). A combination of 100% para-coumaric acid and 60% ferulic acid content of hemp dusts can be recovered with the same pilot plant. In this pilot plant, the HCA extraction yield can be increased by subjecting the dry residue to a second extraction with a polar solvent, confirming twin-screw extrusion process to be a valuable tool for hemp lignocellulosic waste valorization, as it guarantees the extraction of considerable amounts of valuable products with a lower consumption of chemicals than conventional ones (Candy et al., 2017).

12.5.1.3 Conversion of hurds to biofuels and other valuable biomolecules

Because of their high carbohydrate content, hurds have been exploited for fermentable sugars production and further conversion to biofuels (Barta et al., 2010) or valuable molecules such as poly-3-hydroxybutirate (P-3-HB), which is considered a green alternative to petroleum plastics and exhibits physical and functional properties similar to polypropylene (Ji et al., 2021). The techniques useful for hurds valorization were presented in Section 12.4.1.

12.6 Valorization of seed processing by-products and wastes

12.6.1 Valorization of threshing residues and plant residues

Threshing residues, produced during hemp seed panicles threshing and seed cleaning, are composed of leaves, stalk residues, inflorescence leaflets, and female flower bracts wrapping hemp seed at maturity. These tissues are elective for the extraction of cannabinoids or other products of hemp metabolism, as they are particularly rich in glandular trichomes with the ability to synthesize and secrete cannabinoids and terpenoids in different amounts and chemical varieties (Romero, Peris, Vergara, & Matus, 2020). The main cannabinoids produced by glandular trichomes are the acid forms of tetrahydrocannabinol (THCA) and CBD, and in less quantity, of cannabinol (CBN), cannabigerol (GBG), cannabichromene (GBC), and tetrahidrocannabivarin (THCV), all with a broad therapeutic potential (Jin et al., 2020). Hemp terpenoids are responsible of the characteristic hemp fragrance (Sommano, Chittasupho, Ruksiriwanich, & Jantrawut, 2020). They may contribute to the entourage effects of cannabis-based medicinal extracts (Russo, 2011). Flavonoids, produced by almost all green tissues of *Cannabis sativa* (Bautista, Yu, & Tian, 2021), share a wide range of biological effects with cannabinoids and terpenoids (Russo, 2011). It has been estimated that a double purpose (seed and fiber) hemp crop could potentially generate threshing residues up to 2 tons/ha (Andre, Larondelle, & Evers, 2010; Ascrizzi et al., 2019), but the recovery of bioactive compounds can vary depending on the extraction technique, hemp strain, cultivation techniques, environmental conditions, harvesting time, and moisture content (Andre et al., 2010; Ascrizzi et al., 2019; Bautista et al., 2021; Calzolari et al., 2017).

Crop residues consisting of stem and leaves that remain from the "threshing" of the CBD/medical hemp crop are also considered "threshing residues" (Serna-Loaiza et al., 2020). This residual fraction could be another source of cannabinoids due to the extensive distribution of glandular trichomes on the green tissues of hemp plant, contributing to the overall cannabinoids yield, increasing crop profitability and reducing crop wastes, which have a great impact on crop environmental sustainability. The various products of hemp secondary metabolism require different physicochemical techniques to be extracted from plant tissues, according to their different chemical nature.

12.6.1.1 Valorization of threshing residues by cannabinoids extraction

Apart from traditional techniques for the extraction of cannabinoids from hemp plant tissues and products, mainly based on the use of a large amount of organic solvents like hexane or methanol, more environmentally friendly and sustainable extraction techniques are emerging ensuring safe and high quality products, based on the use of alternative solvents (water and green solvents) and the reduction of energy consumption (Chemat, Vian, & Cravotto, 2012). Among them, pressurized liquid extraction (PLE), supercritical fluid extraction (SFE), and ultrasound-assisted extraction (UAE) have been applied for the extraction of cannabinoids and other valuable biomolecules for threshing residue valorization (Calzolari et al., 2017; Nuapia et al., 2020; Vági et al., 2019). These techniques allow higher extraction efficiency and selectivity with a reduction in organic solvents consumption than traditional ones. Furthermore, they reduce molecule degradation and are more easily automatable (Chemat et al., 2012).

Pressurized liquid extraction

PLE is also known as enhanced solvent extraction (ESE), accelerated fluid extraction (ASE), and high-pressure solvent extraction (HSPE) (Rehman, Khan, & Niaz, 2020). In PLE, high pressure (up to 300 psi) is applied to keep a solvent in liquid form beyond its boiling point, which in turn facilitates the extraction process. PLE leads to decreased solvents requirement and reduced extraction time due to high temperature and pressure. In fact, the higher temperature increases the analyte solubility and decreases solvent viscosity and surface tension, thus increasing the extraction rate (Rehman et al., 2020). Because the insoluble matrix remains inside the extraction cell, PLE does not involve a filtration step; hence, it can be applied for the automatic processes as online coupled extraction/separation followed by analysis (Fathordoobady, Singh, Kitts, & Pratap Singh, 2019). By varying the temperature (50–100°C), pressure (50–150 bar), extraction time, and the number of cycles ($n=2$), PLE has been tested on a laboratory scale for cannabinoids extraction from hemp threshing residues using ethanol as a solvent at three different extraction setups containing different solvent volume: flasks (10 mL), speed extractor (22 mL), and autoclave reactor (500 mL) (Serna-Loaiza et al., 2020), whereby the optimal extraction conditions were identified providing valuable inputs for hemp-based biorefineries.

Supercritical fluid extraction

SFE is a separation technology that uses supercritical fluid solvents for extraction. A supercritical fluid (SCF) is a substance that can be either liquid or gas used in a state above the critical point, where gases and liquids can coexist (Fortunati, Luzi, Puglia, & Torre, 2016). A supercritical fluid is generated when a substance is heated at constant pressure; at a critical point, the two phases (liquid and gas) coalesce into a fluid and acquire the same properties. When temperature exceeds the critical temperature, a supercritical phase is obtained, and the substance is then called a supercritical fluid (Ikan & Crammer, 2003). Supercritical fluids can show the properties of both liquids (dissolution) and gases (penetration) simultaneously, making them a very good solvent. Carbon dioxide (CO_2) is the most commonly used supercritical fluid, with other choices including ethanol. Compared with traditional Soxhlet extraction, SFE uses supercritical fluid to provide a broad range of useful properties. It eliminates the use of organic solvents, which reduces the problems of their storage, disposal, and environmental concerns (Yang & Hu, 2014). By using low temperatures, SFE has benefits over conventional methods, especially regarding isolation of heat-sensitive and easily oxidized compounds. Carbon dioxide is a safe, non-toxic solvent with no hazardous residues, which makes it acceptable as an effective solvent for food and pharmaceutical industries (Fathordoobady et al., 2019).

A multistep extraction process has been developed combining SFE-CO_2, PLE, and EAE for hemp threshing residue valorization in a pilot scale plant (Kitrytė et al., 2018). Grinded and sieved threshing residues (0.2 mm) were subjected to consecutive extraction steps, generating a liquid extract and a semisolid residue that was subjected to subsequent extraction and/ or enzymatic treatment, globally leading to the recovery of 51.7 g of extractable substance from 100 g of hemp threshing residues (Kitrytė et al., 2018). Optimized SFE-CO_2 (46.5 MPa, 70°C, 120 min) produced 8.3 g/100 g DW of lipophilic fraction containing > 93% of initial CBD and CBDA amount from plant material. PLE-Ac (100°C, 45 min) and PLE-EtOH/H_2O (100°C, 45 min, EtOH/H_2O 4:1 v/v) produced 4.3 and 18.9 g/100 g DW flavonoid fractions, respectively. PLE residue treatment with cellulolytic enzyme mixture (Viscozyme L) additionally increased the release of mono- and disaccharides up to 94% (Kitrytė et al., 2018).

Subcritical fluid extraction

Subcritical extraction, carried out at lower pressure and temperature conditions than supercritical extraction, requires longer times but is able to preserve some types of more easily degradable biomolecules like terpenes, essential oils, and other compounds. Subcritical CO_2 extraction allows the extraction of the widest range of different cannabinoids; cannabinoid acidic forms such as CBDA, CBGA, and CBNA; terpenes; vitamins; and fatty acids from hemp tissues (Fathordoobady et al., 2019).

Supercritical and subcritical CO_2 extraction allowed the production of CBD and CBDA-rich extracts from hemp threshing residues in a pilot plant equipped with two combined separators (Vági et al., 2020). Threshing residues belonging to two fiber hemp cultivars (THC < 0.2%) grinded at particle size ranging from 8 mm (in the pellet form) to 0.33 mm were subjected to subcritical CO_2 extraction (8 MPa, 27°C) with ethanol as cosolvent (10%) and supercritical CO_2 extraction (8–45 MPa, 45°C) in step-wise extraction mode or with extract fractionation. Under these conditions, supercritical CO_2 extraction yielded 2.5–9.2 g/100 g of lipophilic raw extracts containing 0.2–1.8 g of cannabinoids per 100 g of threshing residues. Extract from the first separator produced by supercritical CO_2 extraction with fractionation contained 30 times more CBD than the extract from the second separator. By applying ethanol as cosolvent, the total extraction yield increased by 30% of the main cannabinoids yield. Subcritical CO_2 extraction resulted in a lower yield of lipophilic raw extract (2.5 g/100 g dry threshing residue) with a high CBD content (0.9 g/100 g dry threshing residue) with 18% less solvent consumption and at mild temperature and pressure conditions, thus resulting in an overall less expansive cannabinoids-rich extract production (Vági et al., 2020).

Ultrasound-assisted extraction

Ultrasound-assisted extraction is a technology that uses ultrasound waves to accelerate the extraction process, increase the extraction yield, improve extraction processes without using toxic and environmentally dangerous solvents, enhance the extraction performance of alternative "green" solvents, and raise the extraction of heat-sensitive compounds (Tiwari, 2015). Therefore, it is a key technology in achieving the objective of sustainable "green" chemistry associated with a high reproducibility and higher purity of the final product in comparison to conventional extraction methods (Chemat et al., 2017). Ultrasound acts through different independent or combined physical and chemical mechanisms such as cavitation (the most dominant), agitation, vibration, pressure, shock waves, shear forces, microjets, compression and rarefaction, acoustic streaming, and radical formation, which causes fragmentation, erosion, capillarity, detexturation, and sonoporation of treated material (Agarwal, Máthé, Hofmann, & Csóka, 2018; Tiwari, 2015). Effects of applied ultrasound are associated with frequency. Low frequencies (20–100 kHz) cause a dominant change in the treated material by physical effects, while higher frequencies (200–500 kHz) are dominantly associated by chemical effects (Tiwari, 2015). The extraction effect of ultrasounds depend on their intensity, the sonication time, the extraction medium properties (e.g., type, viscosity, and surface tension), the properties of the treated material (e.g., particle size), and the extraction conditions (e.g., temperature and pressure) (Fathordoobady et al., 2019; Tiwari, 2015). UAE has been extensively used for cannabinoid extraction from hemp matrices, and it is also used within the European Union method for the quantitative determination of Δ^9-tetrahydrocannabinol in hemp varieties (Reg. UE N. 73/2009, Annex I). Bioactive compounds (CBD, CBG, cannflavin A, Δ^9-THC) have been extracted from hemp threshing residues at different harvesting times by ultrasound-assisted methanol extraction (Calzolari et al., 2017). The highest yield of cannabinoids was associated with the late sampling time (2.5% and 2.9%), whereas the THC level decreased from the first to the last sampling time, as espected. Cannflavin A was present in all evaluated extracts at a variable concentration mainly depending on the cultivar, but its level showed to be affected by high air temperatures at harvesting time. Therefore, harvesting time postponement has been proposed for a better exploitation of threshing residue by CBD and cannflavin A extraction (Calzolari et al., 2017).

12.6.1.2 Valorization of threshing residue by essential oil extraction

Hemp essential oil is mainly composed of terpenoids and flavonoids, although their relative concentration and overall chemical composition can vary depending on hemp variety, growing conditions, harvesting time, tissue processing, and extraction method. Beyond the conventional hydrodistillation and steam distillation, there are other environmentally friendly techniques to assist the extraction of hemp essential oil, such a microwave-assisted extraction (MAE) and its variants (Aramrueang, Asavasanti, & Khanunthong, 2019). Zheljazkov et al. (2020) demonstrated how the utilization of simple processing steps such as grinding and fractionation during distillation can modify hemp essential oil profile and its antimicrobial activity. Generally, it was found that grinding material prior to distillation increased the fraction of monoterpenes in hemp essential oil, twofold to 85% during the first 10 minutes of distillation. Grinding before extraction caused also a slight but significant decrease in essential oil CBD content, but raised the β-pinene and myrcene concentrations. On the other hands, high concentrations of δ-3-carene, limonene, β-caryophyllene, α-(E)-bergamotene, (Z)-β-farnesene, α-humulene, caryophyllenyl alcohol, germacrene D-4-ol, spathulenol, caryophyllene oxide, humulene epoxide 2, β-bisabolol, α-bisabolol, sesquiterpenes, and CBD were obtained from non-grinded material.

Hydrodistillation

Hydrodistillation has long been used for essential oils and bioactive compounds extraction from plant materials without using organic solvents. It can be performed before dehydration of plant materials in three different ways: by water distillation, by water and steam distillation, and by direct steam distillation (Aramrueang et al., 2019). Hydrodistillation involves different physicochemical processes such as hydrodiffusion, hydrolysis, and decomposition by heat. The plant material is packed into a boiler and filled with a sufficient amount of water and subjected to boiling. Alternatively, direct steam is injected into the plant material. Hot water and steam act as extraction media for bioactive compounds from plant tissues. Indirect cooling by water condenses the vapor mixture of water and oil. Condensed mixture flows from the condenser to a separator, where oil and bioactive compounds are automatically separated from the water (Oreopoulou, Tsimogiannis, & Oreopoulou, 2019).

Hydrodistillation has been used for essential oil extraction from hemp threshing residues (Ascrizzi et al., 2019). Threshing resides were dried in a ventilated oven at 35–40°C; hydrodistillation was carried out in standard apparatus for 2 h and GC-MS analysis was used for essential oil composition characterization. Up to 100 different compounds mainly belonging to the sesquiterpene hydrocarbons ($40.1 \pm 6.4\%$) monoterpene hydrocarbons ($28.2 \pm 12.4\%$), oxygenated sesquiterpenes ($20.8 \pm 7.9\%$), cannabinoids ($3.1 \pm 2.0\%$), apocarotenoids ($1.0 \pm 0.3\%$), oxygenated monoterpenes, and non-terpene aldehydes (both below 1%) were identified. Among sesquiterpenes, the most abundant were β-caryophyllene and its oxidized derivatives, and α-humulene. The most abundant monoterpene hydrocarbons were α- and β-pinene, and myrcene. Extraction yield at the early harvest time was revealed to be slightly higher than that of the late harvest time (Ascrizzi et al., 2019).

Steam distillation

Steam distillation is utilized for essential oils extraction at temperatures near 100°C, followed by subsequent condensation to form an immiscible liquid from which the essential oil can be separated in a clarifier. Steam distillation is carried out by passing dry steam through the plant material, whereby the steam volatile compounds are volatilized, condensed, and collected in receivers. Steam distillation is considered a traditional technology for essential oil extraction (Pushpangadan & George, 2012). Essential oil produced by steam distillation of fresh hemp inflorescences has been evaluated for its efficacy against crop pests as filariaris vector *Culex quinquefasciatus*, the peach-potato aphid *Myzus persicae*, the housefly *Musca domestica*, and the tobacco catworm *Spodoptera littoralis* (Benelli et al., 2018). The essential oil was mainly composed by monoterpene and sesquiterpene hydrocarbons, with (E)-caryophyllene (45.4%), myrcene (25.0%), and α-pinene (17.9%) as the most abundant compounds. Toxicity test against the previously mentioned crop pests revealed its high toxicity for *Myzus persicae* aphids (LC_{50} of 3.5 ml/L) and *Musca domestica* flies (43.3 µg/adult insect), moderate toxicity toward *Spodoptera littoralis* larvae (152.3 µg/larva), and scarce toxicity against *Culex quinquefasciatus* larvae (LC_{50} of 252.5 ml/L) and adult insects ($LC_{50} > 500$ µg/cm^2). Furthermore, toxicity tests demonstrated its harmlessness against non-target invertebrate species (Benelli et al., 2018). These results demonstrated another possible way for hemp crop residue valorization as a source of environmentally friendly insecticides against aphids and houseflies, useful for an Integrated Pest Management system or organic farming.

Microwave-Assisted Extraction

MAE is a green extraction technique that offers many advantages over traditional ones: reduction of extraction time, reduction of solvent and energy consumption, possibility of simultaneous extraction of multiple samples, and improvement of sample throughput (Llompart, Garcia-Jares, Celeiro, & Dagnac, 2019). It is based on the breaking of the cell matrix structure induced by nonionizing electromagnetic waves with frequencies in the range from 300 MHz to 300 GHz. During microwave heating, water from the treated sample matrix evaporates, causing an increase in the intracellular pressure, which results in cell wall rupture facilitating the leaching of high-value compounds (Gomez, Tiwari, & Garcia-Vaquero, 2020). The application of Microwave-Assisted Distillation (MAHD) to hemp volatile fraction extraction could overcome the numerous drawbacks of steam and hydrodistillation methods due to their harsh conditions that causes thermal decomposition of volatile compounds, affecting essential oil quality (Gunjević et al., 2021). A fast and cost-effective microwave-assisted cascade protocol has been set up in a pilot scale reactor for the recovery of phytocannabinoids and terpenes from fresh HBRs (inflorescences, leaves, and stalks). MAHD for the recovery of the terpene fraction was carried out in a multimode commercial MW reactor at a maximum delivered power of 1800 W, with matrix-to-liquid ratio ranging from 1/0.5 to 1/1.5 kg/L. Cannabinoids extraction was carried out by microwave-assisted hydrodiffusion and gravity in the same commercial MW reactor, with some process modifications, to evaluate the possibility of overcoming the disadvantages of the classical extraction approach (Gunjević et al., 2021). More than 2.5 kg per cycle of *Cannabis* plant material was efficiently processed by MAHD in a 12 L vessel. The optimized MAHD procedure yielded $0.35 \pm 0.02\%$ w/w of hydrodistillate. The hydrodistilled oil was extremely rich in the characteristic *Cannabis* terpenes (α-pinene, β-myrcene, β-ocimene, *E*-caryophyllene, α-humulene, caryophyllene oxide, and β-selinene). Residual plant material, unaltered from MAHD protocol in terms of cannabinoid content, revealed to be suitable for subsequent cannabinoid recovery. Furthermore, the heating water surrounding the plant material in the vessel resulted rich in polyphenols. The pilot scale MW reactor provided a terpenes rich hydrodistillate, an enriched polyphenols fraction from the undistilled water, and phytocannabinoids with a high level of decarboxylation degree in the MADH solid residue. Furthermore, the absence of solvents strengthened the sustainability of the whole process (Gunjević et al., 2021).

12.6.2 Valorization of seed hulls

Hull contains different complex polysaccharides in varying proportion such as cellulose (22.0%–36.7%) and lignin (16.0%–19.5%), with xylose (5.7%–17.1%) as the most dominant monosaccharide within the non-cellulosic polysaccharides, implying that xylan is the next most abundant polysaccharide after cellulose. The xylan content of hull is of interest for its industrial valorization, because it represents a source for the production of xylan oligosaccharides for prebiotics, which are commonly produced from peanut shells. On the basis of dietary fiber content, the hemp hull should not be ignored but processed and used as a renewable resource for different bio-based solutions (Schultz et al., 2020). Different research demonstrated the unique polyphenol profile of hemp seed, characterized by the abundance of HCA. A total of 26 phenylpropionamides, including HCA amides and lignanamides (Leonard, Zhang, Ying, & Fang, 2021), mainly located in hull fraction, makes it a suitable feedstock for obtaining value-added extracts and/or fortification of food products. It has been shown that cannabisin B and *N-trans*-caffeoyltyramine are the major phenolic compounds in hemp seed hull with proven bioactivity, significantly higher than that of secoisolariciresinol from flaxseed and isoflavones from soybean (Chen et al., 2012). On the basis of outcomes from the enzymatic hydrolysis of hemp bran, Setti et al. (2020) evidenced that hemp bran is highly lignified, made of pectin that surrounds the hemicellulose-cellulose network. Moreover, the presence of starch and arabinoxylanesas as well as the presence of protein moieties linked to cellulose fibers have also been confirmed (Setti et al., 2020).

12.6.3 Valorization of hemp seed cake

Hemp seed cake can be utilized directly in the production of value-added products or as a source of bioactive compounds, mainly proteins for their isolation and production. However, this chapter will not cover the valorization of hemp seed cake as a protein source as it is the scope of another chapter.

Oilseed cake contains soluble conjugated or esterified polyphenols and insoluble polyphenols bounded to structural carbohydrates, lignin, and proteins through ester or glycosidic bonds, so that their extraction from oilseed cakes by conventional extraction methods is hindered (Corrales, Toepfl, Butz, Knorr, & Tauscher, 2008; Teh & Birch, 2014). Apart from micro- and macronutrients and the previously mentioned compounds, hemp seed cake contains residual oil (Pap et al., 2020). The content of the residual oil can be used for the assessment of the de-oiling process efficiency by comparing the oil content of the starting raw material with the residual oil content in the extraction meals, pellets, and/or expeller cakes (International Standards Organization, 2015). The residual content of hemp oil in hemp seed cake can be a beneficial from the nutritive point of view due the abundance of polyunsaturated fatty acids (PUFAs). However, due to their susceptibility to oxidation and formation of oxidative degradation products, deterioration of the chemical, sensory, and nutritional properties of the meals occurs rapidly after oil pressing, which limits their direct utilization (Pap et al., 2020; Yun & Surh, 2012).

To assist hemp seed cake valorization, comprising mainly of the extraction of target compounds and the improvement of their bioavailability, the disruption of plant matrix is required as a pretreatment. It is based on different physical phenomena that facilitate the mass transfer and release of the intracellular bioactive compounds from hemp seed by-products. In that sense, different conventional and innovative technologies are available (Náthia-Neves & Alonso, 2021; Teh et al., 2014).

12.6.3.1 Pre-extraction techniques

The efficient and safe extraction of bioactive compounds from the matrix of interest is an important issue that has captured the attention of food and pharmaceutical scientists and industry experts. Therefore, a variety of techniques that use different physical and physicochemical principles to alter the sample matrix and facilitate the release of target compounds have been emerging that can be used either as standalone techniques (e.g., milling, drying, puffing/extruding, and mechanical pressing) or coupled with extraction processes (e.g., heating, ultrasonication, MAE, sub-critical fluid extraction, SFE, and PLE) (Ummat, Sivagnanam, Rajauria, O'Donnell, & Tiwari, 2021; Wu, Ju, Deng, & Xi, 2017).

Milling/grinding

The simplest example of hemp hull and/or press cake processing is by milling/grinding, resulting in a powdered ingredient that can be utilized either for food fortification or as a starting feedstock for a variety of processes (Pojić, 2021; Schultz et al., 2020). Hemp oilseed cake most commonly comprise the mixture of cotyledon and hull particles, which requires further separation by milling and suitable fractionation technique (Pojić et al., 2014). Milling disrupts the embryonic cell walls of the oilseed cake releasing the protein and other bioactive compounds from plant tissue (Rommi et al., 2015). Traditional milling procedures implies the utilization of roller or abrasive milling procedures, while certain novel processing milling technologies have emerged to improve the quality of the milling products compliant with the requirements of the desired end purpose (Tiwari & Pojić, 2020). To obtain coarse material, utilization of burr, blade, and ball mills are recommended. Hammer, pin, and turbo mills are suitable for obtaining fine milling materials, while obtaining fine and ultrafine materials with particle sizes of $d_{50} < 21\,\mu m$ and $d_{50} < 12\,\mu m$, respectively, require the utilization of jet- (Lazaridou, Vouris, Zoumpoulakis, & Biliaderis, 2018) and/or air-flow mills ($d_{50} \sim 7\,\mu m$) (Rommi et al., 2015). To achieve the targeted milling results, a combination of listed milling technologies can be applied (Delrue & Van De Watering, 2008). Generally, the adjustment of median particle size of the ground material is achieved not only by the selection of the milling equipment but also by the number of the passings of press cake through the mill and the selection of the number of revolutions (Rajabzadeh, Jafari, & Legge, 2015; Rommi et al., 2015). However, Rommi et al. (2015) observed the negative effects of ultrafine milling by pin disc mill on protein extraction, indicating low solubility and low recovery yields of proteins ascribed to protein denaturation due to heat and/or shear forces developed during milling. On the other hand, different milling treatments (ball-milling, jet milling, high pressure micronization) can induce the redistribution of insoluble fibers to soluble as demonstrated by Chau, Wang, and Wen (2007) and Wu, Chien, Lee, and Chau (2007). Along with proteins, milled oilseed cakes contain lipid, ash, fiber, and a variety of bioactives as well as anti-nutritional factors (Laudadio, Bastoni, Introna, & Tufarelli, 2013; Pojić et al., 2014).

Fractionation

At the end of milling, protein bodies can still remain attached to other macronutrients (e.g., starch, fiber), so further processing by suitable fractionation technique is required (Laudadio et al., 2013). In additional to traditional fractionation by sieving and air classification, electrostatic separation is emerging (Tiwari & Pojić, 2020). Air classification is based on the difference in size and density of fractions to be separated, while electrostatic separation is based on charging the material to be separated by tribo-electrification, its introduction into an electric field, and separation of particles according to the acquired charge (Tiwari & Pojić, 2020).

Extrusion technology

Leonard, Zhang, Ying, Xiong, and Fang (2021) demonstrated the potential of extrusion technology for the improvement of total phenolic content, DPPH/ABTS radical scavenging activity, in vitro enzyme (α-glucosidase and AChE) inhibition, and phenylpropionamide content of hemp seed hull by lowering the screw speed from 300 to 150 rpm and controlling the temperature in the range of 80°C to 100°C. However, the demonstrated extrusion conditions appeared to be useful for extrusion of hemp husk and couldn't be unambiguously applied to other fiber-rich materials. The control of the extrusion conditions should be such as to provide the balance between the formation of Maillard reaction products and degradation of phenolic compounds. Simultaneously with retaining the high level of polyphenolics and preservation of their in vitro bioavailability and bioaccessibility, extrusion processing causes the redistribution of insoluble to soluble fibers generally being of more desirable health benefits when compared to insoluble ones, provided that the conditions of extrusion are appropriate (Zhong, Fang, Wahlqvist, Hodgson, & Johnson, 2021). Zhong et al. (2021) demonstrated that the most significant effects on redistribution of insoluble to soluble fibers had the total moisture in the barrel (with negative influence), the barrel temperature (with positive influence), and the interactive effects of the screw speed and total moisture in the barrel. This phenomenon has been demonstrated by the examples of extruded cereal brans, pea, soybean, and lupin seed coats (Zhong et al., 2018, 2021). Generally the composition of extruded products and physicochemical properties of extrudates depend on the characteristics of raw material, mixing and conditioning, barrel temperature, processing pressure, screw speed, moisture content, flow rate, energy input, residence type, and screw configuration, whereby the most critical extrusion parameters are applied temperature, screw speed, and moisture content, having the ability to positively, but also negatively, influence the content of bioactive compounds in the extrudates. Although extrusion as applied by Leonard, Zhang, Ying, Xiong, and Fang (2021) increased the content of total phenolics, certain extrusion studies resulted in reduced content of measurable bioactive compounds in different types of food products. High barrel temperature and high moisture content of material may cause the decarboxylation of phenolic compounds during extrusion and promote polymerization of phenols and tannins, resulting in reduced extractability and antioxidant activity (Brennan, Brennan, Derbyshire, & Tiwari, 2011).

12.6.3.2 Extraction technologies

Ultrasound technology

Teh and Birch (2014) demonstrated the utilization of ultrasonic treatment to assist the extraction of phenolics and flavonoids from defatted hemp seed cake. Sonication of defatted hemp seed cake (at 200 W) not only contributed to the increased polyphenol extraction yields but also enhanced antioxidant capacity of the hemp seed cake extracts during 20 min of treatment. It was shown that the efficiency of the sonication treatment to assist the extraction of the polyphenols can be further increased by simultaneous heating up to 70°C. However, the elevated temperatures up to 70°C have proven to be inadequate for the extraction of flavonoids, which appeared to be degraded in the extraction medium due to elevated temperature and ultrasonication, so that the temperatures between 40°C and 50°C are recommended. The potential of ultrasound to assist the extraction of phenolic compounds from other types of oilseed cakes have been demonstrated (Zardo, de Espíndola Sobczyk, Marczak, & Sarkis, 2019).

Microwave technology

Teh et al. (2014) for the first time demonstrated the potential of microwave technology to assist in the extraction of the total phenolics and flavonoids from hemp seed cake and revealed the optimal combination of conditions to maximize their yields and antioxidant capacity. Microwave treatment, if performed for 5 min with liquid-to-solid ratio of 6 mL/g at electromagnetic power of 700 W, appeared to yield optimal results. It was demonstrated that the higher microwave power was not favorable because it may cause phenolic and flavonoid degradation. The potential of microwaves to assist the extraction of phenolic compounds from other types of oilseed cakes have been demonstrated by Náthia-Neves and Alonso (2021) who reported that microwave-induced microstructural changes of oilseed by-products were affected by the applied irradiation power. An increase in microwave power generally exhibits a positive effect on the extraction yield up to a point to which the temperature increase does not cause a thermal degradation of the recovered compound of interest.

Pulsed electric field technology

Pulsed electric field (PEF)-assisted extraction is based on the electroporation induced by subjecting the sample to electric pulses of short duration (μs to ms) of moderate electric voltage (typically 0.5–20 kV/cm) between two electrodes. High electric voltage (5–50 kV/cm) treatment is applied for preservation and enzyme and microbial inactivation, whilst low to mild electric voltage is recommended for the enhancement of secondary metabolite extraction from plant-based matrixes. By dielectric disruption of cell membrane, its permeabilization increases, which consequently increases the diffusivity of the intracellular substances and mass transfer rate, leading to improved extraction efficiency (Kumari, Tiwari, Hossain, Brunton, & Rai, 2018).

Along with the potential of microwave technology to assist the extraction of the total phenolics and flavonoids from hemp seed cake, Teh et al. (2014) demonstrated the potential of PEF technology for the same purpose. PEF treatment of defatted hemp seed cake, if performed at voltage of 30 V and frequency of 30 Hz, the ethanol concentration of 10% during 10 s yielded maximal total phenolic content and antioxidant capacity. Similar to the microwave power, higher electroporation voltage and frequency appeared to not be favorable because it may cause phenolic and flavonoid degradation.

12.7 Valorization of inflorescence processing waste

Inflorescence processing wastes belonging to cannabinoids extraction or essential oil extraction are composed by plant tissues in a semisolid state that could be subjected to further valorization with a bio-refinery approach. As demonstrated by Kitrytė et al. (2018), hemp extraction residues belonging to a multistep extraction by SFE-CO_2 and PLE can be further processed by enzymatic-assisted extraction for the recovery of monosaccharides and other water-soluble valuable molecules. Monosaccharides could be converted to biofuels by fermentation, thus implementing extraction yield and process sustainability.

12.8 Challenges and opportunities

Although hemp cultivation in the world has grown considerably in the last decade, there are still few concrete examples of complete and profitable supply chains based on industrial hemp cultivation. Much must still be done from a legislative, technological, and industrial point of view to allow the development and spreading of new models of agro-industrial chains capable of exploiting all the potential of this crop in a sustainable way. A modern agro-industrial supply chain must necessarily satisfy criteria of economic and environmental sustainability as well as circularity; to achieve these objectives, it is necessary to make available hemp varieties able of feeding as many markets as possible, minimizing the production of waste, or to imagine new strategies for waste valorization that contribute to the sustainability of the supply chain, increasing its profitability.

12.8.1 Hemp genotypes

When talking about industrial hemp, generally it refers to hemp genotypes characterized by THC content lower than 0.2%–0.3% (according to local legislation) and a good yield in stems (fiber) and/or seeds (dual-purpose hemp). Eighty-one varieties of hemp that meet these requirements are currently registered in the European variety register (EU-Plant variety database, 2021). These are mainly monoecious varieties with flowers of both sexes on the same individual, therefore producing seed and fiber from the same crop. In recent years, new hemp varieties specialized in the production of high levels of CBD have entered the market with the consequent birth of a new industrial chain for the production of CBD. These varieties are generally dioecious (with separate sexes), and the cultivation techniques used for the production of CBD renders the quantity and quality of the fiber produced by their stems very low. The quantity and type of waste produced by these two types of supply chains are decidedly different; the use of dual-purpose varieties has a decisive effect on reducing the production of agricultural and process waste. On the other hand, the currently available dual-purpose germplasm doesn't fulfill the requirements to sustain a modern agro-industrial chain. Some genetic traits, such as monoecy and fruit abscission, should be subject to more stringent genetic selection to ensure greater uniformity of the crop, greater crop yield, and a reduction in waste production. The changing global climate also makes it necessary to select varieties capable of maintaining high production and quality levels (low THC levels) even in conditions of high temperature and low water availability.

12.8.2 Technologies for waste valorization

As explained in this chapter, hemp seed and stems production and processing can generate different types of waste. According to the natural and chemical composition, wastes can be used for the extraction of bioactive molecules, energy and biofuel production, or conversion into molecules with high added value. Over the past 10 years, thanks to numerous public and private economic investments, a large amount of research has been carried out aimed at chemical characterization of hemp waste and the identification of green technologies for their valorization. However, much of these activities have been carried out on a laboratory scale and on waste types that do not always correspond to those that are actually produced by the agro-industrial chain. The identification of the technologies that best meet the needs of the supply chain and the scale-up of the technologies most suitable for the valorization of the supply chain wastes will certainly be an important challenge for the future. Likewise, it is still unknown if up-scaled processes would be sufficiently cost-effective, economically viable, and free of potential hazards related to the emergence of new contaminants for the environment but also the pathogens, antinutritional factors, and allergens in the cases where processed by-products are used as food raw materials. The supporting legislation is either insufficiently clear or still inexistent.

Not less important will be the assessment of the environmental impact of these technologies before their inclusion in the production chain. The quantification of the potential impacts on the environment and human health associated with the different products and wastes of the industrial hemp supply chain will constitute a guide for the concrete development of a sustainable and profitable agro-industrial hemp chain.

12.9 Conclusion

Valorization of different waste streams and by-products generated in agricultural hemp production and along different industrial hemp value chains (e.g., food, medicine, fiber) has been motivated by the global needs to increase the sustainability of agricultural and industrial production and environmental protection. Although by nature hemp is a multipurpose crop utilized within different value chains, valorization of its wastes and by-products additionally contributes to the creation of new opportunities for the development of knowledge-based bioeconomy and creation of new value chains. A variety of innovation and developments in science and technology has been proposed and activities performed to utilize hemp by-products, encompassing the selection of feedstocks, pretreatments, technological processes, refinement, purification, and conversion of by-product fractions efficiently and economically into high-value molecules, and/or biobased products. Such endeavors require an extensive multidisciplinary and cross-sectoral research to adjust and optimize the procedures and processes to obtain materials and products of desirable quality. Because the valorization of agri-food wastes and by-products in general and hemp-based wastes and by-products in particular represents an issue of ecological concern, the framework of the Quintuple Helix innovation model (Carayannis, Barth, & Campbell, 2012) can be utilized to efficiently address the listed challenges.

Acknowledgments

M. Pojić would like to acknowledge the financial support of the Ministry of Education, Science and Technological Development of the Republic of Serbia (Contract No. 451-03-68/2022-14/ 200222).

References

Abramovič, H. (2015). Antioxidant properties of hydroxycinnamic acid derivatives: A focus on biochemistry, physicochemical parameters, reactive species, and biomolecular interactions. In V. R. Preedy (Ed.), *Coffee in health and disease prevention* (pp. 843–852). Academic Press. https://doi.org/10.1016/B978-0-12-409517-5.00093-0.

Acosta-Estrada, B. A., Gutiérrez-Uribe, J. A., & Serna-Saldívar, S. O. (2014). Bound phenolics in foods, a review. *Food Chemistry*, 152, 46–55. https://doi.org/10.1016/j.foodchem.2013.11.093.

Adesina, I., Bhowmik, A., Sharma, H., & Shahbazi, A. (2020). A review on the current state of knowledge of growing conditions, agronomic soil health practices and utilities of hemp in the United States. *Agriculture*, 10(4), 129. https://doi.org/10.3390/agriculture10040129.

Agarwal, C., Máthé, K., Hofmann, T., & Csóka, L. (2018). Ultrasound-assisted extraction of cannabinoids from *Cannabis sativa* L. optimized by response surface methodology. *Journal of Food Science, 83*(3), 700–710.

Agbor, V., Zurzolo, F., Blunt, W., Dartiailh, C., Cicek, N., Sparling, R., et al. (2014). Single-step fermentation of agricultural hemp residues for hydrogen and ethanol production. *Biomass and Bioenergy, 64*, 62–69. https://doi.org/10.1016/j.biombioe.2014.03.027.

Akaraonye, E., Keshavarz, T., & Roy, I. (2010). Production of polyhydroxyalkanoates: The future green materials of choice. *Journal of Chemical Technology and Biotechnology, 85*(6), 732–743. https://doi.org/10.1002/jctb.2392.

Álvarez, C., Mullen, A. M., Pojić, M., Hadnađev, T. D., & Papageorgiou, M. (2021). Classification and target compounds. In C. M. Galanakis (Ed.), *Food waste recovery* (2nd ed., pp. 21–49). Academic Press. https://doi.org/10.1016/B978-0-12-820563-1.00024-X.

Amaducci, S., Zatta, A., Raffanini, M., & Venturi, G. (2008). Characterisation of hemp (*Cannabis sativa* L.) roots under different growing conditions. *Plant and Soil, 313*(1), 227–235. https://doi.org/10.1007/s11104-008-9695-0.

Andre, C. M., Larondelle, Y., & Evers, D. (2010). Dietary antioxidants and oxidative stress from a human and plant perspective: A review. *Current Nutrition and Food Science, 6*(1), 2–12.

Aramrueang, N., Asavasanti, S., & Khanunthong, A. (2019). Leafy vegetables. In Z. Pan, R. Zhang, & S. Zicari (Eds.), *Integrated processing technologies for food and agricultural by-products* (pp. 245–272). Academic Press. https://doi.org/10.1016/B978-0-12-814138-0.00010-1.

Ascrizzi, R., Ceccarini, L., Tavarini, S., Flamini, G., & Angelini, L. G. (2019). Valorisation of hemp inflorescence after seed harvest: Cultivation site and harvest time influence agronomic characteristics and essential oil yield and composition. *Industrial Crops and Products, 139*. https://doi.org/10.1016/j.indcrop.2019.111541, 111541.

Asquer, C., Melis, E., Scano, E. A., & Carboni, G. (2019). Opportunities for green energy through emerging crops: Biogas valorization of *Cannabis sativa* L. residues. *Climate, 7*, 142. https://doi.org/10.3390/cli7120142.

Bandyopadhyay-Ghosh, S., Ghosh, S. B., & Sain, M. (2015). The use of biobased nanofibres in composites. In O. Faruk, & M. Sain (Eds.), *Biofiber reinforcements in composite materials* (pp. 571–647). Woodhead Publishing. https://doi.org/10.1533/9781782421276.5.571.

Barta, Z., Oliva, J. M., Ballesteros, I., Dienes, D., Ballesteros, M., & Réczey, K. (2010). Refining hemp hurds into fermentable sugars or ethanol. *Chemical and Biochemical Engineering Quarterly, 24*(3), 331–339.

Bautista, J. L., Yu, S., & Tian, L. (2021). Flavonoids in *Cannabis sativa*: Biosynthesis, bioactivities, and biotechnology. *ACS Omega, 6*(8), 5119–5123. https://doi.org/10.1021/acsomega.1c00318.

Benelli, G., Pavela, R., Petrelli, R., Cappellacci, L., Santini, G., Fiorini, D., et al. (2018). The essential oil from industrial hemp (*Cannabis sativa* L.) by-products as an effective tool for insect pest management in organic crops. *Industrial Crops and Products, 122*, 308–315. https://doi.org/10.1016/j.indcrop.2018.05.032.

Brennan, C., Brennan, M., Derbyshire, E., & Tiwari, B. K. (2011). Effects of extrusion on the polyphenols, vitamins and antioxidant activity of foods. *Trends in Food Science and Technology, 22*(10), 570–575. https://doi.org/10.1016/j.tifs.2011.05.007.

Caldeira, C., Vlysidis, A., Fiore, G., De Laurentiis, V., Vignali, G., & Sala, S. (2020). Sustainability of food waste biorefinery: A review on valorisation pathways, techno-economic constraints, and environmental assessment. *Bioresource Technology, 312*. https://doi.org/10.1016/j.biortech.2020.123575, 123575.

Callaway, J. C., & Pate, D. W. (2009). Hempseed oil. In R. A. Moreau, & A. Kamal-Eldin (Eds.), *Gourmet and health-promoting specialty oils* (pp. 185–213). AOCS Press. https://doi.org/10.1016/B978-1-893997-97-4.50011-5.

Calzolari, D., Magagnini, G., Lucini, L., Grassi, G., Appendino, G. B., & Amaducci, S. (2017). High added-value compounds from *Cannabis* threshing residues. *Industrial Crops and Products, 108*, 558–563. https://doi.org/10.1016/j.indcrop.2017.06.063.

Candy, L., Bassil, S., Rigal, L., Simon, V., & Raynaud, C. (2017). Thermo-mechano-chemical extraction of hydroxycinnamic acids from industrial hemp by-products using a twin-screw extruder. *Industrial Crops and Products, 109*, 335–345. https://doi.org/10.1016/j.indcrop.2017.08.044.

Carayannis, E. G., Barth, T. D., & Campbell, D. F. (2012). The Quintuple Helix innovation model: Global warming as a challenge and driver for innovation. *Journal of Innovation and Entrepreneurship, 1*, 2. https://doi.org/10.1186/2192-5372-1-2.

Cayuela, M. L., Sánchez-Monedero, M. A., Roig, A., Hanley, K., Enders, A., & Lehmann, J. (2013). Biochar and denitrification in soils: When, how much and why does biochar reduce N_2O emissions? *Scientific Reports, 3*(1), 1732. https://doi.org/10.1038/srep01732.

Chan, K. Y., Van Zwieten, L., Meszaros, I., Downie, A., & Joseph, S. (2007). Agronomic values of greenwaste biochar as a soil amendment. *Soil Research, 45*(8), 629–634. https://doi.org/10.1071/SR07109.

Chau, C.-F., Wang, Y.-T., & Wen, Y.-L. (2007). Different micronization methods significantly improve the functionality of carrot insoluble fibre. *Food Chemistry, 100*(4), 1402–1408.

Chemat, F., Rombaut, N., Sicaire, A.-G., Meullemiestre, A., Fabiano-Tixier, A.-S., & Abert-Vian, M. (2017). Ultrasound assisted extraction of food and natural products. Mechanisms, techniques, combinations, protocols and applications. A review. *Ultrasonics Sonochemistry, 34*, 540–560. https://doi.org/10.1016/j.ultsonch.2016.06.035.

Chemat, F., Vian, M. A., & Cravotto, G. (2012). Green extraction of natural products: Concept and principles. *International Journal of Molecular Sciences, 13*, 8615–8627. https://doi.org/10.3390/ijms13078615.

Chen, T., He, J., Zhang, J., Li, X., Zhang, H., Hao, J., et al. (2012). The isolation and identification of two compounds with predominant radical scavenging activity in hempseed (seed of *Cannabis sativa* L.). *Food Chemistry, 134*(2), 1030–1037. https://doi.org/10.1016/j.foodchem.2012.03.009.

Corrales, M., Toepfl, S., Butz, P., Knorr, D., & Tauscher, B. (2008). Extraction of anthocyanins from grape by-products assisted by ultrasonics, high hydrostatic pressure or pulsed electric fields: A comparison. *Innovative Food Science and Emerging Technologies, 9*(1), 85–91. https://doi.org/10.1016/j.ifset.2007.06.002.

Crini, G., Lichtfouse, E., Chanet, G., & Morin-Crini, N. (2020). Applications of hemp in textiles, paper industry, insulation and building materials, horticulture, animal nutrition, food and beverages, nutraceuticals, cosmetics and hygiene, medicine, agrochemistry, energy production and environment: A review. *Environmental Chemistry Letters, 18*(5), 1451–1476. https://doi.org/10.1007/s10311-020-01029-2.

Das, L., Liu, E., Saeed, A., Williams, D. W., Hu, H., Li, C., et al. (2017). Industrial hemp as a potential bioenergy crop in comparison with kenaf, switchgrass and biomass sorghum. *Bioresource Technology, 244*, 641–649. https://doi.org/10.1016/j.biortech.2017.08.008.

Delrue, R., & Van De Watering, C. G. (2008). *Reduction of fibre content infibre-containing oilseeds* (European Patent EP1908355A1). European Patent Office. https://patentscope.wipo.int/search/en/detail.jsf?docId=EP14899835.

Fathordoobady, F., Singh, A., Kitts, D. D., & Pratap Singh, A. (2019). Hemp (*Cannabis sativa* L.) extract: Anti-microbial properties, methods of extraction, and potential oral delivery. *Food Reviews International, 35*(7), 664–684. https://doi.org/10.1080/87559129.2019.1600539.

Ferreira, A. F. (2017). Biorefinery concept. In M. Rabaçal, A. Ferreira, C. Silva, & M. Costa (Eds.), *Biorefineries. Lecture Notes in Energy: 57. Biorefineries - Targeting energy, high value products and waste valorisation*. Cham: Springer. https://doi.org/10.1007/978-3-319-48288-0_1.

Finnan, J., & Burke, B. (2013). Nitrogen fertilization to optimize the greenhouse gas balance of hemp crops grown for biomass. *GCB Bioenergy, 5*(6), 701–712. https://doi.org/10.1111/gcbb.12045.

Fortunati, E., Luzi, F., Puglia, D., & Torre, L. (2016). Extraction of lignocellulosic materials from waste products. In D. Puglia, E. Fortunati, & J. M. Kenny (Eds.), *Multifunctional polymeric nanocomposites based on cellulosic reinforcements* (pp. 1–38). William Andrew Publishing. https://doi.org/10.1016/B978-0-323-44248-0.00001-8.

Gandolfi, S., Ottolina, G., Consonni, R., Riva, S., & Patel, I. (2014). Fractionation of hemp hurds by organosolv pretreatment and its effect on production of lignin and sugars. *ChemSusChem, 7*(7), 1991–1999. https://doi.org/10.1002/cssc.201301396.

Gandolfi, S., Ottolina, G., Riva, S., Fantoni, G. P., & Patel, I. (2013). Complete chemical analysis of carmagnola hemp hurds and structural features of its components. *BioResources, 8*(2), 2641–2656.

Genesio, L., Miglietta, F., Baronti, S., & Vaccari, F. P. (2015). Biochar increases vineyard productivity without affecting grape quality: Results from a four years field experiment in Tuscany. *Agriculture, Ecosystems and Environment, 201*, 20–25. https://doi.org/10.1016/j.agee.2014.11.021.

Gomez, L., Tiwari, B., & Garcia-Vaquero, M. (2020). Emerging extraction techniques: Microwave-assisted extraction. In M. D. Torres, S. Kraan, & H. Dominguez (Eds.), *Sustainable seaweed technologies: Cultivation, biorefinery, and applications* (pp. 207–224). Elsevier. https://doi.org/10.1016/B978-0-12-817943-7.00008-1.

Gopal, P. M., Sivaram, N. M., & Barik, D. (2019). Paper industry wastes and energy generation from wastes. In D. B. T.-E. from T. O. W. for H, & P. G. Barik (Eds.), *Energy from toxic organic waste for heat and power generation* (pp. 83–97). Woodhead Publishing. https://doi.org/10.1016/B978-0-08-102528-4.00007-9.

Goyal, H. B., Seal, D., & Saxena, R. C. (2008). Bio-fuels from thermochemical conversion of renewable resources: A review. *Renewable and Sustainable Energy Reviews, 12*(2), 504–517.

Gunjević, V., Grillo, G., Carnaroglio, D., Binello, A., Barge, A., & Cravotto, G. (2021). Selective recovery of terpenes, polyphenols and cannabinoids from *Cannabis sativa* L. inflorescences under microwaves. *Industrial Crops and Products, 162*, 113247. https://doi.org/10.1016/j.indcrop.2021.113247.

Gunnarsson, I. B., Kuglarz, M., Karakashev, D., & Angelidaki, I. (2015). Thermochemical pretreatments for enhancing succinic acid production from industrial hemp (*Cannabis sativa* L.). *Bioresource Technology, 182*, 58–66. https://doi.org/10.1016/j.biortech.2015.01.126.

Henniges, U., Hasani, M., Potthast, A., Westman, G., & Rosenau, T. (2013). Electron beam irradiation of cellulosic materials – Opportunities and limitations. *Materials, 6*(5), 1584–1598. https://doi.org/10.3390/ma6051584.

Hu, X., & Gholizadeh, M. (2019). Biomass pyrolysis: A review of the process development and challenges from initial researches up to the commercialisation stage. *Journal of Energy Chemistry, 39*, 109–143. https://doi.org/10.1016/j.jechem.2019.01.024.

Huber, G. W., Iborra, S., & Corma, A. (2006). Synthesis of transportation fuels from biomass: Chemistry, catalysts, and engineering. *Chemical Reviews, 106*(9), 4044–4098. https://doi.org/10.1021/cr068360d.

Ikan, R., & Crammer, B. (2003). Organic chemistry, compound detection. In R. A. Meyers (Ed.), *Encyclopedia of physical science and technology* (3rd ed., pp. 459–496). Academic Press. https://doi.org/10.1016/B0-12-227410-5/00541-X.

International Standards Organization. (2015). *Oilseed meals – Determination of oil content – Rapid extraction method.* (ISO 22630).

Islam, M. N., & Rahman, F. (2019). Production and modification of nanofibrillated cellulose composites and potential applications. In G. Koronis, & A. Silva (Eds.), *Green composites for automotive applications* (pp. 115–141). Woodhead Publishing. https://doi.org/10.1016/B978-0-08-102177-4.00006-9.

Ji, A., Jia, L., Kumar, D., & Yoo, C. G. (2021). Recent advancements in biological conversion of industrial hemp for biofuel and value-added products. *Fermentation, 7*(1), 6. https://doi.org/10.3390/fermentation7010006.

Jin, D., Dai, K., Xie, Z., & Chen, J. (2020). Secondary metabolites profiled in *Cannabis* inflorescences, leaves, stem barks, and roots for medicinal purposes. *Scientific Reports, 10*(1), 3309. https://doi.org/10.1038/s41598-020-60172-6.

Jönsson, L. J., Alriksson, B., & Nilvebrant, N.-O. (2013). Bioconversion of lignocellulose: Inhibitors and detoxification. *Biotechnology for Biofuels, 6*(1), 16. https://doi.org/10.1186/1754-6834-6-16.

Khattab, M. M., & Dahman, Y. (2019). Production and recovery of poly-3-hydroxybutyrate bioplastics using agro-industrial residues of hemp hurd biomass. *Bioprocess and Biosystems Engineering, 42*(7), 1115–1127. https://doi.org/10.1007/s00449-019-02109-6.

Kim, D., Yoo, G. C., Schwarz, J., Dhekney, S., Kozak, R., Laufer, C., et al. (2021). Effect of lignin-blocking agent on enzyme hydrolysis of acid pretreated hemp waste. *RSC Advances, 2021*(11), 22025–22033. https://doi.org/10.1039/D1RA03412J.

Kim, H. S., Oh, Y. H., Jang, Y.-A., Kang, K. H., David, Y., Yu, J. H., et al. (2016). Recombinant *Ralstonia eutropha* engineered to utilize xylose and its use for the production of poly(3-hydroxybutyrate) from sunflower stalk hydrolysate solution. *Microbial Cell Factories, 15*, 95. https://doi.org/10.1186/s12934-016-0495-6.

Kitrytė, V., Bagdonaitė, D., & Rimantas Venskutonis, P. (2018). Biorefining of industrial hemp (Cannabis sativa L.) threshing residues into cannabinoid and antioxidant fractions by supercritical carbon dioxide, pressurized liquid and enzyme-assisted extractions. *Food Chemistry, 267*, 420–429. https://doi.org/10.1016/j.foodchem.2017.09.080.

Kornpointner, C., Sainz Martinez, A., Marinovic, S., Haselmair-Gosch, C., Jamnik, P., Schröder, K., et al. (2021). Chemical composition and antioxidant potential of Cannabis sativa L. roots. *Industrial Crops and Products, 165*. https://doi.org/10.1016/j.indcrop.2021.113422, 113422.

Kraszkiewicz, A., Kachel, M., Parafiniuk, S., Zając, G., Niedziółka, I., & Sprawka, M. (2019). Assessment of the possibility of using hemp biomass (*Cannabis sativa* L.) for energy purposes: A case study. *Applied Sciences, 9*(20), 4437. https://doi.org/10.3390/app9204437.

Kreuger, E., Prade, T., Escobar, F., Svensson, S.-E., Englund, J.-E., & Björnsson, L. (2011). Anaerobic digestion of industrial hemp—Effect of harvest time on methane energy yield per hectare. *Biomass and Bioenergy, 35*(2), 893–900. https://doi.org/10.1016/j.biombioe.2010.11.005.

Kreuger, E., Sipos, B., Zacchi, G., Svensson, S.-E., & Björnsson, L. (2011). Bioconversion of industrial hemp to ethanol and methane: The benefits of steam pretreatment and co-production. *Bioresource Technology, 102*(3), 3457–3465. https://doi.org/10.1016/j.biortech.2010.10.126.

Kuglarz, M., Alvarado-Morales, M., Karakashev, D., & Angelidaki, I. (2016). Integrated production of cellulosic bioethanol and succinic acid from industrial hemp in a biorefinery concept. *Bioresource Technology, 200*, 639–647. https://doi.org/10.1016/j.biortech.2015.10.081.

Kuglarz, M., Gunnarsson, I. B., Svensson, S. E., Prade, T., Johansson, E., & Angelidaki, I. (2014). Ethanol production from industrial hemp: Effect of combined dilute acid/steam pretreatment and economic aspects. *Bioresource Technology, 163*, 236–243. https://doi.org/10.1016/j.biortech.2014.04.049.

Kumar, R., Hu, F., Sannigrahi, P., Jung, S., Ragauskas, A. J., & Wyman, C. E. (2013). Carbohydrate derived-pseudo-lignin can retard cellulose biological conversion. *Biotechnology and Bioengineering, 110*(3), 737–753.

Kumari, B., Tiwari, B. K., Hossain, M. B., Brunton, N. P., & Rai, D. K. (2018). Recent advances on application of ultrasound and pulsed electric field technologies in the extraction of bioactives from agro-industrial by-products. *Food and Bioprocess Technology, 11*(2), 223–241. https://doi.org/10.1007/s11947-017-1961-9.

Ladakis, D., Papapostolou, H., Vlysidis, A., & Koutinas, A. (2020). Inventory of food processing side streams in European Union and prospects for biorefinery development. In M. R. Kosseva, & C. Webb (Eds.), *Food industry wastes* (2nd ed., pp. 181–199). Academic Press. https://doi.org/10.1016/B978-0-12-817121-9.00009-7.

Laudadio, V., Bastoni, E., Introna, M., & Tufarelli, V. (2013). Production of low-fiber sunflower (*Helianthus annuus* L.) meal by micronization and air classification processes. *CyTA – Journal of Food, 11*(4), 398–403. https://doi.org/10.1080/19476337.2013.781681.

Lazaridou, A., Vouris, D. G., Zoumpoulakis, P., & Biliaderis, C. G. (2018). Physicochemical properties of jet milled wheat flours and doughs. *Food Hydrocolloids, 80*, 111–121. https://doi.org/10.1016/j.foodhyd.2018.01.044.

Lee, C. H., Khalina, A., Lee, S. H., & Liu, M. (2020). A comprehensive review on bast fibre retting process for optimal performance in fibre-reinforced polymer composites. *Advances in Materials Science and Engineering, 2020*, 6074063. https://doi.org/10.1155/2020/6074063.

Leonard, W., Zhang, P., Ying, D., & Fang, Z. (2021). Lignanamides: Sources, biosynthesis and potential health benefits—A minireview. *Critical Reviews in Food Science and Nutrition, 61*, 1404–1414. https://doi.org/10.1080/10408398.2020.1759025.

Leonard, W., Zhang, P., Ying, D., Xiong, Y., & Fang, Z. (2021). Extrusion improves the phenolic profile and biological activities of hempseed (*Cannabis sativa* L.) hull. *Food Chemistry, 346*, 128606.

Llompart, M., Garcia-Jares, C., Celeiro, M., & Dagnac, T. (2019). Extraction/microwave-assisted extraction. In P. Worsfold, C. Poole, A. Townshend, & E. Miró (Eds.), *Encyclopedia of analytical science* (3rd ed., pp. 67–77). Academic Press. https://doi.org/10.1016/B978-0-12-409547-2.14442-7.

Lyu, G., Wu, S., & Zhang, H. (2015). Estimation and comparison of bio-oil components from different pyrolysis conditions. *Frontiers in Energy Research, 3*, 28. https://doi.org/10.3389/fenrg.2015.00028.

Major, J., Rondon, M., Molina, D., Riha, S. J., & Lehmann, J. (2010). Maize yield and nutrition during 4 years after biochar application to a Colombian savanna oxisol. *Plant and Soil, 333*, 117–128. https://doi.org/10.1007/s11104-010-0327-0.

Majumder, S., Neogi, S., Dutta, T., Powel, M. A., & Banik, P. (2019). The impact of biochar on soil carbon sequestration: Meta-analytical approach to evaluating environmental and economic advantages. *Journal of Environmental Management, 250*. https://doi.org/10.1016/j.jenvman.2019.109466, 109466.

Marris, E. (2006). Black is the new green. *Nature, 442*(7103), 624–626. https://doi.org/10.1038/442624a.

Matassa, S., Esposito, G., Pirozzi, F., & Papirio, S. (2020). Exploring the biomethane potential of different industrial hemp (*Cannabis sativa* L.) biomass residues. *Energies, 13*(13), 3361. https://doi.org/10.3390/en13133361.

Mathew, S., & Abraham, T. E. (2004). Ferulic acid: An antioxidant found naturally in plant cell walls and feruloyl esterases involved in its release and their applications. *Critical Reviews in Biotechnology, 24*(2–3), 59–83. https://doi.org/10.1080/07388550490491467.

Miao, C., Hui, L., Liu, Z., & Tang, X. (2013). Evaluation of hemp root bast as a new material for papermaking. *BioResources, 9*(1), 132–142.

Myhre, G., Shindell, D., Bréon, F.-M., Collins, W., Fuglestvedt, J., Huang, J., et al. (2013). Anthropogenic and natural radiative forcing. In T. F. Stocker, D. Qin, G.-K. Plattner, M. Tignor, S. K. Allen, J. Boschung, … P. M. Midgley (Eds.), *Climate change 2013: The physical science basis* (pp. 659–740). Cambridge University Press.

Náthia-Neves, G., & Alonso, E. (2021). Valorization of sunflower by-product using microwave-assisted extraction to obtain a rich protein flour: Recovery of chlorogenic acid, phenolic content and antioxidant capacity. *Food and Bioproducts Processing, 125*, 57–67. https://doi.org/10.1016/j.fbp.2020.10.008.

Nitsos, C., Rova, U., & Christakopoulos, P. (2018). Organosolv fractionation of softwood biomass for biofuel and biorefinery applications. *Energies, 11*(1), 50. https://doi.org/10.3390/en11010050.

Nuapia, Y., Tutu, H., Chimuka, L., & Cukrowska, E. (2020). Selective extraction of cannabinoid compounds from *Cannabis* seed using pressurized hot water extraction. *Molecules, 25*(6), 1335. https://doi.org/10.3390/molecules25061335.

Oreopoulou, A., Tsimogiannis, D., & Oreopoulou, V. (2019). Extraction of polyphenols from aromatic and medicinal plants: An overview of the methods and the effect of extraction parameters. In R. R. Watson (Ed.), *Polyphenols in plants isolation, purification and extract preparation* (pp. 243–259). Academic Press. https://doi.org/10.1016/B978-0-12-813768-0.00025-6.

Pap, N., Hamberg, L., Pihlava, J.-M., Hellström, J., Mattila, P., Eurola, M., et al. (2020). Impact of enzymatic hydrolysis on the nutrients, phytochemicals and sensory properties of oil hemp seed cake (*Cannabis sativa* L. FINOLA variety). *Food Chemistry, 320*. https://doi.org/10.1016/j.foodchem.2020.126530, 126530.

Papirio, S., Matassa, S., Pirozzi, F., & Esposito, G. (2020). Anaerobic co-digestion of cheese whey and industrial hemp residues opens new perspectives for the valorization of agri-food waste. *Energies, 13*(11), 2820. https://doi.org/10.3390/en13112820.

Pari, L., Baraniecki, P., Kaniewski, R., & Scarfone, A. (2015). Harvesting strategies of bast fiber crops in Europe and in China. *Industrial Crops and Products*, *68*, 90–96. https://doi.org/10.1016/j.indcrop.2014.09.010.

Peck, P., Bennett, S. J., Bissett-Amess, R., Lenhart, J., & Mozaffarian, H. (2009). Examining understanding, acceptance, and support for the biorefinery concept among EU policy-makers. *Biofuels, Bioproducts and Biorefining*, *3*(3), 361–383. https://doi.org/10.1002/bbb.154.

Pedrazzi, S., Morselli, N., Puglia, M., & Tartarini, P. (2020). Energy and emissions analysis of hemp hurd and vine pruning derived pellets used as fuel in a commercial stove for residential heating. *Tecnica Italiana-Italian Journal of Engineering Science*, *64*(2–4), 361–638.

Pojić, M. (2021). Smart functional ingredients. In S. Pathania, & B. Tiwari (Eds.), *Food formulation* (pp. 5–26). Wiley. https://doi.org/10.1002/9781119614760.ch2.

Pojić, M., Mišan, A., Sakač, M., Dapčević Hadnađev, T., Šarić, B., Milovanović, I., et al. (2014). Characterization of by-products originating from hemp oil processing. *Journal of Agricultural and Food Chemistry*, *62*(51), 12436–12442. https://doi.org/10.1021/jf5044426.

Polotti, G. (2020). Perspectives from industry. In D. Moscatelli, & M. Sponchioni (Eds.), *Advances in chemical engineering* (pp. 259–330). Academic Press. https://doi.org/10.1016/bs.ache.2020.07.003. 56(1).

Polova, Z., & Bruskova, P. (2018). *Roadmap report: HEMP value chain*. DanuBioValNet http://www.ipe.ro/RR%20 Hemp.pdf.

Pushpangadan, P., & George, V. (2012). Basil. In K. V. Peter (Ed.), *Handbook of herbs and spices* (2nd ed., pp. 55–72). Woodhead Publishing. https://doi.org/10.1533/9780857095671.55.

Rajabzadeh, A. R., Jafari, M., & Legge, R. L. (2015). *Solvent-free approach for separation of constituent fractions of pulses, grains, oilseeds, and dried fruits* (US Patent US20150140185 A1). U.S. Patent and Trademark Office. https://patentscope.wipo.int/search/en/detail.jsf?docId=US133593405&_cid=P20-KTV8VX-16819-1.

Rehman, M. U., Khan, F., & Niaz, K. (2020). Introduction to natural products analysis. In A. S. Silva, S. F. Nabavi, M. Saeedi, & S. M. Nabavi (Eds.), *Recent advances in natural products analysis* (pp. 3–15). Elsevier. https://doi.org/10.1016/B978-0-12-816455-6.00001-9.

Romero, P., Peris, A., Vergara, K., & Matus, J. T. (2020). Comprehending and improving cannabis specialized metabolism in the systems biology era. *Plant Science*, *298*. https://doi.org/10.1016/j.plantsci.2020.110571, 110571.

Rommi, K., Holopainen, U., Pohjola, S., Hakala, T. K., Lantto, R., Poutanen, K., et al. (2015). Impact of particle size reduction and carbohydrate-hydrolyzing enzyme treatment on protein recovery from rapeseed (*Brassica rapa* L.) press cake. *Food and Bioprocess Technology*, *8*(12), 2392–2399. https://doi.org/10.1007/s11947-015-1587-8.

Rondon, M. A., Lehmann, J., Ramírez, J., & Hurtado, M. (2007). Biological nitrogen fixation by common beans (Phaseolus vulgaris L.) increases with bio-char additions. *Biology and Fertility of Soils*, *43*(6), 699–708. https://doi.org/10.1007/s00374-006-0152-z.

Rovetto, L. J., & Aieta, N. V. (2017). Supercritical carbon dioxide extraction of cannabinoids from *Cannabis sativa* L. *Journal of Supercritical Fluids*, *129*, 16–27. https://doi.org/10.1016/j.supflu.2017.03.014.

Russo, E. B. (2011). Taming THC: Potential cannabis synergy and phytocannabinoid-terpenoid entourage effects. *British Journal of Pharmacology*, *163*(7), 1344–1364. https://doi.org/10.1111/j.1476-5381.2011.01238.x.

Ryz, N. R., Remillard, D. J., & Russo, E. B. (2017). *Cannabis* roots: A traditional therapy with future potential for treating inflammation and pain. *Cannabis and Cannabinoid Research*, *2*(1), 210–216.

Salami, A., Raninen, K., Heikkinen, J., Tomppo, L., Vilppo, T., Selenius, M., et al. (2020). Complementary chemical characterization of distillates obtained from industrial hemp hurds by thermal processing. *Industrial Crops and Products*, *155*. https://doi.org/10.1016/j.indcrop.2020.112760, 112760.

Sastri, V. (2010). Other polymers: Styrenics, silicones, thermoplastic elastomers, biopolymers, and thermosets. In V. R. Sastri (Ed.), *Plastics in medical devices: Properties, requirements, and applications* (pp. 217–262). William Andrew Publishing. https://doi.org/10.1016/B978-0-8155-2027-6.10009-1.

Schluttenhofer, C., & Yuan, L. (2017). Challenges towards revitalizing hemp: A multifaceted crop. *Trends in Plant Science*, *22*(11), 917–929. https://doi.org/10.1016/j.tplants.2017.08.004.

Schultz, C. J., Lim, W. L., Khor, S. F., Neumann, K. A., Schulz, J. M., Ansari, O., et al. (2020). Consumer and health-related traits of seed from selected commercial and breeding lines of industrial hemp, *Cannabis sativa* L. *Journal of Agriculture and Food Research*, *2*. https://doi.org/10.1016/j.jafr.2020.100025, 100025.

Searle, S., & Malins, C. (2013). *Availability of cellulosic residues and wastes in the EU*. International Council on Clean Transportation: Washington, DC, USA.

Serna-Loaiza, S., Adamcyk, J., Beisl, S., Kornpointner, C., Halbwirth, H., & Friedl, A. (2020). Pressurized liquid extraction of cannabinoids from hemp processing residues: Evaluation of the influencing variables. *Processes*, *8*(11), 1334. https://doi.org/10.3390/pr8111334.

Setti, L., Samaei, S. P., Maggiore, I., Nissen, L., Gianotti, A., & Babini, E. (2020). Comparing the effectiveness of three different biorefinery processes at recovering bioactive products from hemp (*Cannabis sativa* L.) byproduct. *Food and Bioprocess Technology, 13*(12), 2156–2171.

Shin, S.-J., & Sung, Y. J. (2008). Improving enzymatic hydrolysis of industrial hemp (*Cannabis sativa* L.) by electron beam irradiation. *Radiation Physics and Chemistry, 77*(9), 1034–1038. https://doi.org/10.1016/j.radphyschem.2008.05.047.

Sipos, B., Kreuger, E., Svensson, S.-E., Reczey, K., Björnsson, L., & Zacchi, G. (2010). Steam pretreatment of dry and ensiled industrial hemp for ethanol production. *Biomass and Bioenergy, 34*, 1721–1731.

Sommano, S. R., Chittasupho, C., Ruksiriwanich, W., & Jantrawut, P. (2020). The *Cannabis* terpenes. *Molecules (Basel, Switzerland), 25*(24), 5792. https://doi.org/10.3390/molecules25245792.

Stalmach, A. (2014). Bioavailability of dietary anthocyanins and hydroxycinnamic acids. In R. R. Watson, V. R. Preedy, & S. Zibadi (Eds.), *Polyphenols in human health and disease* (pp. 561–576). Academic Press. https://doi.org/10.1016/B978-0-12-398456-2.00042-6.

Sung, Y. J., & Shin, S.-J. (2011). Compositional changes in industrial hemp biomass (*Cannabis sativa* L.) induced by electron beam irradiation pretreatment. *Biomass and Bioenergy, 35*(7), 3267–3270. https://doi.org/10.1016/j.biombioe.2011.04.011.

Tang, K., Struik, P. C., Amaducci, S., Stomph, T.-J., & Yin, X. (2017). Hemp (*Cannabis sativa* L.) leaf photosynthesis in relation to nitrogen content and temperature: Implications for hemp as a bio-economically sustainable crop. *GCB Bioenergy, 9*(10), 1573–1587. https://doi.org/10.1111/gcbb.12451.

Taofiq, O., González-Paramás, A. M., Barreiro, M. F., & Ferreira, I. C. F. R. (2017). Hydroxycinnamic acids and their derivatives: Cosmeceutical significance, challenges and future perspectives, a review. *Molecules, 22*(2), 281. https://doi.org/10.3390/molecules22020281.

Teh, S.-S., & Birch, E. J. (2014). Effect of ultrasonic treatment on the polyphenol content and antioxidant capacity of extract from defatted hemp, flax and canola seed cakes. *Ultrasonics Sonochemistry, 21*(1), 346–353. https://doi.org/10.1016/j.ultsonch.2013.08.002.

Teh, S.-S., Niven, B. E., Bekhit, A. E.-D. A., Carne, A., & Birch, E. J. (2014). The use of microwave and pulsed electric field as a pretreatment step in ultrasonic extraction of polyphenols from defatted hemp seed cake (*Cannabis sativa*) using response surface methodology. *Food and Bioprocess Technology, 7*(11), 3064–3076. https://doi.org/10.1007/s11947-014-1313-y.

Tiwari, B. K. (2015). Ultrasound: A clean, green extraction technology. *TrAC Trends in Analytical Chemistry, 71*, 100–109. https://doi.org/10.1016/j.trac.2015.04.013.

Tiwari, U., & Pojić, M. (2020). Introduction to cereal processing: Innovative processing techniques. In M. Pojić, & U. Tiwari (Eds.), *Innovative processing technologies for healthy grains* (pp. 9–35). Wiley.

Tongwane, M. I., & Moeletsi, M. E. (2018). A review of greenhouse gas emissions from the agriculture sector in Africa. *Agricultural Systems, 166*, 124–134.

Tran, T. T. A., Le, T. K. P., Mai, T. P., & Nguyen, D. Q. (2019). Bioethanol production from lignocellulosic biomass. In *Alcohol fuels—current technologies and future prospect* IntechOpen. https://doi.org/10.5772/intechopen.86437.

Uddin, M. N., Techato, K., Taweekun, J., Rahman, M. M., Rasul, M. G., Mahlia, T. M. I., et al. (2018). An overview of recent developments in biomass pyrolysis technologies. *Energies, 11*(11), 3115. https://doi.org/10.3390/en11113115.

Ummat, V., Sivagnanam, S. P., Rajauria, G., O'Donnell, C., & Tiwari, B. K. (2021). Advances in pre-treatment techniques and green extraction technologies for bioactives from seaweeds. *Trends in Food Science and Technology, 110*, 90–106. https://doi.org/10.1016/j.tifs.2021.01.018.

Vaccari, F. P., Baronti, S., Lugato, E., Genesio, L., Castaldi, S., Fornasier, F., et al. (2011). Biochar as a strategy to sequester carbon and increase yield in durum wheat. *European Journal of Agronomy, 34*(4), 231–238. https://doi.org/10.1016/j.eja.2011.01.006.

Vági, E., Balázs, M., Komóczi, A., Kiss, I., Mihalovits, M., & Székely, E. (2019). Cannabinoids enriched extracts from industrial hemp residues. *Periodica Polytechnica Chemical Engineering, 63*(2), 357–363.

Vági, E., Balázs, M., Komoczi, A., Mihalovits, M., & Székely, E. (2020). Fractionation of phytocannabinoids from industrial hemp residues with high-pressure technologies. *Journal of Supercritical Fluids, 164*. https://doi.org/10.1016/j.supflu.2020.104898, 104898.

Valentim, B., Guedes, A., Rodrigues, S., & Flores, D. (2011). Case study of igneous intrusion effects on coal nitrogen functionalities. *International Journal of Coal Geology, 86*(2–3), 291–294.

Vandenbossche, V., Candy, L., Evon, P., Rouilly, A., & Pontalier, P.-Y. (2019). Extrusion. In F. Chemat, & E. Vorobiev (Eds.), *Green food processing techniques* (pp. 289–314). Academic Press. https://doi.org/10.1016/B978-0-12-815353-6.00010-0.

Vasić, K., Knez, Ž., & Leitgeb, M. (2021). Bioethanol production by enzymatic hydrolysis from different lignocellulosic sources. *Molecules*, 26(3), 753. https://doi.org/10.3390/molecules26030753.

Wu, S., Chien, P., Lee, M., & Chau, C. (2007). Particle size reduction effectively enhances the intestinal health-promotion ability of an orange insoluble fiber in hamsters. *Journal of Food Science*, 72(8), S618–S621.

Wu, K., Ju, T., Deng, Y., & Xi, J. (2017). Mechanochemical assisted extraction: A novel, efficient, eco-friendly technology. *Trends in Food Science and Technology*, 66, 166–175. https://doi.org/10.1016/j.tifs.2017.06.011.

Yang, Y., & Hu, B. (2014). Bio-based chemicals from biorefining: Lipid and wax conversion and utilization. In B. Waldron (Ed.), *Advances in biorefineries: Biomass and waste supply chain exploitation* (pp. 693–720). Woodhead Publishing. https://doi.org/10.1533/9780857097385.2.693.

Yoo, C. G., Meng, X., Pu, Y., & Ragauskas, A. J. (2020). The critical role of lignin in lignocellulosic biomass conversion and recent pretreatment strategies: A comprehensive review. *Bioresource Technology*, 301. https://doi.org/10.1016/j.biortech.2020.122784, 122784.

Yun, J.-M., & Surh, J. (2012). Fatty acid composition as a predictor for the oxidation stability of Korean vegetable oils with or without induced oxidative stress. *Preventive Nutrition and Food Science*, 17(2), 158–165.

Zardo, I., de Espíndola Sobczyk, A., Marczak, L. D. F., & Sarkis, J. (2019). Optimization of ultrasound assisted extraction of phenolic compounds from sunflower seed cake using response surface methodology. *Waste and Biomass Valorization*, 10(1), 33–44.

Zhang, J., Zhou, H., Liu, D., & Zhao, X. (2020). Pretreatment of lignocellulosic biomass for efficient enzymatic saccharification of cellulose. In A. Yousuf, D. Pirozzi, & F. Sannino (Eds.), *Lignocellulosic biomass to liquid biofuels* (pp. 17–65). Academic Press. https://doi.org/10.1016/B978-0-12-815936-1.00002-2.

Zhao, J., Xu, Y., Wang, W., Griffin, J., Roozeboom, K., & Wang, D. (2020a). Bioconversion of industrial hemp biomass for bioethanol production: A review. *Fuel*, 281. https://doi.org/10.1016/j.fuel.2020.118725, 118725.

Zhao, J., Xu, Y., Wang, W., Griffin, J., & Wang, D. (2020b). Conversion of liquid hot water, acid and alkali pretreated industrial hemp biomasses to bioethanol. *Bioresource Technology*, 309. https://doi.org/10.1016/j.biortech.2020.123383, 123383.

Zhao, J., Xu, Y., Wang, W., Griffin, J., & Wang, D. (2020c). High ethanol concentration (77 g/L) of industrial hemp biomass achieved through optimizing the relationship between ethanol yield/concentration and solid loading. *ACS Omega*, 5, 21913–21921.

Zheljazkov, V. D., Sikora, V., Semerdjieva, I. B., Kačániová, M., Astatkie, T., & Dincheva, I. (2020). Grinding and fractionation during distillation alter hemp essential oil profile and its antimicrobial activity. *Molecules*, 25(17), 3943. https://doi.org/10.3390/molecules25173943.

Zhong, L., Fang, Z., Wahlqvist, M. L., Hodgson, J. M., & Johnson, S. K. (2021). Multi-response surface optimisation of extrusion cooking to increase soluble dietary fibre and polyphenols in lupin seed coat. *LWT – Food Science and Technology*, 140. https://doi.org/10.1016/j.lwt.2020.110767, 110767.

Zhong, L., Fang, Z., Wahlqvist, M. L., Wu, G., Hodgson, J. M., & Johnson, S. K. (2018). Seed coats of pulses as a food ingredient: Characterization, processing, and applications. *Trends in Food Science and Technology*, 80, 35–42. https://doi.org/10.1016/j.tifs.2018.07.021.

13

Industrial hemp in animal feed applications

Ondřej Šťastník, Eva Mrkvicová, and Leoš Pavlata

Department of Animal Nutrition and Forage Production, Faculty of AgriSciences, Mendel University in Brno, Brno, Czech Republic

13.1 Introduction

Given the fact the growth of human population is expected, the need to increase the production of animal products (meat, milk, eggs) is proportionally unavoidable (Shariat Zadeh, Kheiri, & Faghani, 2020). By 2050 an increase in the consumption of animal products by 60%–70% is expected (Makkar, Tran, Heuze, & Ankers, 2014), correlating with the increase in the human population, which is estimated to reach 9 billion people. The increase in animal-based food consumption will require overconsumption of natural resources, with feed being the most demanding in terms of arable land area, climate change, and food-feed-fuel competition (Makkar et al., 2014). The growing global need to find alternative and sustainable protein sources supports research on the unconventional feeds. Moreover, the most common protein sources in animal nutrition are soybean meal and fish meal. These conventional sources are no longer sustainable and their utilization will be further limited by increasing prices (Veldkamp & Bosch, 2015), so that new and sustainable protein sources for animal feeds are necessary. One of the alternatives may be industrial hemp. Although hempseeds have been mainly used for animal nutrition, there is a growing interest for their utilization in human nutrition and hemp-based products (i.e., oil, seed cakes, meal, flour, and protein powder). Hemp products can be marketed as a new, natural, and excellent source of nutrients, containing all the essential amino and fatty acids necessary for maintaining a good health (Andre, Hausman, & Guerriero, 2016).

On the other side, hemp (*Cannabis* genus) is often associated with negative psychogenic effects caused by delta-9-tetrahydrocannabinol (THC) content. In this respect, European Food Safety Authority (EFSA) issued a Scientific Opinion on the safety of hemp (*Cannabis* genus) for use as animal feed (EFSA, 2011). This document identifies four different types of feed materials derived from the hemp plant—hempseed, hempseed meal/cake, hempseed oil, and whole hemp plant (including hemp hurds, fresh or dried). Other products may be hemp flour (ground dried hemp leaves) and hemp protein isolate from seeds. The hemp varieties allowed for cultivation in Europe need not to exceed 0.3% THC (in dry matter). Consequently, the EFSA Panel on Additives and Products or Substances used in Animal Feed (FEEDAP Panel) did not find any option for the use of whole hemp plant-derived feedstuff in animal nutrition. In this contrast, hempseed feeding was considered safe for the consumers (EFSA, 2011).

13.2 Industrial hemp as a sustainable animal feed raw material

Hemp (*Cannabis sativa* L.) is an annual, up to 2 m tall plant with palm split leaves and multi-sex flowers (Padua, Bunyaprafatsara, & Lemmens, 1999). Industrial hemp plant by-products such as leaves, fodder, residual plant fibers (Kleinhenz et al., 2020), and hemp hurds may be used for animal nutrition purposes. Hemp hurds are hemp straw (with 96.3% dry matter; DM) characterized by high fiber content (90% neutral detergent fiber (NDF), 78.9% acid detergent fiber (ADF), in DM), whereas the content of crude protein (3.2% in DM) and ether extract (0.8% in DM) is low (EFSA, 2011). The utilization of so-called hemp herbal biomass, composed of flowers with a proportion of seeds and the upper part of the stem, is also possible in the form of pellets (Fig. 13.1). The listed hemp by-products are cellulose-containing plant materials and therefore will be more suitable for ruminants, especially for cattle (Kleinhenz et al., 2020).

However, whole hempseeds, hemp by-products (i.e., hempseed meal/cake), and hemp-seed oil are the most widely used in non-ruminant animal nutrition. Generally, hempseeds contain about 25% crude protein, 30% ether extract, and 22 MJ/kg gross energy content (Callaway, 2004). The apparent metabolizable energy for hempseeds for pigeons was found to be 18 MJ/kg (Hullar, Meleg, Fekete, & Romvari, 1999), while for hempseed cake for chickens

FIG. 13.1   Hemp herbal biomass pellets (author's archive).

was 10.1 MJ/kg (Kalmendal, 2008). This composition makes industrial hempseed a suitable raw material for animal nutrition (protein feed). A comparative overview of nutritive value of hempseeds and other raw materials that are commonly used in animal nutrition is given in Table 13.1. In terms of nutrients, hempseed is most similar to rapeseed. However, hempseed has the indisputable advantage of a minimal content of anti-nutritional compounds.

The two major proteins present (edestin and albumin) are easily digestible and contain all essential amino acids except lysine and sulfur-containing amino acids (Callaway, 2004; Leizer, Ribnicky, Poulev, Dushenkov, & Raskin, 2000; Wang, Tang, Yang, & Gao, 2008). Identification and characterization of hempseed proteins showed that edestin, rich in valuable amino acids (i.e., glutamic acid and aspartic acid), is the main protein component in hempseed protein isolate (Wang et al., 2008).

Wang et al. (2008) also showed that the proportion of essential amino acids to the total amino acids for hemp protein isolate was significantly higher than that of soy protein isolate. In addition, in vitro digestibility of hemp protein isolate was determined to be 88%–91% (Wang et al., 2008), being higher than digestibility of soybean protein isolate (71%). This can be caused by absence of trypsin inhibitor in hemp protein (EFSA, 2011). Another protein structure, rich in methionine and cysteine, was found in hempseed and characterized as albumin (Odani, 1998). In regard to protein quality, lysine is the first limiting amino acid in hemp protein for several animals (House, Neufeld, & Leson, 2010). Since the main role of lysine is its participation in protein synthesis, it is required for growth and performance (Wijtten, Prak, Lemme, & Langhout, 2004). Although methionine is typically the first limiting amino acid in poultry feeds (due to the incorporation of sulfur into feathers), lysine is considered the most important amino acid for the ideal protein concept in feed formulation as it is used as the reference amino acid (Emmert & Baker, 1997). Lysine is the second most limiting amino acid after methionine for poultry formulas and is considered the most limiting amino acid for poultry growth (Khwatenge, Kimathi, Taylor-Bowden, & Nahashon, 2020). Lysine is the first limiting amino acid in swine nutrition management, since all typical swine diets are based on cereal grains being deficient in lysine (Lewis, 2001; NRC, 2012).

In terms of fat composition, unhulled (natural) hempseeds contain 25%–34% lipids, while dehulled seeds contain 42%–47% lipids. Hemp oil, obtained after the pressing of seeds, consists of 75%–80% polyunsaturated fatty acids (PUFAs), of which 53%–60% is linoleic acid (C18:2, n-6), 15%–25% α-linolenic acid (C18:3, n-3), and 3%–6% γ-linolenic acid (C18:3, n-6). Oleic acids (C18:1, n-9) are present in the quantity of 8%–15%. Hempseed oil is also a rich source of tocopherols, which is contained in 1500 mg/kg (Callaway, 2004; Gunstone & Harwood, 2007). More detailed chemical composition of hempseeds and related products is given in Table 13.2.

TABLE 13.1 Comparison of industrial hempseeds with other commonly used raw materials in animal nutrition (in dry matter).

	Hempseed	Wheat	Maize	Soybean	Rapeseed
Crude protein (%)	23	14	11	40	23
Ether extract (%)	34	2	4	20	45
Crude fiber (%)	26	3	3	7	8

TABLE 13.2   Industrial hempseed, hempseed cakes, and hemp herbal biomass chemical composition (in dry matter basis).

Nutrient	Hempseed	Hempseed oil	Hempseed cakes	Hemp herbal biomass
Gross energy (MJ/kg)	22–25.4[a,b]	39.66[c]	20.4[a]	17.6[a]
Organic matter (g/kg)	944.0[a]	999.0[c]	927.5[a]	836.2[a]
Crude protein (g/kg)	229.0–250.0[a,b,d]	0.1[c]	298–330[a,d,e]	157.7[a]
Aspartic acid (g/kg)	27.8[b]		29.2[a]	22.5[a]
Threonine (g/kg)	8.8[b]		9.6[a]	6.3[a]
Serine (g/kg)	12.7[b]		13.6[a]	10.0[a]
Glutamic acid (g/kg)	45.7[b]		45.2[a]	35.5[a]
Proline (g/kg)	11.5[b]		15.9[a]	11.6[a]
Glycine (g/kg)	11.4[b]		12.5[a]	9.7[a]
Alanine (g/kg)	12.8[b]		12.0[a]	8.7[a]
Valine (g/kg)	12.8[b]		14.6[a]	10.4[a]
Isoleucine (g/kg)	9.8[b]		11.8[a]	7.2[a]
Leucine (g/kg)	17.2[b]		19.9[a]	13.6[a]
Tyrosine (g/kg)	8.6[b]		10.0[a]	5.8[a]
Phenylalanine (g/kg)	11.7[b]		13.8[a]	7.7[a]
Histidine (g/kg)	7.1[b]		9.2[a]	6.2[a]
Lysine (g/kg)	10.3[b]		12.8[a]	9.3[a]
Arginine (g/kg)	31.0[b]		39.6[a]	30.2[a]
Ether extract (g/kg)	250–375.0[a,b,d]	998.9[c]	96.94–120.9[a,d,e]	68.88[a]
Palmitic acid (C16:0; %)		5[b]		
Stearic acid (C18:0; %)		2[b]		
Oleic acid (C18:1; %)		9[b]		
Linoleic acid (C18:2 n-6; %)		53–60[b–d,f]		
α-Linolenic acid (C18:3, n-3; %)		15–25[b,d,f]		
γ-Linolenic acid (C18:3, n-6; %)		3–6[b,f]		
Oleic acid (C18:1, n-9; %)		8–15[b,f]		
Stearidonic acid (18:4, n-3; %)		2[b]		
PUFA (%)		84[b]		
n6/n3 ratio		2.5[b]		
Nitrogen-free extract (g/kg)	119.0–276.0[a,b]		207.1[a]	320.9[a]

TABLE 13.2 Industrial hempseed, hempseed cakes, and hemp herbal biomass chemical composition (in dry matter basis)—cont'd

Nutrient	Hempseed	Hempseed oil	Hempseed cakes	Hemp herbal biomass
Crude ash (g/kg)	56.0[a,b]	1.0[c]	72.46[a]	163.84[a]
Ca (g/kg)	1.45[b]		2.5–30.8[a,e]	31.3[a]
P (g/kg)	11.60[b]		9.6–162.6[a,e]	–
Mg (g/kg)	4.83[b]		5.8[a]	4.5[a]
Na (g/kg)	0.12[b]		0.01[a]	0.05[a]
K (g/kg)	8.59[b]		10.1[a]	11.3[a]
Fe (mg/kg)	140.0[b]		108.1[a]	834[a]
Zn (mg/kg)	70[b]		81.54[a]	69[a]
Mn (mg/kg)	70[b]		98.2[a]	403[a]
Cu (mg/kg)	20[b]		17.16[a]	13[a]
Crude fiber (g/kg)	255.0–280.0[a,b,d]		325.53–430.0[a,d]	288.73[a]
ADF (g/kg)	303.0[a]		334.1–420.64[a,e]	391.95[a]
aNDF (g/kg)	351.0[a]		478.11–491.2[a,e]	407.05[a]
ADL (g/kg)			128.6[e]	
Tocopherols (mg/kg)		1500[f]		
Chlorophyll a (µg/g)	–		7.14[a]	0.99[a]
Chlorophyll b (µg/g)	–		10.37[a]	1.98[a]
Total carotene (mg/kg)	–	31.4–125.0[g]		
β-Carotene (mg/kg)	2–8[h]		18.72[a]	11.33[a]
Lutein (mg/kg)	15–34[h]		ND[a]	ND[a]
Zeaxanthin (mg/kg)	2–5[h]		–	–
Cyanidine-3-glucoside (mg/kg)	–		46.62[a]	203.42[a]
Cannabidiol (mg/kg)	–		170[a]	150[a]
Tetrahydrocannabinol	–		ND[a]	ND[a]
Cannabinol	–		ND[a]	ND[a]

[a] *Šťastník (2018), Stastnik, Pavlata, and Mrkvicova (2020), unpublished data.*
[b] *Callaway (2004, modified).*
[c] *Zeman et al. (1995).*
[d] *EFSA (2011).*
[e] *Halle and Schöne (2013).*
[f] *Gunstone and Harwood (2007).*
[g] *Aladić et al. (2014).*
[h] *Irakli et al. (2019, modified).*

ADF, acid detergent fiber; *ADL*, acid detergent lignin; *aNDF*, neutral detergent fiber; *ND*, non-detectable; *PUFA*, polyunsaturated fatty acids.

In addition to essential macro and micronutrients, industrial hemp contains compounds such as plant sterols and phytocannabinoids (i.e., cannabinoids). Cannabinoids are substances found in hemp stems, leaves and flowers. Hemp contains more than 60 (phyto) cannabinoids that have anti-inflammatory, analgesic effects (Hohmann & Suplita, 2006; Jhaveri et al., 2008; Rea, Roche, & Finn, 2007), antiischemic (Lamontagne, Lepicier, Lagneux, & Bouchard, 2006), antipsychotic, anxiolytic (Crippa et al., 2011), and antiepileptic effects (Mortati, Dworetzky, & Devinsky, 2007). Moreover, antimicrobial, immunomodulatory, antihypertensive, and antioxidant effects are also described (Potter, Clark, & Brown, 2008). The two main cannabinoids that can be contained on hemp seeds surface due to its contact with the resin secreted by the epidermal glands located on flowers and leaves are tetrahydrocannabinolic acid (THCA) and cannabidiolic acid (CBDA). THCA, when dried or heated, converts to the psychoactive cannabinoid THC. Similarly, decarboxylation of CBDA yields cannabidiol (CBD). By definition, "industrial hemp," the hemp of commerce which can be used for medicinal purposes, food, feed, or fiber content, contains high levels of CBD and less than 0.3% THC on a dry matter basis (Kogan, Hellyer, & Robinson, 2016). According to EU (European Commission, 2014) Commission Regulation (No. 809/2014 of 17 July 2014 laying down rules for the application of Regulation (EU) No. 1306/2013 of the European Parliament and of the Council with regard to the integrated administration and control system, rural development measures, and cross compliance), production of industrial hemp is permitted if the content of the principal psychoactive constituent THC is less than 0.2%.

To ensure the best possible use of the gene pool of modern hybrid combinations, enabling the rapid growth of animals, feed mixtures with a high concentration of nutrients must be provided. To formulate such diets, the inclusion of feed lipids must be considered, since fat is the most concentrated source of energy. Hemp oil may be part of feed mixtures for animals, although it is primarily intended for human consumption. Generally, lipids with a predominance of polyunsaturated fatty acids are commonly included in the starter diets. Overall, fat in the starter poultry and pig diets should be dosed at a level of 1%–2%. Later, oil content in feed mixtures can be increased to 5%–8%. According to Gakhar, Goldberg, and House (2010) and Goldberg, Ryland, Gakhar, House, and Aliani (2010), hemp oil may be used up to 12% in laying hen's diets without deteriorating effects on performance parameters, flavor, and aroma profiles of cooked eggs. It seems that hemp oil can be a suitable animal feed not only as an energy source but also for its antioxidant properties. The antioxidant properties of hempseed oil may be caused by presence of cannabinoids, tocopherols, and phenolic compounds, mainly flavonoids that are contained in greater quantities compared with other phenolic compounds. The oil obtained from Finola seeds contained significant amounts of polyphenols, especially flavonoids such as flavanones, flavanols, flavonols, and isoflavones (Smeriglio et al., 2016). Flavonoids are effective antioxidants because of their free radical scavenging and metal ion chelating properties (Kandaswami & Middleton, 1994), with protective role against free oxygen radicals and lipid peroxidation (Smeriglio et al., 2016). Furthermore, the hemp oil has an intensive green color due to chlorophyll content (Aladić et al., 2014). Chlorophylls are fat soluble pigments, having powerful prooxidant properties that may lower oil quality by oxidation processes (Liang, Appukuttan Aachary, & Thiyam-Holländer, 2015). Thus the content of chlorophyll a and chlorophyll b, as well as total carotene content, was determined in oil. The hemp oil obtained by supercritical CO_2 has three times higher chlorophyll content and four times higher total carotenoid content compared to cold press oil (Aladić et al., 2014). Teh and Birch (2013) published that chlorophyll content as mg of pheophytin a/kg of cold-pressed hemp oil was 75.21. Liang et al. (2015) reported the total chlorophyll content of cold-pressed hempseed oil to be 98.6 mg/kg (59.22 mg/kg chlorophyll *a* and 39.45 mg/kg chlorophyll *b*), whereas the supercritical fluid extracted hempseed oil had twice the amount of total chlorophyll content of 228.79 mg/kg (193.50 mg/kg chlorophyll *a* and 35.39 mg/kg chlorophyll *b*).

Industrial hemp, like other crops, contains anti-nutritional compounds that need to be considered when administering hemp-based feeds, due to their ability to reduce the nutrient bioavailability, digestibility of proteins, and mineral absorption. In particular, phytic acid (inositol hexaphosphate) content in the hempseeds (i.e., cakes) may be over 5% (Russo & Reggiani, 2015). Phytic acid is the main organic form of phosphorus present in plant seeds. Its presence reduces protein digestibility and increases the excretion of endogenous nitrogen, amino acids, and minerals, in particular bivalent cations (Cowieson, Acamovic, & Bedford, 2004). The approximate content of other anti-nutritional compounds was found to be 69.4 g/kg for phytic acid, 3.50 g/kg for condensed tannins, 0.11 g/kg for cyanogenic glycosides, 22.7 unit/mg for trypsin inhibitors, and 0.59 g/kg for saponins determined for six hemp varieties (Russo & Reggiani, 2015).

For the most part, the feed quality is determined by the quality of the plant material constituting animal feed. The quality of the plant material and its seeds (i.e., expellers) are determined by many factors, such as applied agricultural technology during plant cultivation, soil quality, climate and weather during cultivation and harvest, seed treatment, storage conditions, and seed processing conditions. These factors affect the feedstock in terms of the composition of basic nutrients and the composition of functional and/or anti-nutritional compounds (Stastnik et al., 2020).

13.3 The technological procedures for conversion of industrial hemp material into animal feed

13.3.1 Milling

The technology for obtaining edible hemp oil results in the generation of hempseed cakes that can be further utilized in both human and animal diets. Briefly, the mechanical pressing of hempseeds to extract oil results in generation of a by-product—defatted hempseed cake, which can be further grinded to obtain the desired particle size. The milled powder is sifted using various screens to obtain products marketed as high-value protein powders. The by-product that does not pass through the sieves is the hempseed protein meal, which is normally considered a waste product (Girgih, Alashi, He, Malomo, & Aluko, 2014).

Generally, the seed cakes (i.e., expellers or pomace) are protein feed with residual 10% ether extract (oil, fat) and relatively high level of crude fiber. Since it is created as a by-product, hempseed cake represents more economical and renewable source of proteins (Hessle, Eriksson, Nadeau, Turner, & Johansson, 2008). Fig. 13.2 shows the change of nutrient content of hempseeds before and after processing by oil expelling. Obtained seed cakes should be further subjected to grinding/milling (via hammer mill or a roller mill) before mixing into feed mixture.

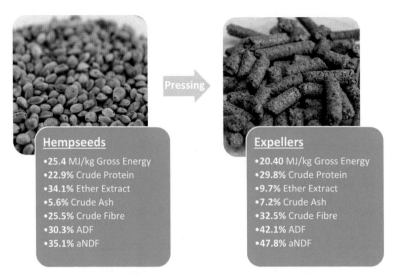

FIG. 13.2 Hempseeds and hempseed expellers of Carmagnola variety and appropriate nutrients content in dry matter. *Photos from author's archive; adapted from Šťastník, O. (2018). The use of milk thistle and hemp in poultry nutrition (Doctoral dissertation (in Czech)). Mendel University in Brno, Czech Republic. Retrieved from https://theses.cz/id/m4p13r/; Stastnik, O., Pavlata, L., & Mrkvicova, E. (2020). The milk thistle seed cakes and hempseed cakes are potential feed for poultry. Animals, 10(8), 1384; unpublished data.*

13.3.2 The pelleting process

Production of pellets typically consists of drying, milling, conditioning, pelleting, and cooling process. To prepare the feeds for pelleting, small size particles are required obtained by milling. For optimizing the milling process, feeds can be dried before milling (Nielsen, Mandø, & Rosenørn, 2020). Hammer mill with appropriate screen opening or a roller mill with suitable roller gap is commonly used for milling. After milling, the feed material is conditioned by hot water or steam, to increase the moisture content of the feeds and to preheat the material up to 60–70°C before pelleting. After conditioning, the feeds are prepared for pelleting. Ring die pellet mills are commonly used for producing pellets. In the pellet mill (Fig. 13.3), feeds are continuously distributed on the inner die surface in front of each roller. The die is equipped with press channels, in which the milled feeds are compressed into pellets. The die rotates, and friction between the die, feeds, and rollers causes the rollers to rotate. A narrow gap between a die and a roller causes the compression of feed material and presses it into press channels to the die. Each time a press channel passes a roller, feed is compressed and pressed into the channel. As material is pressed into the channels, cylinders of compressed material are extruded from the outside of the die, where blades cut them into pellets (Nielsen et al., 2020). The temperature of the pelleting process varies according to the ring die and type of pellet mill and pelleted material. The process temperature can range from 60°C to 100°C with 12%–18% moisture in pelleted material.

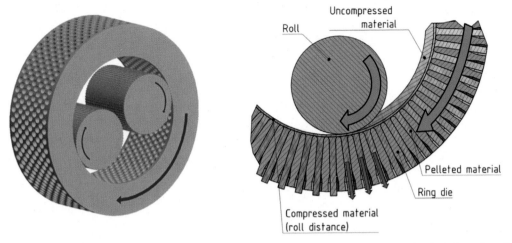

FIG. 13.3 Pellet press ring die—rotating ring die with two rollers (left); cross-section of a ring die and a roller (right). *Taken from Nielsen, S. K., Mandø, M., & Rosenørn, A. B. (2020). Review of die design and process parameters in the biomass pelleting process. Powder Technology, 364, 971–985, with permission.*

13.4 Utilization of industrial hemp by-products as animal feed

Whole hempseeds or hempseeds in combination with hemp oil or hemp by-products (such as seed cakes) have been included into feed mixtures for animals (Eriksson & Wall, 2012; Halle & Schöne, 2013; Khan, Durrani, Chand, & Anwar, 2010; Neijat, Gakhar, Neufeld, & House, 2014). More detailed informations about use industrial hemp products in animal feed brings (Table 13.3). In the experiment of Halle and Schöne (2013), hempseed cakes were used which, according to the analysis, contained (in 91% dry matter) 281 g/kg crude protein, 110 g/kg of ether extract, 447 g/kg of neutral detergent fiber (aNDF), 304 g/kg of acid detergent fiber (ADF), and 117 g/kg of lignin. Serrapica et al. (2019) determined a similar nutrient content of hempseed expellers. Šťastník et al. (2019) reported cannabidiol content of 170 mg/kg, whereas Halle and Schöne (2013) reported the active substances content of THC and CBD below the detection limit of 0.005%. The mineral content of hempseed cakes published by Halle and Schöne (2013) was 28 g/kg calcium and 148 g/kg of phosphorus. An overview of the hempseed and hempseed cake nutrient composition is presented in Table 13.2. The cannabinoid concentration of industrial hemp plant and by-products is presented in Table 13.4.

The diverse physiological effects of cannabinoids reported are due to the existence of specific receptors distributed in some organs and systems of the animals and human body (Alvarado, del Campo Sánchez, & Salcedo, 2017). The hypothesis states that exogenous cannabinoids and the *endo*-cannabinoid system (ECS) increase feed intake and improve weight gain in animals by activating central cannabinoid receptors (CB1) (Di Marzo, Piscitelli, & Mechoulam, 2011). In addition, ECS activation has been found in obese people (Engeli et al., 2005), which is related to adipogenesis, lipogenesis, hepatic steatosis, and increased insulin resistance (Di Marzo et al., 2011). However, a mouse model has shown that a CB1 receptor antagonist is able to effectively reduce weight and thereby reduce possible metabolic risk factors (Tam et al., 2010).

TABLE 13.3 Industrial hemp products in animal feed.

Type of animal	Type of hemp-based feed	Level of inclusion in a diet	The size of the population tested	The duration of feeding experiment	Observations	References
Lactating sows	Hempseed	2% and 5%	40	Last 10 days of gestation	Hempseeds improved the overall antioxidant status of the lactating sows and their progeny	Palade et al. (2019)
Fattened pigs	Hemp oil	3.4 g of C18:3 n-3 (α-linoleic acid)/kg of feed	36	50–105 kg of live weight	Hemp oil improved quantity of α-linoleic acid in the meat	Mourot and Guillevic (2015)
Lohmann white laying hens	Hemp oil or hemp omega	4% and 8%	40	6 weeks	Performance was not affected; eggs and meat were enriched with n-3 PUFAs and gamma-linolenic acid	Jing et al. (2017)
Ross 308	Hemp oil or hemp omega	3% and 6%	150 (75 male, 75 female)	21 days		
Lohmann LSL-Classic laying hens	Hempseeds and hemp oil	10%, 20%, or 30% and 4.5% or 9%, respectively	48	12 weeks	Feed intake, weight gain, laying, and egg weight were not affected	Neijat et al. (2014)
Ross 308	Hempseed expellers	5% and 15%	150	25 days	Hempseed expellers appear to affect the color and odor of broiler chicken's meat	Šťastník et al. (2019)
Swedish Red dairy cows	Hempseed cakes	0, 143, 233, or 318 g/kg of dry matter	40	5 weeks	The increasing hempseed cake levels resulted in significant effects on the milk yields and energy-corrected milk	Karlsson et al. (2010)
Alpine dairy goats	Hempseed	6.4%	18	100 and 150 days in milk	Increased availability of iron in blood	Reggiani and Russo (2016)
Carpathian goats	Hempseed oil	4.79% of dry matter (e.g., 93 g/day)	10	31 days	Modified the goat milk's fatty acid profile with a higher ratio of PUFAs on SFAs without affecting liver function	Cozma et al. (2015)
Californian rabbits	Hemp herbal biomass	10%	20	91 days	Positive effect of dietary administered hemp herbal biomass on the rabbit health was not confirmed	Horakova et al. (2020)
Common carp (Cyprinus carpio L.)	Hempseed cakes	5%, 10%, 15%	200	64 days	Diet with 10% of hempseed cakes exhibited a positive effect on feed conversion ratio and growth rate	Maly et al. (2018)

TABLE 13.4 The cannabinoid concentration of industrial hemp plant components.

Cannabinoid	Whole plant	Leaves	Stalks	Hemp flower	Seed heads	Cleanings	Extracted flower
Cannabinol (μg/g)	9	31	4	27	11	7	21
Δ^9-Tetrahydrocannabinol (μg/g)	186	573	31	664	275	158	301
Δ^9-Tetrahydrocannabinolic acid A (μg/g)	626	4609	119	3379	1228	458	16
Δ^8-Tetrahydrocannabinol (μg/g)	ND	ND	ND	ND	ND	ND	ND
Cannabichromene (μg/g)	192	417	49	513	68	140	ND
Cannabidiol (μg/g)	721	3347	132	3509	262	463	8062
Tetrahydrocannabivarin (μg/g)	30	2	ND	1	303	2	ND
Cannabidiolic acid (μg/g)	4870	36,920	1705	32,900	3184	5309	1960
Cannabigerolic acid (μg/g)	519	1788	362	1938	285	654	154
Cannabichromenic acid (μg/g)	851	4041	500	2916	411	663	ND
Cannabigerol (μg/g)	67	293	28	230	23	79	ND

Adapted from Kleinhenz, M. D., Magnin, G., Lin, Z., Griffin, J., Kleinhenz, K. E., Montgomery, S., & Coetzee, J. F. (2020). Plasma concentrations of eleven cannabinoids in cattle following oral administration of industrial hemp (Cannabis sativa). Scientific Reports, 10.

13.4.1 Hemp in pig nutrition

There are very few available researches in the application of industrial hemp products in pig nutrition. One of them is the study of Palade et al. (2019), which emphasized the potential of hempseed in sow's nutrition. The diets containing 2% and 5% of hempseed improved the overall antioxidant status of the lactating sows and their progeny. Moreover, Mourot and Guillevic (2015) carried out research to diversify the sources of n-3 fatty acid-rich lipids for animal feed. In this study, 36 fattened pigs (between 50 and 105 kg of live weight) received isolipidic diets containing palm oil, rapeseed oil, or hemp oil (providing 0.6, 1.9, and 3.4 g/kg of feed, respectively, of C18:3 n-3 (α-linolenic acid, ALA). The quantity of ALA deposited in the meat was statistically higher in the group of pigs supplemented with hemp oil. Based on these results, hemp oil may be an interesting source of ALA to improve the nutritional quality of pork.

13.4.2 Hemp as poultry feed

The effects of hempseed supplementation in broiler and laying hens were assessed by meat and egg quality, respectively (Konca, Yuksel, Yalcin, Beyzi, & Kaliber, 2019; Vispute et al., 2019). Hempseeds did not negatively affect the growth performance of broilers, but positively influenced the serum lipid profile which was reflected in a reduction of triglycerides, low-density lipoproteins (LDL), and total cholesterol. The improved gut health due to reducing the number of coliform bacteria was also reported (Vispute et al., 2019). In laying hens, both raw and heat-treated hempseeds improved the egg quality, egg fatty acid profile, and hen's performance values. Moreover, the heat treatment modified the flavor of hempseeds restoring the feed consumption at the control group (Konca et al., 2019). The improvement of poultry products quality (i.e., egg and meat) was also confirmed in two separate experiments by Jing, Zhao, and House (2017). The authors tested the efficacy of inclusion of hemp oil and commercial feed product containing hemp oil in poultry diets on the production performance and fatty acid profile of egg yolks and muscle tissues. The results showed that inclusion of hemp oil up to 8% in layer diets and up to 6% in broiler diets provided either by hemp oil or commercial hemp oil-based feed did not negatively affect the overall performance of birds, but resulted in the enrichment of n-3 PUFAs and gamma-linolenic acid in eggs and meat.

From the fattening point of view, the live weight of chickens decreased with the increasing content of hempseed cake in the diet (Stastnik et al., 2015; Šťastník et al., 2019). It is important to realize that seed expellers contain approximately 35% of crude fiber. High levels of fiber in poultry nutrition cause digestive disorders and consequently worsen poultry performance. For this reason, seed expellers must be limited in feed for chickens and hens. Halle and Schöne (2013) confirmed the decreasing trend of chicken's live weight with increasing crude fiber content in diets. Only few publications which recommended the proportion of crude fiber in mixtures for poultry can be found. Crude fiber at the level of 1%–4% for young poultry is recommend by González-Alvarado, Jiménez-Moreno, Lázaro, and Mateos (2007) and Mateos, Jiménez-Moreno, Serrano, and Lázaro (2012) and up to 7% for laying hen's diet (Albiker, Bieler, & Zweifel, 2015). The high content of predominantly soluble fibers in poultry diet can cause deterioration of performance parameters, due to slower passage of feed through the digestive tract and reduced digestibility of nutrients (Honzík, 2015). The hempseed cakes (like all seed cakes) contain high amount of crude fiber (see Table 13.2), which should be taken into account when including seed cakes into the feed mixtures for young poultry and pigs. Higher proportions of seed cakes in diets may reduce the growth and performance parameters of animals mainly due to the crude fiber content. When 5% of hempseed cake was included in the broiler chicken diet, the fiber content of feed mixture was 3.92%. With the inclusion of 15% seed cakes, the fiber content was 5.98%. When 40% of the hempseed cake was included, the fiber content in the broiler chicken's diet was 11.86%. According to our results (Šťastník, 2018), the inclusion of 5% of seed cakes in the diet worsened the growth parameters of broiler chickens. In this regard, EFSA (2011) recommended the maximum hempseed expellers incorporation to the feed mixtures in dose of 3% for fattening poultry, 5%–7% for laying poultry, 2%–5% hempseed and hempseed cake in pig's diets, 5% hempseed cake for ruminants, and 5% hempseed for fish diets.

On the other side, the inclusion of hempseed cakes (25 g/kg) and hemp herbal biomass (10 g/kg) in broiler chicken's diet had no effect on mean live weight and carcass yield. Moreover, experimental diets had no influence on the number of gut microorganisms—*Escherichia coli*, *Lactobacillus* spp., and *Enterococcus* spp. (Stastnik et al., 2016). Eriksson and Wall (2012) in an experiment with Ross 308 in organic mode found a significantly higher chicken's live weight (1194 g) in experimental group fed with 200 g/kg of hempseed cakes compared to the control group fed without hempseed expellers (1071 g) at 35 days of age. According to Di Marzo (2008), exogenous cannabinoids (ingested through a diet containing hemp products) and the endocannabinoid system increased feed intake and improved weight gain of animals by activating central cannabinoid receptors (CB1 receptor).

The inclusion of different proportions of hempseeds (up to 30% in the diet) and hemp oil (up to 9% in the diet) did not affect the performance of laying hens (feed intake, weight gain, laying intensity, and egg weight) (Neijat et al., 2014). These results are consistent with the previous studies on the use of hemp products in laying hen nutrition (Gakhar, Goldberg, Jing, Gibson, & House, 2012; Silversides & Lefrancois, 2005). Neijat et al. (2014) also found significantly lower aspartate aminotransferase (AST) and gamma-glutamyl transferase (GGT) enzyme activities in laying hens fed by feed containing 10% and 20% of hempseeds compared to the control group. The enzymes GGT and AST are considered physiological indicators of liver health. High gamma-glutamyltransferase enzyme activity in avian blood plasma is associated with hyperplasia and bile duct tumors (Harr, 2002). The creatine kinase and electrolytes in the blood plasma were not affected by the experimental treatment. It might appear that the inclusion of hempseeds in a diet had a positive effect on the health of the liver tissue of the laying hens compared to the control group (Neijat et al., 2014). The similar results were reported by Afzali, Barani, and Hosseini Vashan (2015) and Barani, Afzali, and Hosseini Vashan (2015). On the other hand, the increased immunoglobulin G titer was reported for groups of chickens fed with different proportions of hempseeds and hempseed extrudates as compared to the control group (Afzali et al., 2015; Barani et al., 2015). Skřivan, Englmaierová, Taubner, and Skřivanová (2020) reported the improvement of cockerel performance, meat and bone quality, and deposition of alpha tocopherol in the liver due to the dietary supplementation with 40 g/kg hempseed and 60 g/kg extruded flaxseed. These results may indicate the effect of the endocannabinoid system on animal performance, although the study of Neijat et al. (2014) did not state the presence of cannabinoids on hempseeds used in a trial. Therefore the question on the quantity of cannabinoids administered to the animals remained open. On the other hand, the experiment of Halle and Schöne (2013) with laying hens that was fed (inter alia) with hempseed expellers at 5%, 10%, and 15% level demonstrated significantly lower performance (feed intake, egg production, and consumption egg feed) at a level of 15% hempseed expellers in a diet. This reduction in performance parameters could be (i) due to the presence of a low content or absence of cannabinoids in the seed cakes, as emphasized by the authors (THC and CBD were found below the detection limit of 0.005%) or (ii) due to the high content of crude fiber in hen's diets (in connection with the increasing content of seed cakes in the diets) which may deteriorate performance parameters, as stated before. Unfortunately, the authors did not state the crude fiber content in experimental diets. Further increase of the content of seed cakes in the diets led to a reduction in the proportion of yolk and an increase in the proportion of egg white (Halle & Schöne, 2013), for which the mentioned proteins and the quality of their amino acids could be responsible.

According to Irakli et al. (2019), the seven cultivars of hempseeds contained three main carotenoids known as lutein, zeaxanthin, and β-carotene. Lutein was the primary carotenoid ranging from 1.5 (Futura) to 3.4 mg/100 g (Santhica), with a mean value of 2.1 mg/100 g for all cultivars, while zeaxanthin content ranged from 0.2 (Tygra, Finola, and Fedora) to 0.5 mg/100 g (Futura) and β-carotene content from 0.2 (Bialo, Futura, and Finola) to 0.8 mg/100 g (Tygra). The expellers from Carmagnola variety contained 18.72 mg/kg of β-carotene, while lutein content was not determined (Šťastník, 2018; Stastnik et al., 2020). The total carotenoid content reported by Aladić et al. (2014) was 3.14 mg/100 g and 12.50 mg/100 g in hempseed oil (hemp cultivar was not specified) produced by a cold-pressing process and supercritical fluid extraction, respectively. It is therefore apparent that carotenoids are transferred from seeds to the oil. So, if the main carotenoid in hempseed is lutein and the major carotenoid of egg yolk is also lutein, it is appropriate to use hemp oil (or raw hempseeds) in layer diets not only as energy source but also to intensify the color of the egg yolk. Zaheer (2017) stated that egg yolk is highly available and cheap source of lutein and zeaxanthin. However, the profile of carotenoids in egg yolk is highly dependent on the layer diet and therefore can be manipulated by the selection of ingredients in poultry diet.

Yalcin, Konca, and Durmuscelebi (2018) carried out two separate experiments with Japanese quail (*Coturnix coturnix japonica*), conducted to investigate the effects of hempseed on meat quality (experimental diets were administered for 5 weeks) and egg fatty acid (FA) composition (experimental diets were administered for 6 weeks). In both experiments, groups were the same: control (without hempseeds), and with 5%, 10%, and 20% hempseeds in diets. It was found that breast meat cooking loss was significantly lower in a group supplemented with 20% of hempseeds. Hempseed inclusion to diets caused a linear and cubic increase in redness of hip meat. Palmitoleic and oleic FAs decreased in breast meat with hempseed addition, whereas the content of linoleic and linolenic acid contents linearly increased with the increasing dietary hempseed ratio. Hempseed addition yielded lower palmitoleic and oleic FAs in quail eggs. Unfortunately, the authors did not monitored whether the inclusion of hempseeds in a diet affected the yolk color.

Some researchers examined the administration of the extracts of hemp bioactive substances to animals. In this regard, Konieczka et al. (2020) studied the activity of supercritical CO_2 CBD extract and selenium nanoparticles (Nano-Se) in modulating the broiler chickens response to *Clostridium perfringens* infection. Although the infected chickens exhibited no clinical manifestations, a potential hazard remained due to possible pathogen transfer to the food chain.

It was shown that both CBD extract, containing 12% CBD, 0.49% tetrahydrocannabinol, and 0.38% tetrahydrocannabinolic acid, and Nano-Se affected the beneficial responses of chickens to *C. perfringens* infection, manifested by the upregulated expression of genes determining gut barrier function. Likewise, both CBD extract and Nano-Se promoted shifts in gut bacterial enzyme activity to increased energy uptake in infected broilers and upregulated potential collagenase activity. No opposite effect of CBD extract and Nano-Se in mediating the host response to infection was observed, whereas an additive effect was evidenced on the upregulation of gene determining gut integrity. The authors suggested that understanding the action mechanisms of CBD extract and Nano-Se is of great interest for developing a preventive strategy for *C. perfringens* infection in broilers.

13.4.3 Hemp as ruminant feed

Studies about the use of hempseeds in animal nutrition were mainly focused on the quantity and quality of animal production improvement. Ruminants may utilize industrial hemp by-products as they can digest cellulose plant materials in their rumens. The whole hemp plant (including stalk and leaves) could be (due to its high fiber content) a suitable feedstuff for ruminants (and horses), and daily amounts of 0.5–1.5 kg whole hemp plant dry matter could likely be incorporated in the daily ration of dairy cows (EFSA, 2011). For now, the question remains what role may have rumen microflora to metabolism and degradation of cannabinoids administered to ruminants either by direct exposure through specific cannabinoids or hemp (by-)products. Cannabidiolic acid (CBDA) is a precursor of cannabidiol (CBD) (Wang et al., 2016) and present in relatively high concentrations ($7.389\% \pm 3.404$ in Carmagnola variety) in industrial hemp (Glivar et al., 2020). Cannabidiolic acid is converted to CBD via a decarboxylation reaction (Wang et al., 2016). In cattle, CBDA is absorbed from the rumen, where rumen microbes could potentially degrade or metabolize cannabinoids causing alterations in the cannabinoids available for absorption and they could play a significant role in the conversion of fatty acids through biohydrogenation (Kleinhenz et al., 2020). In this regard, Kleinhenz et al. (2020) studied the plasma pharmacokinetics of cannabinoids and their metabolites in cattle (castrated male Holstein calves) after a single oral exposure to industrial hemp flowers. All calves received a single oral dose of 35 g of industrial hemp flowers to achieve a target dose of 5.4 mg/kg cannabidiolic acid (CBDA). The cannabinoids CBDA, tetrahydrocannabinolic acid-A (THCA-A), cannabidivarinic acid (CBDVA), and cannabichromenic acid (CBCA) were detected in all blood samples of calves after industrial hemp dosing. The maximum concentration of CBDA (72.7 ng/mL) was found at 14 h after hemp administration. The half-life of CBDA was 14.1 h. These results showed that acidic cannabinoids (especially CBDA) are readily absorbed from the rumen and available for distribution throughout the body. Similarly to this, high levels of THC metabolites were found in milk of buffaloes grazed at natural pastures in the north of Pakistan, where *Cannabis sativa* L. with high levels of THC spontaneously grows. However, the consumption of such milk, especially by the children, may cause a risk associated with the possible exposure to THC (Ahmad & Ahmad, 1990).

13.4.3.1 Cattle

In feedlot cattle (60 individually penned steers), different amounts of hempseeds (0%, 9%, or 14%) included in the diet did not show any negative effects on either the matter intake or rate or efficiency of weight gain. The positive effects on beef tissue quality related to the relatively higher content of conjugated linoleic acids (CLA) were recorded. However, the increase in saturated fatty acids (SFAs) was also reported (Gibb, Shah, Mir, & McAllister, 2005). Another trial conducted with 16 Swedish Red breed steers, whose diet included hempseed cakes (0.2 and 1.4 kg as fed), showed the increased ratio of polyunsaturated to monounsaturated fatty acid (PUFAs/MUFAs) in both fresh and cooked meat from bovine *M. longissimus dorsi* in comparison to a diet containing soybean meal (Turner, Hessle, Lundström, & Pickova, 2008).

Karlsson, Finell, and Martinsson (2010) evaluated the effects of the increasing proportion of hempseed cakes in the diet of dairy lactating cows on milk production and composition. Four experimental diets (based on a ratio of 494:506 g/kg of dry matter between silage and concentrate mixture) were formulated to contain the increasing concentrations of hempseed cakes: 0, 143, 233, or 318 g/kg of dry matter. No effects in dry matter intake were observed, but significant linear increases in crude protein, fat, and NDF intakes were observed with the increasing proportion of hempseed cakes in the diets. The increasing hempseed cake levels resulted in significant quadratic effects on the milk yields and energy-corrected milk, with the highest value for the group supplemented with 143 g/kg hempseed cake. The milk protein and fat percentage decreased linearly ($P < .05$) with the increasing level of seed cakes in the diet. Furthermore, there was a significant ($P < .001$) linear increase in milk urea concentrations with the enhancement of expellers inclusion due to the increase of crude protein intakes. A linear decrease in crude protein efficiency (milk protein yield/crude protein intake) was also observed.

13.4.3.2 Sheep

Mustafa, McKinnon, and Christensen (1999) carried out a digestibility experiment with sheep using hemp meal (5.2% of lipids on dry matter basis) at different amounts of inclusion (0, 50, 100, 150, 200 g/kg of dry matter) as a replacement of canola meal. The authors maintained isonitrogenous diets between experimental groups, based on barley. Sheep voluntary dry matter intake was not affected by the hemp meal inclusion levels. Total tract dry matter and organic matter digestibility values were similar across experimental groups, suggesting that digestibility of hemp meal is equal to that of canola meal. Based on the obtained results, it was suggested that hemp meal can be used up to 20% on dry matter with no negative effects on nutrient utilization by sheep (Mustafa et al., 1999). Mierliță (2016) carried out an experiment using Turcana ewes ($n = 30$) divided into three groups consisting of a control diet based on hay and supplemented with mixed concentrates and two experimental diets designed to provide the same amount of fat using hempseeds (180 g/day) or hempseed cake (480 g/day). The three diets were isoenergetic and isonitrogenous, and the two diets with hempseeds/hempseed cake had the same amounts of PUFA. Diets supplemented with hempseeds and hempseed cakes caused the increase in milk yield and milk fat content, but decrease in milk lactose. The hemp feeding increased the PUFA content (especially n-3 fatty acids) in ewes' milk and improved the n-6/n-3 ratio. Total CLA content doubled in the milk of the ewes that received hempseed and increased by 2.4 times with the hemp cake inclusion. The addition of 180 g/day hempseeds in Turcana dairy sheep diet increased the quality of milk for human consumption. Therefore the pasture grazing combined with hempseeds exhibited a synergistic effect on milk production in ewes and milk fat without altering the milk protein content (Mierliță, 2016).

13.4.3.3 Goats

Reggiani and Russo (2016) observed that the replacement of 6.4% (on dry matter basis) of corn and soybean with hemp or flax seeds (diets were isonitrogenous) had an effect on the increase of iron content in blood of Alpine lactating goats ($n = 18$). These results are surprising given the high phytate content of seeds of these plants (Russo & Reggiani, 2014, 2015). The authors speculated that some substances (i.e., inulin) present in hemp or flax seeds can stimulate the absorption of iron (Reggiani & Russo, 2016). A possible cause of the iron increase in goats' blood may be due to the higher content of Fe in hempseed. In agreement with these studies, by utilization of hempseed oil as a dietary supplement, it was possible to modify the goat milk's fatty acid profile with a higher PUFA/SFA ratio without affecting liver function (Cozma et al., 2015).

By enriching lactating ewes diet with 5% hempseeds (on dry matter basis) for 35 days, significant increase in milk lactose was observed (from 4.69% to 5.84%), but without changes in total fat, proteins, caseins, and urea. Consequently, no changes in total fat, proteins, or ash were detected in the derived cheeses. The higher lactose concentration of milk used for cheese making affected the metabolism of bacteria present in the cheese matrix, which were responsible for the biochemical processes that affected cheese sensory properties during ripening (Ianni et al., 2020). Luo et al. (2018) found an increased concentration of fat and lactose in colostrum and milk obtained from dairy ewes in early lactation that were supplemented with soy isoflavones. They demonstrated a role of isoflavones in the increased concentration of serum prolactin, which improves mammary gland function. From this point of view, it seems that isoflavones (daidzein and genistein) in cows may be responsible for increased proliferation and development of mammary epithelial cells, milk fat and protein, and milk yield (Liu et al., 2013; Liu, Li, & Feng, 2012). As mentioned earlier isoflavones are commonly present in hempseeds (Smeriglio et al., 2016). On the other side, the lactose concentration in milk may be increased by feeding a diet rich in PUFAs to dairy ewes (Hamer, Coleman, & Relling, 2018).

The major greenhouse gases released from the livestock sector are carbon dioxide (CO_2), methane (CH_4), and nitrous oxide (N_2O). Methane emission originating from the enteric fermentation of ruminant livestock is associated with the highest effect on global warming. Methane emission reduces feed efficiency and is associated with dietary energy loss. Dietary modification is directly linked to changes in the rumen fermentation pattern and types of end products. Studies showed that changing fermentation pattern is one of the most effective ways of methane mitigation. Desirable dietary changes provide twofold benefits—improve production and reduce greenhouse gas emissions (Haque, 2018). In this regard, Wang, Kreuzer, Braun, and Schwarm (2017) demonstrated the potential of oilseeds (i.e., safflower (*Carthamus tinctorius*), poppy (*Papaver somniferum*), hemp (*Cannabis sativa*), and camelina (*Camelina sativa*)) used as feeds that may be used to mitigate CH_4 from ruminant fermentation and hence decrease the carbon footprint. It was shown that hempseeds were the most effective in suppressing the methane yield (mL/g DM, up to 21%) in relation to other tested oilseeds.

All obtained information encourages the use of hempseeds and by-products as a component of feed in animal diets. Despite the positive effects on the quality of products, transcriptomic data related to the hempseed supplementation is lacking. In support of this, the molecular mechanisms, signaling pathways, and metabolic processes behind the hempseed supplementation in ewes were studied (Iannaccone et al., 2019). The transcriptome signature induced by 5% hempseed supplementation of ewes was reported (Iannaccone et al., 2019). It was found that hempseed supplementation influenced the expression of 314 genes, of which 271 were up-regulated and related to complexes and ATP synthase of the electron transport chain that is associated with thermogenesis. Since the perinatal mortality caused by hypothermia is one of the major concerns in the sheep agriculture, the better adaptation to cold stress and increase of cold tolerance with hempseed supplementation is of great practical importance. Moreover, hempseed supplementation affected glycolysis and galactose metabolism pathways, whereby the phosphorylation of galactose is associated with the higher milk lactose content, making the dairy production more profitable.

13.4.4 Hemp as other animals feed (rabbits, fish, pets)

Horakova, Stastnik, Pavlata, and Mrkvicova (2020) studied the dietary effects of hemp herb inclusion in Californian rabbits' diet. The hemp herbal biomass of Carmagnola variety contained (in dry matter) 18.5% of crude protein, 28.9% crude fiber, 6.9% ether extract, and 16.4% of crude ash. The cannabidiol content (CBD) was 0.15%, while THC and cannabinol (CBN) content were below the limit of detection. At the 151 days of rabbits age, no statistically significant differences in live weight between control group and experimental group were found. Final average live weight of rabbits in experimental and control groups were 3.33 and 3.46 kg, respectively. Since no differences were found in feed intake, feed conversion ratio, and mean weight gain, a positive effect of dietary administered hemp herbal biomass on the rabbit health was not confirmed.

In a study of Maly, Mares, Palisek, Sorf, and Postulkova (2018), the effect of hemp by-product addition in common carp (*Cyprinus carpio* L.) diet was studied. The experimental diets were enriched with the hempseed cake of the Finola variety at the proportions of 5%, 10%, and 15% and compared to the commercial feed mixture as a control. The experimental feed mixtures were supplemented with rapeseed oil to achieve isoenergetic diet. The diet containing 10% of hempseed cake exhibited a positive effect on feed conversion ratio and growth rate. Moreover, the diet with hempseed cakes altered hematological and biochemical indicators, such as the hematocrit value, mean corpuscular hemoglobin concentration, and glucose content. The mean corpuscular hemoglobin concentration and glucose content in the blood of fish decreased with the increasing amount of hempseed cakes in diets. In terms of FA, no significant effect on the FA profile was found. The presumed improvement of the n-6 to n-3 ratio in favor of n-3 acids was not observed, so that the nutrient quality of the fish meat was not improved (Maly et al., 2018). The FA profile of hempseed cakes can be influenced by hempseed processing and the presence of residual fat (approximately 10% fat content). Moreover, the FA profile could be also influenced by a chosen industrial hemp variety.

Kogan et al. (2016) conducted a study to find out the consumers' perceptions of hemp products administered to their pets (i.e., dogs, cats). The survey was conducted with an aim to determine which hemp products (Cannabis (THC > 0.3%), hemp isolate (THC < 0.3%), CBD/hemp broad or full spectrum) pet owners are purchasing, what are the reasons for their purchases, and to perceive the value of these products on pets' health. The majority of total responses (58.8%) indicated the favoring of hemp products for dogs, especially for the treatment of seizures, cancer, anxiety, and arthritis (77.6%). Fewer participants indicated the current use of hemp products for cats (11.93%), mainly for the treatment of cancer, anxiety, and arthritis. The effectiveness of hemp products for pets was rated as moderately or very helpful, especially for pain relief, sleeping disorders, reduction of inflammation, and anxiety relief. The most frequently reported side effects were sedation and over-active appetite. The fact that pet owners turned to hemp for the treatment of medical conditions may suggest that, similar to human medicine, many are not satisfied with more conventional modes of health care. The similar study was conducted to explore the motivations and expectations of using hemp products to treat chronic pain in humans and dogs (Wallace, Kogan, Carr, & Hellyer, 2020). Human patients ($n = 313$) and dog owners ($n = 204$) reported similar motivations for using hemp products to treat chronic pain, emphasizing the product naturalness, their preference over conventional medication, and their belief in the effectiveness of treatment for chronic pain. Similar proportions of human patients and dog owners reported that the use of hemp products fulfilled their expectations (86% vs 82%, respectively). It was shown that the administration of hemp products helped in reducing pain; increasing relaxation; and improving sleep, coping, functionality, and overall well-being (Wallace et al., 2020).

Although cannabidiol (CBD)-rich hemp extracts are increasingly used in veterinary medicine, little information is available on the amount of serum cannabinoids. Therefore a study that examined pharmacokinetics of CBD, CBDA, THC, and THCA in oral forms of CBD-rich hemp extracts administered to dogs was performed (Wakshlag et al., 2020). In addition, the metabolized psychoactive component of THC which is 11-hydroxy-19-tetrahydrocannabinol (11-OH-THC) and CBD metabolites which are 7-hydroxycannabidiol (7-OH-CBD) and 7-nor-7-carboxycannabidiol (7-COOH-CBD) were assessed to better understand the pharmacokinetic differences between different extract formulations. The results of this study suggested different absorption or elimination of CBDA and THCA in comparison with CBD or THC, respectively, emphasizing the role of sunflower lecithin used as a base in providing the superior absorption and/or retention of CBDA and THCA. This study showed that CBDA and THCA are readily absorbed and retained in dogs with some differences observed in CBDA absorption and/or retention depending on the medium used to deliver the oral treatment. The finding of mid-dosing concentrations of 75 ng/mL of CBD and CBDA or greater suggested potential for therapeutic use when it was delivered at 1 mg/kg of body weight with feed. A more interesting finding is the retention of THCA in the serum between 10 and 25 ng/mL. The exact functions of CBDA and THCA physiologically suggest similar therapeutic benefits like CBD. Moreover, CBDA and THCA may have the potential to work synergistically with CBD (Wakshlag et al., 2020). These synergistic properties known as the "entourage effect" are currently thought to be the primary reason that lower CBD content in whole hemp extract dosing can be therapeutic when compared to purified CBD (Huntsman et al., 2019; Pamplona, da Silva, & Coan, 2018; Wakshlag et al., 2020). Another study carried out by Deabold, Schwark, Wolf, and Wakshlag (2019) tried to determine the single-dose oral pharmacokinetics of CBD and to provide a preliminary assessment of safety and adverse effects during 12-week administration using a hemp-based product in healthy dogs and cats. Eight of each species were provided a 2 mg/kg total CBD concentration orally twice daily for 12 weeks with screening of single-dose pharmacokinetics in six of each species. Serum chemistry and complete blood counts results showed no clinically significant alterations. In healthy dogs and cats, an oral CBD-rich hemp supplement administered every 12 h was not detrimental based on CBC or biochemistry values. It was observed that cats absorb or eliminate CBD differently than dogs, showing lower serum concentrations but with side effects manifested through excessive licking and headshaking during oil administration. Based on these and other recent data, CBD-rich products appear to be safe in healthy dogs, while more trials with cats are needed to fully understand cannabinoids' benefits and mechanism of absorption (Deabold et al., 2019).

13.5 Challenges and opportunities

The inclusion of hemp by-products, e.g., hempseed cakes to the nutrition of non-ruminant animals, has several disadvantages. The main one is the high fiber content in seed cakes which reduces the utilization of nutrients. On the other side, the inclusion of hemp oil or hempseeds in poultry and pigs feed appears to be advantageous. The hempseeds or hempseed oil is especially advantageous for laying hens due to lutein content—the main carotenoid present in hemp, but also the main carotenoid in egg yolk. The question remains how much carotenoids can be transferred from hempseed (i.e., oil) to egg yolk. In any case, hemp oil is a good source of carotenoids, which intensify the color of the egg yolk. Moreover, carotenoids ensure the supply of antioxidants to laying hens, but also to the consumers due to eggs consumption. It should be noted that the intensity of the yolk color is highly subjective from consumers' point of view and do not indicate the quality of the eggs. CBD products have been used experimentally and commercially only in the nutrition of pets, where pet owners are willing to pay a higher price. For large-scale utilization of industrial hemp active substances in livestock nutrition, these products must be of reasonable prices.

13.6 Conclusion

The growing global need to find alternative and sustainable protein sources supports research on the unconventional feeds. Moreover, the most common protein sources in animal nutrition are soybean meal and fish meal. These conventional sources are no longer sustainable and will be further limited by increasing prices, so that new and sustainable protein sources for animal feeds are necessary. One of the alternatives may be industrial hemp. In this regard, the hemp varieties allowed for cultivation in Europe need not to exceed 0.2% (i.e., 0.3%) delta-9-tetrahydrocannabinol (THC, in dry matter).

Industrial hemp plant by-products such as leaves, fodder, residual plant fibers, and hemp hurds may be used for animal nutrition purposes. Hempseed and hempseed cake may be used as feedstuffs in nutrition of all animal species. Several specific restrictions (i.e., crude fiber content for poultry or polyunsaturated fatty acids content for pigs) may limit the content of these feeds in the feed mixtures. The maximum incorporation rates to the animal diets may be 3% in fattening poultry, 5%–7% in laying poultry, 2%–5% hempseed and hempseed cake in pig's diets , 5% hempseed cake in ruminants and 5% hempseed in fish diet. The whole hemp plant (including stalk and leaves) would be (due to its high fiber content) a suitable feedstuff for ruminants (and horses), and daily amounts of 0.5–1.5 kg whole hemp plant dry matter could likely be incorporated in the daily ration of dairy cows.

In the light of the earlier mentioned findings, it seems more appropriate to apply the extract of the bioactive substances complex to the poultry diets than addition of expellers or other forms of plants processing. The seed expellers mostly worsened the broiler's performance parameters with higher doses in diets. Therefore it is better to include smaller proportions of these by-products in the diets of non-ruminant animals. However, most of the works using the extracts had positive results to animal's performance. But when applying the extract to the diets it is necessary to consider the higher costs of feed production. According to many studies, industrial hemp (i.e., by-products) supplementation could be a good feeding strategy in ruminant animals to improve the bioactive compounds in milk and dairy products. The content of n-3 fatty acids and isomers of conjugated linoleic acid increased in milk and cheese obtained with hemp addition. Further studies are required to determine the effects of including hemp products in dairy rations. In addition, the nutritional value of milk and dairy products (i.e., fatty acids profile, vitamins, bioactive substances) could be determined to know the possible nutraceutical effects of hempseed products.

The potential of hemp-based diet to mitigate the emission of methane originating from the enteric fermentation of ruminant livestock is of high importance for sustainable livestock production and decrease of carbon footprint.

References

Afzali, N., Barani, M., & Hosseini Vashan, S. J. (2015). In *The effect of different levels of extruded hempseed (Cannabis sativa L.) on performance, plasma lipid profile and immune response of broiler chicks Proceedings of the 20th European symposium on poultry nutrition (ESPN)* (p. 196). Prague, Czech Republic.

Ahmad, G. R., & Ahmad, N. (1990). Passive consumption of marijuana through milk: A low level chronic exposure to delta-9-tetrahydrocannabinol (THC). *Journal of Toxicology. Clinical Toxicology, 28*(2), 255–260.

Aladić, K., Jokić, S., Moslavac, T., Tomas, S., Vidović, S., Vladić, J., et al. (2014). Cold pressing and supercritical CO_2 extraction of hemp (*Cannabis sativa*) seed oil. *Chemical and Biochemical Engineering Quarterly, 28*(4), 481–490.

Albiker, D., Bieler, R., & Zweifel, R. (2015). Crude fibre in layer feed influences performance and plumage of LSL hybrid. In *20th European symposium on poultry nutrition (ESPN)* (pp. 394–396). Prague, Czech Republic.

Alvarado, R. I. N., del Campo Sánchez, R. M., & Salcedo, V. V. (2017). Therapeutic properties of cannabinoid drugs and marijuana in several disorders: A narrative review. *Salud Mental, 40*(3), 111–118.

Andre, C. M., Hausman, J. F., & Guerriero, G. (2016). Cannabis sativa: The plant of the thousand and one molecules. *Frontiers in Plant Science, 7*, 19.

Barani, M., Afzali, N., & Hosseini Vashan, S. J. (2015). The effect of hempseed (*Cannabis sativa* L.) on performance, some blood biochemical parameters and immune response of broiler chickens. In *Proceedings of the 20th European symposium on poultry nutrition (ESPN)* (p. 198). Prague, Czech Republic.

Callaway, J. C. (2004). Hempseed as a nutritional resource: An overview. *Euphytica, 140*(1), 65–72.

Cowieson, A. J., Acamovic, T., & Bedford, M. R. (2004). The effects of phytase and phytic acid on the loss of endogenous amino acids and minerals from broiler chickens. *British Poultry Science, 45*(1), 101–108.

Cozma, A., Andrei, S., Pintea, A., Miere, D., Filip, L., Loghin, F., et al. (2015). Effect of hemp seed oil supplementation on plasma lipid profile, liver function, milk fatty acid, cholesterol, and vitamin A concentrations in Carpathian goats. *Czech Journal of Animal Science, 60*(7), 289–301.

Crippa, J. A. S., Derenusson, G. N., Ferrari, T. B., Wichert-Ana, L., Duran, F. L., Martin-Santos, R., et al. (2011). Neural basis of anxiolytic effects of cannabidiol (CBD) in generalized social anxiety disorder: A preliminary report. *Journal of Psychopharmacology, 25*(1), 121–130.

Deabold, K. A., Schwark, W. S., Wolf, L., & Wakshlag, J. J. (2019). Single-dose pharmacokinetics and preliminary safety assessment with use of CBD-rich hemp nutraceutical in healthy dogs and cats. *Animals, 9*(10), 832.

Di Marzo, V. (2008). The endocannabinoid system in obesity and type 2 diabetes. *Diabetologia, 51*(8), 1356–1367.

Di Marzo, V., Piscitelli, F., & Mechoulam, R. (2011). Cannabinoids and endocannabinoids in metabolic disorders with focus on diabetes. In *Diabetes—Perspectives in drug therapy* (pp. 75–104). Springer.

EFSA Panel on Additives and Products or Substances Used in Animal Feed (FEEDAP). (2011). Scientific opinion on the safety of hemp (Cannabis genus) for use as animal feed. *EFSA Journal, 9*(3), 2011.

Emmert, J. L., & Baker, D. H. (1997). Use of the ideal protein concept for precision formulation of amino acid levels in broiler diets. *Journal of Applied Poultry Research, 6*(4), 462–470.

Engeli, S., Böhnke, J., Feldpausch, M., Gorzelniak, K., Janke, J., Bátkai, S., et al. (2005). Activation of the peripheral endocannabinoid system in human obesity. *Diabetes, 54*(10), 2838–2843.

Eriksson, M., & Wall, H. (2012). Hemp seed cake in organic broiler diets. *Animal Feed Science and Technology, 171*(2–4), 205–213.

European Commission. (2014). Commission Regulation (EC) No. 809/2014 of 17 July 2014 laying down rules for the application of Regulation (EU) No 1306/2013 of the European Parliament and of the Council with regard to the integrated administration and control system, rural development measures and cross compliance. *Official Journal of the European Union*, L227, 69–124.

Gakhar, N., Goldberg, E., & House, J. D. (2010). Safety of industrial hemp as feed ingredient in the diets of laying hens and its impact on their performance. *Journal of Animal Science*, 88, 121.

Gakhar, N., Goldberg, E., Jing, M., Gibson, R., & House, J. D. (2012). Effect of feeding hemp seed and hemp seed oil on laying hen performance and egg yolk fatty acid content: Evidence of their safety and efficacy for laying hen diets. *Poultry Science*, 91(3), 701–711.

Gibb, D. J., Shah, M. A., Mir, P. S., & McAllister, T. A. (2005). Effect of full-fat hemp seed on performance and tissue fatty acids of feedlot cattle. *Canadian Journal of Animal Science*, 85(2), 223–230.

Girgih, A. T., Alashi, A., He, R., Malomo, S., & Aluko, R. E. (2014). Preventive and treatment effects of a hemp seed (*Cannabis sativa* L.) meal protein hydrolysate against high blood pressure in spontaneously hypertensive rats. *European Journal of Nutrition*, 53(5), 1237–1246.

Glivar, T., Eržen, J., Kreft, S., Zagožen, M., Čerenak, A., Čeh, B., et al. (2020). Cannabinoid content in industrial hemp (*Cannabis sativa* L.) varieties grown in Slovenia. *Industrial Crops and Products*, 145, 112082.

Goldberg, E., Ryland, D., Gakhar, N., House, J. D., & Aliani, M. (2010). Sensory characteristics of table eggs from laying hens fed diets containing hemp oil or hemp seed. *Journal of Animal Science*, 88, 99.

González-Alvarado, J. M., Jiménez-Moreno, E., Lázaro, R., & Mateos, G. G. (2007). Effect of type of cereal, heat processing of the cereal, and inclusion of fiber in the diet on productive performance and digestive traits of broilers. *Poultry Science*, 86(8), 1705–1715.

Gunstone, F. D., & Harwood, J. L. (2007). Occurence and characterisation of oils and fats. In F. D. Gunstone, J. L. Harwood, & A. J. Dijkstra (Eds.), *The lipid handbook with CD-ROM* (pp. 37–141). Boca Raton, FL: CRC Press.

Halle, I., & Schöne, F. (2013). Influence of rapeseed cake, linseed cake and hemp seed cake on laying performance of hens and fatty acid composition of egg yolk. *Journal für Verbraucherschutz und Lebensmittelsicherheit*, 8(3), 185–193.

Hamer, L., Coleman, D. N., & Relling, A. E. (2018). The effects of supplementing increasing doses of EPA and DHA fatty acids to ewes in late gestation on ewe performance and milk production and offspring performance and plasma metabolites. *Journal of Animal Science*, 96, 257–258.

Haque, M. (2018). Dietary manipulation: A sustainable way to mitigate methane emissions from ruminants. *Journal of Animal Science and Technology*, 60, 15.

Harr, K. E. (2002). Clinical chemistry of companion avian species: A review. *Veterinary Clinical Pathology*, 31(3), 140–151.

Hessle, A., Eriksson, M., Nadeau, E., Turner, T., & Johansson, B. (2008). Cold-pressed hempseed cake as a protein feed for growing cattle. *Acta Agriculturae Scandinavica Section A: Animal Science*, 58(3), 136–145.

Hohmann, A. G., & Suplita, R. L. (2006). Endocannabinoid mechanisms of pain modulation. *AAPS Journal*, 8(4), E693–E708.

Honzík, Z. (2015). Vláknina ve výživě nosnic: Význam výběru správného zdroje vlákniny. *Drůbežář*, 2015(2), 18–20 (in Czech).

Horakova, L., Stastnik, O., Pavlata, L., & Mrkvicova, E. (2020). The use of hemp herb in diet for growing rabbits. *MendelNet*, 27, 102–106.

House, J. D., Neufeld, J., & Leson, G. (2010). Evaluating the quality of protein from hemp seed (*Cannabis sativa* L.) products through the use of the protein digestibility-corrected amino acid score method. *Journal of Agricultural and Food Chemistry*, 58(22), 11801–11807.

Hullar, I., Meleg, I., Fekete, S., & Romvari, R. (1999). Studies on the energy content of pigeon feeds I. Determination of digestibility and metabolizable energy content. *Poultry Science*, 78, 1757–1762.

Huntsman, R. J., Tang-Wai, R., Alcorn, J., Vuong, S., Acton, B., Corley, S., et al. (2019). Dosage related efficacy and tolerability of cannabidiol in children with treatment-resistant epileptic encephalopathy: Preliminary results of the CARE-E study. *Frontiers in Neurology*, 10, 716.

Iannaccone, M., Ianni, A., Contaldi, F., Esposito, S., Martino, C., Bennato, F., et al. (2019). Whole blood transcriptome analysis in ewes fed with hemp seed supplemented diet. *Scientific Reports*, 9(1), 1–9.

Ianni, A., Di Domenico, M., Bennato, F., Peserico, A., Martino, C., Rinaldi, A., et al. (2020). Metagenomic and volatile profiles of ripened cheese obtained from dairy ewes fed a dietary hemp seed supplementation. *Journal of Dairy Science*, 103(7), 5882–5892.

Irakli, M., Tsaliki, E., Kalivas, A., Kleisiaris, F., Sarrou, E., & Cook, C. M. (2019). Effect of genotype and growing year on the nutritional, phytochemical, and antioxidant properties of industrial hemp (*Cannabis sativa* L.) seeds. *Antioxidants, 8*(10), 491.

Jhaveri, M. D., Elmes, S. J. R., Richardson, D., Barrett, D. A., Kendall, D. A., Mason, R., et al. (2008). Evidence for a novel functional role of cannabinoid CB2 receptors in the thalamus of neuropathic rats. *European Journal of Neuroscience, 27*(7), 1722–1730.

Jing, M., Zhao, S., & House, J. D. (2017). Performance and tissue fatty acid profile of broiler chickens and laying hens fed hemp oil and HempOmegaTM. *Poultry Science, 96*(6), 1809–1819.

Kalmendal, R. (2008). *Hemp seed cake fed to broilers* (Master thesis). Uppsala: Swedish University of Agricultural Sciences.

Kandaswami, C., & Middleton, E. (1994). Free radical scavenging and antioxidant activity of plant flavonoids. In *Free radicals in diagnostic medicine* (pp. 351–376). Springer.

Karlsson, L., Finell, M., & Martinsson, K. (2010). Effects of increasing amounts of hempseed cake in the diet of dairy cows on the production and composition of milk. *Animal, 4*(11), 1854–1860.

Khan, R. U., Durrani, F. R., Chand, N., & Anwar, H. (2010). Influence of feed supplementation with *Cannabis sativa* on quality of broilers carcass. *Pakistan Veterinary Journal, 30*(1), 34–38.

Khwatenge, C. N., Kimathi, B. M., Taylor-Bowden, T., & Nahashon, S. N. (2020). Expression of lysine-mediated neuropeptide hormones controlling satiety and appetite in broiler chickens. *Poultry Science, 99*(3), 1409–1420.

Kleinhenz, M. D., Magnin, G., Lin, Z., Griffin, J., Kleinhenz, K. E., Montgomery, S., et al. (2020). Plasma concentrations of eleven cannabinoids in cattle following oral administration of industrial hemp (*Cannabis sativa*). *Scientific Reports, 10*, 12753.

Kogan, L. R., Hellyer, P. W., & Robinson, N. G. (2016). Consumers' perceptions of hemp products for animals. *Journal of American Holistic Veterinary Medical Association, 42*, 40–48.

Konca, Y., Yuksel, T., Yalcin, H., Beyzi, S. B., & Kaliber, M. (2019). Effects of heat-treated hempseed supplementation on performance, egg quality, sensory evaluation and antioxidant activity of laying hens. *British Poultry Science, 60*(1), 39–46.

Konieczka, P., Szkopek, D., Kinsner, M., Fotschki, B., Juśkiewicz, J., & Banach, J. (2020). Cannabis-derived cannabidiol and nanoselenium improve gut barrier function and affect bacterial enzyme activity in chickens subjected to C. *perfringens* challenge. *Veterinary Research, 51*(1), 1–14.

Lamontagne, D., Lepicier, P., Lagneux, C., & Bouchard, J. F. (2006). The endogenous cardiac cannabinoid system: A new protective mechanism against myocardial ischemia. *Archives des Maladies du Coeur et des Vaisseaux, 99*(3), 242–246.

Leizer, C., Ribnicky, D., Poulev, A., Dushenkov, S., & Raskin, I. (2000). The composition of hemp seed oil and its potential use as an important source of nutrition. *Journal of Nutraceuticals, Functional and Medical Foods, 2*(4), 35–53.

Lewis, A. J. (2001). Amino acids in swine nutrition. In A. J. Lewis, & L. L. Southern (Eds.), *Swine nutrition* (pp. 131–150). Boca Raton, FL: CRC Press.

Liang, J., Appukuttan Aachary, A., & Thiyam-Holländer, U. (2015). Hemp seed oil: Minor components and oil quality. *Lipid Technology, 27*(10), 231–233.

Liu, D. Y., He, S. J., Jin, E. H., Liu, S. Q., Tang, Y. G., Li, S. H., et al. (2013). Effect of daidzein on production performance and serum antioxidative function in late lactation cows under heat stress. *Livestock Science, 152*(1), 16–20.

Liu, C. L., Li, Z. Q., & Feng, X. J. (2012). Effects of daidzein or genistein on proliferation and antioxidation of mammary epithelial cell of dairy cow *in vitro*. *Advanced Materials Research, 343*, 649–654.

Luo, Z., Yu, S., Zhu, Y., Zhang, J., Xu, W., & Xu, J. (2018). Effect of various levels of isoflavone aglycone-enriched fermented soybean meal on redox status, serum hormones and milk quality in ewes. *South African Journal of Animal Science, 48*(4), 673–682.

Makkar, H. P. S., Tran, G., Heuze, V., & Ankers, P. (2014). State of-the-art on use of insects as animal feed. *Animal Feed Science and Technology, 197*, 1–33.

Maly, O., Mares, J., Palisek, O., Sorf, M., & Postulkova, E. (2018). Use of by-products from hemp processing in the nutrition of common carp (*Cyprinus carpio* L.). *MendelNet, 25*, 165–170.

Mateos, G. G., Jiménez-Moreno, E., Serrano, M. P., & Lázaro, R. P. (2012). Poultry response to high levels of dietary fiber sources varying in physical and chemical characteristics. *Journal of Applied Poultry Research, 21*(1), 156–174.

Mierliță, D. (2016). Fatty acid profile and health lipid indices in the raw milk of ewes grazing part-time and hemp seed supplementation of lactating ewes. *South African Journal of Animal Science, 46*(3), 237–246.

Mortati, K., Dworetzky, B., & Devinsky, O. (2007). Marijuana: An effective antiepileptic treatment in partial epilepsy? A case report and review of the literature. *Reviews in Neurological Diseases, 4*(2), 103–106.

Mourot, J., & Guillevic, M. (2015). Effect of introducing hemp oil into feed on the nutritional quality of pig meat. *OCL Oilseeds and Fats Crops and Lipids, 22*(6), D612.

Mustafa, A. F., McKinnon, J. J., & Christensen, D. A. (1999). The nutritive value of hemp meal for ruminants. *Canadian Journal of Animal Science, 79*(1), 91–95.

Neijat, M., Gakhar, N., Neufeld, J., & House, J. D. (2014). Performance, egg quality, and blood plasma chemistry of laying hens fed hempseed and hempseed oil. *Poultry Science, 93*(11), 2827–2840.

Nielsen, S. K., Mandø, M., & Rosenørn, A. B. (2020). Review of die design and process parameters in the biomass pelleting process. *Powder Technology, 364*, 971–985.

NRC. (2012). Proteins and amino acids. In *Nutrient requirements of swine* (11 rev. ed., pp. 15–44). Washington, DC: The National Academies Press.

Odani, S. (1998). Isolation and primary structure of a methionine- and cystine-rich seed protein of *Cannabis sativa*. *Bioscience, Biotechnology, and Biochemistry, 62*(4), 650–654.

Padua, L. S., Bunyaprafatsara, N., & Lemmens, R. H. M. J. (1999). Medicinal and poisonous plants. In J. L. C. H. Valkenburg, & M. Bunyapraphatsara (Eds.), *Plant resources of South-East Asia* (pp. 167–175). New Delhi, India: Backhuys Publishers.

Palade, L. M., Habeanu, M., Marin, D. E., Chedea, V. S., Pistol, G. C., Grosu, I. A., et al. (2019). Effect of dietary hemp seed on oxidative status in sows during late gestation and lactation and their offspring. *Animals, 9*(4), 194.

Pamplona, F. A., da Silva, L. R., & Coan, A. C. (2018). Potential clinical benefits of CBD-rich cannabis extracts over purified CBD in treatment-resistant epilepsy: Observational data meta-analysis. *Frontiers in Neurology, 9*, 759.

Potter, D. J., Clark, P., & Brown, M. B. (2008). Potency of Δ^9–THC and other cannabinoids in cannabis in England in 2005: Implications for psychoactivity and pharmacology. *Journal of Forensic Sciences, 53*(1), 90–94.

Rea, K., Roche, M., & Finn, D. P. (2007). Supraspinal modulation of pain by cannabinoids: The role of GABA and glutamate. *British Journal of Pharmacology, 152*(5), 633–648.

Reggiani, R., & Russo, R. (2016). Beneficial effect of supplementation of flax and hemp seeds in the diet of alpine goats on the iron content in blood. *Journal of Scientific Research and Reports, 10*(2), 1–5.

Russo, R., & Reggiani, R. (2014). Valutazione di farine di lino e canapa per l'alimentazione animale: Contenuto di fattori antinutrizionali. In *Progetto Velica-Da antiche colture materiali e prodotti per il futuro. Edizioni* (pp. 15–23). Roma, Italy: CNR. (in Italian). Available from https://www.researchgate.net/publication/265852753_Valutazione_di_farine_di_lino_e_canapa_per_l'alimentazione_animale_contenuto_di_fattori_antinutrizionali.

Russo, R., & Reggiani, R. (2015). Evaluation of protein concentration, amino acid profile and antinutritional compounds in hempseed meal from dioecious and monoecious varieties. *American Journal of Plant Sciences, 6*(1), 14.

Serrapica, F., Masucci, F., Raffrenato, E., Sannino, M., Vastolo, A., Barone, C. M. A., & Di Francia, A. (2019). High fiber cakes from mediterranean multipurpose oilseeds as protein sources for ruminants. *Animals (Basel), 9*(11). https://doi.org/10.3390/ani9110918.

Shariat Zadeh, Z., Kheiri, F., & Faghani, M. (2020). Productive performance, egg-related indices, blood profiles, and interferon-Y gene 383 expression of laying Japanese quails fed on *Tenebrio molitor* larva meal as a replacement for fish meal. *Italian Journal of Animal Science, 19*(384), 274–281.

Silversides, F. G., & Lefrancois, M. R. (2005). The effect of feeding hemp seed meal to laying hens. *British Poultry Science, 46*(2), 231–235.

Skřivan, M., Englmaierová, M., Taubner, T., & Skřivanová, E. (2020). Effects of dietary hemp seed and flaxseed on growth performance, meat fatty acid compositions, liver tocopherol concentration and bone strength of cockerels. *Animals, 10*(3), 458.

Smeriglio, A., Galati, E. M., Monforte, M. T., Lanuzza, F., D'Angelo, V., & Circosta, C. (2016). Polyphenolic compounds and antioxidant activity of cold-pressed seed oil from finola cultivar of *Cannabis sativa* L. *Phytotherapy Research, 30*(8), 1298–1307.

Šťastník, O. (2018). *The use of milk thistle and hemp in poultry nutrition* (Doctoral dissertation (in Czech)). Czech Republic: Mendel University in Brno. Retrieved from https://theses.cz/id/m4p13r/.

Šťastník, O., Jůzl, M., Karásek, F., Fernandová, D., Mrkvicová, E., Pavlata, L., et al. (2019). The effect of hempseed expellers on selected quality indicators of broiler chicken's meat. *Acta Veterinaria Brno, 88*(1), 121–128.

Stastnik, O., Karasek, F., Stenclova, H., Burdova, E., Kalhotka, L., Trojan, V., et al. (2016). The effect of hemp by-products feeding on gut microbiota and growth of broiler chickens. In *Proceedings of the 23rd international Ph.D. students conference—MendelNet* (pp. 289–293). Brno, Czech Republic: Mendel University.

Stastnik, O., Karasek, F., Stenclova, H., Trojan, V., Vyhnanek, T., Pavlata, L., et al. (2015). The effect of hempseed cakes on broiler chickens performance parameters. In *Proceedings of the 22nd international PhD students conference—MendelNet* (pp. 157–160). Brno, Czech Republic: Mendel University.

Stastnik, O., Pavlata, L., & Mrkvicova, E. (2020). The milk thistle seed cakes and hempseed cakes are potential feed for poultry. *Animals, 10*(8), 1384.

Tam, J., Vemuri, V. K., Liu, J., Bátkai, S., Mukhopadhyay, B., Godlewski, G., et al. (2010). Peripheral CB1 cannabinoid receptor blockade improves cardiometabolic risk in mouse models of obesity. *Journal of Clinical Investigation, 120*(8), 2953–2966.

Teh, S. S., & Birch, J. (2013). Physicochemical and quality characteristics of cold-pressed hemp, flax and canola seed oils. *Journal of Food Composition and Analysis, 30*(1), 26–31.

Turner, T., Hessle, A., Lundström, K., & Pickova, J. (2008). Influence of hempseed cake and soybean meal on lipid fractions in bovine *M. longissimus dorsi*. *Acta Agriculturae Scandinavica Section A: Animal Science, 58*(3), 152–160.

Veldkamp, T., & Bosch, G. (2015). Insects: A protein-rich feed ingredient in pig and poultry diets. *Animal Frontiers, 5*(2), 45–50.

Vispute, M. M., Sharma, D., Mandal, A. B., Rokade, J. J., Tyagi, P. K., & Yadav, A. S. (2019). Effect of dietary supplementation of hemp (*Cannabis sativa*) and dill seed (*Anethum graveolens*) on performance, serum biochemicals and gut health of broiler chickens. *Journal of Animal Physiology and Animal Nutrition, 103*(2), 525–533.

Wakshlag, J. J., Schwark, W. S., Deabold, K. A., Talsma, B. N., Cital, S., Lyubimov, A., et al. (2020). Pharmacokinetics of cannabidiol, cannabidiolic acid, Δ^9-tetrahydrocannabinol, tetrahydrocannabinolic acid and related metabolites in canine serum after dosing with three oral forms of hemp extract. *Frontiers in Veterinary Science, 7*, 505.

Wallace, J. E., Kogan, L. R., Carr, E. C., & Hellyer, P. W. (2020). Motivations and expectations for using cannabis products to treat pain in humans and dogs: A mixed methods study. *Journal of Cannabis Research, 2*(1), 1–12.

Wang, S., Kreuzer, M., Braun, U., & Schwarm, A. (2017). Effect of unconventional oilseeds (safflower, poppy, hemp, camelina) on in vitro ruminal methane production and fermentation. *Journal of the Science of Food and Agriculture, 97*(11), 3864–3870.

Wang, X. S., Tang, C. H., Yang, X. Q., & Gao, W. R. (2008). Characterization, amino acid composition and in vitro digestibility of hemp (*Cannabis sativa* L.) proteins. *Food Chemistry, 107*(1), 11–18.

Wang, M., Wang, Y. H., Avula, B., Radwan, M. M., Wanas, A. S., van Antwerp, J., et al. (2016). Decarboxylation study of acidic cannabinoids: A novel approach using ultra-high-performance supercritical fluid chromatography/photodiode array-mass spectrometry. *Cannabis and Cannabinoid Research, 1*(1), 262–271.

Wijtten, P. J. A., Prak, R., Lemme, A., & Langhout, D. J. (2004). Effect of different dietary ideal protein concentrations on broiler performance. *British Poultry Science, 45*(4), 504–511.

Yalcin, H., Konca, Y., & Durmuscelebi, F. (2018). Effect of dietary supplementation of hemp seed (*Cannabis sativa* L.) on meat quality and egg fatty acid composition of Japanese quail (*Coturnix coturnix japonica*). *Journal of Animal Physiology and Animal Nutrition, 102*(1), 131–141.

Zaheer, K. (2017). Hen egg carotenoids (lutein and zeaxanthin) and nutritional impacts on human health: A review. *CyTA Journal of Food, 15*(3), 474–487.

Zeman, L., Šimeček, K., Krása, A., Šimek, M., Lossmann, J., Třináctý, J., et al. (1995). *Katalog krmiv*. Pohořelice VÚVZ. ISBN: 80-901598-3-4 (in Czech).

14

Consumer trends and the consumption of industrial hemp-based products

Hannah Lacasse and Jane Kolodinsky

Department of Community Development and Applied Economics, University of Vermont, Burlington, VT, United States

14.1 Introduction

Hemp-based products have been reintroduced to the market landscape over the last few decades. From milk alternatives, hemp hearts, and food grade oils to insulation and plastics, the opportunity for hemp products is vast and growing. This chapter discusses the role of hemp in the global market landscape and emerging research that characterizes consumers and informs marketing communication objectives. However, hemp is not entirely new; its involvement in human history is long and tangled. Before we can evaluate the modern hemp market, a necessary first step is to understand hemp's historic role.

14.2 A history of hemp

Hemp has been present throughout human history. Remnants of hemp cloth were found in ancient Mesopotamia as early as 8000 BCE (Ministry of Hemp, 2019). The crop originated in Eastern and Central Asia (Fike, 2019), with China having cultivated hemp for over 6000 years—the longest period of hemp cultivation (The Thistle, 2000). As hemp made its way across the world, it gained cultural significance in religious and spiritual practices (Ministry of Hemp, 2019; Robinson, 1996) and contributed to the necessities of daily life through food, clothing, paper, and medicine (Ministry of Hemp, 2019; The Thistle, 2000; van Roekel, 1994). Hemp oil and seeds became a source of health and healing in Chinese and Indian medicine as early as 2300 BCE (Fike, 2019; Robinson, 1996). Hempseed has been used as a food source for centuries and historically served as a staple for the poor and lower class (Fike, 2019; Robinson, 1996), providing a source of nutrition as raw, cooked, or pressed into oil (Callaway, 2004; Matthäus & Brühl, 2008).

In 1500 BCE, hemp was introduced to Europe (Fike, 2016), where it eventually became highly valued for its use as a salt-resistant canvas and durable rope for sailing ships and naval expeditions (Meijer, van der Werf, Mathijssen, & van den Brink, 1995). Throughout North America in the 16th and 17th centuries, hemp cultivation was highly encouraged, and sometimes mandatory, in order to supply the paper, rope, and textile industries (Robinson, 1996). Despite hemp's historical importance, the crop's significance began to fall in the 20th century for a variety of reasons: development of the cotton synthetic fiber industries, technological advances in the shipping industry, and the emphasized connection between the crop and marijuana and associated public health concerns (Micu, 2021; Raymunt, 2020).

Hemp production was prohibited in most of North America through the Marijuana Tax Act of 1937 in the United States and the Opium and Narcotic Act of 1938 in Canada because of its ties to marijuana (Alberta Government, 2015; Fortenbery & Bennett, 2004; Tourangeau, 2015). Both countries saw a brief resurgence in production when World War II required hemp-based supplies, but then resumed prohibition when the war ended (Kolosov, 2009; The Thistle, 2000). Many European countries also began to outlaw hemp production in the 1920s (Bewley-Taylor, Blickman, & Jelsma, 2014). A hemp commission in India was created in the 1890s to determine the public health threat of hemp and whether it caused "lunacy" when used (Ayonrinde, 2020).

India ultimately decided to pursue regulation and taxation of the crop rather than outright prohibition, but there is speculation that this decision was less motivated by minimal public health concern and more due to political and economic interests of the commission members (Hall, 2019). Similarly, there is skepticism regarding the United States' concern for the health effects of hemp. Though 1937 restrictions were framed as protection of public health and safety, there is conjecture that investors in the upcoming plastic, pulp–paper, and petroleum industries pushed for hemp prohibition to limit competition (Robinson, 1996). In addition to economic and ethical rationale, motives to prohibit hemp were also fraught with racist and xenophobic tensions. Prohibition efforts across the world were likely used to scapegoat people of color, immigrants, and other minority groups (Bewley-Taylor et al., 2014; Gray, 1998; Robinson, 1996).

As fear of marijuana fueled prohibition of hemp production around the world, hemp declined further with the introduction of new, more cost-effective technologies and fibers: the cotton gin reduced labor costs in the south, steamboats replaced sailboats and the market for marine cordage and cloth, and abaca and jute were imported at a lower price (Fortenbery & Bennett, 2004; Meijer et al., 1995). As a result, global production of hemp tow waste, or fiber, and hempseed both experienced overall declines from the 1960s until the 1990s (Fig. 14.1).

14.3 A renewed interest in hemp

Despite barriers to production in the 19th and 20th centuries, attitudes toward hemp have shifted (Cherney & Small, 2016). With the distinction between hemp and marijuana becoming clearer to the public, as well as a growing demand for sustainable products and CBD in

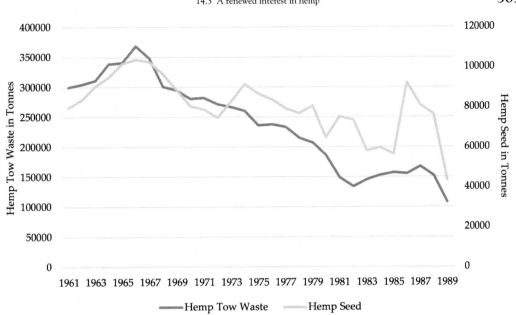

FIG. 14.1 Global production of hemp tow waste and hempseed from 1961 to 1989 (Food and Agriculture Organization of the United Nations, 2020).

general, production for hemp returned (Mark et al., 2020). Bans on hemp production were lifted across Europe in the 1990s (Johnson, 2018). In 1994 Canada legalized hemp production for research purposes and fully legalized production in 1998 (Mark et al., 2020). The United States followed a similar path, though was slow to follow suit. It wasn't until 2014 that the United States legalized hemp production for research purposes and then fully legalized the crop's production in 2018 (Mark et al., 2020).

As such, hemp has experienced a revival in the last several years. Chinese production doubled from 2015 to 2017 to 125,000 metric tons and Canadian hemp acreage has been on the rise since 2010 (Mark et al., 2020). There is a resurgence in consumer interest toward hemp and CBD appears to be the catalyst (Adesina, Bhowmik, Sharma, & Shahbazi, 2020). A search of news articles throughout the world using Nexis Uni, a database of news, law, and business sources, shows that discourse regarding hemp and CBD had been relatively static in 2015 and 2016 but began to rise from 2017 to 2019 (Fig. 14.2). We find similar trends when analyzing hemp and CBD using Google Trends, which pulls a representative sample of normalized search data and provides the term's relative search interest based on the proportion of searches on all topics for that time period (Google Trends, 2021). When viewing searches for hemp and CBD from 2015 to 2018, we see that, like news sources, search interest is relatively low through 2016 and rises steadily from 2017 to 2019 (Figs. 14.3 and 14.4). During this time, people appear to be seeking information on hemp.

However, CBD is not the only hemp product influencing this growing market. Hemp CBD made up just 23% of hemp-based product sales in 2017, with lower but present interest in personal care products, industrial applications, food products, and textiles (Hemp Business Journal, 2018). Industry predicts that fiber sales will escalate in the next few years (Hemp Industry Daily,

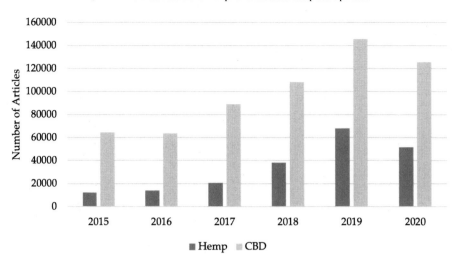

FIG. 14.2   Mentions of "hemp" and "CBD" in news outlets across the world from 2015 to 2020 (LexisNexis, 2021).

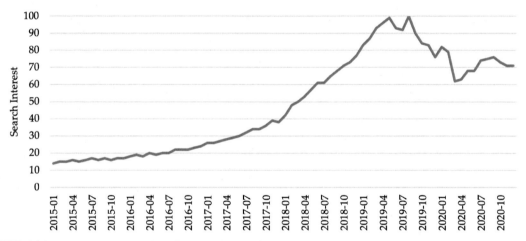

FIG. 14.3   Global Google searches of "hemp" from 2015 to 2020 (Google Trends, 2021).

2021) as well as demand for hempseed (Adesina et al., 2020). A survey of Vermont consumers also finds that, though use of CBD is most common, there are many other types of hemp products in use, including clothing, personal care products, rope, and food products (Fig. 14.5). It is clear that attention should be paid to the variety of products that hemp can produce.

14.4 Hemp products

Throughout its existence, hemp has been processed into thousands of different products, with more coming onto the market seemingly every day. Hemp's versatility lies in the ability to process each part of the plant, including the stalk, seeds, and flower.

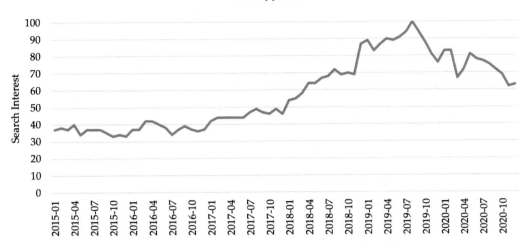

FIG. 14.4   Global Google searches of "CBD" from 2015 to 2020 (Google Trends, 2021).

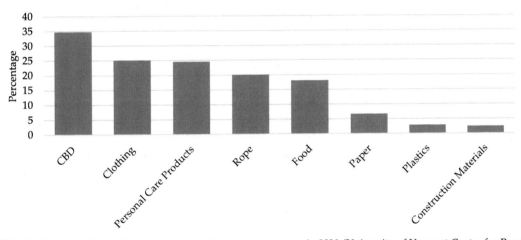

FIG. 14.5   Use of hemp-based products by Vermont consumers in 2020 (University of Vermont Center for Rural Studies, 2020).

Hempseeds can be pressed for oil and used in a variety of products (Johnson, 2018; Schluttenhofer & Yuan, 2017). Hempseed is lauded for its healthy oil and protein content (Adesina et al., 2020; Pihlanto, Mattila, Mäkinen, & Pajari, 2017; Schultz et al., 2020) and can be processed into food products like flour, pasta, hempseed butter, protein powders and bars, milk alternatives, and granola (Borkowska & Bialkowska, 2019; Brzyski & Fic, 2017; Johnson, 2018). The oil can also be applied to cosmetics, like soaps and shampoos, as well as an ingredient for medicinal and therapeutic products (Adesina et al., 2020; Johnson, 2018; Thompson, Berger, & Allen, 1998). The by-product of the pressing process, also known as hempseed cake, can be used as a source of animal feed (Schluttenhofer & Yuan, 2017).

The stalk of the hemp plant can be processed into multiple types of products using the bast and hurd fibers. Bast fibers are harvested from the outer stalk, while the hurds are sourced from the stalk's inner core (Johnson, 2018). These fibers can then be processed into traditional hemp products like clothing, rope, and paper. Clothing, rope, and other textiles made from hemp are stronger than cotton and resistant to water and saline damage (Kraenzel et al., 1998; Montford & Small, 1999). Hemp paper has been used for thousands of years and is competitive with wood pulp paper in terms of lifecycle length and recyclability (Bouloc & van der Werf, 2013; Miritoiu et al., 2019). Hurd fibers can also be used as animal bedding (Schluttenhofer & Yuan, 2017).

More recent innovations for hemp fiber include auto parts, composites, insulation materials, concrete alternatives, and plastics. Hemp-based building composites are comparable to the strength and insulation of commercial product alternatives, with a lower environmental impact and fewer negative human health effects (Sassoni, Manzi, Motori, Montecchi, & Canti, 2014; Väisänen, Batello, Lappalainen, & Tomppo, 2018). Concrete alternatives made from hemp are feasible for green building (Jami, Karade, & Singh, 2019; Maalouf et al., 2018). Hemp plastic innovations have been manufactured for long-term use in the automotive industry (Pervaiz & Sain, 2003), as well for daily, single use as a biocomposite (Brzyski & Fic, 2017).

The hemp female flower produces cannabinoids—a class of terpenophenolic compounds, with CBD as the major cannabinoid and a nonpsychoactive component (Adesina et al., 2020). There are over 100 known cannabinoids but two are most common: tetrahydrocannabinol (THC) and cannabidiol (CBD) (Aizpurua-Olaizola et al., 2016). THC has psychoactive properties that are associated with marijuana and occur in the plant at higher concentrations than do hemp (Hložek et al., 2017). Hemp primarily produces CBD, a nonpsychoactive cannabinoid that can be used for therapeutic, medicinal, and recreational purposes (Andre, Hausman, & Guerriero, 2016; World Health Organization, 2018). CBD has been used to aid ailments such as arthritis, epilepsy, inflammation, anxiety, and pain (Evans Schultes, Klein, Plowman, & Lockwood, 1974; Pertwee, 2004; Rosenberg, Tsien, Whalley, & Devinsky, 2015; World Health Organization, 2018).

14.5 Communicating hemp: The need for a lasting narrative

Regulatory changes reduced barriers to hemp production and set the stage for rising demand, sending a global spotlight to the crop. Though hemp has been a part of human history for thousands of years, release from production prohibition across the world reintroduced hemp to the market in a way it had not seen in decades. Excitement surrounding the opportunity to bolster farmer incomes and profit off of thousands of potential product alternatives fueled dialogue and demand for hemp. But as regulatory, financial, and logistic realities set in for the industry (Mark et al., 2020), the dazzle of the crop may be waning.

Dialogue surrounding hemp in the media is likely to have been a source of publicity and information conveyance to consumers over the last few years. But the buzz of this "new" industry appears to be fading. Figs. 14.2, 14.3 and 14.4 also show that hemp and CBD as mentioned in the news and in internet searches have declined since 2019. Both terms saw a small resurgence in April 2020 and may be attributed to the global COVID-19 pandemic. There is conjecture that sales for CBD products fared relatively well during this time, particularly via online shopping, as consumers sought to cope with the stress associated with quarantining, isolation, and generally living through this global disaster (Convenience Store News, 2021;

Prosser, 2020; Williams, 2021; Wright, 2020). However, this trend may reflect an increase in current users of CBD rather than new users (Wright, 2020). After a relative peak in April (Figs. 14.3 and 14.4), search interest continues to veer downward, indicating that less consumers are searching for information regarding hemp and CBD.

Given historic regulatory controversy, hemp's reentry into the market received more attention than most new products would. Producers who depend on media buzz to carry the advertising weight of hemp-based products may soon find that this reliance cannot be sustained. Though demand for hemp-based products is growing (Mark et al., 2020; Wheeler, Merten, Gordon, & Hamadi, 2020), these search results show that public attention toward hemp may be phasing out, limiting the ease of information access consumers and producers once had.

Just as consumers can learn about new-to-market hemp products, they can quickly forget. Consumer forgetting behavior occurs when they are unable to, or imperfectly, remember the quality or information regarding a product that they had once learned (Song & Fai Tso, 2020; Zhao, Zhao, & Helsen, 2011). Forgetting can take place when advertising subsides (Kogan, Herbon, & Venturi, 2020), or in the case of hemp, when the media moves on to other headlines. For hemp to succeed, the industry must pay attention to the hemp consumer and keep hemp-based products in their line of sight. Otherwise, consumers will forget and move onto the next innovation, or stick with the conventional product alternative.

To market hemp, we need to understand who the hemp consumer is and what qualities about hemp products appeal to them. A study by Ellison (2021) revealed that economics and marketing were the most prioritized areas of research for hemp stakeholders. Despite this expressed need for market research, there is limited peer-reviewed literature on the hemp consumer market (Mark et al., 2020). The remainder of this chapter will outline the available gray and peer-reviewed literature on consumer perceptions of hemp and make suggestions for future research.

Preliminary research exists that evaluates hemp consumption based on consumer demographics. Demographic characteristics of hemp consumers provide stakeholders with a starting point in such a nascent market. Similar to other literature regarding sustainable products or behaviors (Bhaskaran & Hardley, 2002; de Medeiros, Ribeiro, & Cortimiglia, 2014; Ha-Brookshire & Norum, 2011; Hustvedt & Bernard, 2008; Panzone, Hilton, Sale, & Cohen, 2016; Tan, Johnstone, & Yang, 2016; Verain et al., 2012; Verbeke, 2005), findings for demographic influence on hemp product consumption are varied. Higher incomes are associated with consumption of hemp-based cereal brands and hemp nuts (Kim & Mark, 2018). Older consumers are more likely to consume CBD (New Frontier Data, 2020) but less likely to consume hemp foods (Kim & Mark, 2018). Higher levels of education are associated with hemp nuts but not hemp-based cereal brands (Kim & Mark, 2018). Men are more likely than women to consume CBD products (New Frontier Data, 2020). However, two studies of Vermont consumers do not find any significant association between general hemp consumption, or hemp CBD consumption, and demographic variables (Kolodinsky & Lacasse, 2020; Kolodinsky, Lacasse, & Gallagher, 2020). The variation in findings points to a need for hemp consumption research that is tailored to specific products and more holistic in its ability to describe variation of consumption of those products.

A frequently cited limitation of the hemp industry is the cost of production (Amaducci et al., 2015; Bouloc & van der Werf, 2013; Cherney & Small, 2016). Hemp products are more expensive than their conventional alternatives, so it is necessary to determine what qualities make the product worth purchasing or what is inherent about a consumer that makes the product appealing. Hemp retailers, for example, identify quality as their most important differentiator (Hemp Industry Daily, 2021). But it is critical that producer and retailer perceptions of consumer values are accurate reflections of the consumer.

Literature regarding the consumer perceptions of hemp products is limited but beginning to emerge, particularly for CBD and food products. There appears to be low consumer likeness for hemp food based on studies with pork and bread made with hemp ingredients (Hayward & McSweeney, 2020; Zając et al., 2019), indicating the importance of palatability. Perceptions of hemp-based food products that incorporate a cannabis leaf in its branding include associations with marijuana and drugs (del Pozo, 2020), highlighting the need for intentional marketing depending on the intended conveyance of hemp's use. Perceived health benefits also appear influential in consumer likeness for hemp products, including food and CBD for both pet and human uses (Kogan, Hellyer, & Robinson, 2016; Metcalf, Wiener, & Saliba, 2021; Wheeler et al., 2020).

14.5.1 Consumer perceptions of hemp: Considerations from the United States

Hemp has been inextricably linked to marijuana since the early 1900s and was only recently released from prohibitive regulation in 2014 (*Legitimacy of Industrial Hemp Research*, 2014). For nearly 80 years, hemp had been coupled with marijuana under US regulation and classified as a Schedule I substance (*Marihuana Tax Act of 1937*, 1937). Though it has since been determined that the two crops are genetically distinct, particularly in their ability to induce a psychoactive response (World Health Organization, 2018), it would be logical for the general population to maintain this association. Lusk (2017) found that one-third of surveyed respondents believed hemp and marijuana to be the same. If this finding is representative of US consumers, it may influence the potential viability of a hemp product market. It is, therefore, necessary to contextualize the trajectory of marijuana support in the United States and how that may be reflected by hemp consumers.

There is an abundance of literature that analyzes perceptions toward marijuana, especially among youth, and how those perceptions change after policy implementation by way of medicinal or recreational marijuana legalization. These studies are frequently framed by early adopters of marijuana legislation, such as California and Colorado, and compare public opinion to that of nonmedicinal marijuana states. Policy change appears to relate strongly to perceptions of and support for marijuana. Literature comparing perceived risk of marijuana use finds that perceived risk decreases within a state before and after policy change, as well as when comparing states with marijuana legislation to those without (Khatapoush & Hallfors, 2004; Maxwell, 2016; Miech et al., 2015; Schuermeyer et al., 2014). Studies also find higher support for, or lower disapproval of, use of marijuana in states where medicinal or recreational marijuana has been legalized compared to nonmedicinal marijuana states (McGinty, Niederdeppe, Heley, & Barry, 2017; Miech et al., 2015; Schuermeyer et al., 2014).

Although the impacts of policy implementation on marijuana use are mixed, they offer important insight to hemp use. Incidents of marijuana use have been found to increase after policy change compared to nonmedicinal marijuana states among both youth and adults (Maxwell, 2016; Miech et al., 2015; Schuermeyer et al., 2014). However, other studies focused on youth outcomes did not find a significant impact on marijuana-related behavior afterward

(Khatapoush & Hallfors, 2004; Maxwell, 2016). In a study of Nielsen Homescan data, Kim and Mark (2018) found that consumers in hemp-legalized states were more likely to purchase hemp-based cereal products than consumers in nonhemp states. This difference between use impacts of marijuana and hemp may be due to the fact that the hemp product at hand is food based and the connotation of drugs is lessened. However, Kim and Mark (2018) also found that hemp states were *less* likely to consume hemp nut products compared to nonhemp states. These findings highlight the importance of evaluating consumer perceptions across the many products for which hemp can be processed, particularly when assessing the connotation of marijuana or when attitudes might be shaped by mistaken associations with marijuana.

The importance of positive public opinion is cited as a motivator for state-based marijuana legalization (Cruz, Queirolo, & Boidi, 2016; Johns, 2015). Given that many referendums were led by grassroots efforts, these findings imply that public support for marijuana came before legalization (Cruz et al., 2016). Increased use of marijuana has been found to precede state policy implementation as well (Maxwell, 2016; Schwadel & Ellison, 2017). However, policy change does not appear to be the singular driver of rising marijuana support.

There appear to be two other determinants of support, or lack thereof, for marijuana use and legalization. The first are the economic benefits that accompany legalization. A study by McGinty et al. (2017) reveals that Americans find pro-legalization arguments surrounding economic benefits as more persuasive compared to antilegalization arguments that frame marijuana as a public health risk. Though not the primary factor for city adoption of marijuana policy in Colorado, Johns (2015) cites the ability of economic benefits to change perceptions of former opposition and the importance of understanding community perceptions of the extent of these economic outcomes. Therefore, consumer perceptions of how hemp contributes to, or detracts from, the local economy may be an important factor in their consumption of hemp-based products.

Finally, a major correlate to marijuana support appears to be political affiliation. Democrats are more likely to support marijuana legalization and pro-legalization arguments than Republicans (Denham, 2019; McGinty et al., 2017; Schwadel & Ellison, 2017). It appears that this trend is historic, with studies finding higher Democrat support as early as 1973 (Schwadel & Ellison, 2017). However, this disparity appears to widen in the 21st century, with Republicans expressing significantly less support than both Democrats and Independents beginning in 2004 (Denham, 2019; Schwadel & Ellison, 2017). These findings indicate growing partisanship and politicization of marijuana policy across the United States (Denham, 2019).

Though hemp policy has received bipartisan support and conversations surrounding hemp aren't as heavily politicized as conversations about marijuana, the crops' close ties make political affiliation a likely determinant of hemp acceptance and use (Malone & Gomez, 2019; Steenstra, 2018). This link has been confirmed in studies of Vermont consumption of hemp-based products, where Independent and Progressive respondents were more likely to support and use hemp products compared to Republican respondents (Kolodinsky et al., 2020; Kolodinsky & Lacasse, 2020).

These findings indicate that hemp may continue to be associated with marijuana by the public, translating to political disparities regarding its approval. Though this case study focuses on the policy environment of the United States, it could have important implications for other countries across the world who have similar historic tensions between hemp and marijuana. Given the context of marijuana support and recommendations from hemp consumer behavior findings, a greater understanding of consumer perceptions of hemp's relationship to marijuana, as well as the role of political affiliation and associated ideology toward the hemp industry, may be of consequence to comprehensively understand the hemp consumer.

14.6 Considerations for consumer behavior research

The literature reviewed previously provides a starting point for understanding consumer perceptions of hemp products. The variation of findings indicates that marketing requirements for hemp will likely vary by type of product (Kim & Mark, 2018; Kolodinsky & Lacasse, 2020). However, the current literature does not provide a comprehensive picture for the potential consumption of the thousands of products that can be manufactured from hemp. As the excitement of hemp begins to subside, stakeholders across the supply chain require a more complete understanding of the hemp consumer in order to market hemp and advance this new industry. There is a myriad of opportunities for consumer behavior research regarding hemp to move the industry forward.

Whether consumers seek hemp products because they contain hemp—and why that is—is an important determination for creating effective marketing strategies in this industry. There may be several reasons why consumers actively seek hemp-based products or ultimately choose those products over alternatives. Hemp can be associated with environmentally friendly or sustainable choices and consumers may consider products to contribute to those perceived efforts (Lagoa et al., 2020; Li & Kallas, 2021). Or they may view hemp as a durable and high-quality ingredient. On the other hand, hemp may not at all be the motivator for the purchase decision. An analysis of the buying hierarchy as applied to hemp may be a useful tool for understanding the decision-making process of the hemp consumer (Kolodinsky, 1997; Mole, Halbrendt, Wang, & Kolodinsky, 1997).

The buyer hierarchy is a set of stages that lead a consumer to purchase a given product and include exposure, awareness, knowledge, attitude, intention to purchase, and purchase (Barry & Howard, 1990; Shimp & Andrews, 2013). Though some consumers may follow each step of this model to reach a purchase decision, many others will enter at different points and even go up and down this figurative ladder before choosing a product. The type of relationship a consumer has with the buyer hierarchy of a product can be described through their involvement in product choice. Involvement can be influenced by risk, past experience, how frequently the product is purchased, and the time available for the consumer to consider the decision at hand. High risk and low frequency purchases are each likely to result in a more involved decision and may result in the consumer navigating each step of the buyer hierarchy. Consumers with more experience with a product, or less time to make a decision, are less likely to be involved in the purchase decision.

Distinguishing between public, private, and luxury goods also explain a consumer's relationship with the buyer hierarchy and could be a particularly helpful tool in analyzing different hemp-based products. An individual's purchase of a private good can be seen or known by others. Private consumption indicates that a good purchased will not be known by other consumers. Luxury goods are those that are purchased because a consumer wants it, not because they need it. Goods that are consumed publicly are more likely to have more involved decisions, since others will know and see the purchase (Bearden & Etzel, 1982).

Given the variety of hemp products on the market, this type of consumption categorization may provide an additional understanding of the involvement of a hemp purchase decision.

When determining the target market for hemp consumers, the role of product, price, place, and promotion may be a helpful starting point (Schwartz, 2000). Product is the market offering, price is the amount of money a consumer must pay to receive the product, place is the location where the product originated and can also be where the product is available for purchase, and promotion includes strategies that motivate a consumer to purchase, such as discounts. Each variable can relate to the other at different magnitudes. Determining the optimum relationship between them may result in consumer choice of that product. To fully understand the relationship between these variables, it is likely necessary to conduct analysis based on a given hemp product category.

Though a main takeaway from the review of literature is that distinguishing between hemp product types is necessary, there may be a need for further disaggregation as products become saturated in the market. Understanding how consumers come to a decision regarding a specific product, say CBD for example, may reveal more about their values and perceptions. Howard and Sheth's (1969) Theory of Buyer Behavior seeks to model consumer choice of a specific brand given all the possible brands a consumer is aware of (Howard & Sheth, 1969). Similar to a study conducted by Kolodinsky et al. (2020), elaborations of this model can reveal not only which brands a consumer is aware of but also why the ultimate brand choice is made.

Hemp has been present throughout human history and its demonstrated versatility as an input for thousands of different products make it a compelling target for consumer behavior studies. Though currently limited, literature surrounding hemp demand continues to emerge and will play a critical role in directing the industry amid a new wave of hemp interest.

References

Adesina, I., Bhowmik, A., Sharma, H., & Shahbazi, A. (2020). A review on the current state of knowledge of growing conditions, agronomic soil health practices and utilities of hemp in the United States. *Agriculture, 10*(4), 1–15. https://doi.org/10.3390/agriculture10040129.

Aizpurua-Olaizola, O., Soydaner, U., Öztürk, E., Schibano, D., Simsir, Y., Navarro, P., et al. (2016). Evolution of the cannabinoid and terpene content during the growth of *Cannabis sativa* plants from different chemotypes. *Journal of Natural Products, 79*(2), 324–331. https://doi.org/10.1021/acs.jnatprod.5b00949.

Alberta Government. (2015). *Industrial hemp Enterprise.* Alberta Agriculture and Forestry. Agdex 153/830-1 https://open.alberta.ca/dataset/644b036a-b04e-48e1-a45b-cf530aa61b01/resource/56707132-5dd0-4c95-a82d-17bf49db3e17/download/Agdex-153-830-1-Nov2015.pdf.

Amaducci, S., Scordia, D., Liu, F. H., Zhang, Q., Guo, H., Testa, G., et al. (2015). Key cultivation techniques for hemp in Europe and China. *Industrial Crops and Products, 68*, 2–16. https://doi.org/10.1016/j.indcrop.2014.06.041.

Andre, C. M., Hausman, J.-F., & Guerriero, G. (2016). *Cannabis sativa*: The plant of the thousand and one molecules. *Frontiers in Plant Science, 7*(February), 1–17. https://doi.org/10.3389/fpls.2016.00019.

Ayonrinde, O. A. (2020). Cannabis and psychosis: Revisiting a nineteenth century study of "Indian hemp and insanity" in colonial British India. *Psychological Medicine, 50*(7), 1164–1172. https://doi.org/10.1017/S0033291719001077.

Barry, T. E., & Howard, D. J. (1990). A review and critique of the hierarchy of effects in advertising. *International Journal of Advertising, 9*(2), 121–135.

Bearden, W. O., & Etzel, M. J. (1982). Reference group influence on product and brand purchase decisions. *Journal of Consumer Research, 9*(2), 183. https://doi.org/10.1086/208911.

Bewley-Taylor, D., Blickman, T., & Jelsma, M. (2014). *The rise and decline of cannabis prohibition.* Transnational Institute. https://doi.org/10.1086/470120.

Bhaskaran, S., & Hardley, F. (2002). Buyer beliefs, attitudes and behaviour: Foods with therapeutic claims. *Journal of Consumer Marketing, 19*(7), 591–606. https://doi.org/10.1108/07363760210451410.

Borkowska, B., & Bialkowska, P. (2019). Evaluation of consumer awareness of hemp and its applications in different industries. *Scientific Journal of Gdynia Maritime University, 110*(19), 7–16. https://doi.org/10.26408/110.01.

Bouloc, P., & van der Werf, H. M. G. (2013). The role of hemp in sustainable development. In P. Bouloc, S. Allegret, & L. Arnaud (Eds.), *Hemp: Industrial production and uses* (pp. 278–289). CABI. http://www.cabi.org/cabebooks/ebook/20133324485.

Brzyski, P., & Fic, S. (2017). The application of raw materials obtained from the cultivation of industrial hemp in various industries. *Economic and Regional Studies, 10*(1), 100–113. https://doi.org/10.2478/ers-2017-0008.

Callaway, J. C. (2004). Hempseed as a nutritional resource: An overview. *Euphytica, 140*(1–2), 65–72. https://doi.org/10.1007/s10681-004-4811-6.

Cherney, J. H., & Small, E. (2016). Industrial hemp in North America: Production, politics and potential. *Agronomy, 6*(58), 1–24. https://doi.org/10.3390/agronomy6040058.

Convenience Store News. (2021). "COVID commencers" driving growth in CBD category. *Convenience Store News.* January 5 https://csnews.com/covid-commencers-driving-growth-cbd-category.

Cruz, J. M., Queirolo, R., & Boidi, M. F. (2016). Determinants of public support for marijuana legalization in Uruguay, the United States, and El Salvador. *Journal of Drug Issues, 46*(4), 308–325. https://doi.org/10.1177/0022042616649005.

de Medeiros, J. F., Ribeiro, J. L. D., & Cortimiglia, M. N. (2014). Success factors for environmentally sustainable product innovation: A systematic literature review. *Journal of Cleaner Production, 65*, 76–86. https://doi.org/10.1016/j.jclepro.2013.08.035.

del Pozo, C. (2020). *Exploring college students' interpretations and implications of the use of cannabis leaves on packaging of foods with hemp-derived ingredients* (Master's thesis). Colorado State University. ProQuest Dissertations Publishing.

Denham, B. E. (2019). Attitudes toward legalization of marijuana in the United States, 1986-2016: Changes in determinants of public opinion. *International Journal of Drug Policy, 71*, 78–90. https://doi.org/10.1016/j.drugpo.2019.06.007.

Ellison, S. (2021). Hemp (Cannabis sativa L.) research priorities: Opinions from U.S. hemp stakeholders. *GCB Bioenergy, 13*(4), 562–569. https://doi.org/10.1111/gcbb.12794.

Evans Schultes, R., Klein, W. M., Plowman, T., & Lockwood, T. E. (1974). Cannabis: An example of taxonomic neglect. *Botanical Museum Leaflets Harvard University, 23*(9), 337–367.

Fike, J. (2016). Industrial hemp: Renewed opportunities for an ancient crop. *Critical Reviews in Plant Sciences, 35*(5–6), 406–424. https://doi.org/10.1080/07352689.2016.1257842.

Fike, J. (2019). The history of hemp. In *Industrial hemp as a modern commodity crop* (pp. 2–25). ASA, CSSA and SSSA.

Food and Agriculture Organization of the United Nations. (2020). *FAOSTAT statistical database.*

Fortenbery, T. R., & Bennett, M. (2004). Opportunities for commercial hemp production. *Review of Agricultural Economics, 26*(1), 97–117. https://doi.org/10.1111/j.1467-9353.2003.00164.x.

Google Trends. (2021). *Google trends data.*

Gray, M. (1998). The devil weed and Harry Anslinger. In *Drug crazy* Random House. https://books.google.co.uk/books?id=a8g7QptAUtwC&pg=PA66&lpg=PA66&dq=%E2%80%9CThe+reign+of+tears+is+over.+The+slums+will+soon+be+a+memory.+We+will+turn+our+prisons+into+factories+and+our+jails+into+storehouses+and+corncribs.+Men+will+walk+upright+now,+women+will+#v=onepage&q=%E2%80%9CThe%20reign%20of%20tears%20is%20over.%20The%20slums%20will%20soon%20be%20a%20memory.%20We%20will%20turn%20our%20prisons%20into%20factories%20and%20our%20jails%20into%20storehouses%20and%20corncribs.%20Men%20will%20walk%20upright%20now%2C%20women%20will&f=false.

Ha-Brookshire, J. E., & Norum, P. S. (2011). Willingness to pay for socially responsible products: Case of cotton apparel. *Journal of Consumer Marketing, 28*(5), 344–353. https://doi.org/10.1108/07363761111149992.

Hall, W. (2019). The Indian hemp drugs commission 1893–1894. *Addiction, 114*(9), 1679–1682. https://doi.org/10.1111/add.14640.

Hayward, L., & McSweeney, M. B. (2020). Acceptability of bread made with hemp (*Cannabis sativa* subsp. sativa) flour evaluated fresh and following a partial bake method. *Journal of Food Science, 85*(9), 2915–2922. https://doi.org/10.1111/1750-3841.15372.

Hemp Business Journal. (2018). *The U.S. Hemp Industry grows to $820mm in sales in 2017.* Hemp Business Journal. https://www.hempbizjournal.com/size-of-us-hemp-industry-2017/.

Hemp Industry Daily. (2021). *Annual hemp & CBD industry Factbook.* Hemp Industry Daily.

Hložek, T., Uttl, L., Kadeřábek, L., Balíková, M., Lhotková, E., Horsley, R. R., et al. (2017). Pharmacokinetic and behavioural profile of THC, CBD, and THC+CBD combination after pulmonary, oral, and subcutaneous administration in rats and confirmation of conversion in vivo of CBD to THC. *European Neuropsychopharmacology, 27*(12), 1223–1237. https://doi.org/10.1016/j.euroneuro.2017.10.037.

Howard, J. A., & Sheth, J. N. (1969). The theory of buyer behavior. *Journal of the American Statistical Association, 65*(331), 1406. https://doi.org/10.2307/2284311.

Hustvedt, G., & Bernard, J. C. (2008). Consumer willingness to pay for sustainable apparel: The influence of labelling for fibre origin and production methods. *International Journal of Consumer Studies, 32*(5), 491–498. https://doi.org/10.1111/j.1470-6431.2008.00706.x.

Jami, T., Karade, S. R., & Singh, L. P. (2019). A review of the properties of hemp concrete for green building applications. *Journal of Cleaner Production, 239.* https://doi.org/10.1016/j.jclepro.2019.117852, 117852.

Johns, T. L. (2015). Managing a policy experiment. *State and Local Government Review, 47*(3), 193–204. https://doi.org/10.1177/0160323x15612149.

Johnson, R. (2018). *Hemp as an agricultural commodity.* Congressional Research Service. www.crs.gov.

Khatapoush, S., & Hallfors, D. (2004). "Sending the wrong message": Did medical marijuana legalization in California change attitudes about and use of marijuana? *Journal of Drug Issues, 34,* 751–770.

Kim, G., & Mark, T. (2018). Who are consuming hemp products in the U.S.? Evidence from Nielsen Homescan data. *SSRN Electronic Journal,* 1–28. https://doi.org/10.2139/ssrn.3176016.

Kogan, K., Herbon, A., & Venturi, B. (2020). Direct marketing of an event under hazards of customer saturation and forgetting. *Annals of Operations Research, 295*(1), 207–227. https://doi.org/10.1007/s10479-020-03723-4.

Kogan, L. R., Hellyer, P. W., & Robinson, N. G. (2016). Consumers' perceptions of hemp products for animals. *Journal of the American Holistic Veterinary Medical Association, 42,* 40–48. http://www.drugabuse.gov/.

Kolodinsky, J. (1997). Marketing of hemp products—The consumer is key. In *Flax and other bast plants symposium proceedings.*

Kolodinsky, J., & Lacasse, H. (2020). Consumer response to hemp: A case study of Vermont residents from 2019 to 2020. *GCB Bioenergy, 00,* 1–9. https://doi.org/10.1111/gcbb.12786.

Kolodinsky, J., Lacasse, H., & Gallagher, K. (2020). Making hemp choices: Evidence from Vermont. *Sustainability, 12,* 1–14. https://doi.org/10.3390/su12156287.

Kolosov, C. A. (2009). Evaluating the public interest: Regulations of industrial hemp under the controlled substances act. *UCLA Law Review, 38.* http://files/85/HS_STAT_2018_Population_Estimates_Bulletin.pdf.

Kraenzel, D. G., Petry, T. A., Nelson, B., Anderson, M. J., Mathern, D., & Todd, R. (1998). *Industrial hemp as an alternative crop in North Dakota* (Agricultural Economics Report 402, issue number 402). https://doi.org/10.22004/ag.econ.23264.

Lagoa, N. C., Marcon, A., Duarte Ribeiro, J. L., Fleith de Medeiros, J., Brião, V. B., & Antoni, V. L. (2020). Determinant attributes and the compensatory judgement rules applied by young consumers to purchase environmentally sustainable food products. *Sustainable Production and Consumption, 23,* 256–273. https://doi.org/10.1016/j.spc.2020.06.003.

Legitimacy of industrial hemp research. (2014).

LexisNexis. (2021). *Nexis Uni.*

Li, S., & Kallas, Z. (2021). Meta-analysis of consumers' willingness to pay for sustainable food products. *Appetite, 163*(September 2020), 105239. https://doi.org/10.1016/j.appet.2021.105239.

Lusk, J. L. (2017). Consumer research with big data: Applications from the food demand survey (foods). *American Journal of Agricultural Economics, 99*(2), 303–320. https://doi.org/10.1093/ajae/aaw110.

Maalouf, C., Ingrao, C., Scrucca, F., Moussa, T., Bourdot, A., Tricase, C., et al. (2018). An energy and carbon footprint assessment upon the usage of hemp-lime concrete and recycled-PET façades for office facilities in France and Italy. *Journal of Cleaner Production, 170,* 1640–1653. https://doi.org/10.1016/j.jclepro.2016.10.111.

Malone, T., & Gomez, K. (2019). Hemp in the United States: A case study of regulatory path dependence. *Applied Economic Perspectives and Policy, 41*(2), 199–214. https://doi.org/10.1093/aepp/ppz001.

Marihuana Tax Act of 1937. (1937). (testimony of U.S. Congress).

Mark, T., Shepherd, J., Olson, D., Snell, W., Proper, S., & Thornsbury, S. (2020). *Economic viability of industrial hemp in the United States: A review of state pilot programs United States Department of Agriculture.* U.S. Department of Agriculture, Economic Research Service. www.ers.usda.gov.

Matthäus, B., & Brühl, L. (2008). Virgin hemp seed oil: An interesting niche product. *European Journal of Lipid Science and Technology, 110*(7), 655–661. https://doi.org/10.1002/ejlt.200700311.

Maxwell, J. C. (2016). What do we know about the impact of the laws related to marijuana? *Journal of Addiction Medicine*, 10(1), 3–12. https://doi.org/10.1117/12.2549369.Hyperspectral.

McGinty, E. E., Niederdeppe, J., Heley, K., & Barry, C. L. (2017). Public perceptions of arguments supporting and opposing recreational marijuana legalization. *Preventive Medicine*, 99, 80–86. https://doi.org/10.1016/j.ypmed.2017.01.024.

Meijer, W. J., van der Werf, H. M., Mathijssen, E. W., & van den Brink, P. W. (1995). Constraints to dry matter production in fibre hemp (*Cannabis sativa* L.). *European Journal of Agronomy*, 4(1), 109–117. https://doi.org/10.1016/S1161-0301(14)80022-1.

Metcalf, D. A., Wiener, K. K. K., & Saliba, A. (2021). Comparing early hemp food consumers to non-hemp food consumers to determine attributes of early adopters of a novel food using the food choice questionnaire (FCQ) and the Food Neophobia Scale (FNS). *Future Foods*, 3(December 2020), 100031. https://doi.org/10.1016/j.fufo.2021.100031.

Micu, A. (2021). The rise and fall of hemp—And how we can make it great again. *ZME Science*. January 22 https://www.zmescience.com/science/hemp-history-future-feature-91352342/.

Miech, R. A., Johnston, L., O'Malley, P. M., Bachman, J. G., Schulenberg, J., & Patrick, M. E. (2015). Trends in use of marijuana and attitudes toward marijuana among youth before and after decriminalization: The case of California 2007–2013. *International Journal of Drug Policy*, 26(4), 336–344. https://doi.org/10.1016/j.drugpo.2015.01.009.

Ministry of Hemp. (2019). *History of hemp in the U.S.*. Ministry of Hemp. https://ministryofhemp.com/hemp/history/.

Miritoiu, C. M., Stanescu, M. M., Burada, C. O., Bolcu, D., Padeanu, A., & Bolcu, A. (2019). Comparisons between some composite materials reinforced with hemp fibers. *Materials Today: Proceedings*, 12, 499–507. https://doi.org/10.1016/j.matpr.2019.03.155.

Mole, M., Halbrendt, C., Wang, Q., & Kolodinsky, J. (1997). Willingness to pay for hemp based products: Evidence from a consumer survey. *Symposium Magazine*, 60.

Montford, S., & Small, E. (1999). Measuring harm and benefit: The biodiversity friendliness of Cannabis sativa. *Global Biodiversity*, 8(4), 2–13. http://files/1062/Measuring_harm_and_benefit_Th.pdf.

New Frontier Data. (2020). *Frequency of use among U.S. CBD consumers*. https://newfrontierdata.com/cannabis-insights/frequency-of-use-among-u-s-cbd-consumers/.

Panzone, L., Hilton, D., Sale, L., & Cohen, D. (2016). Socio-demographics, implicit attitudes, explicit attitudes, and sustainable consumption in supermarket shopping. *Journal of Economic Psychology*, 55, 77–95. https://doi.org/10.1016/j.joep.2016.02.004.

Pertwee, R. G. (2004). Pharmacological and therapeutic targets for Δ9-tetrahydrocannabinol and cannabidiol. *Euphytica*, 140(1–2), 73–82. https://doi.org/10.1007/s10681-004-4756-9.

Pervaiz, M., & Sain, M. M. (2003). Sheet-molded polyolefin natural fiber composites for automotive applications. *Macromolecular Materials and Engineering*, 288(7), 553–557. https://doi.org/10.1002/mame.200350002.

Pihlanto, A., Mattila, P., Mäkinen, S., & Pajari, A. M. (2017). Bioactivities of alternative protein sources and their potential health benefits. *Food & Function*, 8(10), 3443–3458.

Prosser, D. (2020). UK demand for CBD products soars amid Covid-19 pandemic. *Forbes*. May 11 https://www.forbes.com/sites/davidprosser/2020/05/11/uk-demand-for-cbd-products-soars-amid-covid-19-pandemic/?sh=247ee6916e07.

Raymunt, M. (2020). *Hemp cultivation in Europe: Key market details and opportunities*. Hemp Industry Daily.

Robinson, R. (1996). *The great book of hemp*. Park Street Press.

Rosenberg, E. C., Tsien, R. W., Whalley, B. J., & Devinsky, O. (2015). Cannabinoids and epilepsy. *Neurotherapeutics*, 12(4), 747–768. https://doi.org/10.1007/s13311-015-0375-5.

Sassoni, E., Manzi, S., Motori, A., Montecchi, M., & Canti, M. (2014). Novel sustainable hemp-based composites for application in the building industry: Physical, thermal and mechanical characterization. *Energy and Buildings*, 77, 219–226. https://doi.org/10.1016/j.enbuild.2014.03.033.

Schluttenhofer, C., & Yuan, L. (2017). Challenges towards revitalizing hemp: A multifaceted crop. *Trends in Plant Science*, 22(11), 917–929. https://doi.org/10.1016/j.tplants.2017.08.004.

Schuermeyer, J., Salomonsen-Sautel, S., Price, R. K., Balan, S., Thurstone, C., Min, S. J., et al. (2014). Temporal trends in marijuana attitudes, availability and use in Colorado compared to non-medical marijuana states: 2003-11. *Drug and Alcohol Dependence*, 140, 145–155. https://doi.org/10.1016/j.drugalcdep.2014.04.016.

Schultz, C. J., Lim, W. L., Khor, S. F., Neumann, K. A., Schulz, J. M., Ansari, O., et al. (2020). Consumer and health-related traits of seed from selected commercial and breeding lines of industrial hemp, *Cannabis sativa* L. *Journal of Agriculture and Food Research*, 2(January). https://doi.org/10.1016/j.jafr.2020.100025, 100025.

Schwadel, P., & Ellison, C. G. (2017). Period and cohort changes in Americans' support for marijuana legalization: Convergence and divergence across social groups. *Sociological Quarterly*, 58(3), 405–428. https://doi.org/10.1080/00380253.2017.1331715.

Schwartz, D. G. (2000). Concurrent marketing analysis: A multi-agent model for product, price, place and promotion. *Marketing Intelligence & Planning*, 18(1), 24–30. https://doi.org/10.1108/02634500010308567.

Shimp, T. A., & Andrews, J. C. (2013). *Advertising, promotion, and other aspects of integrated marketing communications* (9th ed.). South-Western Cengage Learning. https://doi.org/10.1300/J057v01n01_07.

Song, L., & Fai Tso, G. K. (2020). Consumers can learn and can forget—Modeling the dynamic decision procedure when watching TV. *Journal of Management Science and Engineering*, 5(2), 87–104. https://doi.org/10.1016/j.jmse.2020.05.002.

Steenstra, E. (2018). Hemp farming legalized across the United States by the 2018 farm bill. *Vote Hemp*. December 20.

Tan, L. P., Johnstone, M.-L., & Yang, L. (2016). Barriers to green consumption behaviours: The roles of consumers' green perceptions. *Australasian Marketing Journal; AMJ*, 24(4), 288–299. https://doi.org/10.1016/j.ausmj.2016.08.001.

The Thistle. (2000). *The People's history*. MIT - The Thistle. https://www.mit.edu/~thistle/v13/2/history.html.

Thompson, E. C., Berger, M. C., & Allen, S. N. (1998). *Economic impact of industrial hemp in Kentucky*. http://gatton.gws.uky.edu/cber/cber.htm.

Tourangeau, W. (2015). Re-defining environmental harms: Green criminology and the state of canada's hemp industry. *Canadian Journal of Criminology and Criminal Justice*, 57(4), 528–554. https://doi.org/10.3138/cjccj.2014.E11.

University of Vermont Center for Rural Studies. (2020). Vermonter Poll 2020 data.

Väisänen, T., Batello, P., Lappalainen, R., & Tomppo, L. (2018). Modification of hemp fibers (*Cannabis sativa* L.) for composite applications. *Industrial Crops and Products*, 111, 422–429. https://doi.org/10.1016/j.indcrop.2017.10.049.

van Roekel, G. J. (1994). Hemp pulp and paper production. *Journal of the International Hemp Association*, 1, 12–14. http://www.internationalhempassociation.org/jiha/iha01105.html.

Verain, M. C. D., Bartels, J., Dagevos, H., Sijtsema, S. J., Onwezen, M. C., & Antonides, G. (2012). Segments of sustainable food consumers: A literature review: Segments of sustainable food consumers. *International Journal of Consumer Studies*, 36(2), 123–132. https://doi.org/10.1111/j.1470-6431.2011.01082.x.

Verbeke, W. (2005). Consumer acceptance of functional foods: Socio-demographic, cognitive and attitudinal determinants. *Food Quality and Preference*, 16(1), 45–57. https://doi.org/10.1016/j.foodqual.2004.01.001.

Wheeler, M., Merten, J. W., Gordon, B. T., & Hamadi, H. (2020). CBD (Cannabidiol) product attitudes, knowledge, and use among young adults. *Substance Use and Misuse*, 55(7), 1138–1145. https://doi.org/10.1080/10826084.2020.1729201.

Williams, P. (2021). Demand for CBD soars in UK during COVID-19 as country's market now estimated at £690 million. *Hemp Grower*. May 6 https://www.hempgrower.com/article/demand-cbd-soars-uk-covid-19-estimated-690-million-pounds/.

World Health Organization. (2018). *Cannabiodiol (CBD): Critical review report*. Expert Committee on Drug Dependence. https://www.who.int/medicines/access/controlled-substances/CannabidiolCriticalReview.pdf.

Wright, M. (2020). The CBD industry surprising changes and growth due to COVID-19. *ASD Market Week*. August 5 https://asdonline.com/blog/retail-news/the-cbd-industry-surprising-changes-and-growth-due-to-covid-19/.

Zając, M., Guzik, P., Kulawik, P., Tkaczewska, J., Florkiewicz, A., & Migdał, W. (2019). The quality of pork loaves with the addition of hemp seeds, de-hulled hemp seeds, hemp protein and hemp flour. *LWT*, 105(October 2018), 190–199. https://doi.org/10.1016/j.lwt.2019.02.013.

Zhao, Y., Zhao, Y., & Helsen, K. (2011). Consumer learning in a turbulent market environment: Modeling consumer choice dynamics after a product-harm crisis. *Journal of Marketing Research*, 48(2), 255–267. https://doi.org/10.1509/jmkr.48.2.255.

Index

Note: Page numbers followed by *f* indicate figures and *t* indicate tables.